D1327652

LIVERPOOL HOPE
UNIVERSITY COLLEGE

LIBRARY

PO BOX 95
LIVERPOOL L16 9LB

Landform Monitoring, Modelling and Analysis

British Geomorphological Research Group
Symposia Series

LIVERPOOL HOF   Y COLLEGE

Landform Monitoring, Modelling and Analysis

Edited by

Stuart N. Lane

Keith S. Richards

Department of Geography,
University of Cambridge, UK

and

Jim H. Chandler

Department of Civil Engineering,
Loughborough University of Technology, UK

JOHN WILEY & SONS

Chichester • New York • Weinheim • Brisbane • Singapore • Toronto

Copyright © 1998 by John Wiley & Sons Ltd,
Baffins Lane, Chichester,
West Sussex PO19 1UD, England

National   01243 779777
International (+44) 1243 779777
e-mail (for orders and customer service enquiries): cs-books@wiley.co.uk
Visit our Home Page on http://www.wiley.co.uk
or http://www.wiley.com

All rights reserved. No part of this publication may be reproduced, stored in a retrieval system, or transmitted, in any form or by any means, electronic, mechanical, photocopying, recording, scanning or otherwise, except under the terms of the Copyright, Designs and Patents Act 1988 or under the terms of a licence issued by the Copyright Licensing Agency, 90 Tottenham Court Road, London UK W1P 9HE, without the permission in writing of the publisher.

Other Wiley Editorial Offices

John Wiley & Sons, Inc., 605 Third Avenue,
New York, NY 10158-0012, USA

WILEY-VCH Verlag GmbH, Pappelallee 3,
D-69469 Weinheim, Germany

Jacaranda Wiley Ltd, 33 Park Road, Milton,
Queensland 4064, Australia

John Wiley & Sons (Asia) Pte Ltd, 2 Clementi Loop #02-01,
Jin Xing Distripark, Singapore 129809

John Wiley & Sons (Canada) Ltd, 22 Worcester Road,
Rexdale, Ontario M9W 1L1, Canada

LIVERPOOL HOPE
UNIVERSITY COLLEGE

Order No./Invoice No. 51·35

LO17001713/8499635

Accession No.
201906

Class No.
910·514 LAN

Control No.
ISBN·

Catal.
10/7/98

Library of Congress Cataloging-in-Publication Data

Landform monitoring, modelling, and analysis / [edited by] Stuart N. Lane,
  Keith S. Richards, Jim H. Chandler.
    p. cm. — (British Geomorphological Research Group symposia series)
  Includes bibliographical references and index.
  ISBN 0-471-96977-X
  1. Geomorphology—Maps. 2. Digital mapping. I. Lane, Stuart N.
  II. Richards, K. S. III. Chandler, Jim H. IV. Series.
GB400.42.M3L36 1997
551.41'022'3—dc21   97-5546
                                                                   CIP

British Library Cataloguing in Publication Data

A catalogue record for this book is available from the British Library

ISBN 0-471-96977-X

Typeset in 10/12pt Times from authors' disks by Mayhew Typesetting, Rhayader, Powys
Printed and bound in Great Britain by Bookcraft (Bath) Ltd

This book is printed on acid-free paper responsibly manufactured from sustainable forestry, in which at least two trees are planted for each one used for paper production.

Contents

List of Contributors

M. G. Anderson Department of Geography, University of Bristol, University Road, Bristol, BS8 1SS, UK

N. S. Arnold Department of Geography, University of Cambridge, Downing Place, Cambridge, CB2 3EN, UK

R. Barker Department of Geography, University of Portsmouth, Buckingham Building, Lion Terrace, Portsmouth, Hampshire, PO1 3HE, UK

P. D. Bates Department of Geography, University of Bristol, University Road, Bristol, BS8 1SS, UK

K. Beven Centre for Research on Environmental Systems, Lancaster University, Lancaster, LA1 4YQ, UK

J-M. Bonvin Grande Dixence SA., Rue des Creusets 41, Sion, Valais, Switzerland

M. Bray Department of Geography, University of Portsmouth, Buckingham Building, Lion Terrace, Portsmouth, Hampshire, PO1 3HE, UK

J. H. Chandler Department of Civil Engineering, University of Loughborough, Loughborough, LE11 3TU, UK

C. D. Clark Sheffield Centre for Earth Observation Science, Department of Geography, University of Sheffield, Winter Street, Sheffield, S10 2TN, UK

M. A. R. Cooper Engineering Surveying Research Centre, Department of Civil Engineering, The City University, London, EC1V OHB, UK

L. Copland Department of Earth and Atmospheric Sciences, University of Alberta, Edmonton, Alberta, T6G 2E3, Canada

W. E. Dietrich Deptartment of Geology and Geophysics, University of California, Berkeley, CA 94720, USA

L. Dixon Department of Geography, University of Portsmouth, Buckingham Building, Lion Terrace, Portsmouth, Hampshire, PO1 3HE, UK

I. S. Evans Earth Surface Systems Research Group, Department of Geography, University of Durham, Science Laboratories, South Road, Durham, DH1 3LE, UK

P. Farres Department of Geography, University of Portsmouth, Buckingham Building, Lion Terrace, Portsmouth, Hampshire, PO1 3HE, UK

R. I. Ferguson Department of Geography, University of Sheffield, Winter Street, Sheffield, S10 2TN, UK

J. C. Gallant   Centre for Resource and Environmental Studies, The Australian National University, Canberra, ACT 0200, Australia

D. J. Gilvear   Department of Environmental Science, University of Stirling, Stirling, Scotland, FK9 4LA, UK

J. Hardisty   School of Geography and Earth Resources, University of Hull, Hull, HU6 7RX, UK

J. Hooke   Department of Geography, University of Portsmouth, Buckingham Building, Lion Terrace, Portsmouth, Hampshire, PO1 3HE, UK

M. Horritt   Department of Geography, University of Bristol, University Road, Bristol, BS8 1SS, UK

B. P. Hubbard   Centre for Glaciology, Institute of Earth Studies, University of Wales, Aberystwyth, Dyfed, SY23 3DB, UK

R. Inkpen   Department of Geography, University of Portsmouth, Buckingham Building, Lion Terrace, Portsmouth, Hampshire, PO1 3HE, UK

R. Lamb   Centre for Research on Environmental Systems, Lancaster University, Lancaster, LA1 4YQ, UK. Currently at the Institute of Hydrology, Wallingford, Oxfordshire, UK

S. N. Lane   Department of Geography, University of Cambridge, Downing Place, Cambridge, CB2 3EN, UK

R. J. Martin   School of Mathematics and Statistics, University of Sheffield, Winter Street, Sheffield, S10 2TN, UK

M. J. McCullagh   Department of Geography, University of Nottingham, University Park, Nottingham, NG7 2RD, UK

A. Merel   Department of Geography, University of Portsmouth, Buckingham Building, Lion Terrace, Portsmouth, Hampshire, PO1 3HE, UK

R. Middleton   School of Geography and Earth Resources, University of Hull, Hull, HU6 7RX, UK

A. M. Milner   Department of Geography, University of Birmingham, Edgbaston, Birmingham, B15 2TT, UK

D. R. Montgomery   Department of Geological Sciences, University of Washington, Seattle, WA 98195-1310, USA

S. Myrabo   Institute of Geophysics, University of Oslo, 1022 Blindern, N-0315 Oslo, Norway

D. Payne   Department of Geography, University of Portsmouth, Buckingham Building, Lion Terrace, Portsmouth, Hampshire, PO1 3HE, UK

J. S. Pethick   Cambridge Coastal Research Unit and Department of Geography, University of Cambridge, Downing Place, Cambridge, CB2 3EN, UK

K. S. Richards Department of Geography, University of Cambridge, Downing Place, Cambridge, CB2 3EN, UK

H. Rouse School of Geography and Earth Resources, University of Hull, Hull, HU6 7RX, UK

M. J. Sharp Department of Earth and Atmospheric Sciences, University of Alberta, Edmonton, Alberta, T6G 2E3, Canada

A. Shelford Department of Geography, University of Portsmouth, Buckingham Building, Lion Terrace, Portsmouth, Hampshire, PO1 3HE, UK

K. Sullivan Weyerhaeuser Technology Center, Weyerhaeuser Company, Tacoma, WA 98477, USA

O. Turpin Department of Geography and Sheffield Centre for Earth Observation Science, University of Sheffield, Winter Street, Sheffield, S10 2TN, UK

D. Twigg Department of Civil Engineering, Loughborough University of Technology, Loughborough, LE11 3TU, UK

C. P. Vencatasawmy Sheffield Centre for Earth Observation Science, Department of Geography and School of Mathematics and Statistics, University of Sheffield, Winter Street, Sheffield, S10 2TN, UK

T. M. Waters Department of Environmental Science, University of Stirling, Stirling, FK9 4LA, UK

D. Whyatt School of Geography and Earth Resources, University of Hull, Hull, HU6 7RX, UK

I. C. Willis Department of Geography, University of Cambridge, Downing Place, Cambridge, CB2 3EN, UK

J. P. Wilson Department of Earth Sciences, Montana State University, Bozeman, MT 59717-0348, USA

S. M. Wise Department of Geography, University of Sheffield, Winter Street, Sheffield, S10 2TN, UK

Preface

Geomorphologists played a major role in terrain research in the 1960s and 1970s as morphometric descriptions of both specific landforms and general land form were developed, and as the potential of the computer was realised for storing and manipulating terrain data. With growth in both "pure" and "applied" aspects of terrain analysis, further technical development, and growing input from other disciplines, this research has increasingly overlapped with the domain of Geographical Information Systems and Remote Sensing. Furthermore, the development of terrain analytical approaches has begun, more clearly than in the past, to reflect the needs of allied disciplines such as hydrology, rather than simply those of geomorphology itself. The relative importance of geomorphologists in developing this research area has therefore reduced, despite the centrality of terrain description and analysis as a geomorphological objective. However, innovation and inter-disciplinarity have accordingly increased, bringing new ideas and approaches.

A British Geomorphological Research Group (BGRG) conference with a spatial analytical theme was held in Cambridge in the early 1970s, and the three of us therefore suggested that the Annual Conference a quarter of a century later (in September 1995) should be held in Fitzwilliam College, Cambridge, with the related theme of terrain monitoring, modelling and analysis, and with the aim of examining recent and anticipated developments in this field. This proved to be a successful conference because its technical focus was of interest to geomorphologists of all persuasions, because its concern with land form was seen to be central to the discipline, and because it was clearly timely to review the technical developments in the field. As conference organisers and proceedings editors, we had all found ourselves concerned with terrain analysis, but approaching it from different perspectives. J.H.C. trained as a land surveyor, but undertook research in a Department of Civil Engineering where analytical photogrammetric methods were developed and applied to solve specific problems, including the geomorphological one of recreating the three-dimensional morphology of a landslide evolving through time using dated digital elevation models. Collaboration between K.S.R and J.H.C developed following the 1990 BGRG spring field meeting, resulting in a Natural Environment Research Council research grant and research studentship completed by S.N.L. This employed photogrammetric and digital elevation modelling methods to help understand and ultimately model the behaviour of dynamic gravel-bed rivers. This fusion of engineering surveying expertise with interest in specific geomorphological problems has resulted in fruitful collaboration. The stimulus for both the conference and this volume was our perception of the success of this collaboration, and a desire to diffuse both (i) the developing terrain measurement techniques, and (ii) new perspectives on the way in which terrain information may be used to answer geomorphological problems.

With these objectives in mind, this proceedings volume contains both reviews of existing material and significant original research papers. All of the papers that follow have been peer-reviewed to the same standards as an international journal, and we are particularly grateful to the following who freely gave of their time, often at short notice, for this purpose: Frank Ahnert, Keith Beven, James Brasington, Sue Brooks, Mike Cooper, Mariza Costa-Cabral, Richard Dikau, Jane Drummond, Ian Evans, Jon French, Ian Heywood, Rob Lamb, Dave Montgomery, Clive Oppenheimer, Tony Parsons, Richard Pike, Gareth Rees, David Sear, Martin Sharp, Geoff Smith, Tom Spencer, John Thornes, John West, Ian Willis and Steve Wise. Referees were asked to ensure that all papers provided one of the following: (i) an original review of relevant technical developments or their application to geomorphological problems; (ii) a significant advance in a particular methodological issue relevant to landform research; or (iii) a novel way of using existing methods to address a particular geomorphological problem.

The organisation of the conference was helped significantly by postgraduate students from the Department of Geography, University of Cambridge, and notably by Kate Bradbrook, James Brasington, Ben Brock, Justin Butler, David Gaselee, Chris Pyle and Sudhanshu Sinha. Nadine Keating, Gill Renshaw, Jane Robinson and Lynda Williams provided secretarial assistance. The Conference Office and staff at Fitzwilliam, and Caroline Choat in particular, allowed the conference to run smoothly. The following organisations exhibited at and kindly sponsored the conference, allowing coverage of the costs of the key-note speakers, and subsidy of postgraduate attendance: Ashtech Europe Ltd, Catena Verlag Publishers, C.Z. Scientific Instruments Ltd, ERDAS U.K. Ltd, Geotronics Ltd, Intergraph (UK) Ltd, Leica U.K. Ltd and SOKKIA Ltd.

Stuart Lane
Keith Richards
University of Cambridge

Jim Chandler
Loughborough University

July 1996

1 Landform Monitoring, Modelling and Analysis: Land *Form* in Geomorphological Research

STUART N. LANE[1], JIM H. CHANDLER[2] and KEITH S. RICHARDS[1]
[1] *Department of Geography, University of Cambridge, UK*
[2] *Department of Civil Engineering, University of Loughborough, UK*

INTRODUCTION

Pike (1995) has noted that developments in the description of land *form* have lagged behind those in the understanding of *process* in the quantitative study of the earth's surface. This is perhaps surprising given that there are strong substantive and methodological reasons for the study of topography in geomorphology, as well as a purely etymological one. The lag may reflect the fact that developments in landform description have often been driven by applications which have emerged outside geomorphology at unpredictable times, and that these have often had military overtones; for example, terrain trafficability studies (Wood and Snell, 1957), support for basin morphometry studies by the Office of Naval Research in the USA in the 1950s, and the relationship between developments in the digital representation of terrain and in missile guidance systems. With the elaboration of physically-based, semi-distributed modelling in hydrology, different practical requirements for geomorphological data have emerged, whose intellectual aims are more closely allied to those of geomorphology.

In terms of the substantive importance of topographic description for geomorphology, it is possible to detect a growing emphasis on the significance of morphology as a control of geomorphological processes. As Shreve (1972) noted, the fundamental differential equations that describe changes in the form of the landscape represent processes whose operation is determined by the form itself. Thus, while topography drives hydrological process patterns, those patterns change the topography. Anderson and Burt (1978) showed qualitatively that soil moisture and runoff patterns reflect contour curvature, and this is now quantitatively embodied in a range of topographic indexes (Zevenbergen and Thorne, 1987) that underpin hydrological models such as TOPMODEL and TAPES-C (Beven and Kirkby, 1979; O'Loughlin, 1986; Moore *et al.*, 1991). Digital terrain data can be used to identify the locations in landscapes where different slope processes, such as mass movement and surface runoff, occur (Dietrich *et al.*, 1993), where hillslope processes give way to channel processes (Tribe,

---

Landform Monitoring, Modelling and Analysis. Edited by S. N. Lane, K. S. Richards and J. H. Chandler.
© 1998 John Wiley & Sons Ltd.

1991), and where river valleys switch from bedrock to alluvial forms (Montgomery *et al.*, 1996). However, it is more than a question of topography controlling process; the spatially distributed effects of process on topography also change the landscape. Convergence of flow into a hillslope hollow may result in a feedback in which solutional weathering accentuates the hollow, although this is modulated by the changing effects of soil water chemistry, residence times and chemical kinetics. Morphology is thus not only the consequence of past processes, but is also a factor that affects the course of present erosion, and hence the future morphology. This feedback can be considered qualitatively, as in Williams' (1987) analysis of tower karst, not merely as passive products of weathering, but as active factors in relief evolution. In some small-scale and dynamic landforms the feedback is now beginning to be quantified, particularly in fluvial geomorphology. Ashworth and Ferguson (1986) discussed the bi-directional causality of gravel bar construction by bedload transport and bedform influences on patterns of bedload transport, and Lane *et al.* (1995a) have begun to model this spatially-distributed feedback using digital topographic representations of gravel river beds.

The methodological significance of topography follows from its substantive role. If morphology is a critical control upon both process type and its rate of operation, then the detailed specification of morphology should be an important part of any geomorphological investigation. The development of process-based simulation modelling as a means of understanding landform evolution has emphasised the importance of accurate specification of the topographic boundary condition (Richards *et al.*, 1995). Measurement of morphology, and possibly of its changes through time, may be required either to verify model predictions or to update the topographic surface where model predictions have diverged significantly from reality, perhaps because simulation models fail to accommodate possible non-linear topographic responses (Haff, 1996). One crucial development encouraging a resurgence of interest in form has been the replacement of traditional one-dimensional morphometric statistics by two- and three-dimensional representations of topography which allow analysis of the internal spatial distribution of form–process relations. The most pronounced methodological shift, however, is found in the growing recognition that process rates themselves may be better estimated from observations of morphological change, particularly if measurement of process rate is difficult to achieve over relevant time scales (Lane *et al.*, 1995b; Ashmore and Church, 1995). This implies a radical methodological switch, from process study being the key to understanding landforms, to detailed description of forms being the key to assessment of process.

Despite these substantive and methodological cases for measuring form in geomorphology, there remains a sense of *déjà vu*. This volume of essays arises from the 1995 Annual Meeting of the British Geomorphological Research Group (BGRG). One stimulus for this meeting was the fact that 25 years previously a similar BGRG conference had been held in Cambridge, organised by R. J. Chorley, with selected papers appearing in the 1972 volume *Spatial Analysis in Geomorphology*. Chorley noted then that the arrival of spatial analytical methods in geomorphology in the 1960s was late when compared with the advances made in quantitative geology (for example, by Krumbein) in the 1940s and in human geography in the 1950s and

1960s, and could even be described as having been not particularly successful. He identified two reasons for this. The first was a failure to invest geomorphic problems with the appropriate spatial dimensionality, with three-dimensional morphometric problems relating to forms and processes being specified in only a one- or two-dimensional form because of their complexity. The second was that, with some exceptions, automated cartography had focused on the mainly technical problems of data storage and retrieval rather than on the use of such data in spatial model building. Thus the centrality of analysis of form has apparently long been recognised as an etymological necessity for a discipline known as geomorphology, but the scientific weaknesses in the manner of its investigation *per se* remain apparent today, especially in the context of landform evolution. Pike's review in 1995 thus suggests that things have changed little since 1972. Arguing in a similar vein to Chorley, he notes: (i) the subordinate role of topography in process-oriented work; (ii) the complexity of terrain and the consequent difficulties in quantifying it; (iii) a lack of agreement on method for the study of landform; (iv) the inefficiency of manual data gathering; and (v) a reluctance to abandon the qualitative approach to landform description that for a long time seemed adequate for research and teaching.

This volume does not address all of Pike's concerns directly, although many are implicit. Five issues in particular may be highlighted, and are reflected in the following essays. One is the need to raise awareness of developments in technologies used to monitor and generate digital representations of terrain, and hence to answer geomorphological questions in new ways. A second is the growing recognition of the need to consider the quality of data used to represent numerical terrain surfaces. Thirdly, it is necessary to examine the potential of recent technical developments which can now provide landform information at rates and densities that are commensurate with the rates at which processes operate. Fourth is the importance of land *form* for the investigation of process, notably using spatially distributed simulation models which require morphological data as an input in the form of boundary conditions. Finally, there is an increasing recognition that changes in morphology may provide an alternative, and in some cases a more useful, means of estimating process rates. The remainder of this introduction will consider these points further.

TECHNICAL DEVELOPMENTS IN TERRAIN MONITORING

Both Chorley (1972) and Pike (1995) have referred to the complexity of terrain arising from its three-dimensional spatial character, and therefore its four-dimensional time-dependent nature. This complexity implies that improvements in our ability to measure and monitor terrain in three spatial dimensions are critical. Some landform monitoring methods traditionally used by geomorphologists involve parameters and technologies that are inherently one- or two-dimensional (e.g. measuring drainage density from maps or river channel cross-section form by simple levelling). The fact that few of the chapters in this volume utilise such methods reflects the rapid diffusion of the three-dimensional digital representation of topography using digital elevation models (DEMs). There have been some significant changes in the technology available to acquire DEMs and the impact and implications of these are assessed by Cooper (Chapter 2). Modern surveying methods are dependent upon computerised methods but vary widely in their degree of sophistication and the complexity of infrastructure necessary for their implementation. At the risk of generalisation, they may also be classed in terms of their applicability to different scales of geomorphological investigation: micro-scale studies of areas less than 10 km², meso-scale studies of less than 100 km², and macro-scale studies which may be global in extent.

Micro-scale Studies: the Modern Total Station/Electronic Tacheometer

The simplest method for three-dimensional survey of small areas involves a theodolite and electromagnetic distance measurement (using an EDM). However, there has been a significant increase in the sophistication of such instruments, commonly known as total stations or electronic tacheometers (Schofield, 1994; Cooper, Chapter 2). Modern systems include a lightweight EDM capable of measuring distances of greater than 2 km reliably, in a wide range of weather conditions and using a minimum number of prisms, as long as lines of sight are not restricted. Quoted accuracies of the latest generation EDMs are typically $\pm(3 \text{ mm} + 5 \text{ ppm})$. Significantly, EDMs based on pulsed laser rather than traditional phase difference measurement even offer the potential of distance measurement without a prism. Although this capability is restricted to certain surfaces and relatively short distances (300 m), it offers the possibility of automated measurement of complex natural surfaces.

Modern total stations are equipped with a dual-axis compensator and include the ability to measure, store and correct for the systematic error source known as collimation error (see Cooper, Chapter 2). If the instrument is working within acceptable limits, these features enable single face readings to be taken without loss of accuracy, greatly increasing the speed of use. Digital measurement of both the horizontal and vertical measuring circles enables such systematic errors to be removed, but also provides the ability to store measured data. Storage and transfer of measured and manually entered attribute data now enable most aspects of subsequent data processing to be automated. The transformation of raw measured data to three-dimensional coordinates based upon a datum is typically carried out using a PC and one of the variety of survey data-processing packages. Output from such a package can remain a conventional paper plot at the required scale, but increasingly the preferred product is digital three-dimensional coordinates to be used subsequently in a digital terrain modelling package.

The most recent (and expensive) total stations (Geodimeter 600 and Leica TCA1800) include a servo-powered and controlled telescope, and consequently a target and prism tracking capability. This combination allows a single user to obtain all required field measurements at very high rates of data acquisition. These developments ensure that the modern total station will remain for some time the obvious technology to generate DEMs in micro-scale studies.

Meso-scale Studies and the Global Positioning System

The total station is less appropriate for meso-scale studies, because of the time required to set up a control network over the whole study area. The most significant advance in survey technology at this scale has been the development of the Global Positioning System (GPS), which has expanded the range of field-based survey by at least an order of magnitude (Fix and Burt, 1995). A massive financial investment by the US Department of Defense has developed and provided the infrastructure for the system, which now consists of 24 operational satellites orbiting at 20 200 km above the earth's surface, with five monitoring stations distributed throughout the world. Usage of GPS now goes well beyond the original requirement to provide military navigational data for ships, aircraft and missiles. A diverse range of GPS equipment of varied cost and accuracy is now available, which can be bewildering to the uninformed geomorphologist. Twigg (Chapter 3) aims to address that shortfall in knowledge by providing a review of the technology, techniques and applications of GPS surveying. The current state of the GPS constellation and the basic position-fixing methods are described, and some of the practical issues which need to be considered when carrying out survey work with GPS are discussed. These include satellite availability and configuration; transformations from WGS84 to different coordinate systems and national mapping systems; and determining heights and height differences using GPS methods, and the accuracy of heights thus derived. These are important constraints on any use of GPS and are discussed further by Cooper (Chapter 2). Twigg (Chapter 3) concludes with three illustrative case studies of the practical problems and benefits of GPS survey.

Macro-scale: Remote Sensing Methods and Continental DEMs

Global-scale terrain data are now becoming relatively widely available, and regional, continental and global terrain analysis is now a possibility, although the appropriate resolutions and accuracies necessarily vary. Ten years ago, Cogley (1985) demonstrated the potential of continental hypsometry based on visual estimation of elevations from maps on a 1° grid, with mean elevation accuracies of about 100 m. Thelin and Pike (1991) developed a 30 arc-second (0.8 km) resolution terrain model based on contour data for the conterminous United States, with elevation accuracies of about 30 m, to produce a hill-shaded relief map which forms the basis for mapping of both structural regions and specific landforms. Today, global digital elevation data are even accessible via the World Wide Web; for example, the global 30 arc-second DEM based on the US Defense Mapping Agency 1:1 million scale contour data. This implies that large-scale terrain analysis can now be conducted with surprising ease but, as noted below, this often occurs without rigorous assessment of data quality.

Analysis of terrain properties has increasingly overlapped with the domain of Geographical Information Systems (GIS) and remote sensing; indeed, integrated GIS might now be regarded as the combination of GIS, remotely sensed data and terrain modelling. A wide range of remote sensing methods has been exploited in geomorphology. The use of aerial photography by geomorphologists has included both qualitative analysis of landforms and recovery of quantitative information by conventional analogue photogrammetry. Investigation of landforms at large spatial scales has tended to focus on qualitative attributes, and Vencatasawmy *et al.* (Chapter 8) illustrate this in assessment of the significance of synthetic aperture radar imagery to detect lineaments in the landscape over Alaska. Similar approaches may be used at sub-continental scales to investigate large-scale ice moulding features associated with ice sheets (Clark, 1994). At a rather smaller but still regional scale, and employing a more quantitative analysis, Gilvear *et al.* (Chapter 9) describe a technique which utilises the variation in grey-scale values in aerial photographs as a method of estimating water depth. This technique is then used as a means of assessing stream morphology recovery rates following gold placer mining in Alaska, and appears to offer potential for clear and shallow streams of low turbidity. However, the most important contribution to the derivation of quantitative terrain data from imaged sources has been the development of image processing methods such as stereo-matching, which has enabled terrain data to be recovered automatically from aerial photography and satellite imagery. These methods transcend particular scales of analysis, and are considered briefly in the next section.

Photogrammetry

Pike (1995) argues that manual data-gathering methods may have hindered progress in the study of land form, but it is possible that recent developments in photogrammetry may well address this problem. Photogrammetric techniques can be applied across the range of spatial scales described above, provided consideration is given to image resolution and scale. Photogrammetry is widely used as a primary method of acquiring three-dimensional topographic data, and most national mapping organisations (including the Ordnance Survey in the UK) compile maps from data acquired photogrammetrically (Cooper, Chapter 2). Photogrammetry has changed markedly since analogue methods were originally developed, and analytical techniques became practicable during the late 1980s. The chapter by Dixon *et al.* (Chapter 4) outlines the advantages of analytical photogrammetry for geomorphology, reviewing the processes and practicalities of carrying out a photogrammetric survey and deriving coordinates using analytical methods. Assuming no change in focal length, photographic scale is dependent upon the distance between the camera and the object. Photogrammetric methods are therefore appropriate to monitor objects of a wide range of size or scale, with commensurate precision. Such flexibility of scale is typified by the use of photogrammetry to measure both the rate of limestone decay using cores only 50 mm in diameter, and the rate of coastal cliff erosion using aerial photographs of a section of cliffs 1.8 km long in Christchurch Bay, UK (Dixon *et al.*, Chapter 4).

Geomorphologists have used photogrammetric methods in the past (Petrie and Price, 1966; El-Ashry and Wanless, 1967; Welch and Jordan, 1983), but the cost of the equipment required for both analogue and analytical photogrammetry has been prohibitively expensive. Developments in digital photogrammetry, in which points are measured using a scanned digital image presented on a computer screen, are now rectifying this situation. Typically, all that is required is access to a high resolution scanner and a powerful PC or UNIX workstation with appropriate software. With the ever-reducing costs of computer power, memory and disc storage, initial investment can be as low as £5000. The other important benefit with digital photogrammetry is automation, particularly DEM generation, which is now both practicable and in an advanced state of development. This automation is based upon sophisticated image correlation or image matching techniques which automatically identify small image patches appearing on two overlapping digital images. Using established photogrammetric methods, these two image measurements are transformed into object coordinates and the process repeated. With appropriate hardware and software this cycle can recur at speeds in excess of 100 points per second, so very dense and consequently accurate DEMs can be generated. Automated DEM acquisition at micro-scales (e.g. Brunsden and Chandler, in press) is best suited to situations in which there are both a lack of vegetation and sharp discontinuities in the land surface. At meso- and macro-scales using satellite imagery, the effect of relief-related shadow is more critical, and may require a correction based on solar elevations. A recent development is the application of stereo-matching methods to the construction of DEMs of river banks from ground photography of river banks, the objective being to compare successive DEMs in order to map the spatio-temporal pattern of bank erosion (Pyle *et al.*, 1997). The precision of automatically generated DEMs was ±12 mm and the locations where individual clasts had been removed from the gravel river bank could be identified. A similar method is currently being used to quantify the three-dimensional form of exposed and sub-aerial river-bed gravels (Lane *et al.*, 1996).

Depending upon requirements, photogrammetry normally requires some form of ground control survey, typically carried out using either a total station or differential GPS as indicated above. Thus, while it is possible to suggest that different methods are applicable at different scales, in reality these methods are often complementary across scales. Furthermore, in terrain analysis at all scales, three-dimensional terrain data are increasingly derived by image processing methods, and the primary concern is the control exercised by the characteristics of the image and image-acquisition methods over the precision and accuracy of the derived terrain data.

CONCEPTUAL ISSUES IN TERRAIN MODELLING

While there remain many practical limitations to land surveying, the technical developments identified above have almost reversed the traditional geomorphological equation. Formerly, to produce a field area map required the tedious and subjective application of methods that resulted in only an approximation of the surface, from which one- and two-dimensional data were extracted (slope profiles and cross-sections). Realistic monitoring of spatial patterns of landform change was inconceivable, and it was process monitoring that seemed to offer most potential in the investigation of that change. Today, there is a real prospect that topography can be represented as a four-dimensional entity (as a time series of DEMs), the weak link is then the expense of process monitoring at a single point or at very few points within this system. As discussed below, topographic changes then become the vehicle for understanding some of the space–time complexity of earth surface processes, rather than *vice versa*. Furthermore, photography, which can inexpensively record information but can be recovered by digital photogrammetry at a later date if required, now offers considerable potential for longer-term assessment of change in landforms than has been possible via process monitoring. This, however, implies that new kinds of geomorphological questions will be asked, new approaches to research design will be needed, and greater attention will be paid to the quality of this digital representation of land form.

Data and Terrain Surface Quality

Many commentators have noted the dangers of using digitised contours to generate DEMs (Fryer *et al.*, 1994; McCullagh, Chapter 5; Wise, Chapter 7), and such concerns are magnified where statistics are used to describe DEMs and their derivatives, and where DEMs are used in process simulations. As Robinson (1994) comments, many users accept DEMs uncritically and there is a need to re-emphasise that all geomorphologists should assess the quality of the DEMs they use. The importance of this is illustrated by Cooper (Chapter 2). Comparison of landform surfaces obtained from two consecutive time periods may imply differences which are not, however, attributable to geomorphological processes. Instead, they may be due to either (or to both) an inexact datum definition or various error sources introduced during the original surveys. Any user of morphological information should attempt to consider the likely magnitude of these errors and its implications for geomorphological conclusions. Error analysis of this kind is not something to which geomorphologists are accustomed, and it is reflected in some of the papers in this volume. However, it is critical if geomorphological conclusions are to be accepted as meaningful.

As Cooper explains, surveyors and photogrammetrists identify three different kinds of error: random error (normal variation expected during the measurement process which may be modelled statistically); systematic error (inexact mathematical models used to relate measurements); and blunders (avoidable mistakes made during measurement). Aspects of data quality can be related to these three types of error: precision can be regarded as a measure of random errors; accuracy is a measure of quality with respect to systematic error; and reliability can be related to the detection of blunders and their impact. There is some difficulty with this trichotomy, partly as a result of the widespread confusion of accuracy and precision in common usage, and even amongst some surveying equipment manufacturers (Twigg, Chapter 3). Despite this, such definitions are important for survey measurement, and can also be significant for derived data. Proper consideration of data quality in terms of accuracy, precision and reliability is a necessary element of a rigorous and scientific geomorphology, as illustrated by some of the studies reported in this volume.

Accuracy may perhaps be regarded as the key component of quality, but in most instances is difficult to estimate. DEM accuracy is often assessed by comparing elevations in a DEM with "true" values of elevation derived by independent means, and by computing the root mean square error (McCullagh, Chapter 7). The issue of accuracy of DEMs derived from widely used contour data is examined by Wise (Chapter 7). It is suggested that aspect derived from contour data is a good means of identifying and measuring erroneous artefacts within DEMs. McCullagh (Chapter 5) also considers factors which affect the accuracy of a DEM, and illustrates the significance of the distribution and density of points and the importance of break lines. Notions of reliability (repeatability) are examined by Evans (Chapter 6), who assesses the variability of statistical moments (mean, standard deviation, skew and kurtosis) describing altitude matrices and their derivatives (slope, aspect, and profile and plan curvature). This study suggests that terrain statistics are repeatable for altitude and gradient, but that third and fourth moments used to describe higher-order derivatives (profile and plan convexity) are variable and dependent on grid size (data density). The impact of digitised contours tagged with the incorrect elevation is also considered by McCullagh (Chapter 5). Fortunately such gross errors are readily visualised in the terrain surface as an artefact with the appearance of a "Roman ramp-and-ditch fortification", which illustrates the importance of visual and qualitative methods of assessing the reliability of a data set. The significance of precision of a particular method and scale of monitoring is perhaps best illustrated by the work of Dixon *et al.* (Chapter 4), in which photogrammetric methods are applied to generate spatial data at a variety of scales. Clearly, a precision of ±0.2 m, adequate for monitoring changes along a section of eroding sea cliffs, is entirely unsuitable for assessing the rate of decay on a limestone rock core.

Some papers in this volume do consider implications of data quality for geomorphological parameters. The topographic surfaces used to define boundary conditions in environmental modelling applications will contain error, and it is important that the effects of such error are assessed, together with the uncertainties in the process components of the model. There is evidently some overlap here with sensitivity analyses in numerical simulations. One demonstration of this sensitivity is provided by the effects on parameter values in hydrological simulations of varying the grid size in the underlying DEM, which is used to generate the area–slope index for TOPMODEL. Brasington and Richards (in press), for example, have simulated runoff data for a small Nepalese catchment, employing separate calibration and validation data periods. As the grid resolution was degraded, the calibrated parameter values changed systematically, demonstrating that although they have conceptual–physical meaning, their numerical values are dependent on DEM properties. In particular, the calibrated hydraulic conductivity parameter increased to compensate for the loss of representation of hillslope hollows as the grid coarsened. This maintained an approximately constant mean of the $\ln(a/T_o\tan\beta)$ distribution function (where a is contributing area, T_o is transmissivity, and β is local slope), which otherwise increases with the grid cell size. This interdependence of grid resolution and parameter value was necessary for the model to maintain its predictions of saturated area, and of the proportions of surface and sub-surface flow. The additional impact of spatially distributed errors in boundary condition specification may be assessed through observing spatially distributed changes in internal model predictions as such errors are introduced. This approach has been adopted by Lane *et al.* (1994) in investigating flow processes within a divided reach of river, and Bates *et al.* (Chapter 13) examine such effects for floodplain flows during overbank flood conditions by assessing model response to small changes in the topographic surface defining the River Culm floodplain. Systematic variations in simulated inundation in both time and space were observed, indicating that the flow model was indeed sensitive to topographic errors. A similar problem is considered by Lane (Chapter 14) in a review of the errors in process estimates obtained from sampling complex gravel-bed surfaces in river channel studies. The interaction between DEM resolution and accuracy in the description of a river's bed topography in hydraulic models, and the value of the roughness coefficient describing the boundary friction, illustrates a similar problem to that of DEM resolution in hydrological modelling as described above. The effects of terrain model quality in numerical modelling are explored further in a subsequent section.

Time and Space Scales

A key criticism of some modern geomorphology is its preoccupation with enquiries over short time scales and small space scales (e.g. Baker and Twidale, 1991), and the resultant "unfulfilled promise" of process studies as the means of acquiring understanding of landform evolution (Douglas, 1982). Indeed, some of the papers in this volume (e.g. Lane, Chapter 14) could be criticised for their overemphasis upon intensive investigation of small areas over short time periods. This is particularly problematic given the importance of transfers of mass between different landforms, and hence the need to understand landform systems rather than just single landforms. However, concerns with landform systems involving multiple process environments are raised by Montgomery *et al.* (Chapter 11), who illustrate the essential coupling between hillslope processes and spatial variation in river channel characteristics that emerges in a holistic study of landform systems at the drainage basin scale. Recognition of the coupling of different scales of landform behaviour (cf. Schumm and Lichty, 1965), and notably of the way in which processes that act over short time and area scales result in long-term landform evolution, can only come from land form data of a high temporal resolution, with a spatial density that is sufficient to describe the landform surface under investigation. Tests of time-based geomorphological hypotheses concerning the existence of equilibrium in landforms also demand data whose resolution in time and space is appropriate for the purpose, that is, commensurate with the time scales over which equilibrium and departures from equilibrium occur. One reason for concentration on small-scale dynamic systems has been the relative ease of obtaining data which illustrate such process–form adjustment. It is therefore a considerable strength of the methods of landform monitoring and modelling emphasised in this volume that they allow re-creation of data for testing these hypotheses. An example is the study of the Black Ven mudslide (Chandler and Cooper, 1989; Chandler and Brunsden, 1995), in which archival oblique and vertical aerial photographs from 1946, 1958, 1969, 1976 and 1988 were used to generate sequential DEMs and consequent DEMs of difference. The study demonstrated that slope angle distribution remains roughly constant through time, despite dramatic cliff recession, and suggests the existence of a secular equilibrium. Opportunities to examine stability and change in landscapes by comparing DEMs derived from historical photography open up new horizons for the geomorphologist, while at the same time illustrating the significance of photographic archives for geomorphological research.

It therefore appears that it is precisely those technological developments described above that are beginning to release geomorphologists from traditional scale restraints, allowing increases in the ease and speed of data acquisition until it becomes possible to acquire data at the density, rate and precision required by the time scale and area scale over which landforms are changing. Methodological development is increasingly allowing direct acquisition by geomorphologists of data of appropriate quality at the large scale (e.g. Dixon *et al.*, Chapter 4), rather than their being constrained to make use of data provided by organisations such as the Ordnance Survey. This gives the geomorphologist increasing control over data quality, both temporal and spatial, and may reduce the cost of data acquisition. It is only with high spatial and temporal resolution data that it becomes possible to understand the extent to which processes operating over small time scales may exert control on longer time scales of change over larger areas. This is illustrated by Lane (Chapter 14) who describes how estimates of bed material transport rate using morphological methods vary as a function of both the spatial density of surface measurement and the length of time between morphological remeasurement.

Process Simulation and the Topographic Boundary Condition

Chorley (1972, p.9) was able to conclude that

> . . . it is also clear that some of the most exciting aspects of this work will involve the construction of spatial simulation models in geomorphology in which a much better balance will be achieved than hitherto between deterministic and stochastic processes . . .

The growth of physically-based and spatially-distributed simulation modelling in geomorphology has perhaps been the key development in the years since 1972. Process–response models in the 1970s were explicitly driven by morphology: a process was represented by a mathematical relationship between material flux and topographic attributes such as upslope area and slope, which were surrogates for discharge, shear stress and stream power (e.g. Ahnert, 1976). Furthermore, the empirical parameters in these relationships were commonly spatially-averaged. Recent developments, with their emphasis upon physical realism in process representation and spatial distribution of parameter values and topographic information, have increasingly allowed process and form to be coupled in a more rigorous manner.

Topography itself provides fundamental process information. The upstream contributing area draining to a point and the local topography are key topographic predictors of hydrological response, an index of contributing area divided by local gradient providing a measure of the propensity of water to accumulate and saturate the soil (Beven and Kirkby, 1979; O'Loughlin, 1986). Recent research illustrates the importance of accurate specification of topography for this purpose, and this is implicit in the papers by Wilson and Gallant (Chapter 10) and Lamb *et al.* (Chapter

12). The paper by Lamb *et al.* is concerned with application of TOPMODEL theory to a small catchment, for which detailed spatial mapping of water table and soil characteristics allows spatially distributed verification of TOPMODEL predictions. The paper illustrates the potential need for a power law form of the topographic–soils index, for reasons that may be attributed to inadequate DEM specification. This may be due to the underestimation of drained areas and the overestimation of effective slope, as measured from the DEM as topographic slope, particularly in the vicinity of catchment boundaries. As the authors note, this may be partly due to the assumptions used in the hydrological analysis (for example, that the water table is sub-parallel to the surface slope), rather than to errors in the terrain analysis *per se*. However, there are several technical issues that still require experimental assessment (Moore *et al.*, 1991), including the data structures required for terrain data (grid-based, Triangular Irregular Networks, or contour-and-slope units), and the algorithms for routing accumulating area. Such assessment may conclude that no universally applicable procedures exist, but rather that appropriate procedures are terrain- and problem-dependent.

Although topography contains important process information, topographic parameters alone should not drive process–response models. Landforms subject to erosional and depositional processes change morphology as a result of spatially distributed sediment flux imbalances, identified as local lack of continuity of sediment transport. These imbalances are predicted from sediment transport relationships in which the topography indirectly influences local discharge (since the contributing area controls overland flow or throughflow, for example) and local shear stresses (through the slope effect). However, the shear stress or fluid power that directly determines the sediment transport will depend on depth and/or velocity of flow, which are also dependent on surface friction. Thus, there are additional feedbacks from surface conditions to transport capacity and sediment flux that are not represented by process–response models in which the transport is driven by topographic variables alone, and which therefore are inherently circular and, despite occasional successes such as Ahnert's (1987) studies of the Kall valley, unlikely to provide reliable time scales.

Morphology is thus best considered as the topographic information required as a boundary condition for process models. Many such applications of distributed models are dependent upon careful specification of a terrain surface (e.g. Dietrich *et al.*, 1993). Bates *et al.* (Chapter 13) note the sensitivity of model predictions of floodplain inundation to accurate topographic specification, which is particularly critical in the case of low-gradient floodplain surfaces where elevation errors may be proportionally more significant (Fryer *et al.*, 1994). However, in addition they note that, with the improving capability of specifying topographic boundary conditions given the technical developments noted above, so the ability of numerical models to handle detailed topographic information is also critical. Bates *et al.* show that one of the computational models that they have used (RMA-2) employs numerical stability criteria that do not allow the incorporation of micro-scale topographic features, even though they may be critical to adequate representation of certain processes. Further model development is thus required to allow more complex topographies to be incorporated, which in turn will allow the more realistic floodplain flow simulation that is necessary in order that the simulated flow variables can drive other processes (such as suspended sediment transport). In essence, numerical models may require further development if they are to be applied to the complex geometries characteristic of most natural river channel and floodplain surfaces; in addition, the necessary computing power will increase as these complex surfaces are represented by dense computational grids or finite element meshes.

Related to these problems is the need to improve understanding and representation of sub-grid-scale processes, particularly those associated with scales of topography smaller than the grid spacings typically used in such models (Lane *et al.*, 1996). Here again, issues of data quality converge with the parameterisation process. In hydraulic modelling, the choice of grid spacing, and hence the scale of roughness parameterisation, needs to make reference to the quality and spatial density of topographic information available. The grid spacing will determine the scale of topographic roughness "seen" by the model, and any smaller scales of roughness will have to be parameterised in the roughness coefficient. However, this often means that the roughness coefficient required is not a simple, conventional grain-scale roughness index, but one which incorporates micro-topographic roughness. Some scales of micro-topographic roughness may be sufficiently understood to be represented in an additive coefficient, but others may not. There is thus a need to consider the resolution of topographic information required in order that the roughness parameterisation can be achieved physically; if this resolution cannot be achieved, the implication is that roughness will need to be optimised. It may be necessary to explore a hierarchical system of data acquisition, which allows analysis of sub-grid processes at successive levels of resolution and investigates the scales of information contained within the topographic data using, for example, fractal analysis. This should provide insights into the aspects of topography represented directly in DEMs, those needing additional parameterisation, and the basis for this parameterisation (Lane *et al.*, 1994). Ultimately, model development might need to have recourse to spatially variable grid structures, with a variability in resolution that reflects topographic characteristics. This is an issue that merits further research using high quality and high density topographic data sets.

Changes in Form and the Estimation of Process Rates

In some instances, monitoring changes in form may provide a more successful basis for developing understanding of landform dynamics than monitoring the processes driving those dynamics. In particular, a considerable challenge for geomorphology remains the acquisition of spatially-distributed information on process rates. All too often, the ease of measurement of a process at a fixed point has resulted in insufficient emphasis on the variation of process rates across landscapes. There are a number of ways in which process rates, both spatially-averaged and spatially-distributed, can be estimated from morphological information, depending upon the degree of feedback between form and the processes involved.

In some cases, all that is available is a series of static topographic surfaces. However, this may allow either qualitative estimation of process behaviour or quantitative estimation of process rates. Static topographic information can be used to provide estimations of spatially distributed process characteristics, as in the modelling of hydrological systems (Wilson and Gallant, Chapter 10; Lamb *et al.*, Chapter 12), and in the derivation of patterns of snow-melt in glacierised catchments (Turpin *et al.*, Chapter 16; Copland, Chapter 17) or spatially distributed bed shear stresses in a divided river (Lane, Chapter 14). In the latter case, the evolution of the topography may be inferred from the spatial patterns of shear stress and the bed velocity vectors. Static applications do not allow explicit estimation of process rates, as they lack a time component. However, with a fixed topography, other information may provide a time component. Thus, Hardisty *et al.* (Chapter 19) combine a DEM of the Humber estuary with a time-series of water surface elevations obtained from tide gauge information to estimate estuary-distributed patterns of tidal velocity through time. This qualitative linkage of terrain data with fluid dynamic information may be supplemented by deriving digital terrain models (DTMs) of elevation differences, which result from subtracting successive DEMs in order to highlight regions experiencing net aggradation and degradation. If the pattern of bed velocity vectors is coupled with measured or calculated sediment inflow information, and with the DTM of difference which defines patterns of scour and fill, it is possible to estimate spatial patterns of sediment routing through the topography. Lane (Chapter 14) shows how this combination of distributed information on patterns of river bed erosion and deposition can be combined with a routing equation to provide spatially-distributed estimation of bedload transport rates. In a similar vein, Willis *et al.* (Chapter 15) illustrate how sequential information on glacier surface topography allows the estimation of distributed patterns of glacier velocity.

However, a fully dynamic approach involves application of a mass transport equation and the continuity equation to cases in which the morphology itself varies as a function of time. This requires an initial topographic boundary condition for which the flow pattern can be modelled, an estimation of sediment flux that permits calculation of the patterns of scour and fill, and up-dating of the boundary condition for the next time step. Successive DEMs may then be used to validate the progress of this physical and distributed simulation of the evolution of a landform. While this fully dynamic modelling may still be some way off, even the quasi-dynamic methods using DTMs of difference provide the geomorphologist with a means of estimating the spatial variation in process rates, something which is extremely difficult to achieve using observations of process *per se*. These approaches capture the essential three-dimensionality of links between form and process. However, they also raise important challenges for geomorphologists in terms of ensuring that the representation of morphology is of sufficient quality. As Lane (Chapter 14) shows, in situations where process rates are highly unsteady, different sampling frequencies will result in very different estimates of process rates from the morphological changes observed over time. While this is no different from the experience of direct process monitoring, it serves to indicate that all of the traditional sampling issues remain, together with new problems that reflect the sea-change that is now possible, in which spatial landform dyamics are used to infer process, rather than point-based process measurements being used to infer landform evolution.

BOOK STRUCTURE

In the planning of the original conference, papers were grouped under four headings: landform monitoring; landform modelling; landform analysis; and geomorphological applications. During the process of reviewing papers and compiling the book it became apparent that such a classification was inappropriate, and most authors were clearly making links across these themes. This is not surprising given the interdependence between data quality issues and the nature of the geomorphological application in which the data are employed. The editors' philosophy is thus that if they divorce issues of data acquisition from issues of data use, geomorphologists may find themselves with either: (i) too much data with too little purpose; or (ii) plenty of ideas, but data of insufficient quality to be able to develop useful interpretation. Thus the book is divided into papers that are predominantly technical in emphasis and papers that are predominantly conceptual and applied in emphasis, although most papers bridge that divide.

It is instructive to make a brief comparison of the contents of this book with those of Chorley's (1972) *Spatial Analysis in Geomorphology.* Evans' (1972) chapter on general geomorphometry in the early volume is one obvious forerunner to the terrain analytical framework of this book. However, like many of the other chapters, it has a predominantly statistical emphasis. Spatial analysis in 1972 was interpreted in terms of trend surface analysis of planation surfaces, point pattern analysis of karst depressions and stream sinks in limestone areas, multiple regression analysis of karst depression shapes, harmonic and spectral analysis of landscape grain and scale, principal component and cluster analyses of geomorphic regions, and the probabilities of drainage network structures. With hindsight, it appears that the second particularly forward-looking chapter was King's (1972) review of a process-based simulation model of coastal spit development. The physico-mathematical basis of a large proportion of the terrain-based modelling in this book is its major difference from the approaches embodied in its precursor, and it would have required some prescience in 1972 to have predicted that the intervening quarter of a century would succeed in filling the gaps between Evans' chapter on elevation matrices on the one hand, and King's pioneering application of a process-based simulation model on the other.

REFERENCES

Ahnert, F., 1976. Brief description of a comprehensive three-dimensional model of landform development. *Zeitschrift für Geomorphologie Supplementband*, **25**, 29–49.

Ahnert, F., 1987. Approaches to dynamic equilibrium in theoretical simulations of slope development. *Earth Surface Processes and Landforms*, **12**, 3–15.

Anderson, M. G. and Burt, T. P., 1978. The role of topography in controlling throughflow generation. *Earth Surface Processes and Landforms*, **3**, 331–344.

Ashmore, P. E. and Church, M., 1995. Sediment transport and river morphology: a paradigm for study. *Gravel-bed Rivers IV Workshop – Gravel-bed Rivers in the Environment*, Gold Bar, Washington.

Ashworth, P. J. and Ferguson, R. I., 1986. Interrelationships of channel processes, changes and sediments in a proglacial braided river. *Geografiska Annaler*, **68A**, 361–371.

Baker, V. R. and Twidale, C. R., 1991. The re-enchantment of geomorphology. *Geomorphology*, **4**, 73–100.

Beven, K. and Kirkby, M. J., 1979. A physically-based variable contributing area model of basin hydrology. *Hydrological Sciences Bulletin*, **2**, 43–69.

Brasington, J. and Richards, K. S., in press. The role of DEM resolution and entropy in the parameterisation and evaluation of hydrologic models. In *Geocomputation*, eds M. J. Kirkby *et al.*, John Wiley & Sons, Chichester.

Brunsden, D. and Chandler, J. H., 1996. The continuing evolution of the Black Ven mudslide, 1946–95. Theories put to the test. In *Advances in Hillslope Processes*, eds S. M. Brooks and M. G. Anderson, John Wiley & Sons, Chichester, 869–96.

Chandler, J. H. and Brunsden, J. H., 1995. Steady state behaviour of the Black Ven mudslide: the application of archival analytical photogrammetry to studies of landform. *Earth Surface Processes and Landforms*, **20**, 255–275.

Chander, J. H. and Cooper, M. A. R., 1989. The extraction of positional data from historical photographs and their application in geomorphology. *Photogrammetric Record*, **13(73)**, 69–78.

Chorley, R. J. (ed.), 1972. *Spatial Analysis in Geomorphology*. Methuen, London, 393pp.

Clark, C. D., 1994. Large-scale ice moulding: a discussion of genesis and geological significance. *Sedimentary Geology*, **91**, 253–268.

Cogley, J. G., 1985. Hypsometry of the continents. *Zeitschrift für Geomorphologie, Supplementband*, **53**, 1–48.

Dietrich, W. E., Wilson, C. J., Montgomery, D. R. and McKean, J., 1993. Analysis of erosion thresholds, channel networks, and landscape morphology using a digital terrain model. *Journal of Geology*, **101**, 259–278.

Douglas, I., 1982. The unfulfilled promise: earth surface processes as a key to landform evolution. *Earth Surface Processes and Landforms*, **7**, 101.

El-Ashry, M. R. and Wanless, H. R., 1967. Shoreline features and their changes. *Photogrammetric Engineering*, **33**, 184–189.

Evans, I. S., 1972. General geomorphometry, derivatives of altitude and descriptive statistics. In *Spatial Analysis in Geomorphology*, ed. R. J. Chorley, Methuen, London, 17–90.

Fix, R. E. and Burt, T. P., 1995. Global Positioning System: an effective way to map a small area or catchment. *Earth Surface Processes and Landforms*, **20**, 817–827.

Fryer, J. G., Chandler, J. H. and Cooper, M. A. R., 1994. On the accuracy of heighting from maps and aerial photographs: implications for process modellers. *Earth Surface Processes and Landforms*, **19**, 577–583.

Haff, P. K., 1996. Limitations on predictive modelling in geomorphology. In *The Scientific Nature of Geomorphology*, eds B. L. Rhoads and C. E. Thorn, John Wiley & Sons, Chichester, 337–358.

King, C. A. M., 1972. Some spatial aspects of the analysis of coastal spits. In *Spatial Analysis in Geomorphology*, ed. R. J. Chorley, Methuen, London, 355–369.

Lane, S. N., Richards, K. S. and Chandler, J. H., 1994. Applications of distributed sensitivity analysis to a model of turbulent open channel flow in a natural river channel. *Proceedings of the Royal Society, Series A*, **447**, 49–63.

Lane, S. N., Richards, K. S. and Chandler, J. H., 1995a. Within-reach spatial patterns of process and channel adjustment. In *River Geomorphology*, ed. E. J. Hickin, John Wiley & Sons, Chichester, 105–130.

Lane, S. N., Richards, K. S. and Chandler, J. H., 1995b. Morphological estimation of the time-integrated bedload transport rate. *Water Resources Research*, **31**, 761–772.

Lane, S. N., Chandler, J. H. and Butler, J. B., 1996. Roughness paarmeterisation for hydraulic models. Paper presented to the *Annual Conference of the RGS-IBG*, Strathclyde, January 1996.

Montgomery, D. R., Abbe, T. B., Buffington, J. M., Peterson, N. P., Schmidt, K. M. and

Stock, J. D., 1996. Distribution of bedrock and alluvial channels in forested mountain basins. *Nature*, **381**, 587–589.

Moore, I. D., Grayson, R. B. and Ladson, A. R., 1991. Digital Terrain Modelling: a review of hydrological, geomorphological and biological applications. *Hydrological Processes*, **5**, 3–30.

O'Loughlin, E. M., 1986. Prediction of surface saturation zones in natural catchments by topographic analysis. *Water Resources Research*, **22**, 794–804.

Petrie, G. and Price, R. J., 1966. Photogrammetric measurements of the ice wastage and morphological changes near the Casement Glacier, Alaska. *Canadian Journal of Earth Sciences*, **3**, 783–798.

Pike, R. J., 1995. Geomorphometry – progress, practice and prospect. *Zeitschrift für Geomorphologie, Supplementband*, **101**, 221–238.

Pyle, C. J., Chandler, J. H. and Richards, K. S., 1997. Digital photogrammetric monitoring of river bank erosion. *Photogrammetric Record*, **15**, 753–64.

Richards, K. S., Arnold, N., Lane, S., Chandra, S., El-hames, A., Mattikalli, N. and Chandler, J., 1995. Numerical landscapes: static, kinematic and dynamic process-form modelling. *Zeitschrift für Geomorphologie, Supplementband*, **101**, 201–20.

Robinson, G. J., 1994. The accuracy of digital elevation models derived from digitised contour data. *Photogrammetric Record*, **14(83)**, 805–814.

Schofield, W., 1994. *Engineering Surveying*. Butterworth-Heinemann, London, 554pp.

Schumm, S. A. and Lichty, R. W., 1965. Time, space and causality in geomorphology. *American Journal of Science*, **263**, 110–119.

Shreve, R. L., 1972. Movement of water in glaciers. *Journal of Glaciology*, **11**, 205–214.

Thelin, G. P. and Pike, R. J., 1991. *Landforms of the conterminous United States – a digital shaded-relief portrayal*. United States Department of the Interior, Map I-2206, 16pp.

Tribe, A., 1991. Automated recognition of valley heads from digital elevation models. *Earth Surface Processes and Landforms*, **16**, 33–49.

Welch, R. and Jordan, T. R., 1983. Analytical non-metric close range photogrammetry for monitoring stream channel erosion. *Photogrammetric Engineering and Remote Sensing*, **49**, 367–374.

Williams, P. W., 1987. Geomorphic inheritance and the development of tower karst. *Earth Surface Processes and Landforms*, **12**, 453–465.

Wood, W. F. and Snell, J. B., 1957. *A quantitative system for classifying landforms*. US Department of the Army, Natick, Massachusetts, Technical Report EP-124.

Zevenbergen, L. W. and Thorne, C. R., 1987. Quantitative analysis of land surface topography. *Earth Surface Processes and Landforms*, **12**, 47–56.

Section 1

TECHNICAL ISSUES

2 Datums, Coordinates and Differences

M. A. R. COOPER

Engineering Surveying Research Centre, The City University, London, UK

ABSTRACT

Digital data in the form of three-dimensional Cartesian coordinates of a large but finite number of points are often used to define the size, shape and location of a surface. Several measurement techniques for producing coordinates of points on a terrain surface are described and compared. Changes in coordinates with time are caused by measurement errors, datum differences and by changes in the terrain itself. Since only the latter are of interest, methods for assessing, reducing or eliminating the effects of the other causes are described and related to specific measurement techniques. The importance in terrain monitoring of the quality of measured and derived data is stressed. It is shown how quality may be defined in terms of accuracy, precision and reliability and that such definitions make possible the design of monitoring schemes to meet specific geomorphological objectives.

INTRODUCTION

Numerical methods of analysis have been used in science and engineering for over 300 years. Empirical numerical data are used to test hypotheses about physical processes. In engineering, *a priori* numerical data are a basis for designing a structure or other artefact so that it can be made economically and will be useful and safe. For the last 30 years, decreasing costs and increasing speed and convenience of computers have led to opportunities for routine use of numerical methods in many disciplines. Development of instrumentation for automated generation, at increasing rates, of empirical numerical data creates demands for faster software and hardware for processing, which in turn offer more opportunities for numerical methods in science and engineering. There is no sign yet of any remission of this state of affairs.

The ease with which empirical data can now be generated tends to divert attention away from the quality of data and towards procedures for processing them and interpreting the results. Automation of measuring instruments, data generation and analysis ought to include automated methods for assessing and indicating the quality of data, whether measured or derived by numerical processing, but this is not often done. Manufacturers of instrumentation generally design and make devices that they understand their customers want. It seems as if users do not consider automated assessments of data quality to be as important as low cost, convenience, speed and durability of a device. Probably the first three of these

Landform Monitoring, Modelling and Analysis. Edited by S. N. Lane, K. S. Richards and J. H. Chandler.
© 1998 John Wiley & Sons Ltd.

attributes would be diminished if the device were to include assessment of data quality.

Undetected errors in empirical data can lead to invalid conclusions about physical processes which the data are intended to represent (Fryer *et al.*, 1994). In terrain monitoring, when three-dimensional (3D) coordinates of surface points define the site topography at one epoch and at another, differences between the coordinates (or between surfaces derived from them) can arise for the following three reasons: (i) the coordinate datums for the two epochs are not identical; (ii) errors are present in the coordinates, arising from errors in measurement and additionally errors in surfaces derived from them through interpolation; and (iii) the topography has changed. Geomorphologists (e.g. Lane, Chapter 14; Willis *et al.*, Chapter 15) are interested in the latter and it is therefore critical that the data that geomorphologists use for surface reconstruction attempt to minimise (i) and (ii) with respect to the magnitude of (iii).

For terrain monitoring based on coordinates (or on data derived from them) at two epochs it is essential to ensure that the coordinate datums are identical and to make a careful assessment of the magnitudes of errors in the two sets of data before coming to any conclusions about changes in the terrain based on differences in spatial data at the two epochs. It is not easy to comply with either of these two prerequisites, but it is unscientific to ignore them. Differences in coordinates will exist, whether or not the terrain has changed.

This paper is an attempt to explain in general terms how technological changes have brought not only increasing speed and automation to the acquisition of spatial data, but also difficulties, particularly when differences between data sets are important. Some of the numerous datums in use for positioning are discussed first. The picture is complicated and, at least in Great Britain, will not become simpler for a few years yet. Discussions follow on types of measurement error and indicators of data quality. General principles of measurement systems for generating 3D coordinates that geomorphologists might use for terrain monitoring are then briefly described, with an indication of their data quality. No detailed mathematical descriptions are given here, but appear in the references. If a geomorphologist decides to use automated systems to produce terrain coordinates for monitoring, some understanding is necessary of the complications which often arise. Such understanding may be acquired by the geomorphologist, or provided by working in cooperation with a topographic or engineering surveyor.

COORDINATE REFERENCE SYSTEMS, THEIR DATUMS AND TRANSFORMATIONS

Coordinates are generally numbers representing distances along axes or angles between lines and planes. Without definitions of the origins of axes, their directions and other reference lines and planes, coordinates are meaningless. The most common reference system for defining 3D positions is the right-handed orthogonal cartesian system (XYZ). Seven elements are necessary to define uniquely a datum for 3D positions: the origins of three axes (three translational elements) the directions of the axes (three rotational elements) and the scale along each of the three axes (only one element if the scale is isotropic).

Many coordinate systems are in use for defining positions of points on or near the surface of the earth. They can be for global, regional, national or local positioning. The most commonly used global reference system is known as WGS84 (1984 World Geodetic System) and is a geocentric, earth-fixed, earth-related reference system used to define coordinates derived from satellite signals, in particular the Global Positioning System (GPS). In 3D Cartesian form its Z-axis is the mean axis of rotation of the earth (between 1900 and 1905) and is directed northwards. Its X-axis lies in the mean equatorial plane and passes through the mean Greenwich meridian; the Y-axis lies in the mean equatorial plane and is directed so as to complete a right-handed orthogonal system. Associated with the WGS84 Cartesian system is an ellipsoidal reference system. The generating ellipse has its origin at the origin of the Cartesian system. Its minor axis, the axis of rotation, lies along the Z-axis. Ellipsoidal coordinates are ϕ, λ and h, where ϕ and λ are the ellipsoidal latitude and longitude respectively, and h is the height above the ellipsoid, reckoned along the ellipsoidal normal which passes through the given point. The shape and size of a reference ellipsoid are defined by its semi-major axis (a) and flattening, $f = (a-b)/a$, where b is its semi-minor axis. Conversion formulae between geocentric Cartesian and ellipsoidal coordinates, $(XYZ) \leftrightarrow (\phi\lambda h)$, are exact and well-known (e.g. Seeber, 1993), but some care is needed to avoid round-off errors. Many similar reference systems are in use. They generally have slightly different ellipsoids, origins and alignments. Regional reference systems include the 1989 European Terrestrial Reference Frame (ETRF89) and 1983 North American Datum (NAD83).

Transformation from one geocentric Cartesian system to another, $(XYZ)_1 \leftrightarrow (XYZ)_2$, is usually carried out by a 3D similarity transformation with seven independent parameters: three datum translation elements, three rotational elements, and a uniform scale change (one element). The formulae for these transformations are well-known (e.g. Hofmann-Wellenhof et $al.$, 1992). The minimum data necessary to determine the transformation elements are 3D coordinates, in both systems, of two points and a coordinate, again in both systems, of a third point non-collinear with the other two. In practice, 3D coordinates of several widely spaced points in both systems are used in a least squares estimation process to evaluate the transformation parameters. The procedure can be simplified if the transformation parameters are known to be small (e.g. Leick, 1990).

Eastings and Northings (EN), or map coordinates are derived by projection of points with ellipsoidal coordinates ($\phi\lambda$) according to orthomorphic projection formulae (e.g. Bomford, 1983). In Great Britain the reference ellipsoid used as the basis of mapping is the Airy 1849 ellipsoid. Formulae are available for the transformation $(EN)_{NG} \leftrightarrow (\phi\lambda)_{Airy}$ between transverse Mercator orthomorphic map coordinates (National Grid coordinates) and ellipsoidal coordinates (Anon, 1995b). The transformation is two-dimensional (2D); heights are treated separately, as will be seen later. Ordnance Survey National Grid coordinates used for mapping are known as OSGB36 coordinates, or $(EN)_{NG,OSGB36}$. They are based on a computation of the primary triangulation of Great Britain which was started in 1936 but not completed until 1953. The measurements were horizontal angles and taped baselines. Very soon afterwards, rapid developments in instrumentation, particularly in electronic distance measurement (EDM) showed that the accuracy of the triangulation begun in 1936 should be improved for scientific purposes by new measurements and recomputations. These gave rise to National Grid coordinates $(EN)_{NG,OSGB70}$ which have been superseded by $(EN)_{NG,OS(SN)80}$ as the Ordnance Survey's Scientific Network coordinates, intended for use in geodesy.

In most countries, systematic national mapping preceded the definition of modern geodetic reference systems derived from satellite and other extra-terrestrial signals. Many map datums are therefore independent of such reference systems and the relationships between a map datum and relevant geodetic datums are not well known. This is so in Great Britain. A method has, however, been devised and published recently for the transformation of WGS84 coordinates (from GPS) to National Grid map coordinates: $[(XYZ)_{WGS84}$ or $(\phi\lambda h)_{WGS84}]\to(EN)_{NG,OSGB36}$, which is accurate only to 2 m (Anon, 1995a). The Ordnance Survey will carry out transformations accurate to 0.2 m for a fee of £60 (+VAT) per point (special rates are negotiable for a large number of points). It must be emphasised that these are essentially 2D transformations; although $(XYZ)_{WGS84}$ and $(\phi\lambda h)_{WGS84}$ are 3D, heights are lost in the transformations, which are in effect $[(XYZ)_{WGS84}$ or $(\phi\lambda h)_{WGS84}]\to(xy)_{OSGRS80}\leftrightarrow(EN)_{NG,OSGB36}$, where OSGRS80 is an intermediate transverse Mercator projection of the 1980 Global Reference System (GRS80) ellipsoid (not the same as the Airy ellipsoid). Worked examples of $(XY)_{OSGRS80}\to(EN)_{NG,OSGB36}$ may be found in Anon (1995b).

It is now necessary to consider how heights of a point in different datums are related, particularly ellipsoidal height h in $(\phi\lambda h)_{WGS84}$ from GPS, and orthometric height H above mean sea level found by spirit levelling. Figure 2.1 is a vertical section through a point P on the topographic surface. A reference ellipsoid is shown. Pp, the normal to the ellipsoid which passes through P, defines p on the ellipsoid. The ellipsoidal coordinates of p are $(\phi\lambda)$; those of P are $(\phi\lambda h)$. The equipotential surface arising from the earth's gravity field and rotation that most closely corresponds to the mean level of the oceans is the geoid, commonly called mean sea level (MSL). The normal to the geoid that passes through P coincides with the vertical direction at P. This is the direction to which survey measurements obtained with theodolites and levels are referenced. The distance of P from the geoid, measured along the geoidal normal, is the orthometric height of P or the height of P above MSL. The angle between the vertical and the ellipsoidal normal through P is the deviation (or deflection) of the vertical. Its component in the plane of the figure is ψ. In a sense, "deviation (or deflection) of the vertical" is a mis-nomer; it is the ellipsoidal normal which deviates from the vertical. The separation of the geoid and ellipsoid is N. An approximate relationship between the ortho-metric height and the ellipsoidal height of P is:

$$h = N + H \tag{2.1}$$

so if h has been evaluated in WGS84, for example, the geoid/ellipsoid separation must be known before the height of P above MSL can be derived from it.

Geodesy provides the means of producing contour maps of the geoid/ellipsoid separation N. The geoid itself is a unique surface but the reference ellipsoid chosen

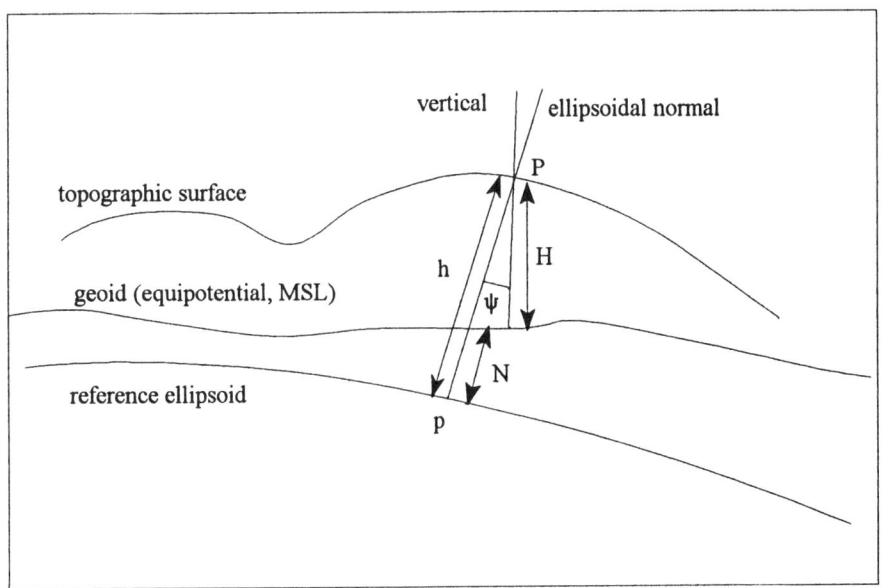

Figure 2.1 Orthometric (MSL) and ellipsoidal heights

in any particular case will determine the form of geoidal contours. Geoid maps can be global, regional or national. They are derived by applications of satellite positioning, astrogeodesy, geopotential modelling and gravimetry, often in combination. Olliver (1992) uses different methods and reference ellipsoids to give several geoid contour maps of Great Britain. Two are of particular interest in the present context. One is derived from GPS observations at 54 triangulation stations and the other from astrogeodetic observations at 192 stations. The geoid/Airy ellipsoid separation N is seen to vary from about 3.5 m to –2 m, mainly from west to east across Great Britain. Contours are at intervals of 0.5 m. Interpolated values could be used in equation (2.1) to convert ellipsoidal heights from GPS to orthomorphic heights and *vice versa*, but only if errors of the order of several decimetres are acceptable. A more recent geoid map of Great Britain by Featherstone and Olliver (1994) is derived from gravimetry, based on GRS80 and intended for scientific use. The measures of absolute geoidal height on which it is based are accurate to better than 10 cm. Geoidal height differences over distances of 100 km are accurate to about 1 cm.

A distinction in the context of geomorphology must be made between the absolute and relative effects of inaccuracies in N. If N is in error by about 1 m, the resultant error in the $(EN)_{OSGB36}$ position of a point transformed from $(XYZ)_{WGS84}$ will be more than 0.5 m and of course the error in H will be about 1 m. If GPS data are used locally to give the positions of specific points on the topographic surface at two different epochs, changes in $(XYZ)_{GPS}$ coordinates for each point should be virtually free from datum uncertainties, but are topographically meaningless. They will include errors from the GPS procedures, but these can be kept small (or at least estimated) by suitable procedures and processing. Coordinate differences $(\Delta X \Delta Y \Delta Z)_{WGS84}$ can be transformed to $(\Delta E \Delta N)_{OSGB36}$ and ΔH with accuracies much greater than absolute coordinates. Variation of N over the area of investigation could be allowed for by selecting at least three bench marks close to the perimeter of the area and obtaining values of N at each using observed $(XYZ)_{WGS84}$ coordinates. Values of N elsewhere could be obtained by linear inter-polation. However, no indication of the measurement errors can be obtained from this procedure. A better procedure is to produce a local geoid map by obtaining by GPS geoidal heights of points close to primary and secondary bench marks distributed throughout the area. Conventional precise levelling of these points gives their orthometric heights. Dodson (1995) reports that this procedure for nine bench marks over an area about 30 km × 15 km gave orthometric heights from GPS ellipsoidal heights to accuracies better than 10 mm. This was in an area where the geoid undulations were not severe, and primary and secondary bench marks were available. If only tertiary bench marks could be used, the results would be significantly worse. Errors arising from datum transformations are additional to the errors of measurement which are inevitably present.

For some sites of limited extent, monitoring can be based on a local datum. A natural or artificial station mark, situated in an area expected to be stable, can be given arbitrary coordinates $(xy)_{LOCAL}$ to define the origin. (At least three such datum marks, located outside the area of interest, should be set up and measure-ments made between them at regular intervals throughout the study to confirm their stability, or otherwise to re-establish the datum.) The direction of one of the coordinate axes can be defined by assigning a value to the bearing of the line from the datum mark to another mark. The other axis is orthogonal to the first; the two axes lie in the horizontal plane through the datum mark. The remaining datum element, scale, is usually defined by the distance measurements and assumed to be unity. Heights are defined by levelling, but a local height datum might be used. Surveys can be computed in the local system, but the curvature of the geoid will have a significant effect on computed positions of points that are more than a few kilometres from the coordinate origin unless some additional correction terms are computed. Since one of the main reasons for using a local datum is its simplicity, it is better to confine the extent of coverage to no more than a kilometre or so from the datum mark. If the geomorphologist can also use the same instruments and methods for successive surveys, and carry out those surveys under similar conditions (e.g. Lane et al., 1994), the derived coordinate differences are less likely to be caused by measurement errors and datum differences than when data from different methods, such as instrumental site surveying, photogrammetry, digitised maps and satellite positioning systems, are used in conjunction with one another.

For the 2D similarity transformation between local coordinates and national map coordinates, for example $(xy)_{LOCAL} \leftrightarrow (EN)_{NG,OSGB36}$ in Great Britain, the values of four transformation parameters must be known: two translational, one rotational and one scale parameter. A unique solution for the four parameters can be obtained if both $(xy)_{LOCAL}$ and $(EN)_{NG,OSGB36}$ coordinates are known for two points, preferably widely spaced. The closeness of the evaluated scale factor to the local projection scale factor (which, for Great Britain, can be found in Anon, 1995b) gives an indication of the accuracy of the transformation, but it is better to have more than two points with both $(xy)_{LOCAL}$ and $(EN)_{NG,OSGB36}$ coordinates and to estimate the transformation parameters by least squares.

MEASUREMENT ERRORS AND INDICATORS OF DATA QUALITY

In surveying and photogrammetry it is customary to identify three different kinds of error: random errors, systematic errors and blunders. Random errors are unavoidable, subject to chance and can be estimated by statistics. Systematic errors are avoidable in theory (and should be reduced to insignificance in practice) and occur through the use of inexact functional models and improperly calibrated equipment. Blunders are unavoidable because they arise through incorrect procedures, usually through human fallibility. Aspects of data quality are related to the three types of error: precision is an indicator of random error, accuracy relates to systematic errors, and reliability to blunders. Internal reliability is an indication of how easy it is to detect a blunder; external reliability is a measure of the effect of an undetected blunder on results – 3D coordinates in the context of this paper. Least squares estimation processes underlie the computation of spatial positioning in many cases, so it is convenient to use them also to indicate the quality of data. Some theory on this aspect of data quality is given by Cooper and Cross (1988, 1991).

Any measurement procedure to be followed in order to produce coordinates is accompanied by a functional model which in general is of the (non-linear) form $F(\ell,x) = 0$, where ℓ represents measured elements (angles, distances, phase differences, photocoordinates etc.) and x represents coordinates. The functional model should include all terms thought to be necessary for adequate accuracy (e.g. temperature and pressure, lens distortion, EDM calibration, and ionospheric delay terms). The datum will be part of the functional model. Measured values substituted for ℓ in function F do not in general give a unique solution for x, so least squares estimation is used. The process is iterative because the functional model is non-linear (except for special cases such as spirit levelling). The weight w_i of a measurement ℓ_i is defined as inversely proportional to its variance: $w_i \propto \sigma_i^{-2} = \sigma_o^{-2}/\sigma_i^{-2}$, where σ_o^{-2} is the variance factor. Least squares processing of the measurements and their variances produces estimates of the coordinates and their variances and covariances (e.g. Cooper, 1987b; Mikhail, 1976). Statistical tests of hypotheses (e.g. Koch, 1987) enable the surveyor to decide, at specific confidence levels, whether or not the precisions of the measurements and of the derived coordinates are satisfactory, systematic errors are present, or a blunder is present in one or more of the measurements. The statistics used for these tests are taken as indicators of the quality of measured and derived data. A fourth indicator of data quality that is relevant to terrain monitoring is the sensitivity of the measurement scheme to change. If one scheme allows the difference in height of a point to be determined to ± 5 mm, for example, then that is a much more sensitive scheme than one which gives the difference to ± 50 mm. Such sensitivities need to be evaluated with respect to the magnitude of the elevation change that is being detected (e.g. Lane, 1994).

An important aspect of these measures of data quality is that a survey scheme can be designed to meet specified criteria, including sensitivity. Indicators of precision and reliability are independent of measured values. They do, however, depend upon what measurements are proposed and what their standard deviations are intended to be. Design is the selection of elements to be measured and the precision with which they should be measured so that the criteria will be met and the scheme is feasible. The coordinate datum is important at the design stage because some of the statistics are datum-dependent (variances and covariances of coordinates are obvious examples).

A very simple assessment of data quality in any particular case is to apply independent checks to measured data on site. How this might be done depends upon circumstances, but the important aspect is that the checks must be independent of (and preferably more accurate than) the procedures used to produce the data that are being assessed. For example points with known $(EN)_{NG}$ coordinates can be included in the current positioning scheme as unknowns and their coordinates evaluated together with those of all other points. Discrepancies $(\Delta N \Delta E)$ at each check point will indicate at least the order of accuracy of the current scheme (and of course the National Grid coordinates). A height difference found by trigonometrical levelling might be checked by spirit levelling. Many opportunities can be taken for independent checks without causing too much delay in the site work.

TECHNIQUES FOR PRODUCING TERRAIN COORDINATES

The most common techniques for producing 3D terrain coordinates are electronic tacheometry, photogrammetry (including the digitising of graphical maps originally produced by photogrammetry) and satellite positioning.

Electronic Tacheometry

Electronic tacheometry will be based on either the national map datum or a local datum. Instrumental tests, checks and adjustments should be carried out according to the manufacturer's handbook and any necessary adjustments made by the user. An error of 60 arc-seconds in horizontal collimation of a tacheometer telescope will give a positional error of about 0.3 m at 1 km unless rectified. The manufacturer should also be asked to carry out regular recalibration of the EDM system. A recent publication by Fédération Internationale des Géomètres (Becker, 1994) gives some simple procedures for routine tests and checks of electro-optical distance meters.

Automated methods for registering, correcting and transforming measured data are carried out by the instrument's microprocessor in response to instructions input by the user via a keypad. The ease with which 3D coordinates can be obtained at the press of a button should not detract attention from data quality. Systematic errors will arise from uncorrected instrumental defects and incorrect mathematical modelling of the atmosphere. In this context, Rüeger (1990) discusses EDM and

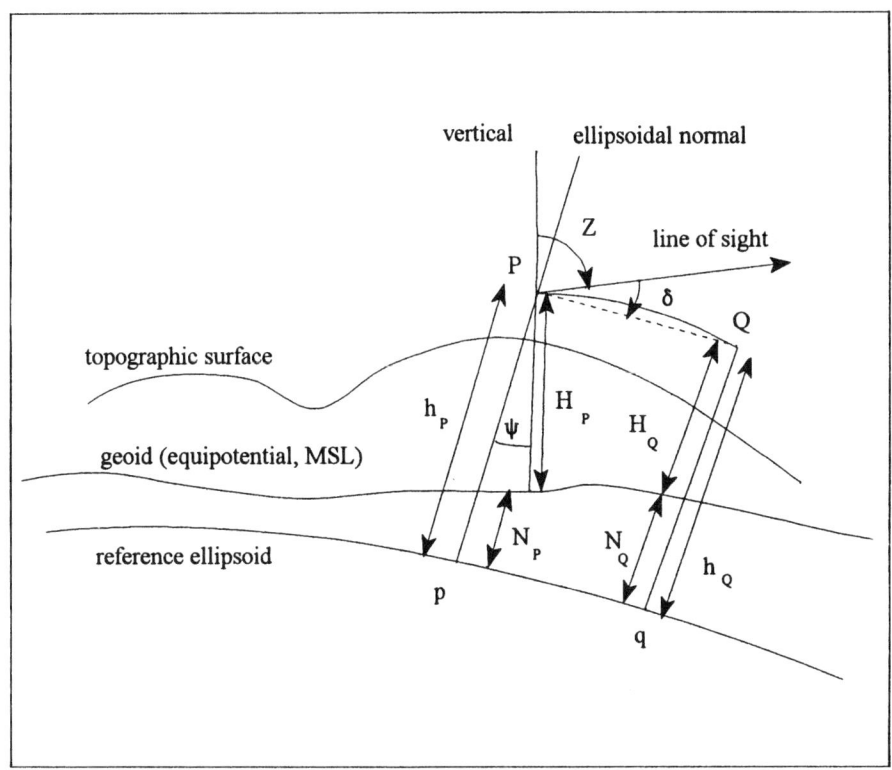

Figure 2.2 Atmospheric refraction and its effect on heights derived from EDM and zenith angles

Cooper (1987a) discusses instrumentation for angular measurement. An example of how significant systematic error could occur in electronic tacheometry is in the trigonometrical height difference between the tacheometer and a prism reflector computed by the tacheometer's microprocessor. Figure 2.2 shows a section along the line from a tacheometer at P to a reflector at Q. The tacheometer records the measured distance $S = PQ$, corrected for the effects of the earth's atmosphere on the velocity of propagation of the EDM signal and the curvature of its path. The zenith angle Z between the vertical through P and the tangent to the line of sight from P to Q is also recorded. The orthometric height difference $\Delta H = H_Q - H_p$ is computed by the tacheometer's microprocessor from:

$$\Delta H = S \cos Z + \frac{(1-k)S^2 \sin^2 Z}{2R} \qquad (2.2)$$

This equation already depends upon approximations: the deviation of the vertical (ψ) is zero; the reference ellipsoid and the geoid are parallel over the line PQ and have constant radius of curvature R. These approximations do not cause significant systematic error over a range of the order of 1 to 2 km, which is the usual maximum with a single prism reflector commonly used for terrain heights. However, the value used for k, the coefficient of atmospheric refraction, is important. It is related to the angle of refraction δ according to $\delta = Sk/2R$, and is a function of the values and vertical gradients of atmospheric temperature and pressure across the line of sight, which are extremely difficult to evaluate in practice. Accordingly, a "standard" value of 0.13 is often used, corresponding to adiabatic conditions and common values of pressure and temperature and of their vertical gradients in temperate climates. Under other conditions, the value of k can be very different from standard. During temperature inversion, the line of sight will be convex towards the centre of the earth and k will be negative; values between 3.0 and +1.5 have been reported by Brunner (1975) and Ashkenazi and Howard (1984). Dodson and Zaher (1985) found negative values of k common for lines up to 1.5 km at different sites in the UK. Lack of knowledge of atmospheric pressure and temperature and their vertical gradients makes accurate trigonometrical levelling very difficult. If a measurement is made over a distance of 1.5 km when the value of k is actually 2.5 but the tacheometer is using the standard value of 0.13, the derived height difference will be in error by about 450 mm, despite the fact that it will be displayed and recorded to 1 mm resolution. Methods for improving accuracy are proposed by Brunner (1975), Ashkenazi and Howard (1984), Dodson and Zaher (1955) and Cooper (1987b).

Processing the measurements by least squares estimation to evaluate 3D coordinates should be commonplace; computer programs are available commercially at reasonable cost, configured for use with a PC. Only those programs that provide the user with adequate information about the accuracy, reliability and precision of the data should be considered for computing coordinates from surveys for terrain monitoring.

Photogrammetry

Photogrammetry has been used for topographic mapping at small, medium and large scales for decades, and it continues to have much potential for geomorphological applications (e.g. Dixon et al., Chapter 4; Lane, Chapter 14; Willis et al., Chapter 15). Routine procedures have been developed by national mapping organisations for standard map production. Photo scales, map scales and vertical contour intervals are interrelated and, until recently, the accuracy of mapping has been determined by the scale of the published map. Assessment of data quality is possible if a "block adjustment" is used to evaluate 3D coordinates of pass-points and tie-points throughout the photographic block based on coordinates of ground control points (e.g. Chandler, 1989; Lane, 1994). The datum is almost always the national map datum. Chandler (1989), Lane (1994) and Fryer et al. (1994) discuss aspects of photogrammetric data quality in relation to geomorphological applications.

The block adjustment is a least squares process using as measurements either 3D stereoscopic model coordinates or 2D photocoordinates. The latter, known as a "bundle adjustment", is the more flexible of the two procedures and gives better indicators of data quality. Its flexibility is shown by Chandler and Cooper (1989) to be particularly useful in geomorphology. Archival photographs taken for different purposes and with different cameras over a 40-year period were used to produce a variety of visual and numerical data about changes in the morphology of a slope. The bundle adjustment allows for aerial and terrestrial photography (from different kinds of cameras) and survey data to be combined in a simultaneous least squares process for 3D coordinates with indicators of their accuracy, reliability and precision. It is also used for very high precision monitoring of structures, when the deformation to be detected is of the same order of magnitude as the measurement errors (typically less than 1 mm on the structure). Rigorous statistical tests of hypotheses about the significance of coordinate differences, using variances and covariances from their cofactor matrix, are then essential for the deduction of deformation at specific confidence levels (e.g. Caspary, 1988).

Traditionally in topographic mapping, data from the block adjustment are used as control for plotting plan positions of features, spot-heights and contours at map scale. The procedure is carried out by operators of stereoscopic photogrammetric analogue plotters. Operators have a high level of coordinated visual and manipulative skill, recently aided by computers in the form of analytical plotters, particularly for the creation of digital terrain models (DTMs; Lane *et al.*, 1994; Dixon *et al.*, Chapter 4). Although an indication of data quality is possible from a bundle adjustment, no such quantitative information about the processes of plotting, cartography and printing is available. Recently, however, digital images instead of photographs and digital 3D data rather than printed maps have brought both advantages and disadvantages in relation to traditional procedures and products. Konecny (1995) gives a comprehensive overview of terrestrial, airborne and spaceborne data acquisition systems for mapping.

For terrain monitoring, digital photogrammetry offers many advantages over other methods, photogrammetric or not, but automated procedures and processes for 3D computerised terrain modelling and visualisation are still under development. Associated automated indication of the quality of the data lags far behind, as can be verified by reference to a recent newsletter (Timmerman, 1995) of l'Organisation Européenne d'Études Photogrammétriques Experimentales (OEEPE). The Organisation's research plan includes 14 topics under investigation, including the development of a quality model and definition of quality concepts and parameters to be used for general (application-independent) standards in photogrammetry; definition of rules for data used in DTMs; how to define accuracy and quality of digital elevation models (DEMs); and other matters of interest to terrain monitoring including standardisation of 3D Geographical Information Systems (GISs). At the 1995 Cambridge Conference for National Mapping Organisations (Ordnance Survey, 1995), very few papers discussed the quality of spatial positioning information in terms that would give confidence to users who are seeking to identify small changes in terrain morphology. Geomorphologists may, in such cases, decide to establish, define and maintain a local datum for such studies. Until OEEPE has published its findings, and national mapping organisations see that application of strict quality control of spatial positioning information for terrain monitoring is profitable, geomorphologists should be vigilant in using mapping information for detecting change. A published map printed at a specific scale has obvious accuracy limitations and a clearly defined datum. Three-dimensional data from digital photogrammetry, however, can be displayed and measurements made at all scales using computer graphics. Even though tens of thousands of points may have been matched automatically and used for surface-fitting, the accuracy limitations are generally not obvious. The datum will probably be the same as the control survey datum, but the next section shows that if GPS has been used for the control, accuracy of MSL heights should be examined particularly closely.

Satellite Positioning

As a generator of coordinates, the US Department of Defense's NAVSTAR-GPS (Navigation Satellite Timing and Ranging Global Positioning System) is unsurpassed in the variety of its applications (see Twigg, Chapter 3). It was designed and put into effect to meet military objectives set by the US Department of Defense in 1978 to provide satellites and signals for world-wide real-time navigation, with positional errors of the order of 10 m, tolerant of interference (intentional or not) and capable of being used when the receiver is travelling at high speeds (e.g. military aircraft).

The signals transmitted by the satellites are now exploited by civilian users in ways and for purposes that were never intended. The result is that commercial receivers and systems vary from simple receivers costing a few hundreds of pounds to systems costing 100 times as much. In the former case, 2D positions are obtained with accuracies of a few hundreds of metres, and in the latter case, differences in position can be determined to accuracies better than 10 mm over 10 km. Clearly, choice of equipment for terrain monitoring is crucial.

Military GPS receivers make use of time information that is coded and transmitted by modulation of carrier waves in the L-band of the frequency spectrum from at least four satellites. In effect, the distance from the receiver to each satellite is measured, so knowing the positions of the satellites (which are transmitted with the timing data) it is possible to compute the position of the receiver. Ranges to three satellites are the minimum necessary to determine $(XYZ)_{\text{WGS84}}$ coordinates of the receiver, but the fourth satellite is needed to reduce timing errors. Civilian users of GPS for surveying do not always have access to the encoded timing data and in any case, the designed 10 m accuracies are too low. Alternative methods of using the GPS signals are therefore needed. Many ways have been devised for treating the signals and processing GPS data for surveying purposes (Wells, 1986).

Two procedures are common to most GPS surveying systems. The modulation containing the coded time information is removed, but the phase of the carrier signal is tracked and used as the basic measurement. This means, in effect, that the integer number of carrier wavelengths in the range from receiver to satellite is not known, but the fractional part of the wavelength in the range is known. There is said to be an "integer ambiguity" which must be resolved. The second common procedure in surveying with GPS is to use differential positioning, where one receiver remains stationary whilst another is transported over the area to be surveyed. The moving receiver can be stopped for a minute or so at each point whose coordinates are required to collect data, or it can collect data continuously, recording every second or so as it moves. Data logged by the stationary receiver and by the moving receiver are generally processed after the site work has finished ("post-processing"). The "baseline vector" between the stationary receiver and the moving receiver at each epoch of data is computed. If at some time during the site work, data were collected at one or more survey stations with $(EN)_{OSGB36}$ coordinates, the baseline vectors $(\Delta X \Delta Y \Delta Z)_{WGS84}$ can be transformed to give OSGB36 map coordinates. Manufacturers' software has been developed to guide the user both during data collection on site and during post-processing. Matters such as integer ambiguity resolution, recovery of loss of lock on satellites, other system operations and computation of map coordinates are largely automated, but allow interaction with the user. Recently, "real-time kinematic positioning" has become possible. This requires a radio link between the moving receiver and a stationary receiver at a point of known coordinates.

Systematic errors arise mainly from ionospheric and tropospheric effects on the signals. These errors affect heights more than plan positions. Precision of positioning is degraded if the satellite configuration or the geometry of the baselines and survey stations used to define the transformation to map datum is weak. Blunders can arise through undetected cycle slips and incorrect resolutions of integer ambiguities. Any good software system will indicate the quality of the data it produces. In any case, the user can devise simple independent checks of data quality. When a survey is partially completed and post-processing has taken place, the next data collection phase should include visits to a few points already surveyed and coordinated in the first phase. Discrepancies between the two sets of coordinates are a good indication of the overall accuracy.

The original registered data are subjected to considerable numerical processing in order to produce coordinates. The degree of complexity of the process can be grasped if it is recalled that the phase of an electromagnetic signal is the "observable" (in fact a phase difference between two signals) and coordinates are the result; both entities are abstract concepts in so far as they cannot be detected by the senses. GPS procedures in surveying are various and developing; even nomenclature is not yet standard. However, some terms are generally accepted. Chapter 7 of Hofmann-Wellenhof *et al.* (1992) is recommended as a guide to the most important aspects of GPS practice for those who are not specialists in the technique, but who think it might be useful and want to understand more about it. The specific chapter reference is particularly valuable because it concludes by listing information to be given in a report of a GPS survey. Such information is a quality audit trail for GPS surveys, but the principle is applicable to all surveying techniques. It is not necessary here to repeat the information, but the following quotation will serve as an epilogue to this keynote paper on terrain monitoring:

When surveys are properly performed and documented, they provide a lasting contribution. . . . Often, data can be used in later years by others to study a particular phenomenon or the work may be included in a larger project. Proper use of measurements can only be made when the survey is thoroughly documented for posterity.

Hofmann-Wellenhof *et al.* (1992)

REFERENCES

Anon, 1995a. *National Grid/ETRF89 Transformation Parameters*. Ordnance Survey, Southampton, 10pp.

Anon, 1995b. *The Ellipsoid and the Transverse Mercator Projection*. Ordnance Survey, Southampton, 25pp.

Ashkenazi, V. and Howard, P. D., 1984. An empirical method for refraction modelling in trigonometrical heighting. *Survey Review*, **27**(213), 311–322.

Becker, J.-M., (ed.) 1994. *Recommended Procedures for Routine Tests and Checks of Electro-Optical Distance Meters (EDM)*. Fédération Internationale des Géomètres (FIG Bureau, c/o RICS, London), 61pp.

Bomford, G., 1983. *Geodesy*, 4th edition reprinted with corrections. Clarendon Press, Oxford, 855pp.

Brunner, F. K., 1975. Coefficients of refraction on a mountain slope. *Unisurv G*, **22**: 81–96 (School of Surveying, University of New South Wales).

Caspary, W. F., 1988. *Concepts of Network and Deformation Analysis*. Monograph 11, School of Surveying, University of New South Wales, 183pp.

Chandler, J. H., 1989. *The acquisition of spatial data from archival photographs and their application to geomorphology*. Unpublished PhD Thesis, City University, London.

Chandler, J. H. and Cooper, M. A. R., 1989. The extraction of positional data from historical photographs and their application to geomorphology. *Photogrammetric Record*, **13**(73), 69–78.

Cooper, M. A. R., 1987a. *Modern Theodolites and Levels*, 2nd edition. BSP Professional Books, Oxford, 258pp.

Cooper, M. A. R., 1987b. *Control Surveys in Civil Engineering*. BSP Professional Books, Oxford (originally Collins), 381pp.

Cooper, M. A. R. and Cross, P. A., 1988. Statistical concepts and their application in photogrammetry and surveying. *Photogrammetric Record*, **12**(71), 637–663.

Cooper, M. A. R. and Cross, P. A., 1991. Statistical concepts and their application in photogrammetry and surveying (continued). *Photogrammetric Record*, **13**(77), 645–678.

Dodson, A. H., 1995. The status of GPS for height determination. *Survey Review*, **33**(256), 66–76.

Dodson, A. H. and Zaher, M., 1985. Refraction effects on vertical angle measurements. *Survey Review*, **28**(217), 169–183.

Featherstone, W. E. and Olliver, J. G., 1994. A new gravimetric determination of the geoid of the British Isles. *Survey Review*, **32**, 464–478.

Fryer, J. G., Chandler, J. H. and Cooper, M. A. R., 1994. On the accuracy of heighting from aerial photographs and maps: implications to process modellers. *Earth Surface Processes and Landforms*, **19**, 577–583.

Hofmann-Wellenhof, B., Lichtenegger, H. and Collins, J., 1992. *GPS Theory and Practice*. Springer Verlag, Vienna, 326pp.

Koch, K.-R., 1987. *Parameter Estimation and Hypothesis Testing in Linear Models*. Springer-Verlag, Berlin, 378pp.

Konecny, G., 1995. Data acquisition for mapping and mapping updates using terrestrial, airborne and spaceborne methods. In *Proceedings, Cambridge Conference for National Mapping Organisations*. Ordnance Survey, Southampton, 24pp.

Lane, S. N., 1994. *Monitoring and modelling morphology, flow and sediment transport in a gravel-bed stream*. Unpublished PhD thesis, University of Cambridge, 373pp.

Lane, S. N., Chandler, J. H. and Richards, K. S., 1994. Developments in monitoring and terrain modelling small-scale river-bed topography. *Earth Surface Processes and Landforms*, **19**, 349–368.

Leick, A., 1990. *GPS Satellite Surveying*. John Wiley & Sons, New York, 352pp.

Mikhail, E. H., 1976. *Observations and Least Squares*. IEP, New York, 497pp.

Olliver, J. G., 1992. Space-derived geoid maps of Great Britain. *Survey Review*, **31**, 310–320.

Ordnance Survey, (pub.) 1995. Proceedings of the Conference for National Mapping Organisations, St Johns College, University of Cambridge, 1995.

Rüeger, J. M., 1990. *Electronic Distance Measurement*, 3rd edition. Springer-Verlag, Berlin, 266pp.

Seeber, G., 1993. *Satellite Geodesy*. Walter de Gruyter, Berlin, 531pp.

Timmerman, J., (ed.) 1995. *OEEPE Newsletter 1*. ITC, Enschede, 12pp.

Wells, D., (ed.) 1986. *Guide to GPS Positioning*. Canadian GPS Associates 1986, Fredericton, *c.* 600pp.

3 The Global Positioning System and its Use for Terrain Mapping and Monitoring

DAVID R. TWIGG
Department of Civil and Building Engineering, Loughborough University, UK

ABSTRACT

A review of the technology, techniques and applications of global positioning system (GPS) surveying is presented, with particular reference to the requirements of terrain mapping and monitoring. Three case studies are described, showing the use of GPS in a control survey, a surface detail survey and an underwater detail survey in Turkey and the UK.

INTRODUCTION

The NAVSTAR Global Positioning System (GPS) is a satellite-based navigation system designed and operated for military purposes by the United States Department of Defense (DoD). The satellites continuously transmit radio signals which enable a single receiver on land, on sea or in the air to determine its instantaneous position 24 hours a day in any weather conditions on a world-wide basis.

Since the early stages in the development of GPS, the transmitted signals have been used by civilians for navigation purposes and it is now possible to purchase hand-held receivers for as little as £200 which give a positional fix to within approximately 100 m. Alongside this basic usage, the transmitted signals have been exploited for surveying and geodetic purposes by surveyors and equipment manufacturers to give the relative position of two receivers to a level of accuracy that was never envisaged for the system. This relative position accuracy is typically at the sub-centimetre level for baselines up to 10 km, but the cost of achieving it can be over £50 000 for a pair of receivers and the associated processing software. With such a vast range of capabilities and costs, it is clearly necessary to decide what level of accuracy is needed for the intended application and to choose equipment and measuring method accordingly.

A good understanding of GPS technology and techniques, and of the extensive jargon currently in use, is essential for anyone involved in either collecting or using

Landform Monitoring, Modelling and Analysis. Edited by S. N. Lane, K. S. Richards and J. H. Chandler.
© 1998 John Wiley & Sons Ltd.

GPS data. This paper reviews the state of GPS surveying, with particular reference to the requirements of terrain mapping and monitoring applications, before presenting three case studies of recent uses of GPS during fieldwork in Turkey and the UK. It should be noted that "accuracy", rather than "precision" or a combination of the two, is adopted throughout the paper as a measure of the quality of results obtained. Accuracy is the term generally adopted in GPS literature and is universally used by equipment manufacturers in their specifications.

GPS SATELLITES AND SIGNALS

The full GPS constellation, which was finally achieved in 1995, some 17 years after the launch of the first system satellite, consists of 24 operational satellites arranged in six orbital planes at heights of approximately 20 000 km above the earth's surface. The orbital planes and the spacing of the satellites in each plane are arranged to give a minimum of four satellites visible at any one time at any world-wide location.

Each satellite transmits two radio signals known as the L1 and L2 carriers, each modulated with one or two codes and a navigation message. The purpose of the codes is to enable a receiver to lock on to the carrier signal in order to determine the travel time of the signal from satellite to receiver and hence calculate the distance between the two.

The Coarse Acquisition code (C/A code) is carried by the L1 signal, is fully accessible by all receivers and its use leads to the Standard Positioning Service (SPS). However, the full capabilities of the C/A code are controlled by the DoD by a process known as Selective Availability (SA). This is an intentional artificial degradation of the satellite clock time and satellite position carried as part of the navigation message. Although it was announced in 1996 that SA is to be phased out within the next decade, it is currently set such that the SPS gives a positional fix within 100 m in plan and 150 m in height, 95% of the time.

The Precise code (P code) is carried by both the L1 and L2 signals and its use leads to the Precise Positioning Service (PPS) which gives a positional fix to within 15 m. Since 1994 the P code has been encrypted, in which form it is referred to as the Y code, and the use of the PPS has been restricted to authorised military users. The encryption process, known as anti-spoofing (AS), prevents the signals from being imitated with false information. It should be noted, however, that equipment manufacturers have developed methods of dealing with AS which enable receivers to make full use of the L2 signal even though they do not have access to the P code itself.

The navigation message is transmitted on both the L1 and L2 signals and contains, amongst other data, orbital parameters for computing satellite positions, satellite clock corrections and ionospheric corrections. The navigation message is regularly updated from the Master Control Station in Colorado Springs which collects data from several monitoring stations located around the world.

GPS POSITIONING TECHNIQUES

Pseudo-ranging and Carrier Phase Observations

The basic principle of GPS positioning is that a receiver's location can be determined as the intersection of measured distances from the unknown receiver station to a number of satellites whose positions are known at the time of measurement. These distances can be determined by two distinct types of observations of the transmitted signals: pseudo-ranges and carrier phases.

Pseudo-ranges

By comparing the received code sequence from a satellite with that generated relative to its own clock, a receiver can measure the travel time of the signal and hence calculate the distance by multiplying this time by the speed of light. However, this distance will be in error due to an unknown time difference between the atomic clock used in the satellite and the less accurate clock used in the receiver, as well as other less significant effects such as atmospheric propagation delay. The uncorrected observed distance is known as a pseudo-range. If pseudo-ranges are observed to three satellites, whose positions are known, then the three unknown coordinates of the receiver can be determined. These coordinates will not be very accurate because the receiver clock error has not been accounted for. If observations are made to four satellites then the measured pseudo-ranges can be used simultaneously to determine both the clock error and receiver coordinates, which are now free from this error. Hence there is a system design specification for a minimum of four satellites to be visible at all times. This so-called navigation solution of the instantaneous position of a single receiver is the most basic GPS method and would be capable, without SA, of producing a position to within about 30 m when used with the C/A code.

Carrier Phases

Accurate distance measurement using pseudo-ranging is limited by the clock used in the receiver, as 0.1 microseconds (a typical clock specification) is equivalent to a distance of 30 m travelled by the signal. One way in which the system has been exploited in order to give accuracies required for surveying use is to make carrier phase shift measurements. This phase observable is the difference between the received carrier phase and the phase of the constant reference frequency generated by the receiver. As phase resolution is about 1% of the wavelength, this is equivalent to an accuracy of about 2 mm in the measured distance, as the wavelengths of the L1 and L2 signals are 0.19 and 0.24 m respectively. This measurement process only gives the fractional part of a wavelength, however, and it is also necessary to determine the whole number of wavelengths, known as resolving the ambiguity, as well as eliminating or reducing the effects due to satellite clock error, receiver clock error and atmospheric delay errors. This is done by considering linear combinations of range equations (which are written in terms of the measured phase difference, integer ambiguity and the above errors) between one or two receivers and one or more satellites.

Thus a single-difference equation giving the range difference for a single receiver observing two satellites simultaneously eliminates the receiver clock error. A single-difference equation for the range difference between two receivers simultaneously measuring to the same satellite eliminates the much smaller satellite clock error. Both these single-differences also greatly reduce the errors due to atmospheric delay. A double-difference equation gives the difference between two single-differences obtained from two receivers measuring to two satellites simultaneously. This type of equation not only eliminates both clock errors and reduces atmospheric effects, but also enables the integer ambiguity to be resolved. Finally, a triple-difference equation can be written as the difference between two double-differences and thus involves two receivers observing two satellites at two different times. Although triple-differences eliminate the ambiguities, they are mainly used to obtain good approximate positions to be used in a double-difference solution. When using phase observables in practice, many observations are made from two receivers to at least four satellites and the resulting single-, double- and triple-difference equations are used in a least squares solution to obtain the ranges and hence receiver coordinates.

Point and Relative Positioning

The designed method of use for GPS is for a single receiver to observe pseudo-ranges to the satellites in order to determine the absolute coordinates of the point position at which the receiver is located. The way the system has been adopted by the surveying community, however, is for phase observations to be made in conjunction with the advantages of relative, or differential, positioning. One receiver is located at a known point and the baseline, or vector, between this and a second receiver at an unknown point is determined in order to give the position of the second relative to the first. The advantage of this approach is that many of the system errors are either eliminated or their effects are greatly reduced.

Several field techniques have evolved, under a variety of names, for relative positioning but they all use the same basic principle. One receiver, the Reference, is located at a known point and records observations continuously for the duration of the survey work. A second receiver, the Rover, moves from point to point and collects enough data in order to calculate each position relative to the Reference. The only difference between the various techniques is the time spent at each point by the Rover, with its consequent effect on the way the ambiguities are resolved and the resulting accuracy. The most common techniques used are rapid-static, stop–go, kinematic and on-the-fly, although it should be noted that different users and different equipment manufacturers use different names for the same methods.

Rapid-static

This will give the most accurate position and the Rover remains on a point for a sufficient period of time for the ambiguities to be resolved and the system accuracy to be achieved. The required time depends on the receiver being used, the baseline length, the number of satellites observed, the satellite geometry and the state of the ionosphere. The minimum time will be about 5 min for a dual frequency (L1/L2) receiver and about 15 min for a single frequency (L1) receiver. The name rapid-static refers to the short observation periods, brought about by improvements in receiver technology and data processing algorithms, compared with those used in the so-called static approach.

Stop–Go

After an initialisation period, during which the ambiguities are resolved, the Rover moves from point to point remaining stationary for a short period at each. The initialisation stage can be carried out by obtaining a rapid-static fix at an unknown point or by occupying a known point for a minimum of four epochs of observations, which is about 20 s for a typical observation recording rate of 5 s used in stop–go surveying. After initialisation, the Rover should remain at each of the subsequent points to be surveyed for a minimum of four epochs. During this moving part of the survey it is essential that the Rover maintains lock on at least four satellites so that the ambiguities can be carried forward. If lock is lost then the Rover must be re-initialised before the survey can continue.

Kinematic

This is very similar to the stop–go technique except that observations are made at fixed time intervals rather than at distinct points during the moving part of the survey. The Rover can therefore move continuously around the area to be surveyed after the initialisation period and provided lock is maintained to at least four satellites.

On-the-fly

This enables a kinematic survey to be carried out without the need for an initialisation period but it does require the use of a dual frequency receiver. The Rover moves continuously from the start of the survey and as soon as the ambiguities are resolved they are carried forward to subsequent points and back to the points surveyed before ambiguity resolution. Once again, the Rover must maintain lock on at least four satellites.

Each of these different relative techniques has its own uses in the context of terrain mapping and monitoring applications and these are summarised in Table 3.1.

Real-time and Post-processed Results

When using a single receiver for the Standard Positioning Service, results are obtained in real time in that the coordinates of a point are computed in the field at almost the same time as the observations to the satellites are made. The standard approach for high accuracy GPS, however, which involves two receivers making code and phase observations over extended periods of time, is for the Reference and

Table 3.1 Relative positioning applications in terrain mapping and monitoring

Observations	Field technique	Typical applications
Code only	Any	GIS applications
Code and phase	Rapid-static	Location of control stations; local transformation parameters; local geoid determination
	Stop–Go	Detail surveys with specific points of interest
	Kinematic	Detail surveys in smooth terrain, e.g. hydrographic surveys (if initialisation is possible); snowfield and ice-cap surveys
	On-the-fly	Hydrographic surveys when initialisation is not possible

Rover to each record the measured data for subsequent down-loading and post-processing in a computer package. The results are only known some time after the fieldwork has been completed (but typically the same day) and it is only at this stage that the surveyor knows whether ambiguity resolution has been successful or not.

One of the recent developments in high accuracy GPS is a real-time processing capability. The data observed at the Reference are continuously transmitted by a radio link to the Rover where they are combined with the Rover observations and processed almost immediately. The results of the survey are obtained as the fieldwork is taking place, but disadvantages of using real-time GPS are a substantial increase in the cost of the system and a limited distance, typically less than 5 km, over which data can be transmitted.

Finally, it should be noted that pseudo-ranges from a single suitable receiver can be recorded and post-processed in order to average out the effects of Selective Availability. The accuracy of a positional fix will be increased from the SPS value of 100 m to within about 10 m if observations are made over a sufficient period of time.

Summary of Techniques and Accuracies

The previous sections have outlined a wide variety of methods by which GPS results can be obtained: positioning can be absolute or relative; observations can be code only or code and phase; receivers can be single frequency (L1) or dual frequency (L1/L2); results can be post-processed or obtained in real time; relative positioning can be rapid-static, stop–go, kinematic or on-the-fly. The accuracies involved for each of these variations are summarised in Table 3.2 and some typical costs involved in achieving them are listed in Table 3.3.

Although a dual frequency receiver may be needed for some high accuracy work, the main advantage of this type of receiver over a single frequency one is not the increase in accuracy but the much shorter time required for rapid-static obser-vations and the initialisation part of stop–go and kinematic surveys. The most appropriate combination for general terrain mapping and monitoring applications in terms of budget requirements and achievable accuracy is listed in Table 3.4.

Table 3.2   GPS positioning accuracies

		Real-time	Post-processing	
Point		100 m (95%)	<10 m	
		L1/L2	L1	L1/L2
Relative	Code	1 m	1 to 2 m	0.5 to 1 m
	Code and phase	Rapid-static 5 to 10 mm to 1 ppm	5 to 10 mm + 2 ppm	5 mm + 1 ppm
		Stop–go 1 cm + 2 ppm	1 to 3 cm + 2 ppm	1 to 2 cm + 1 ppm
		Kinematic 1 cm + 2 ppm	1 to 3 cm + 2 ppm	1 to 2 cm + 1 ppm
		On-the-fly 1 cm + 2 ppm	–	1 to 2 cm + 1 ppm

Notes.   1. The ppm values are parts per million, or mm per km, of the measured baseline.
2. Accuracies in height are typically double the position accuracies listed in the table.

Table 3.3   Some typical 1996 costs for relative GPS positioning

Observations	Signal(s)	Cost
Code	L1	£5000–£10 000
Code and phase	L1	£15 000–£20 000
	L1/L2	£30 000–£40 000
	L1/L2 real-time	£45 000–£55 000

Note. Costs are for two receivers and processing software.

Table 3.4   GPS techniques most suitable for general terrain mapping and monitoring applications

Positioning method	Relative
Observations	Code and phase
Receiver	Single frequency (L1)
Results	Post-processed
Field techniques	Rapid-static for control; stop–go for detail; kinematic for detail

PRACTICAL GPS SURVEYING: SOME GENERAL CONSIDERATIONS

Survey Planning

Although in theory it is possible to carry out GPS surveying 24 hours a day, there are certain weather conditions in which the surveyor would not want to work and there are certain periods when the number of satellites visible and their distribution are not ideal for the intended application and the required accuracy. With most GPS processing packages it is possible to obtain plots of satellite availability for a given location and a given date and time using the latest available navigation message, or almanac, which is always recorded when observations are being made.

Two typical plots centred on Loughborough for 5 September 1995 are given in Figures 3.1 and 3.2. It can be seen from Figure 3.1 that the number of satellites visible during the 24 hour period varies from four to nine. However, it can also be seen that the Geometric Dilution of Precision (GDOP) increases at times corresponding to low numbers of visible satellites. GDOP is a measure of the expected accuracy of the results based on estimated errors for coordinates and time, reflecting the combined satellite and receiver geometry. For good results, the GDOP value should be as low as possible and periods of high GDOP, say greater than 8, should be avoided. Thus Figure 3.1 indicates that periods around 07:30 and 14:00 are not suitable for high accuracy GPS surveying. It should be noted that the plotted GDOP values assume that there are no obstructions such as buildings, trees or mountains reducing the number of visible satellites. Any such obstruction could drastically change the situation to the extent that GPS surveying is not possible at all if less than four satellites are available. Other DOP values, such as the Positional Dilution of Precision (PDOP) shown in Figure 3.1, can be computed but they are simply components of the composite GDOP value.

The sky plot in Figure 3.2 shows the distribution of satellites during the 24 hour period in terms of azimuth and elevation. The dark grey band represents the portion of the sky up to a cut-off angle of 15° above the horizon. Satellites in this band are not usually observed as the signals are affected by the atmosphere to a much greater extent than those from satellites with a higher elevation. The most striking aspect of Figure 3.2 is that, for this location, there are no visible satellites in the northern sector of the sky above the cut-off angle. This has the benefit that any obstructions to the north of the receiver will not affect the observations as the number of available satellites will not be reduced. It also suggests that it is better to hold the antenna to the south of the surveyor's body when walking around during a stop–go survey in order to minimise the possibility of blocking out satellites.

The effect of changing locations is shown in the sky plots of Figures 3.3 and 3.4 for two other sites visited in 1995. Figure 3.3 is for the Oksfjordjokelen area of Arctic Norway and it can be seen that satellites now appear to the north above the 15° cut-off angle but the satellites to the south have reduced maximum elevations. The use of the northern satellites in this situation was essential during recent fieldwork as the base-camp reference receiver for the intended work was located in a valley with an open vista to the north but with steeply rising mountains in the whole of the southern sector. Figure 3.4 is for the Çumra region of central-southern Turkey and it can be seen that for this latitude the southern rising satellites achieve a maximum elevation of about 120° but there are no satellites above the horizon in the northern sector. This situation presented no problems in this particular case as the survey work was being carried out on an extensive elevated plain, but GPS surveying in nearby mountainous regions could be problematic.

Coordinate Systems, Projections and the Geoid

Unless baseline vectors are required for a control survey network adjustment, the results obtained from a relative GPS survey are the coordinates of each position of the roving receiver as determined from the known or assumed coordinates of the

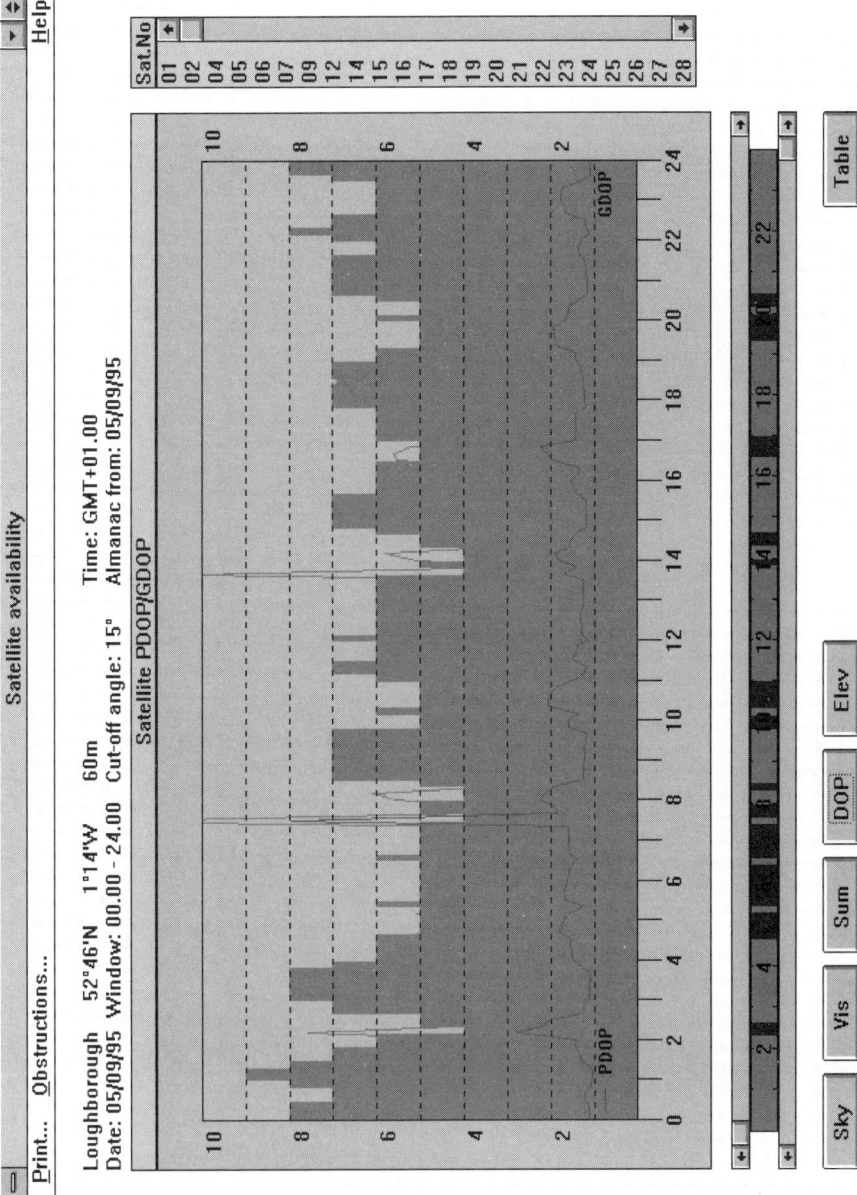

Figure 3.1 Satellite availability over Loughborough, UK (plot obtained from Leica SKI software) (lower line PDOP, upper line GDOP)

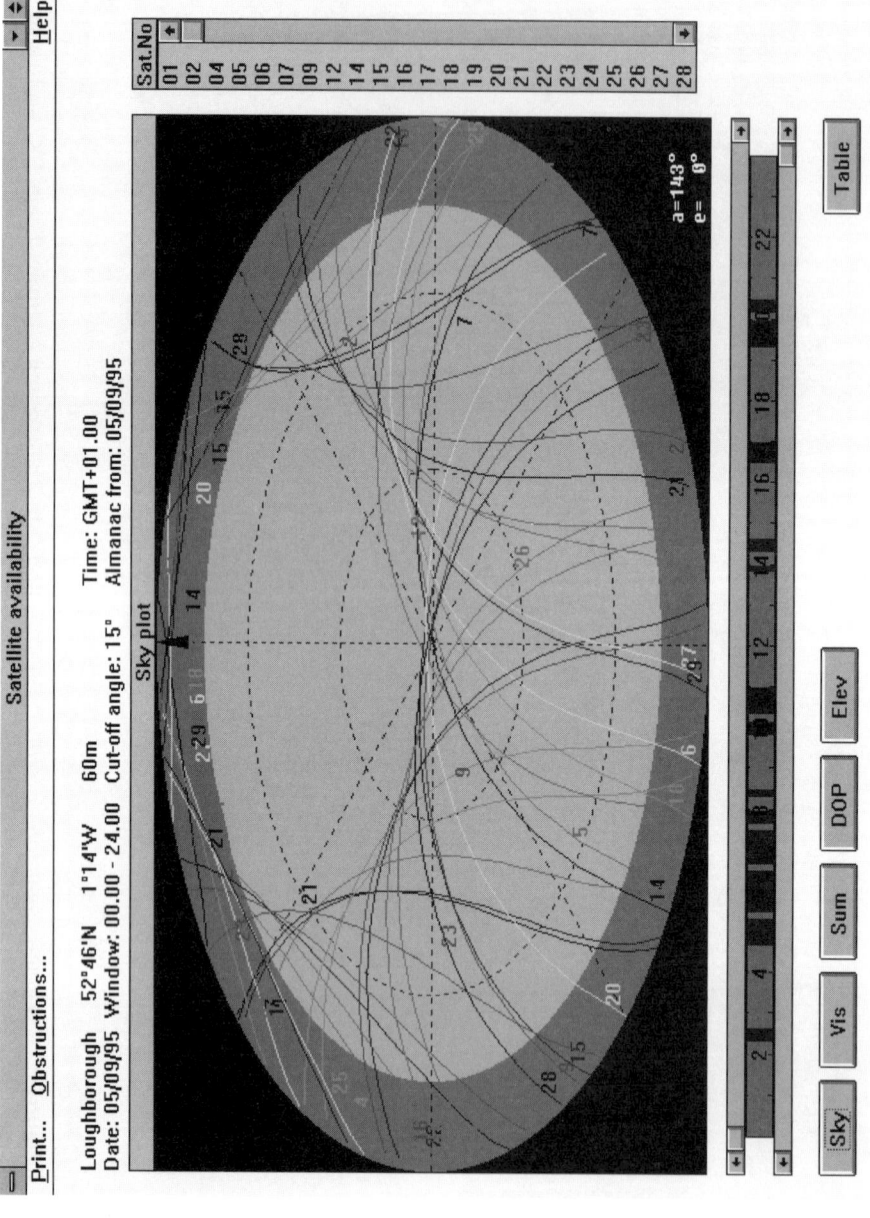

Figure 3.2 Sky plot of satellite availability over Loughborough, UK (plot obtained from Leica SKI software)

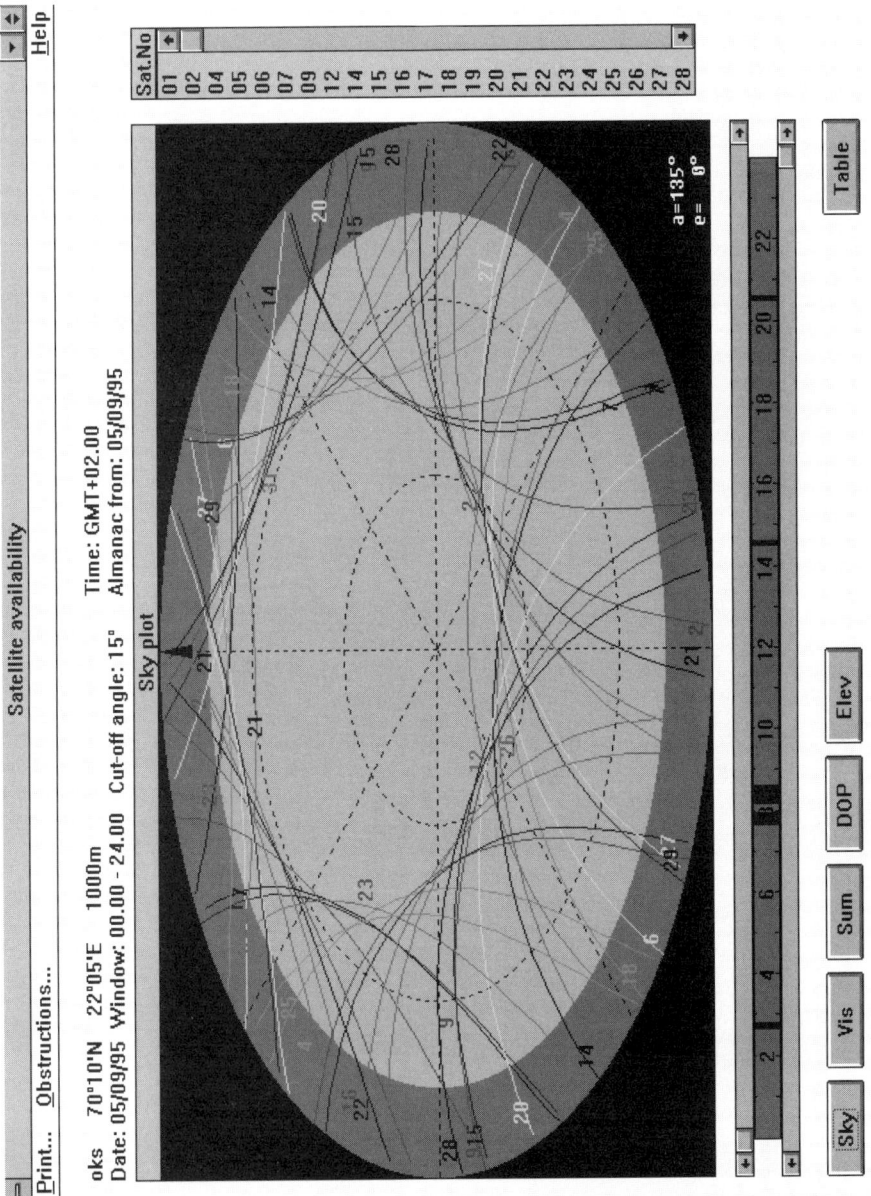

Figure 3.3 Sky plot of satellite availability over Oksfjordjokelen, Norway (plot obtained from Leica SKI software)

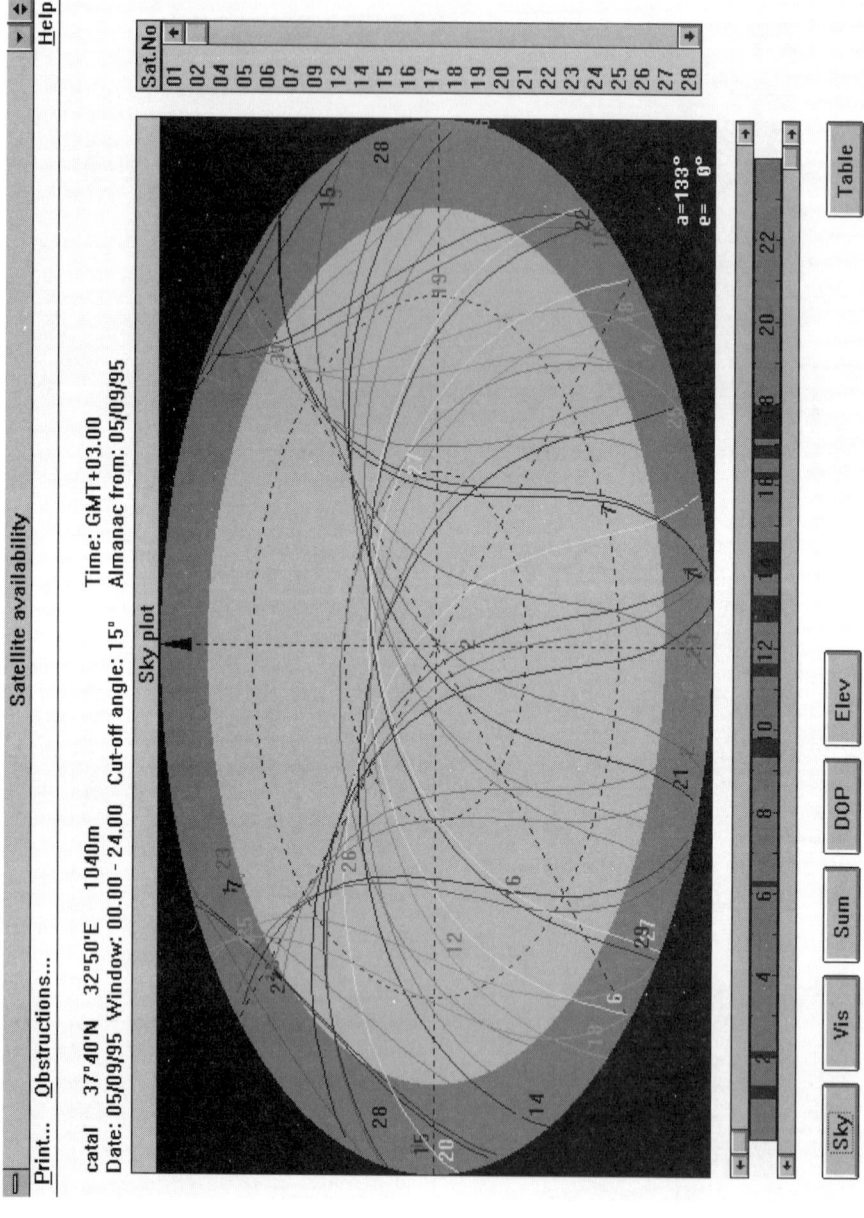

Figure 3.4 Sky plot of satellite availability over Çumra, Turkey (plot obtained from Leica SKI software)

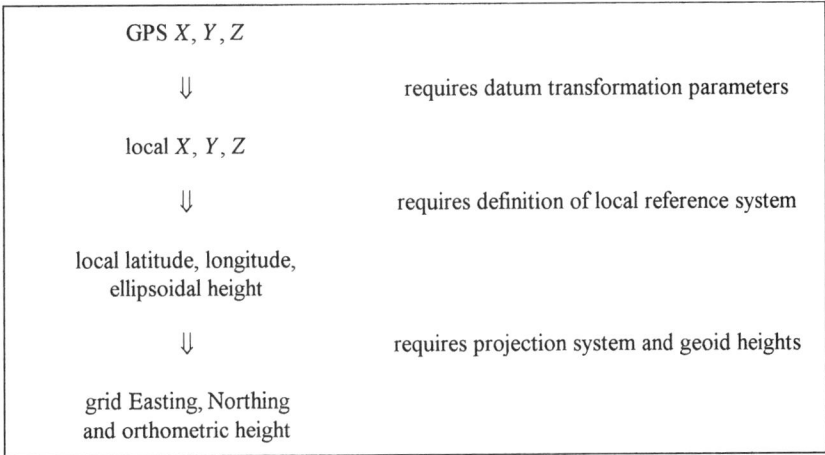

GPS X, Y, Z

⇓   requires datum transformation parameters

local X, Y, Z

⇓   requires definition of local reference system

local latitude, longitude,
ellipsoidal height

⇓   requires projection system and geoid heights

grid Easting, Northing
and orthometric height

Figure 3.5 Transformation of WGS84 coordinates to local grid coordinates

fixed reference receiver. These coordinates are usually given in the cartesian system (i.e. X, Y and Z) of the World Geodetic System of 1984 (WGS84) or geodetic system (i.e. latitude, longitude and ellipsoidal height) based on the ellipsoid used by WGS84 (see Cooper, Chapter 2). Although the X, Y and Z coordinates will rarely be used directly, the geodetic coordinates can be an acceptable final output, particularly if only relative heights are required and the surveyed area is sufficiently small, say 1 km × 1 km, for the ellipsoid and geoid to be considered parallel. It should be remembered here that the geoid is a unique irregular physical surface that coincides with mean sea level, whereas an ellipsoid is an ellipse of revolution chosen to approximate the geoid over an area of interest. This area can be very local as with the Airy ellipsoid chosen for the UK, or it can be world-wide as with the WGS84 ellipsoid.

In general, the survey information obtained from GPS will need to be combined with existing data taken from maps or observed using traditional techniques. In either case the old coordinates will probably be in terms of Eastings and Northings based on a local map projection and heights measured relative to the geoid (i.e. orthometric heights). The GPS coordinates therefore need to undergo several transformations before they can be used, as shown in Figure 3.5 (see Cooper, Chapter 2).

All necessary computations will usually be carried out within the processing software package being used, but it is still necessary to specify the various items listed on the right hand side of Figure 3.5. In general there are seven datum transformation parameters (three translations, three rotations and a scale change) but it is common practice to approximate the transformation for a local area by using only three parameters corresponding to the X, Y and Z shifts of the centre of the WGS84 ellipsoid to the centre of the local ellipsoid. This is possible because reference systems are usually defined such that the rotations and scale change are either zero or very small. Average values of the three transformation parameters in a local area, computed from station coordinates that are known in both systems, are available (for example, the User Guide for the Idrisi for Windows GIS package lists nearly 200 sets for world-wide datums which were obtained from the United States Defense Mapping Agency). These can be used in situations where the utmost accuracy is not required. If published values are not available then they can be derived by measuring WGS84 coordinates by GPS at a station of known local grid coordinates and orthometric heights. These local values can be transformed to Cartesian coordinates via geodetic coordinates provided the geoid height on the local ellipsoid is known at the station. Comparison of the two sets of Cartesian coordinates will give the required transformation parameters. Although only one point is needed in theory, several points would be used in practice with the parameters obtained from a least squares solution.

A local reference system requires eight parameters to define an ellipsoid and its relationship to the geoid: two for the size and shape of the ellipsoid; three for its position relative to the centre of the earth; two for the direction of its minor axis relative to the spin axis of the earth; and one for defining the zero longitude position. The only two of these parameters that need to be specified explicitly are the ellipsoid size and shape, usually in terms of the semi-major axis and either the flattening or eccentricity squared, as the remaining values will have been included in the datum transformation parameters.

Parameters of the projection system used to convert three-dimensional geodetic coordinates based on a particular ellipsoid to two dimensional map, or grid, coordinates are, for a transverse Mercator projection, the latitude of the origin, the longitude of the central meridian, the grid coordinates of the origin (the false Easting and false Northing) and the scale factor on the central meridian. These parameters should be readily available for the map projection being used. The final values that are required for the transformation process are geoid heights. Although these are available in the form of geoid maps, with say 1 m contours, on a national or world-wide basis, it is not generally possible to obtain accurate values for localised areas. If accurate values are required (for example, where the area of the survey warrants correction of the relative ellipsoidal heights by the varying geoid–ellipsoid separation) then the solution in practice is for a local geoid to be derived. This is done by measuring GPS ellipsoidal heights at positions of known ortho-metric height and plotting a contoured plan of the height differences.

A summary list of all the required parameters, together with example values for transforming from WGS84 to OSGB36 coordinates, are given in Table 3.5.

Finally, it should be stressed that if any information is extracted from a published map then it is essential to make a note of the geodetic information printed in the map's margin. This will usually include the ellipsoid and datum (which together define the reference system), the map projection and the vertical datum used for heights. This information is needed not only for combining map data with GPS or any other surveyed data, but also for comparing maps of the same area published at different times. For example the 1979 edition of the 1:25 000 map of Oksfjordjokelen in Arctic Norway is based on a Universal Transverse Marcator (UTM) projection with the international ellipsoid of the European Datum, whereas the 1993 edition of the same map is based on a UTM projection with the WGS84 ellipsoid. The result is

Table 3.5 Required information for coordinate transformations

Stage	Parameters	WGS84 to OSGB36	Notes
Datum Transformation	Delta X Delta Y Delta Z	−375 m 111 m −431m	Average values for England, Scotland and Wales
Ellipsoid	Semi-major axis Flattening	6 377 563.396 m 1/299.324 964 6	Airy ellipsoid
Projection system	Latitude of origin Longitude of origin False Easting False Northing Central scale factor	49° north 2° west 400 000 m −100 000 m 0.999 601 271 7	United Kingdom transverse Mercator
Geoid	Geoid height	−2.0 m to +3.5 m	Depends on location

that there are differences between the two maps in Eastings and Northings of 54 m and 197 m respectively. These substantial movements are not immediately apparent from the printed grid lines.

Need For a Known Point

When computing base lines in high accuracy GPS, the WGS84 coordinates of the reference receiver need to be known to within about 10 m to avoid the introduction of systematic errors. If the local grid coordinates and height of the reference station are known, either by observation or purchasing them from the relevant mapping authority, then the WGS84 coordinates can be derived by the reverse process to that outlined in the previous section (see Figure 3.5). If the local coordinates are unknown and cannot be obtained then it is possible to determine the WGS84 coordinates by processing the data recorded by the reference receiver in order to average out the effects of Selective Availability. Provided sufficient observations are made, for at least an hour, then the resulting coordinates should be within the required 10 m.

CASE STUDIES

Rapid-static GPS and Geoid Determination

The work referred to in this section relates to GPS surveying data obtained in the Çumra district of Turkey as part of the on-going Konya Basin Palaeoenvironmental (KOPAL) Research Programme led by Dr Neil Roberts of Loughborough University. The main aim of the 1995 fieldwork season was to undertake a systematic geoarchaeological and geomorphological survey of the area north of Çumra in order to establish a three-dimensional lithostratigraphic sequence for the Çarşamba alluvial fan. Topographic data were obtained at 25 sites, over an area of 15 km ×

20 km, in order to link stratigraphic information obtained by coring at archaeo-logical sites and by cleaning sections in irrigation ditches. The GPS equipment used for the work was a pair of Leica SR261 single frequency receivers with a quoted accuracy of 5 to 10 mm plus 2 ppm. The data collected were transferred to a laptop computer on a daily basis and processed using the Leica SKI software package. Additional computer processing to produce terrain models and plans was done within the LISCAD and Microstation SiteWorks packages.

Reference Station

The location of each site was determined relative to a single reference station which was positioned, mainly for security reasons, over the old triangulation pillar on the east mound of the Çatalhöyük archaeological site. The UTM Zone 36N coordinates based on the European Datum, as used on Turkish large-scale maps, of this station were unknown and therefore the WGS84 coordinates had to be determined by processing the data observed at the station. As the station was occupied on 18 occasions during the three week fieldwork season, for periods ranging from just under 1 h to over 5 h, it was possible to obtain not only mean values for the station coordinates but also a measure of their precision. The standard errors calculated from the residuals of all observations were about 4 m in plan position and 10 m in height. Leica recommend that several hours of observations are used for single point positioning computations but these results, as well as others obtained in Loughborough, suggest that reasonable results will be obtained provided observa-tions are made for at least an hour. Once the WGS84 coordinates of this station were obtained, they were transformed, as were all subsequently computed coordinates, to UTM36N values using published Cartesian transformation parameters along with the various parameters of the UTM projection based on the international ellipsoid.

Baseline Observation Time

The observation time required to obtain enough data to enable the ambiguities to be resolved and to achieve an accurate result depends on the receiver being used, the baseline length, the number of satellites observed, the satellite geometry and ionospheric disturbance. The recommended times for the single frequency receiver being used were 5 m per kilometre of baseline but not less than 15 min, with a GDOP of less than 8 for four or more satellites above the 15° cut-off angle. The baselines measured from Çatalhöyük ranged from a few hundred metres to over 13 km, which would require measuring times between 15 min and over 1 h. How-ever, as six satellites were being tracked for most of the time with GDOP values of 2 or 3, a compromise was made on the observation period, as the surveying had to be fitted in with the time-consuming coring and sectioning work, and a standard time of 20 min was adopted. This did result in ambiguities not being resolved on two occasions and the baseline measurements had to be repeated. It is suggested that the full recommended time should be adhered to, partly to overcome the inconvenience of repeating work but mainly because there have been a few occa-sions since on baselines measured over short periods in Loughborough where the solution has not been acceptable (with a maximum error of 0.7 m in height) even though the ambiguities have been resolved. The error was only known because several computations had been carried out to obtain checks, but would, of course, be undetectable if only a single baseline measurement had been made. Where there are time constraints over the observation period it may be worthwhile considering the advantage of dual frequency receivers, where the recommended observation times are about 1–2 min per kilometre of baseline with a minimum of 5 min.

Local Geoid Determination

Because the KOPAL Research Programme makes it necessary to link stratigraphic information recorded at sites over a wide area, the accurate determination of heights is essential and due allowance must be made for variations in the geoid–ellipsoid separation. No detailed geoid information was available for the area and observations were therefore made at a number of points of "known" orthometric height in order to determine a local geoid. These heights were obtained from 1:25 000 maps which in Turkey are classified as secret and can only be obtained for inspection from the government representative assigned to each visiting research group. The spot heights chosen (which were located on archaeological mounds, or höyüks) were specified to 0.1 m on the older maps, but only to 1 m on newer ones, and were indicated in the map legend as "survey stations". In reality some of the spot heights were occupied by old survey stations, generally flush with the ground, some by new survey stations, concrete pillars of no fixed design, some by both and some by neither. The quoted heights were assumed to be the ground level at the survey station if there was one or the top of the mound otherwise. The local "geoid" obtained by comparing observed GPS ellipsoidal heights with the orthometric heights at 25 stations, all referenced to the Çatalhöyük east mound station, is shown in Figure 3.6. The surface described by the geoid contours is not very realistic but that is not surprising given the unreliable nature of the orthometric heights. A closer inspection of the data indicates that the old survey stations give a much better trend, but these only cover a small section of the whole area. As the results suggest that there could be a 2 m variation in geoid heights over the region, further work is obviously needed to resolve this issue before the main landscape modelling work can continue. This will involve making traditional trigonometrical height observations in the 1996 field season at some of the new pillars in order to obtain reliable orthometric heights. By way of comparison, in order to show the sort of result that would be expected, a local geoid computed for the Loughborough region is shown in Figure 3.7. This was determined by comparing GPS ellipsoidal heights at 11 stations with orthometric heights determined by levelling to adjacent Ordnance Survey bench marks.

Stop–Go Detail Surveying with GPS

A plan of Çatalhöyük west mound surveyed by GPS during the fieldwork season in Turkey described above is shown in Figure 3.8. This survey was carried out at the request of Professor Ian Hodder of Cambridge University, Project Director of the

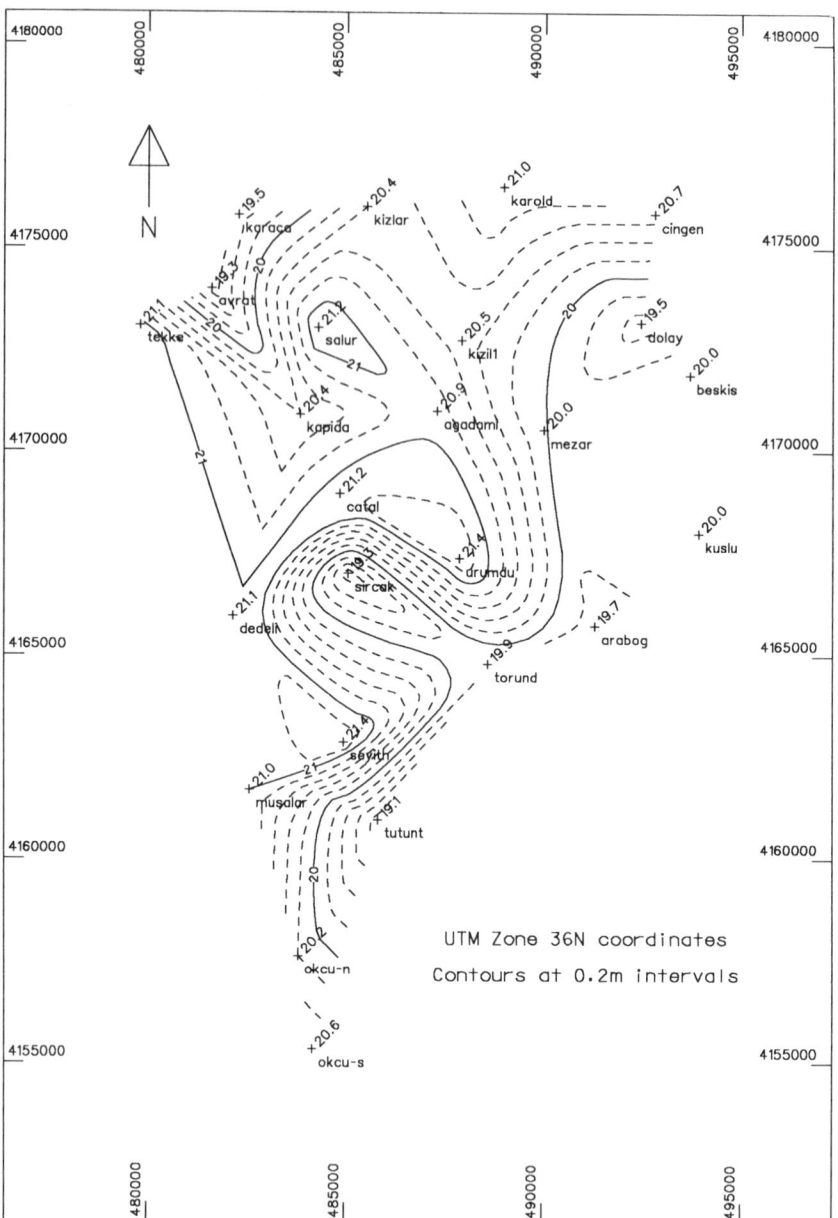

Figure 3.6 Derived geoid contours on the WGS84 ellipsoid from GPS observations at 25 stations with unreliable orthometric heights in the Çumra region of Turkey

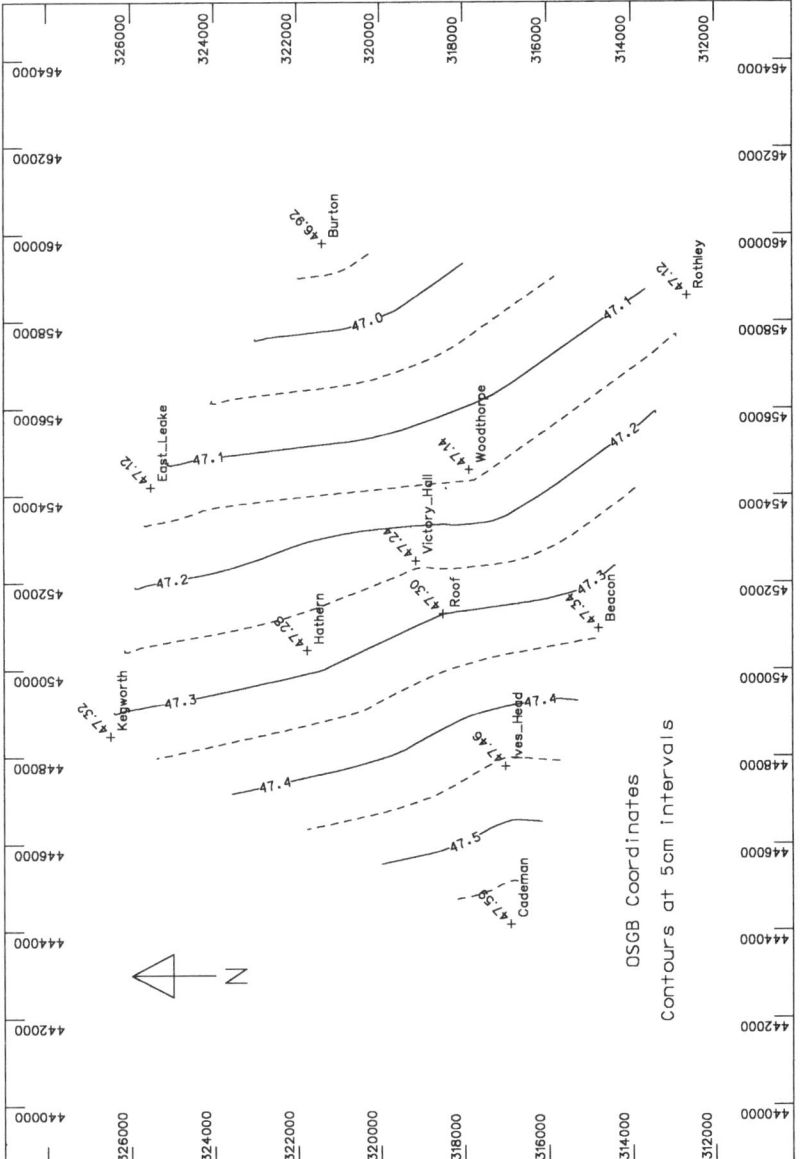

Figure 3.7 Derived geoid contours on the WGS84 ellipsoid from GPS observations at 11 stations with reliable orthometric heights in the Loughborough region of the UK

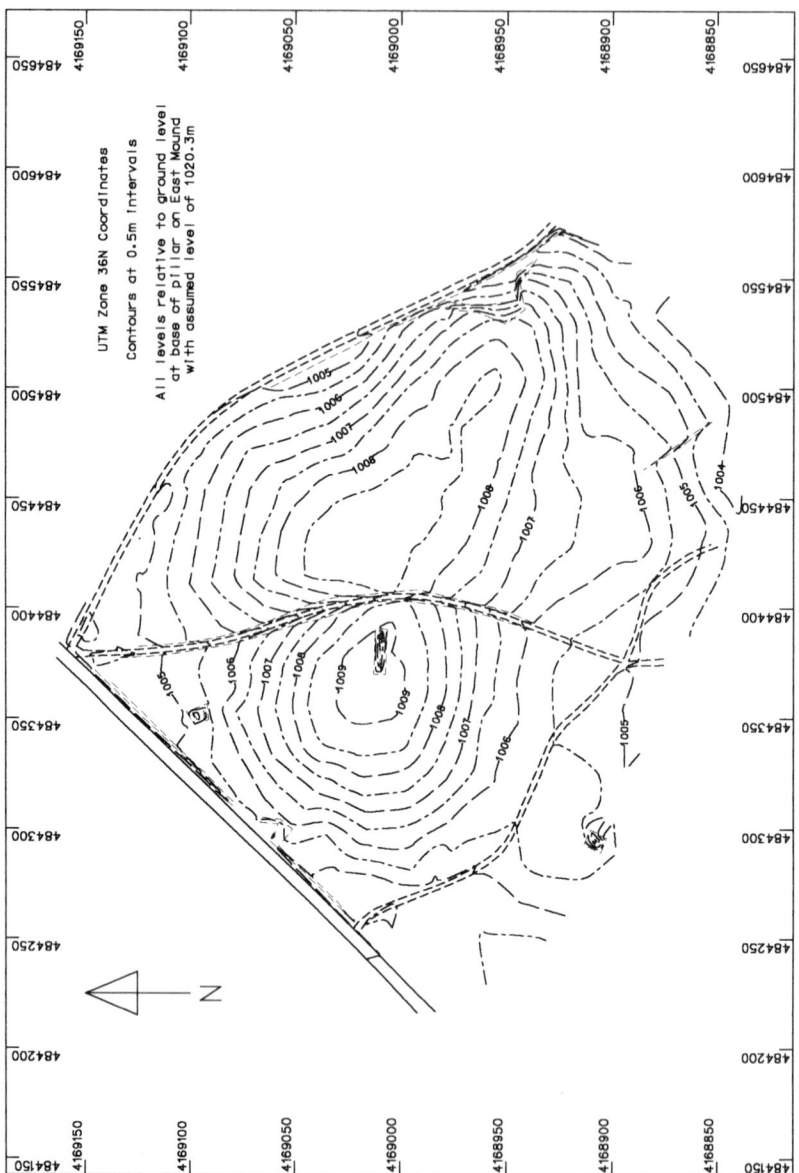

Figure 3.8 Stop–go GPS survey of Çatalhöyük west mound, Turkey

Çatalhöyük Research Trust, to complement the survey of the east mound which was completed by his team in 1994 using total-stations. The original east mound survey, done at a very high resolution to reflect the importance of the site, involved the collection of some 26 000 data points and had taken a team of surveyors several weeks to complete. By contrast, the GPS survey of the west mound was completed by one person in two afternoons and, although the same resolution was not used (with approximately 500 data points collected), all significant features of the topography are indicated.

The reference station used for the survey was Çatalhöyük east mound, whose coordinates were determined from the multiple single point positioning described in the previous section. The stop–go survey on the first afternoon was initialised by a rapid-static determination on a fixed ground mark which was then used as a known station for a quick initialisation on the second afternoon. Another advantage of having a fixed ground mark is that re-initialisation can take place very quickly if lock is not maintained on at least four satellites. The observation rate for the two receivers was set at 5 s so that the Rover only had to be held stationary for about 20 s at each surveyed point. The southeast section of the mound was bordered by tall trees and care had to be taken not to lose lock on too many satellites. Once the observations had been processed, the WGS84 coordinates were transformed to UTM36N values using the previously described procedures. Because the survey was of limited extent, some 300 m × 300 m, it was assumed that the derived relative ellipsoidal heights were equal to relative orthometric heights. Each ellipsoidal height difference was therefore simply subtracted from the known orthometric height of the Çatalhöyük reference station to give the required orthometric heights of surveyed points, without any need for a knowledge of the local geoid.

Although this detail survey was completed by using GPS without any problems, alternative methods should be considered. Advantages of using GPS were that only one person was needed, plan coordinates were obtained directly in the UTM system, and orthometric heights were obtained directly from the reference station. Disadvantages were that very expensive equipment was being used, a substantial amount of equipment (antenna, controller, recorder and battery) had to be carried around the surveyed area, and some sections were almost impossible to survey because of tree coverage. If a total-station had been used to survey the same area, much cheaper equipment would have been used, only the reflector would have been carried around the area, and all parts of the survey would have been easily accessible, albeit with a number of different instrument set-ups. Disadvantages are that two surveyors would have been needed, at least two instrument set-ups would have been required to maintain instrument–reflector line of sight, heights would have to have been transferred by levelling from the nearest known station, and tying the survey into the UTM36N coordinate system would have taken a lot of extra effort. GPS was used for this particular survey because it was the only equipment available at the time. The ideal solution, however, would have been to use GPS to locate two or three control stations from which the detail survey could have been carried out using a total-station. Obviously the pros and cons of both techniques, as well as equipment and personnel availability, should be considered for any individual project.

Kinematic Detail Surveying with GPS

A plan of the bed of Blackbrook Reservoir near Loughborough obtained from a kinematic GPS and echo-sounder survey is shown in Figure 3.9. The survey was completed in four sessions, each lasting about 2 h.

The reference station used for the survey was a roof station at Loughborough University, some 6 km from the reservoir. The coordinates of the station had been determined previously by measuring a GPS baseline from a nearby first-order triangulation pillar with known OSGB36 coordinates. The necessary transformation to WGS84 coordinates was accomplished using published Cartesian transformation parameters along with the various parameters of the OSGB36 projection based on the Airy ellipsoid. An initialisation station was located on the shore of the reservoir during the first session by a rapid-static fix and this point was used to obtain a quick initialisation during each subsequent session. The time interval for recorded points was set at 10 s, a value used to give a reasonable spacing of survey points for the approximate speed of travel of the small boat being used. GPS was used to keep a track of the boat as it zigzagged across the reservoir, and the echo-sounder trace was marked with the GPS epoch number at regular intervals and at turning points so that the two sets of data could be correlated. The processed GPS observations gave the plan positions of the boat and the levels of the echo-sounder, which were very consistent with less than 7 cm variation, from which the echo-sounder depths could be subtracted to give the bed levels. As with the previous stop–go detail survey, the ellipsoidal height differences were assumed to be equal to orthometric height differences as the survey area was relatively small, at about 1 km east–west by 800 m north–south, and the geoid map of Figure 3.7 shows that there is only a variation in geoid heights of about 5 cm per kilometre in the east–west direction and very little in the north-south direction. A few surveyed points did cause problems, however, with the water level appearing to jump by about 1 m. As these points were observed in the vicinity of the dam wall, it was assumed that this was due to the effects of multi-path (the antenna picking up both direct signals and signals reflected off the wall) and consequently were ignored. The above techniques were obviously only used on the open water and further points were obtained by walking round the water line with the GPS in kinematic mode and by surveying the top of the shore, which was generally under a tree canopy, with a total-station and reflector.

Similar bed surveys have been carried out in the past on a fjord in Norway and on another reservoir in Loughborough. In each case the boat was tracked by assuming that it travelled in a straight line at constant speed between two points whose locations were determined either from the intersections of directions measured with theodolites or from directions and distances measured with a total-station. However, when sighting a boat which is not quite stationary, there are problems with both these surveying techniques which could lead to highly inaccurate coordinates. In addition, a limited number of tracks can be observed and the assumption of straight line travel at constant speed is of dubious validity. Unlike the previous stop–go detail survey where the use of GPS was not essential, the accuracy of the reservoir survey was undoubtedly increased substantially by using GPS to keep a constant track on the boat's position.

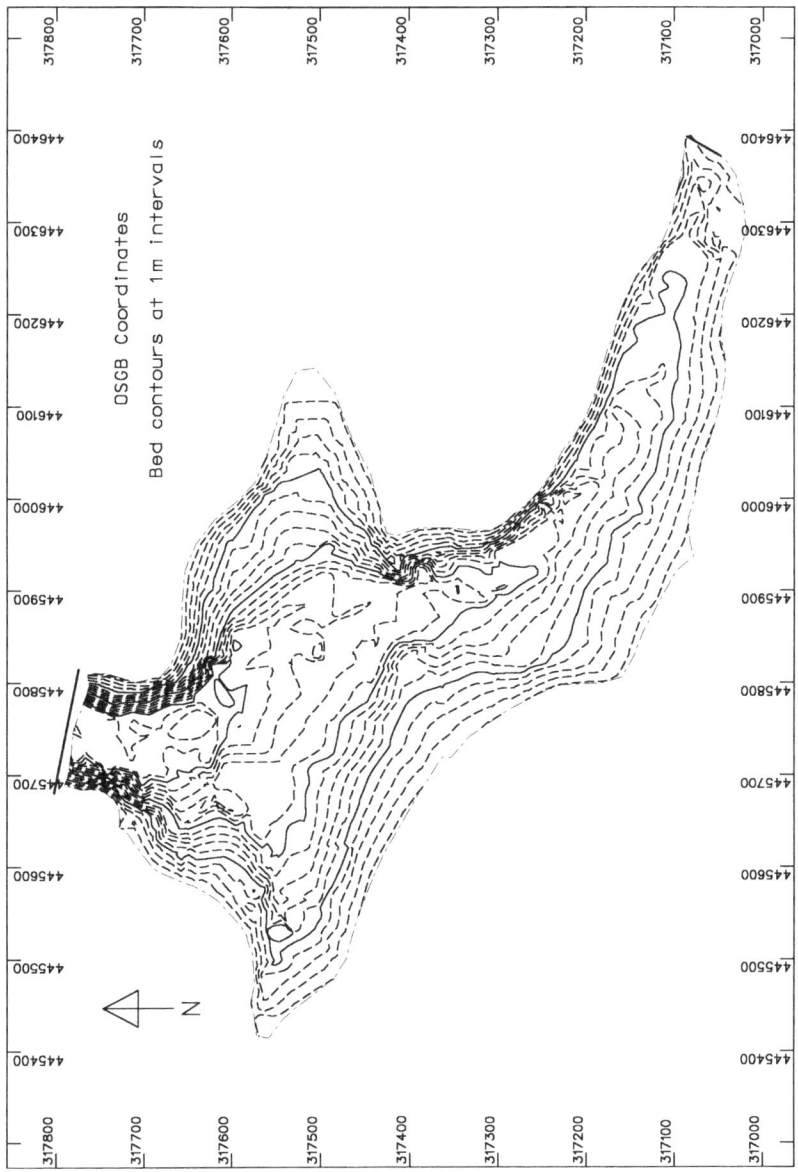

Figure 3.9 Kinematic GPS and echo-sounder survey of Blackbrook Reservoir near Loughborough, UK

CONCLUSIONS

GPS technology has been developed to the stage where it is now possible to obtain an instantaneous position to within 100 m with a £200 hand-held receiver, or to obtain the relative position of two points several kilometres apart to within 1 cm using a system which may cost over £50 000.

Relative GPS positioning can be used for the rapid and accurate collection of topographic data in many terrain mapping and monitoring applications where it would be extremely difficult and time-consuming to employ traditional surveying techniques. However, to make full use of the available technology it is essential that all users of GPS equipment and data have a good understanding of the principles and associated geodetic theory. This is particularly important where data from a variety of sources are to be combined, where the survey is over an extensive area, or where accurate orthometric heights are required and knowledge of the local geoid is limited.

ACKNOWLEDGEMENTS

The KOPAL Research Programme in Turkey was sponsored by the National Geographic Society and the reservoir survey near Loughborough was carried out with the permission of Severn Trent Water Ltd. I would also like to thank Neil Roberts, Hakan Yiğitbaşioğlu, Romola "Chip" Parish, Pete Boyer and John Tibby for their assistance with the fieldwork in Turkey, and final year students Mark Doggett, Robert Evans and Norris Riley for use of GPS data collected in Loughborough with my assistance.

SOURCES OF INFORMATION

Since the early 1970s there have been a large number of articles and books written about the technology, techniques and applications of GPS. As the needs of individual users will vary widely, even in the field of terrain mapping and monitoring, it is considered more appropriate to give a list of information sources rather than the usual list of references. Information is considered under four headings: books, journals, the Internet and equipment manufacturers.

Books

This short list contains two standard surveying texts which include introductory sections on GPS and two texts exclusively devoted to GPS.

Elfick, M. H., Fryer, J. G., Brinker, R. C. and Wolf, P. R., 1994. *Elementary Surveying*, 8th edition (SI adaption). Harper Collins, 510pp.
Hofmann-Wellenhof, B., Lichtenegger, H. and Collins, J., 1994. *GPS Theory and Practice*, 3rd revised edition. Springer-Verlag, 355pp.
Leick, A., 1995. *GPS Satellite Surveying*, 2nd edition. John Wiley & Sons, 560pp.
Schofield, W., 1993. *Engineering Surveying*, 4th edition. Butterworth-Heinemann, 554pp.

Journals

Many GPS related papers can be found in the following journals (in English despite some of the titles).

Bolletino di Geodesia e Scienze Affini
Bulletin Géodésique (until 1995 – now the *Journal of Geodesy*)
Civil Engineering Surveyor
Earth Surface Processes and Landforms
Geomatica
Geomatics Research Australasia
GPS World
Journal of Geodesy (from 1996)
Journal of Geophysical Research
Journal of Surveying Engineering (American Society of Civil Engineers)
Manuscripta Geodaetica (until 1995 – now the *Journal of Geodesy*)
Mapping Awareness
Navigation
Survey Review
Surveying and Land Information Systems

The Internet

There are an enormous number of sites on the Internet containing GPS information of which the four listed below may be found useful because of the wide coverage and extensive links that they provide.

gopher://unbmvsl.csd.unb.ca:70/hPUB.CANSPACE.GPS.INTERNET.SERVICES.HTML
(This site can also be located via "GPS-related services available on the Internet" from http://
   degaulle.hil.unb.ca/Geodesy/CANSPACE.html)
http://www.inmet.com:80/~pwt/gps_gen.htm
http://wwwhost.cc.utexas.edu/ftp/pub/grg/gcraft/notes/gps/gps.html
http://galaxy.einet.net/editors/john_beadles/introgps.htm
The contents pages of many of the journals listed in the previous section can be obtained at
http://www.geod.emr.ca/~craymer/tcg/ and regular updates can be obtained by subscribing to
   the Canadian Space Geodesy Forum – see the CANSPACE html above.
Finally, the journal GPS World has its own home page at http://www.advanstar.com/GEO/
   GPS/ which has a useful Resource section including a link to a very extensive glossary of
   GPS related terminology.

Equipment Manufacturers

Manufacturers can provide extensive literature as well as glossy brochures. The following list contains manufacturers or distributors, with UK contact telephone numbers, of high accuracy GPS equipment.

Ashtech Europe Ltd	Tel: 01993-883533	Fax: 01993-883977
C.Z. Scientific Instruments Ltd	Tel: 0181-9531688	Fax: 0181-9539456
Del Norte Technology	Tel: 01793-784487	Fax: 01793-784409
Geotronics Ltd	Tel: 01480-433555	Fax: 01480-432480
Leica UK Ltd	Tel: 01908-666663	Fax: 01908-609992
Sokkia Ltd	Tel: 01270-250525	Fax: 01270-250533
Toposell Ltd	Tel: 01580-860801	Fax: 01580-860802
Trimble Navigation Europe Ltd	Tel: 01256-760150	Fax: 01256-760148

4 Analytical Photogrammetry for Geomorphological Research

L. F. J. DIXON, R. BARKER, M. BRAY, P. FARRES, J. HOOKE, R. INKPEN, A. MEREL, D. PAYNE and A. SHELFORD
Department of Geography, University of Portsmouth, UK

ABSTRACT

This paper examines the basic principles and constraints on the application of analytical photogrammetry to a variety of geomorphological research problems. Principles of image acquisition, acquisition of ground control, model orientation and data capture are outlined. These principles are applied to two- and three-dimensional landscape analysis using aerial photography and to meso- and micro-scale studies using non-standard terrestrial imagery. In these applications, the potentials and limitations of the principles of analytical photogrammetry are exemplified. These examples illustrate the versatility of this technique for monitoring and understanding change over a range of physical environments and spatial and temporal scales.

INTRODUCTION

Geomorphology involves the measurement, monitoring and analysis of forms and the processes that produce them. The understanding and interpretation of such process–form relationships provides the framework in which landform dynamics can be investigated. There has been a long tradition of process measurement with various attempts to link results to landform evolution (e.g. Brunsden, 1974). Other studies have interpreted surface forms so as to understand better their formative processes (e.g. Brunsden and Jones, 1980). However, complex, variable surface forms have proved inherently difficult to measure over appropriate spatial and temporal scales, thereby forestalling many process–form modelling approaches.

Conventional techniques of measurement of surface form have included various methods of ground survey, from the use of pin frames at one scale (Allamaras *et al.*, 1966) through to micro-erosion meters (Trudgill, 1976) at the other extreme. Changes in surface form have been considered from sequential surveys or from historical sources such as maps (Hooke and Kain, 1982). Such conventional methods, apart from being time-consuming and costly, can potentially contain within them indefinable errors and omissions (e.g. Carr, 1962; Thieler and Danforth, 1994).

Aerial photographs have long provided a source for the mapping and interpretation of morphology (Collin and Chisholm, 1991), but the analogue method used for

Landform Monitoring, Modelling and Analysis. Edited by S. N. Lane, K. S. Richards and J. H. Chandler.
© 1998 John Wiley & Sons Ltd.

previous studies was inflexible in terms of the type of photography that could be used. There is also the problem of acquiring suitable imagery, due to either availability from existing archives or the significant cost of obtaining new photography. Although recent developments have obviated some of these problems, for example the use of laser height scanners at the micro-scale (Bradford and Huang, 1992) and Global Positioning System (GPS) surveys at a larger scale (Ackermann, 1992; Cornelius et al., 1994), these cannot be used retrospectively as archive historical data do not exist.

Developments in analytical photogrammetry, recently brought to the attention of geomorphologists by Lane et al. (1993), offer a potential means of tackling many of these problems. Unlike most other precise measuring techniques, it has the great advantage of being able to utilise the abundant information contained within historical photographic archives (Chandler and Cooper, 1988). Furthermore, it permits quantification of surface form from a much greater variety of image types, overcoming many of the mechanical restrictions of the previous analogue method. Vast improvements in data capture, processing and replicability coupled with increasingly powerful and easy-to-use digital terrain modelling (DTM) methods enable not only the efficient graphical display of information, but also the possibility of sophisticated quantitative analysis of morphological change in two or three dimensions (e.g. Lane et al., 1994). With such advances, calculations of area and volume changes between time series become routine. Morphological data can now be collected at a greater range of scales and models of geomorphological investigation applied.

This paper investigates the practicalities of applying analytical photogrammetry to a variety of contrasting geomorphological research problems. By reference to these diverse applications, it explains some of the basic principles and constraints of the technique. Attention is focused upon the types of analysis possible and the potential for geomorphological interpretation of results. The examples developed progress from two-dimensional landscape analysis (e.g. river channel planform change), through three-dimensional landform scales (e.g. volumetric analysis of coastal cliff erosion), to meso- and micro-scale investigations using non-standard terrestrial image photography (e.g. river bank erosion, soil erosion, stone weathering). These case studies demonstrate clearly the versatility and potential of this technique across a range of image types, physical environments and spatial scales.

THE ANALYTICAL PHOTOGRAMMETRIC TECHNIQUE

Photogrammetry requires stereoscopic coverage of an area, created by acquiring photographs taken from two different positions. In an analogue plotter, the relationships between photographic images and object which existed at the time the photographs were taken is physically recreated at reduced scale by mechanical means (Lane et al., 1993). In the analytical plotter, a mathematical solution is used to establish the relationship between the object and image (Ghosh, 1988). The principal advantages of the analytical technique are detailed by Lane et al. (1993) and are outlined as follows.

1. It is non-destructive and non-invasive, and thus allows replication of measurements over time without any direct contact with the target surface except for the acquisition of ground control.
2. The photographic record of the surface is preserved and so can be returned to at any time for further reference and measurement.
3. Given sufficient control, the same points on any surface, can be "revisited" automatically and therefore changes through time can be accurately monitored.
4. Relationships between height characteristics of surface points and non-morphological attributes can be easily observed and assessed.
5. The technique permits accurate measurements in all three spatial dimensions allowing for the constraints imposed by image quality and ground control.
6. Different image formats can be accommodated including oblique, distorted and exceptionally, historical photography which would not be regarded as strictly stereo in traditional analogue terms.
7. The data are captured in digital form and can therefore easily be transferred between Geographical Information Systems (GIS) and other data analysis computer packages.

One of the main reasons why applications have so far been limited is that the technique requires expensive, highly specialised equipment and skilled personnel. The applications presented in this paper have all made use of a Kern DSR-14 analytical photogrammetric plotter, with Kork software attached for digital storage, retrieval and data transfer. The main technical considerations in using these methods are described below.

Image Acquisition

For simple photogrammetric extraction of precise quantitative data, a pair of overlapping metric photographs is required. The geometric configuration created by this photo-pair is critical and can be defined by the base–distance ratio. This is the ratio of the distance between the two original exposure stations and the average distance to the surface of interest. A base–distance ratio of between 0.1 and 0.25 is considered to be the optimum for ease of measurement, although the ratio can be increased up to 0.4 if greater precision is required (Granshaw, 1980). These ratios are attained by creating a strategy for the photographic mission that achieves a balance between maximising the parallactic angles between intersecting light rays to common points, whilst minimising the distance of the exposure stations from the surface of interest (Wolf, 1983). This type of strategic balance is exemplified by the use of super-wide-angle lenses for aerial photography. This allows a low flying height, maintains the desired parallactic angles but still provides sufficient coverage to make the venture economically viable. Costs are reduced by minimising the number of photographs that have to be acquired and consequently the number of stereo-models that have to be processed.

The mathematical model used in the analytical solution is fundamental to the greater freedom of choice in type of photography used. The analogue method constrained the user to a maximum tilt from the vertical of typically ±5–6°, whereas the analytical solution has no such constraints, thus allowing the use of oblique photography. With two photographs and knowledge of the position and orientation of the two camera exposure stations, the analytical method allows the calculation of the position of points in the object. Inversely, knowledge of the positions of points identifiable on the photographs can be used to calculate the positions of the camera stations. This is important for temporal applications of terrestrial photogrammetry, because provided that there are features or targets of known position (ground control) appearing on the photographs, the camera station locations do not have to remain the same for each site visit. This could be vital if the initial camera station positions were found to be unusable because of subsequent geomorphological activity. The conventional method is to establish the coordinates of a network of targets visible from both camera stations. Two plan and three height points are the minimum that should be visible, but the introduction of additional points will allow a least squares solution which can improve the result. Further detail on the acquisition of ground control and its use in the photogrammetric procedure is discussed later.

Analytical methods also provide increased flexibility in respect of the camera used. Traditional photogrammetric cameras are calibrated to define important geometric properties and advance knowledge of these parameters is vital for analogue photogrammetry. The collinearity equations associated with the analytical method can be extended to incorporate the unknown camera parameters (Lane *et al.*, 1993), allowing medium and even 35 mm formats to be used. Accurate self-calibration requires high redundancy which is best produced using a highly convergent configuration consisting of many photographs and so fully calibrated metric cameras are still preferred for ease of photogrammetric processing (see Tait (1980) for a summary of terrestrial photogrammetric cameras and their properties). Such cameras are also more likely to be equipped with a pressure plate to keep the film truly flat and within the focal plane during exposure. Images taken using cameras without this feature will need further distortions removed (Short, 1992). An important coordinate reference system in the focal plane is provided by fiducial marks which define the centre of collimation of a camera. Metric air cameras and purpose-built metric terrestrial cameras such as the Wild P32 include fiducial marks and these appear on every image. Use of these marks is funda-mental to photogrammetric work, particularly for the task known as Interior Orientation (Wolf, 1983). If a camera is not equipped with such a reference system (e.g. the Pentax 645 used for one of the studies reported here), it is possible to manufacture a glass plate with scribed reference marks. It is also recognised here that the edges or corners of the image format can be used in some circumstances (Short, 1992).

The diapositive is the optimum image type for use in an analytical plotter, but, through adding an additional illumination system, normal paper prints can be used. However, care must be taken with prints as the paper backing may not be flat and could distort over time; the use of the original negatives provides a preferred alternative. The quality of the image is dependent on the properties of the film, and therefore care must be taken in choice of film type and exposure times. For terrestrial photogrammetry, slower, less sensitive films are preferable because of the smaller grain size within the emulsion which allows a greater resolution (Wolf, 1983); an ISO of 100 or less is suggested. A problem with this type of film is that its use can lead to an image of high contrast which reduces the detail visible within shadow and dark areas. A solution to this may be the use of a flash gun or other means of additional illumination to infill such areas.

The longer exposure time required for films with a speed rating of ISO 100 is not a problem when the camera is mounted on a tripod and the target is stationary. For aerial photography, faster films coupled with fast shutter speeds were originally required in order to compensate for the motion of the camera. This led to images of a lower resolution and contrast. However, developments in forward motion compensation (FMC) for aerial cameras has permitted a much greater freedom in choice of film. The effects of shutter speed, length of exposure and film speed become less relevant when the film is transported by the camera to compensate for the continued forward movement of the aircraft (Cox, 1992). Also, to enhance the depth of field characteristics of the image, it is necessary to use a small aperture size (or f setting), which will also reduce the potential amount of lens distortion. Each camera will have its own optimum f setting for best depth of field with the minimum amount of distortion, but this setting will generally be either f-16 or f-22. The taking of test shots is advised, especially in the case of terrestrial applications.

Acquisition of Ground Control

Photogrammetric control consists of any points on the surface of interest for which horizontal and elevation values have been established and that are identifiable on the photographs taken for a particular photogrammetric operation (Combs, 1980). Whatever the scale of the photography, the ground control points must satisfy two requirements: they must be clearly visible on all photographs and located in favourable positions. In order to achieve this, potential control points can be selected from the photographs after they have been taken and coordinates determined in the field. Alternatively, targets can be placed in suitable locations prior to photography. The accuracy to which the coordinates of these points is determined depends on the aim of the study, but generally it can be assumed that the ground control must be surveyed at a height accuracy of at least one-twentieth of the required contour interval (Fryer *et al.*, 1994). It must be remembered that the accuracy of the finished map can be no better than the ground control on which it is based (Wolf, 1983).

For aerial photography, traditional survey methods can be used to fix the positions of features on the ground that are to be used as control. For example, total-station surveys originating from a known Ordnance Survey control point or a fixed hypothetical origin would be of sufficient accuracy for the majority of scales of aerial photography. Developments in differential GPS provide an alternative although initially more expensive method. Here, with a reference receiver set up on a known point, the surveyor is free to fix any point visible on the aerial photographs to a high degree of accuracy using a "roving" receiver. This development is important as "line-of-sight" is not necessary between receivers, the survey is very rapid and is accurate to within 5 mm +1 ppm in horizontal position and

10–15 mm +1 ppm in elevation (figures for Leica 300 series differential GPS system in static mode). Traditional survey methods, although adequate for aerial photographs, are often of insufficient accuracy for close-range terrestrial photogrammetry, especially on vertical surfaces. For example, the measuring of targets on an overhanging river bank face using a total-station in conjunction with a prism would result in inaccurate estimates for the control coordinates. This type of survey would be better achieved using intersecting horizontal and vertical angles from each end of a fixed baseline. Here, the operator simply has to align the cross-hairs of the total-station telescope with the centre of each target from an instrument set up at both ends of the baseline, and record the angles. The coordinates within this local grid system can then be calculated trigonometrically to a high degree of precision, the level of which is dependent on the precision of the baseline distance measurement. The quoted accuracy of a total-station such as the Geodimeter 400 series is $\pm(5$ mm + 5 ppm) and therefore this type of instrument is ideally suited for such a task.

The requirements for ground control used for non-topographic photogrammetry is often more stringent. For example, annual erosion rates on limestone masonry have been recorded for St Paul's Cathedral (using a micro-erosion meter) at 40 μm (Trudgill *et al.*, 1989). At this micro-scale, annual measurement using analytical photogrammetry would be almost impossible. The solution is perhaps to attempt measurements at a larger temporal scale by simulating the erosion conditions at an increased rate. However, this still requires very accurate measurements of the ground control, coupled with a good knowledge of all the potential errors, if any reliable data are to be extracted. For example, factors such as the film not being flat within the camera could lead to a systematic error of 40 μm in the image, operator error in measurement of the fiducial marks and ground control points could be as much as ±5 μm and the height measurements of the surface itself could also have an operator error of a similar value, or worse, depending on the method of extraction (Fryer *et al.*, 1994). Studies have established the very high measurement accuracies achievable using analytical photogrammetry in ideal conditions (Neill, 1994; Stevens *et al.*, 1992), but prospective users should be aware of the potential pitfalls that can be overlooked.

Setting up the Model: The Three-stage Orientation Procedure

Like the analogue solution, analytical photogrammetry utilises three phases when setting up the model: the inner (or interior), relative and absolute (or exterior) orientations. These orientations reconstruct the bundle of rays originally projected onto the photographic negative and allow the creation of the object-model used for all measurement (Ghosh, 1988).

The *inner orientation* can be defined as the restoration of the internal geometry of the camera at the time of exposure. For standard analytical photogrammetry, the fiducial coordinates and principal distances of the images are entered into the computer, perhaps via selection of the appropriate camera calibration table. The stereo-pair is then placed on the photo-carriers and fiducial marks on each photograph are digitised in order to derive machine or comparator coordinates. The analytical plotter will drive to the vicinity of subsequent fiducial marks once the first has been measured, whereupon the operator will take the final measurement. The measured and calibrated coordinates of the fiducials are then used to derive the parameters of a two-dimensional transformation (typically an affine) using least squares estimation. Measurement residuals are displayed and an opportunity is provided for the operator to either accept the measurements and transformation or to remeasure the fiducial locations. Another aspect which is typically considered at this stage is the specification of parameters necessary to model systematic error sources such as lens distortion, atmospheric refraction and earth curvature (Wolf, 1983).

The *relative orientation* involves re-establishing the geometric relationship between two photographs (a stereo-pair) which existed at the instant of exposure. This is achieved by the operator removing what is termed the y-parallax (Ghosh, 1988) at a minimum of six pre-selected positions. The analytical plotter assists this process by driving the measuring mark to the six optimum locations. Additional measurements are recommended and the consequent redundancy helps to derive the best least squares estimates for those parameters used to express the required geometric condition (coplanarity). Again, points can be remeasured, and when the operator is satisfied with the residuals, parameters are stored.

The final process is *absolute orientation.* This is carried out in order to rotate and translate the stereo-model to the desired scale and orientation relative to a horizontal datum by utilising the ground control points. The coordinates of all control points have to be entered and the operator needs to measure the location of each point within the stereoscopic model. For this process a minimum of two plan points and three height points are required (Ghosh, 1988), but more are recommended so that a least squares solution can be used. The stereo-model is effectively scaled, shifted and rotated mathematically to obtain the best fit to the ground control. Again, the operator can accept, add or re-measure any of these points until the residuals are satisfactory and the parameters defining the solution are stored.

The three-stage orientation process described above can be undertaken for each model in approximately 15 min. Once these parameters have been computed and saved, a model can be restored in less than 5 min by simply carrying out the inner orientation procedure. This is a vast improvement on the amount of time required to set up a model on an analogue plotter. Having established the geometric relationship between images and object, the operator can determine the three-dimensional coordinates of points within the stereo-model by simply placing the measuring mark on the ground. These coordinates are then recorded using digital mapping software as either individual points or as strings of consecutive points to define linear features.

Data Capture and Surface Reconstruction

The method of data capture chosen is dictated by the objective of the research, the nature of the surface morphology and the ways in which the data are going to be analysed and displayed. The majority of examples given here require data for the generation of digital terrain models (DTMs) displayed as either contour maps or block diagrams. The alternative methods of data capture are therefore explained with this requirement in mind.

The Grid Method

The Kork digital software used in conjunction with the DSR-14 has the capacity to sample points on the surface of the stereo model according to a user-defined grid. This allows the operator to measure the plan position and height at each grid intersection to produce a uniform density. This can be performed either sequentially working down the rows, along columns, or randomly where the plotter drives the user to a decreasing number of grid intersections. This latter method eliminates the "dragging" effect where the operator may lose accuracy because of possible local, subtle changes in z value. Heighting accuracy has been shown to improve when an operator's measuring mark is relocated away from the previous measured point and such a method was used in the soil surface studies described later. Composite grids can reduce the amount of redundant data especially for areas of mixed relief. A fairly coarse grid can be used for the initial measurement of the whole model followed by denser grids over areas of greater change in relief (Petrie, 1990).

The grid method of data capture requires only a limited amount of reconstruction in order to generate a surface or DTM. The joining together of adjacent grid nodes is generally sufficient using the z values directly. However, the accuracy of the DTM generated from gridded data is highly correlated with sampling interval, that is, with a small sampling interval; or a composite grid, the relative accuracy of the resultant DTM is improved (Li, 1992).

The Contouring Method

This method requires the operator to fix the floating mark at a desired height by locking the z measurement control on the plotter. Contours are digitised at a user-specified interval as the operator traces around the model whilst keeping the floating mark on the ground surface. Additional user-specified points are useful to ensure coverage of any significant areas between contours. The density of points is variable according to the sampling and the contour interval. The cliff erosion study employed this method as it is fast, and naturally concentrates points where morphology is most variable. Its drawback is that the measurement accuracy of individual points is on average half as accurate as spot heighting of natural features (Fryer, 1994; Petrie 1990). The operator will tend to stray above or below the actual surface whereupon the two dots of the floating mark will separate because of excessive x parallax. The operator then reacts and moves the floating mark back to the true contour, and it is this fluctuation above and below the surface that introduces the error (Kumler, 1994). The extent to which it happens depends on the ability of the operator to "read" the terrain, and it is also dependent on the speed at which the operator attempts to digitise the contour.

Irregular data of this type will require some form of interpolation in order to reconstruct the surface. The two most widely used methods of surface reconstruction are regular grids (DTMs or digital elevation models (DEMs)) and triangulated

Table 4.1   Summary of interpolation routines. Source: Keckler (1995)

Interpolation method	Summary
Inverse distance	• weighted average interpolator – weight of one data point diminishes with distance from the grid node • normally an exact interpolator – honours the data points when they coincide with a grid node; can add a smoothing parameter if desired • tends to create concentric rings around data points • fast for medium sized datasets
Kriging	• geostatistical gridding method – has proven to be popular • generates visually appealing contours that attempt to depict trends in the data • uses irregularly spaced data – good for random point photogrammetric method • can be an exact or smoothing interpolator • user can define specific variogram model • can be slow for large data sets, but results are worth the wait
Minimum curvature	• popular in the earth sciences • generates the smoothest possible surface while trying to honour each point • is not an exact interpolator so not all points are honoured exactly • fast for most data sets
Nearest neighbour	• assigns the value of the nearest datum point to each grid node • useful when data are already on a grid, or for filling holes when data values are missing for a grid
Polynomial regression	• used to define large-scale trends and patterns within the data • not really an interpolator as does not attempt to predict unknown z values • Very fast but detail lost within the resultant grid
Radial basis functions	• consist of a range of exact interpolators that attempt to honour the data points • multiquadratic is probably the best for a smooth surface that fits the data • the results are similar to those produced by kriging
Shepard's method	• uses inverse distance weighted least squares method • similar to inverse distance but local fit reduces concentric circle effect • can be an exact or smoothing interpolator
Triangulation with linear interpolation	• uses optimal Delauney triangulation • original data points are connected by lines to create triangles – produces a patchwork of triangle faces over the extent of the grid • is an exact interpolator • good for evenly distributed points; with large enough data sets breaks in slope can be preserved

irregular networks (TINs). Attempts have been made to define which method is conclusively superior (e.g. Kumler, 1994); however, the range of terrain type, sample structure and modelling routine is so vast that attempts to make generalisations about the most appropriate method are fraught with difficulty. Table 4.1 summarises the interpolation options provided by one terrain modelling package, namely Surfer for Windows, and briefly suggests which method is the most suitable for particular types of data.

The Random Point Method

Here, individual points are measured across the stereo-model surface in an evenly distributed manner. It provides the user with the flexibility of being able to increase the density of points around specific features of geomorphic interest such as along breaks of slope. Such a point distribution also improves the quality of DTM block diagram displays, makes the interpolation procedure more accurate and presents a more realistic representation of the morphology (Petrie, 1990). Once again, these irregular data require reconstruction through interpolation. Additional strings depicting breaks in slope, as well as feature-specific points, may be added to both DTMs and TINs. This will have the effect of further improving the accuracy of the resultant model (Lane *et al.*, 1993; Li, 1992; Petrie, 1990).

Revisiting Specific Points

The operating software of the plotter includes a useful macro facility, that allows the data capture routine used on one stereo-model to be identically repeated for another. This is, of course, especially useful when considering a time series of stereo-models covering the same surface.

In summary, the accuracy of the reconstructed surface is dependent on the following general parameters as highlighted by Li (1992):

1. the accuracy, density and distribution of the source data;
2. the terrain characteristics;
3. the method of surface reconstruction (DTM/DEM or TIN);
4. the surface characteristics constructed from the source data.

These factors usually have to be considered within an economic framework and are subject to availability of software and hardware.

Graphical Display of Data

The simplest way to present the photogrammetrically generated data is to create two-dimensional plots directly from the Kork files via a pen plotter. The Kork software allows a variety of cartographic outputs, including the incorporation of text. It is possible to export the strings contained within the Kork files in .DXF format to other Windows-based mapping software such as MapViewer, or to CAD packages such as AutoCad. These software packages permit further manipulation and cartographic design, with the results being viewed in both soft and hard copy. The data can be exported into GIS such as ArcInfo and MapInfo. Such systems have the facility to edit the map coverage, but more importantly can attach further spatial information to the map via a relational database.

It is the interpolated surfaces of the TINs and DTMs that utilise fully the three-dimensional component of the data. Contour maps can be generated at pre-determined intervals and block diagram representations and shaded relief maps/hillshades provide powerful tools for surface visualisation. For gridded surfaces, the density of the rows and columns within the grid can be intensified to provide a more detailed surface, although this requires more processing time. Furthermore, for both the TINs and the DTMs, the z component of the surface can be exaggerated to aid visualisation of the surface and can be viewed in different orientations. Again, this output can be viewed either on the screen or as hard copy on a variety of printers and plotters.

The terrain modelling software, in addition to providing a good visual representation of the surface, allows volumetric and planar area calculations to be carried out between models of the same area that were captured for different periods of time. This permits further insight into the evolution of a surface, for example as it undergoes erosion or deposition events. Surfer for Windows (utilised for the applications discussed later) uses three methods to calculate volumes: (i) the trapezoidal rule; (ii) Simpson's rule; and (iii) Simpson's 3/8 rule (see Press *et al.* (1986) for further information on volume calculations). It is possible to create contour maps and surfaces of change between two data sets by subtracting one surface from another. Additionally, the Surfer software allows the overlay of different layers, thus permitting direct visual comparison of a surface as it evolves through time. It is possible to stack either contour maps, surface plots, surface change plots or a combination of all three in order to compare different surfaces.

APPLICATIONS OF ANALYTICAL DIGITAL PHOTOGRAMMETRY

Planimetric and Three-dimensional Mapping and Analysis

Geomorphological research often requires the accurate spatial representation of landscape features in plan form, and quantification of how these forms change through time. Although such applications can be satisfied using traditional photogrammetric instruments, analytical plotters do offer several advantages, particularly the capacity to utilise different image formats at different scales for the same surface. Furthermore, the digital data output allows efficient transfer of feature maps to a GIS where they can be combined with alternative data sources such as historical maps. Together, these two particular advantages of analytical photogrammetry enhance an ability to determine accurately planimetric form changes over extended time scales. Such advantages have been incorporated into a study of river channel change of an active meandering system covering a 20 km reach of the upper Severn, near Caersws in mid-Wales (Hooke *et al.*, 1994).

Analysis of River Channel Change

Detailed geomorphological maps, of both planimetric and hypsometric characteristics of the river channel and floodplain features, were plotted photogrammetrically from 1:5000 colour vertical air photographs taken in 1992 and 1984. The same ground control was used for both sets of photographs, these points being established by ground survey using a Geodimeter 412 total-station, providing a minimum of six points for each stereo-model. Digital data representing the bank lines were combined with data derived by digitising bank lines on Ordnance Survey (OS) 1:2500 plans for various dates, and tithe maps of *c*.1840. To minimise the errors inherent in such data, common control points were used throughout the study. The data derived from the photographs and maps were fed directly into the GIS package ARC-INFO, and by using "rubber-sheeting" routines, discrepancies between the different data sources were minimised. It should also be noted that in using diverse sources care is needed in the interpretation to ensure that the "same" feature is being mapped. Specifically, in this case, it was important to check that the definition of bank line from the aerial photographs was directly comparable with that derived from the maps. However, here, as in many applications, the accuracy of the respective sources was within acceptable limits *given the levels of activity of the river*. The extension backwards in time made possible by combining historical maps with contemporary air photography adds immense value to the analysis.

The main advantages of using an analytical photogrammetric approach were the ease with which digital data could be generated and directly combined within a GIS to enable additional quantitative analysis. For example, the direction and amount of movement of the bank was used to calculate the areas of erosion and deposition for selected reaches and/or epochs (Figure 4.1). Such data were then used for spatial comparison between different parts of the river course, showing differences in mobility and rates of activity. Significantly, the data derived from the most accurate sources (the photogrammetry) also covered the period for which there were discharge data. Tables 4.2a and 4.2b show that deposition exceeded erosion in the period 1973–1984, but contrasts with higher amounts of erosion for the period 1984–1992. Discharges were higher in the latter period (Table 4.2c) and other evidence from processes and landforms points towards a causative link and a relatively rapid adjustment of channel form to effective discharges. Additionally, the data can be used for more detailed analysis of propagation effects, such as detecting the transmission of waves of erosion or sedimentation along the channel, for relating changes to specific features or characteristics (e.g. bedrock outcrops or insertion of bank protection), and can be used for comparison with results of modelling of meander change.

Although such a two-dimensional analysis is facilitated by an analytical photogrammetric approach to surface feature mapping, the full potential of analytical feature mapping is expressed when non-standard images are used, or when additional heighting information is added. This enables analysis of three-dimensional form which may be related to structural components and features of the surface under consideration.

Analysis of Limestone Decay

The approach of three-dimensional (3D) mapping using terrestrial photography has been applied to the weathering forms created by the exposure of building stone to surface environmental processes (Shelford *et al.*, 1996) to develop causal explanations of the forms. This used samples taken from a number of beds in the south Dorset outcrop areas of the Portland Limestone Formation, aiming to consider how the physical and chemical properties of rock surfaces control the

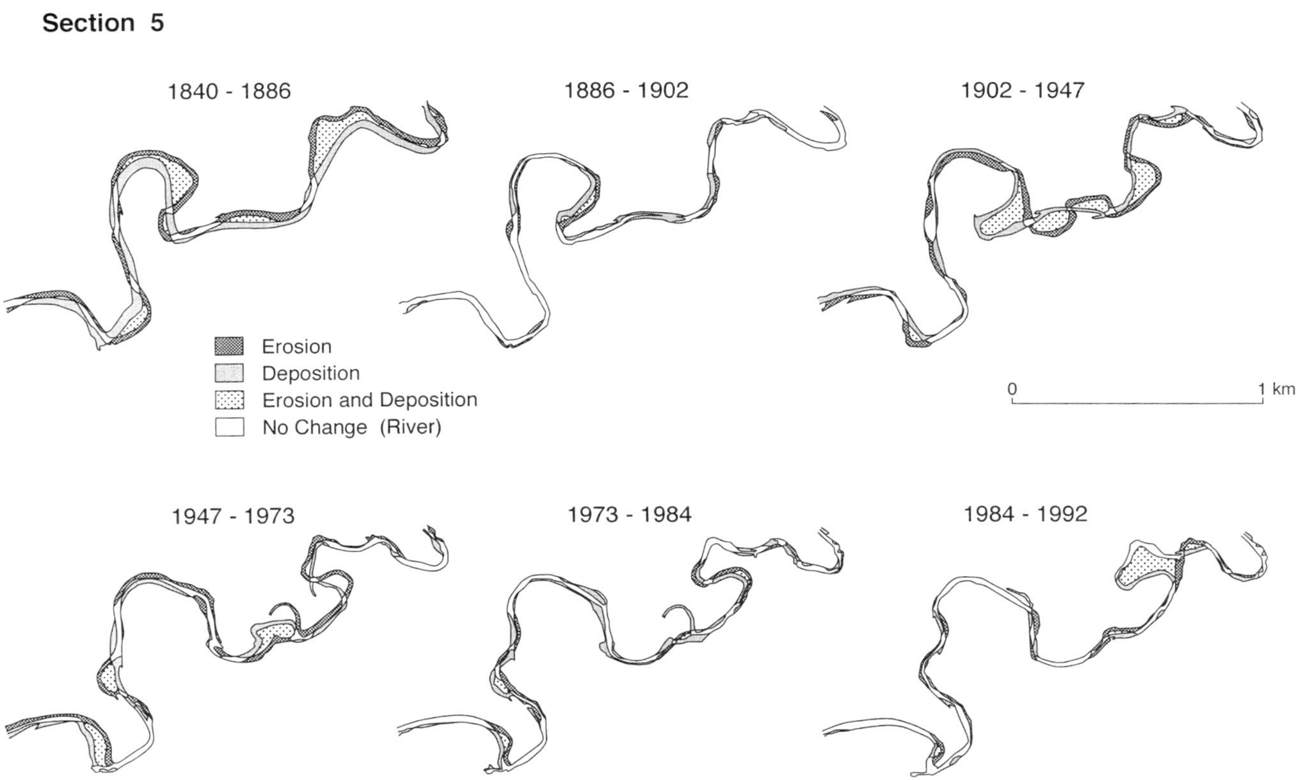

Section 5

1840 - 1886

1886 - 1902

1902 - 1947

0   1 km

1947 - 1973

1973 - 1984

1984 - 1992

Erosion
Deposition
Erosion and Deposition
No Change (River)

Figure 4.1 Zones of erosion, deposition and stability in successive periods on one section of the Upper River Severn, near Caersws, mid-Wales

Table 4.2 (a) Total floodplain areas (in square metres per year) reworked by processes for each period evidence. (b) Ratio of areas of deposition to erosion in different time periods, Upper River Severn, near Caersws. (c) Discharge characteristics of the periods 1973–1983 and 1984–1992 on the Upper River Severn at Abermule gauging station

a

Time period	Erosion	Deposition	Erosion and deposition	Stable
1840–1886	9169	8812	9582	3015
1886–1902	8754	11 677	3450	23 383
1902–1947	6584	6898	4000	4539
1947–1973	8266	9859	8363	9391
1973–1984	7785	14 779	1716	26 957
1984–1992	15 089	11 385	5821	36 390

b

Date	Ratio of deposition to erosion
1840–1886	0.96
1886–1902	1.33
1902–1947	1.04
1947–1973	1.19
1973–1984	1.90
1984–1992	0.75

c

	1973–1983	1984–1992
Mean daily flow (m^3s^{-1})	14.21	14.72
Mean annual maximum daily flow (m^3s^{-1})	109.50	131.62
Mean annual flood (m^3s^{-1})	174.85	192.06
Average no. peaks per year over threshold	1.36	2.89

morphological development of surface weathering forms. Due to the variety of constituent grains within the rocks, including an extensive fossil fauna, the variability of surface produced will make it possible to relate site-specific surface change to the individual grain forms which are present on a variety of scales.

Experimental tablets were prepared from 50 mm cores cut from bedding-oriented samples collected from the field sites. The cores were sliced into 25 mm thick tablets, and the surfaces were lapped with a coarse grit to remove saw marks. Two sets of tablets from each sample were cut, one along the bedding plane and one at right angles to this. Each tablet was firmly mounted on a perspex base, surrounded by six screws of varied height mounted vertically on the slab around the tablet.

These were required as ground control for each individual model. Prior to exposure to a variety of weathering simulations, each sample was photographed in stereo using a vertically mounted Polaroid MP3 land camera with a focal length of 27.5 mm set at a distance of 275 mm above the target surface. A fine grained film coupled with a low angle incident light source maximised clarity of the obtained images. Further sets of stereo photographs were taken after a number of weathering cycles.

Data collection using the DSR took three forms, with the aim of producing accurate two-dimensional (2D) and 3D surface maps to illustrate the progressive change of the surface as it weathers. The example here illustrates the results obtained from applying the technique to a freshly prepared sample, and the results are illustrated in Figures 4.2a, 4.2b and 4.3. Figure 4.2a shows a feature map plotted to highlight dominant surface structure. This illustrates the presence of bivalve shells, fossil moulds and ooids amongst other grain types.

The entire surface of the tablet was measured using random data points, which can be reused at each stage of model creation. These data allowed production of 2D contour maps and 3D surface maps of the tablet (Figure 4.2b), to allow identification of areas of height change, which would be related in turn to the base map to determine the influence of structure on weathering patterns. With this particular sample, owing to the presence of fossil moulds, there is already quite extensive relief. Figure 4.3 shows a surface map of a bivalve mould, where it has been possible to pick up many of the complex elements which form its internal structure, such as gill crenellations and pallial lines. These figures illustrate the results of an early test of the technique which has since highlighted a number of potential errors for such close range photogrammetry in the study of weathering. Further studies have refined the control within these images to allow intra-granular susceptibility to weathering to be observed, in addition to the inter-granular scale.

These examples illustrate the potential of applying photogrammetric techniques to the micro-scale. In conjunction with thin sectioning data and other geological analytical techniques, it is becoming possible to relate accurately the way physical and chemical properties of the stone at a variety of scales influence the type and rate of weathering, and so influence the shape of the resulting surface.

Application to Soil Surfaces

Another example of a similar approach, but in a field monitoring situation, can be seen in the work of soil surface evolution. Once a freshly tilled soil surface is left exposed to a sequence of natural rainfall events it begins to be transformed from a rough discrete soil aggregate surface to a continuous, flat, dense, discrete particle crusted surface. Such surfaces have significance in soil erosion modelling because they define the probability of surface water flow generation from subsequent storm events. One aspect of interest in this study is how the distribution of stones, individual soil aggregates and voids control the pattern and rate of soil crust seal evolution. As with the weathering study, therefore, it is critical to map these non-morphological characteristics of the initial object's surface. A series of 10 m × 1 m

Legend.

■	Fossil mould (general)	M
□	Bivalve shell (1)	B1
▨	Bivalve shell (2)	B2
⠃	Individual ooids	OO
▥	Echinoid fragment	G

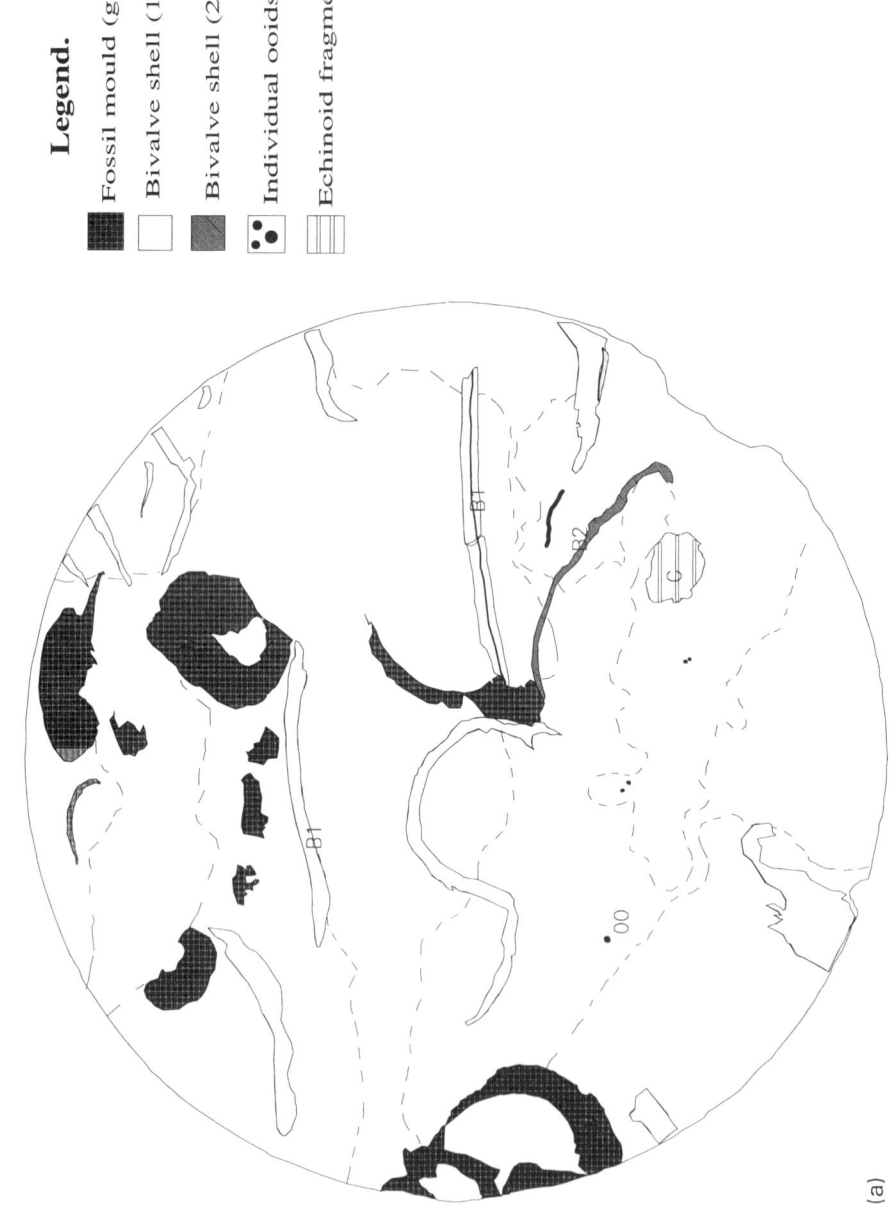

(a)

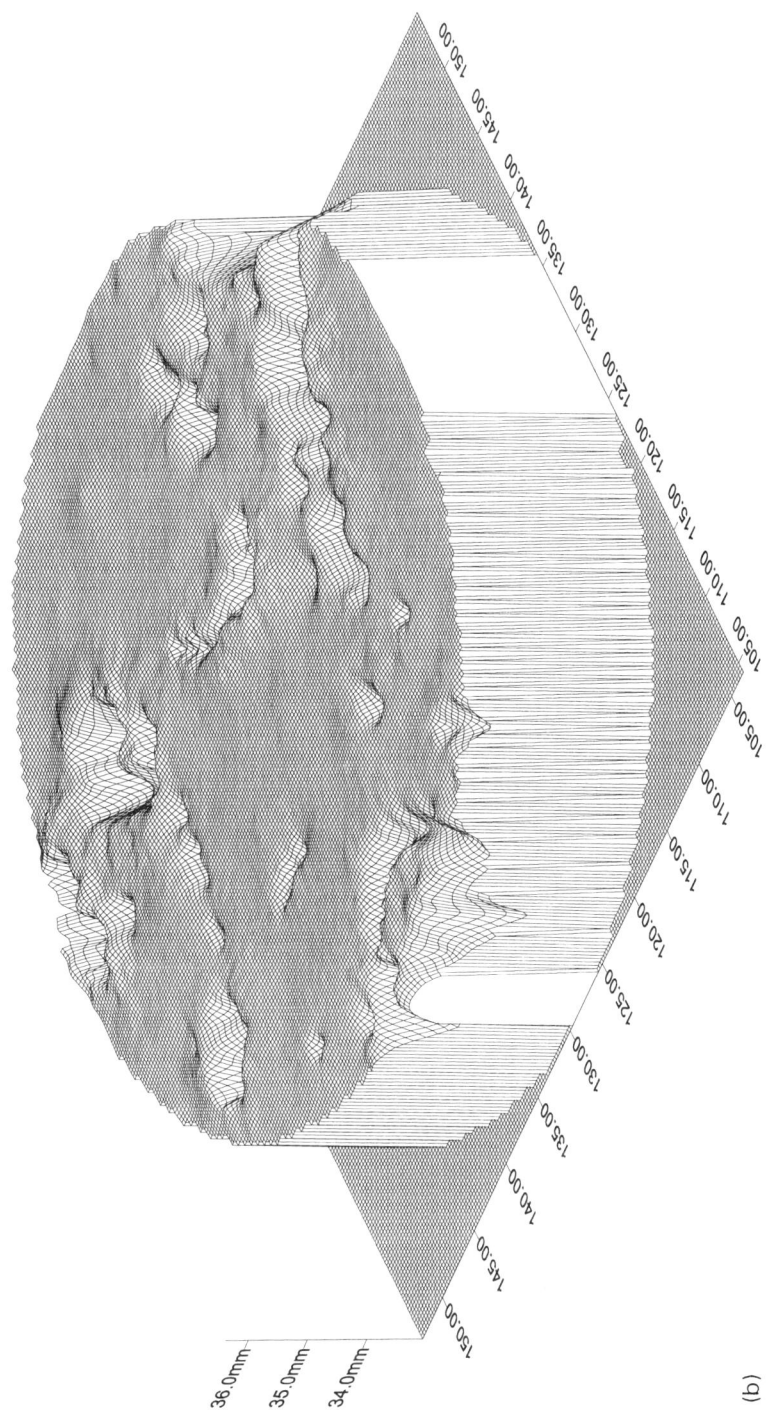

(b)

Figure 4.2 (a) Feature basemap of surface of stone tablet. (b) DTM of stone tablet surface. Note that the Z-scale has been exaggerated in order to aid interpretation

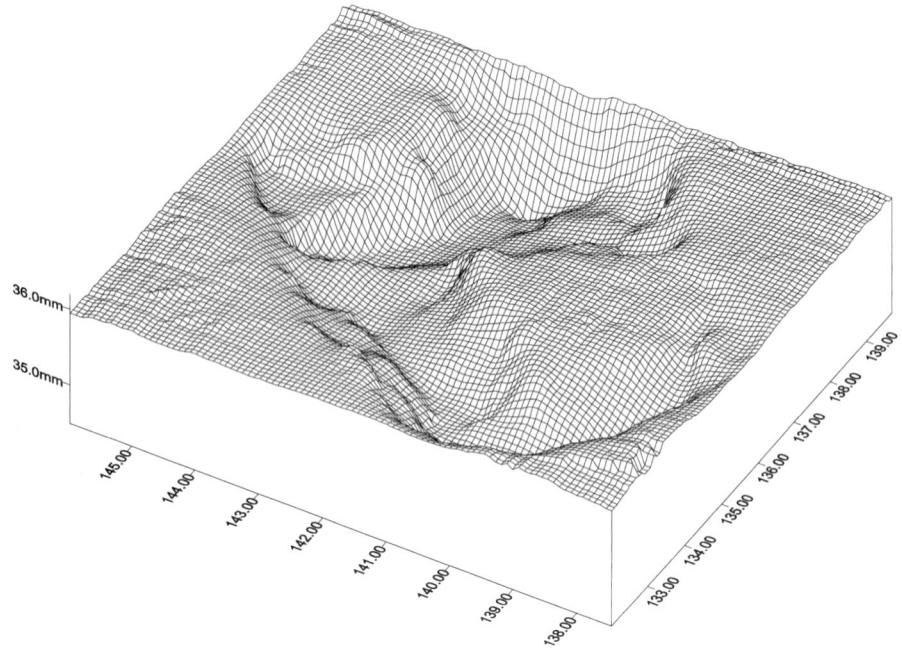

Figure 4.3 DTM of individual bivalve mould

plots were prepared in the field to produce a natural tilled surface, typical of the conditions at the start of a growing season. Photographs were obtained using a Wild P32 camera capable of producing 70 m × 52.5 mm format diapositive images. The camera was mounted on a specially designed frame 2.4 m above the surface, the resulting images each giving a 3 m × 1.5 m coverage of the ground surface. Within the photographs, but not directly on the experimental plot surfaces, control points were established and coordinates determined using a Geodimeter 412 total-station. Such experimental design produced between five and seven control points on each image. Images were obtained fortnightly for a period of six weeks. Figures 4.4a and 4.4b are surface feature maps for one of the replicate soil surfaces and Figures 4.5a and 4.5b are their corresponding DTMs. Such maps can subsequently be used to attempt an explanation of the interactions between non-morphological features and morphological change.

The experiments in both the limestone decay and soil surface examples have an exceptional degree of control over the capture of the images, so one can clearly relate the height at particular points on the object's surface for different time periods, to structures on that surface, with a very high degree of accuracy. The research demonstrates how analytical photogrammetry can therefore produce planimetric feature maps at a scale and accuracy previously impossible and without any contact between the observer and the naturally evolving morphology. Such relationships between features and form may well provide a basis for improved understanding of causal explanations of form evolution.

(a)

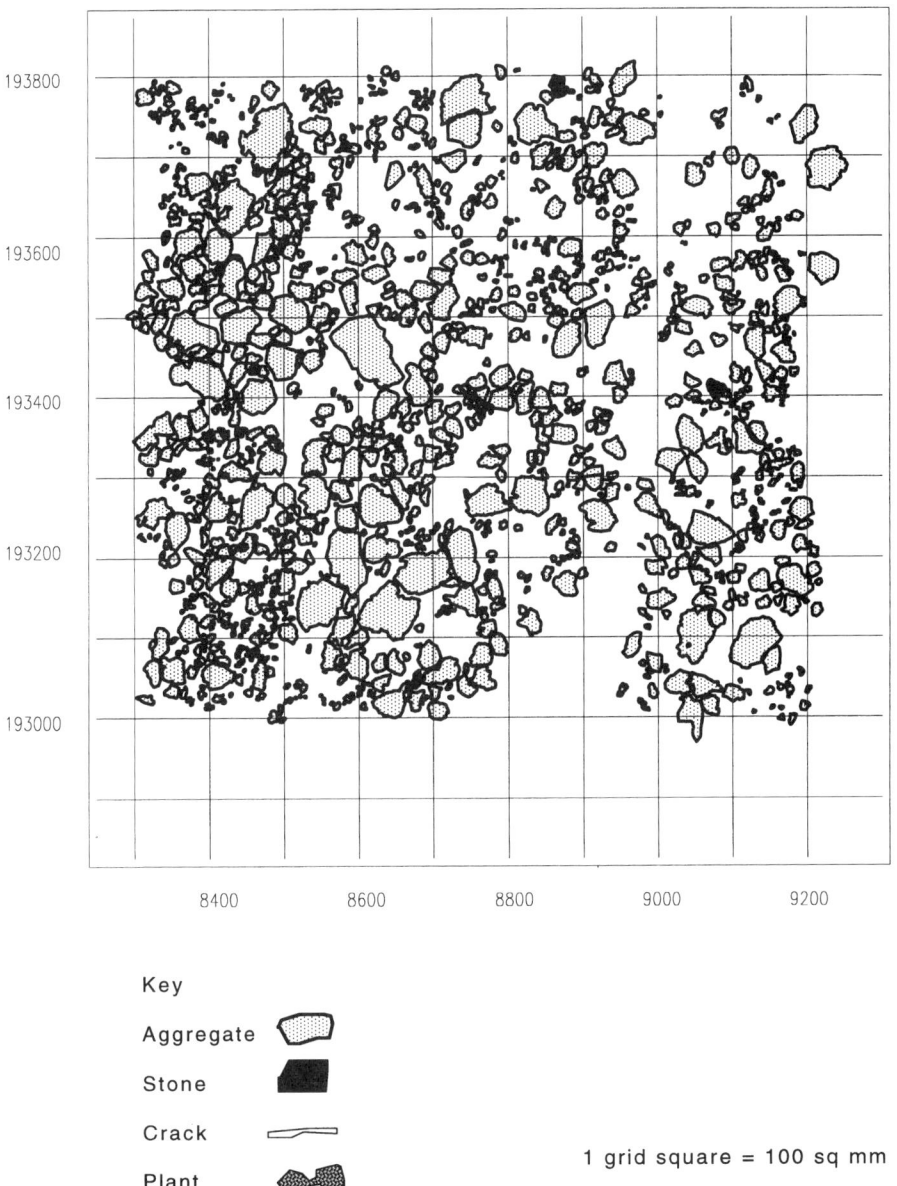

Key

Aggregate

Stone

Crack

Plant

1 grid square = 100 sq mm

Figure 4.4   (a) Feature map of soil surface at Time 0 – initial surface state. (b) Feature map of soil surface at Time 1 – surface after six week's exposure

(b)

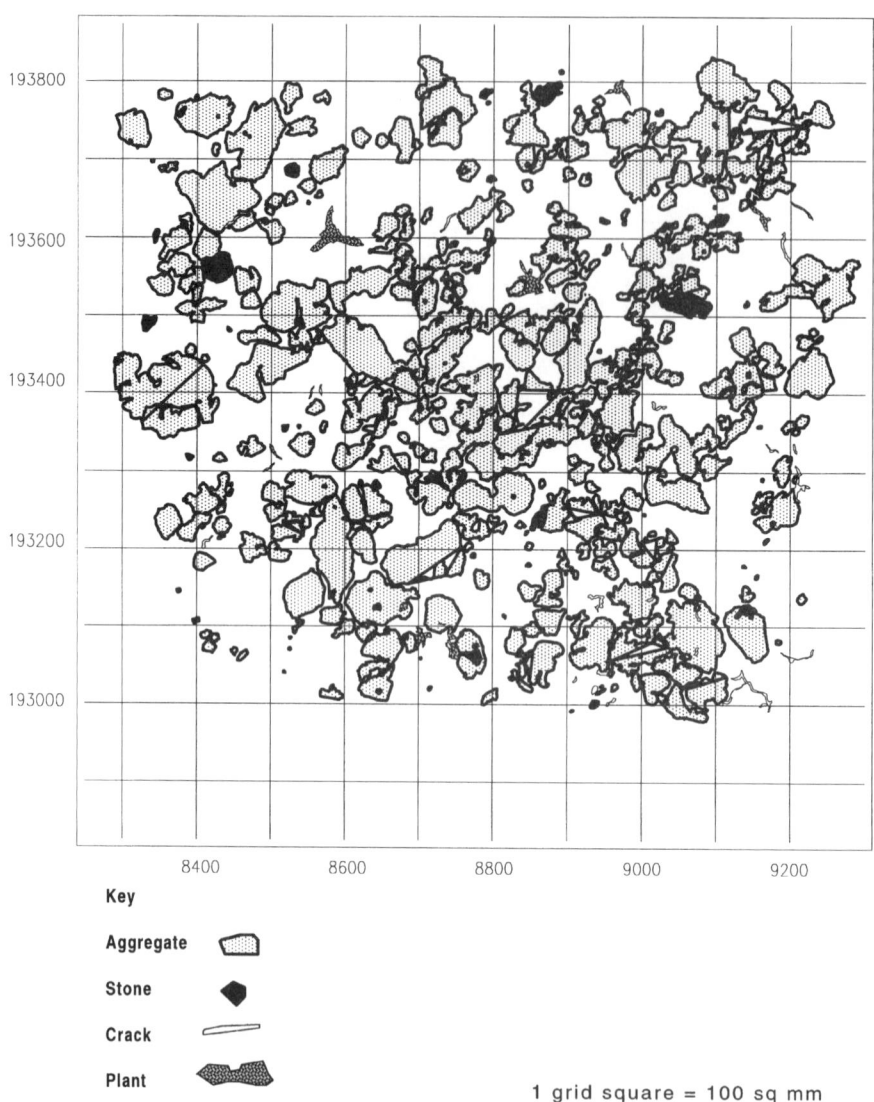

Figure 4.4 (*continued*)

(a)

(b)

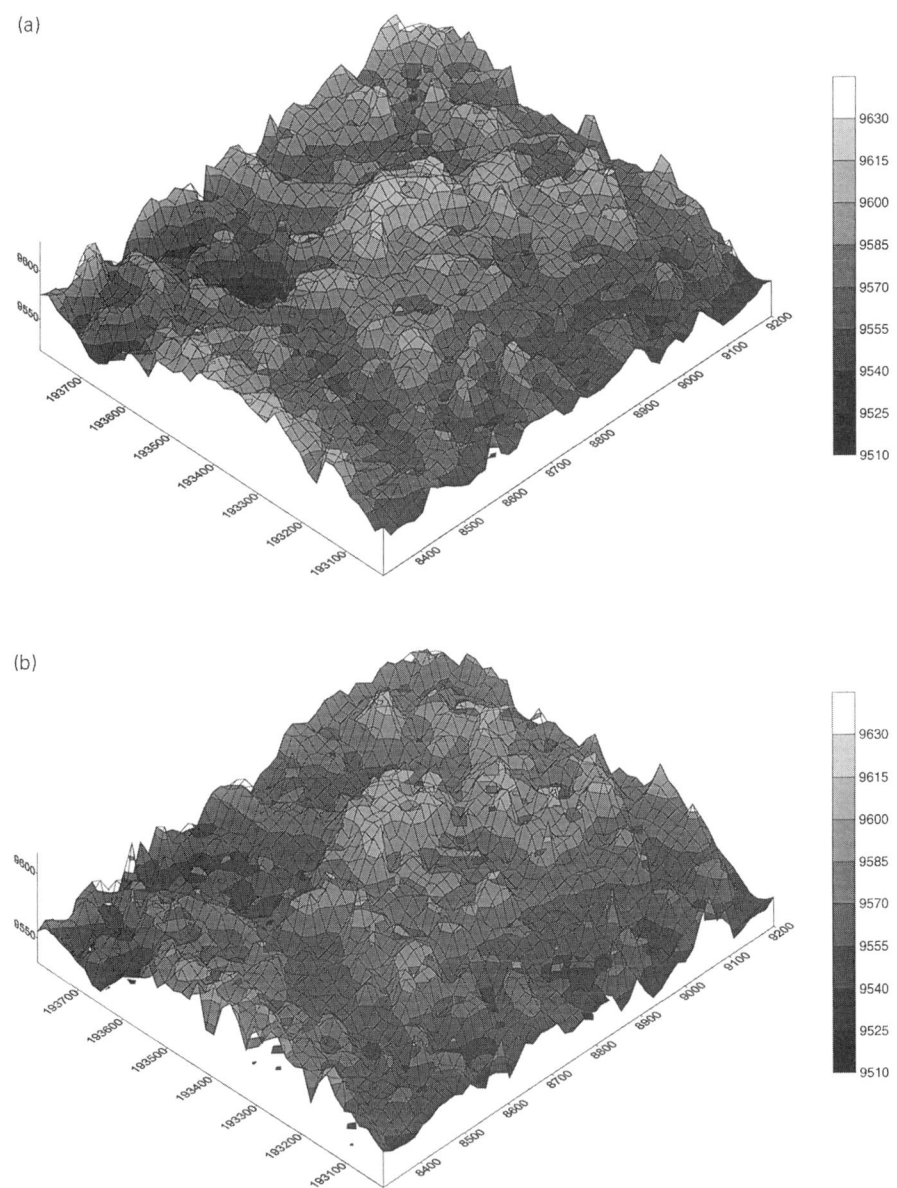

Figure 4.5 (a) DTM of surface at Time 0 – initial surface state. (b) DTM of surface at Time 1 – surface after six week's exposure

Temporal Analysis

The full potential of analytical photogrammetry is fulfilled when 3D data are extracted for a surface and compared over time. This enables changes in form to be analysed in detail and calculations of volumetric change between time periods to be made. For example, the volume of material removed during the initiation and development of a thalweg gully in an arable agricultural field can be measured and used to make an estimate of the weight of soil material removed. In a similar way, river channel dynamics, in terms of volume of material involved in bar migration, can also be determined photogrammetrically (e.g. Lane *et al.*, 1994). Calculations like these provide invaluable inputs for mathematically based models whose structure depends on mass balance continuity statements.

Application of quantification of morphological changes using sequential traditional air photographs is illustrated by cliff retreat along a dynamic section of coast in southern Britain. Aerial photograph analysis has long been a valuable investigative technique in coastal geomorphology (e.g. El Ashry, 1978; Chisholm, 1990), because the extensive, complex and dynamic surface forms typical of coasts are not represented well by conventional maps (Carr, 1962, 1980; Collin and Chisholm, 1991). Analytical photogrammetry and digital data processing provide considerable advantages over conventional analogue plotting and make possible new types of coastal monitoring and reconstruction of historical change (e.g. Chandler and Cooper, 1988; Chandler and Brunsden, 1995).

Coastal Cliff Erosion

The rapidly eroding Tertiary sands and clay cliffs in Christchurch Bay, Hampshire, are an especially dynamic landform that poses significant management problems (e.g. Clark *et al.*, 1976). Since 1976, the New Forest District Council has monitored these cliffs using high quality vertical aerial photography flown annually at a scale of 1:2500. The present study is a preliminary analysis of photographs from 1967, 1976 and 1993 along a 1.8 km frontage of mostly unprotected cliffs to the east of Barton-on-Sea. The work shows how analytical photogrammetry both complements and extends information available from conventional sources.

When setting up the stereo-models, the large scale of the photographs caused limited ground coverage and created some problems in acquiring sufficient reliable ground control. This problem was compounded by the nature of the landscape because significant areas comprised unstable cliffs, beaches and the sea, all locations where ground control is not normally advised. Sea-level could not be used to check heighting and assist absolute orientation within single models, because of inconsistencies produced by wave action. It was therefore necessary to determine by field survey the precise OS coordinates of 50 "permanent" features that could be clearly identified on the 1993 photography. Common points were passed back from the 1993 to the 1976 and 1967 photography. The high quality of photogrammetric resolution ultimately achieved (standard deviations of control points within ±0.2 m for height and plan) was due to the large photographic scale and the capacity of the analytic method to utilise this high quality ground control.

Different strategies were adopted to capture data throughout the complete study area and within specific detailed sites. The research design involved two distinct scales of study: (i) an extensive assessment of erosional change throughout the complete coastal section; and (ii) a detailed intensive analysis focusing on a 220 m section. In the extensive survey, for each epoch, strings of coordinates (X, Y, and Z) representing the positions of the cliff top, 10 m contour (AOD), cliff toe and the 1 m contour (equivalent to mean high water) were collected. In addition, analogous features (cliff top, cliff toe and mean high water) were digitised as strings (x and y) from OS 1:2500 plans for 1870, 1898, 1910, 1932 and 1958. Using a similar research design to that of the River Severn study, these strings were also accommodated within the same grid system as those derived from the photos. All features were plotted in plan to produce a detailed record of historical change. Erosion trends can be determined by sampling recession distances, calculating rates per epoch and plotting results in longshore sequence (Figure 4.6). Retreat is most rapid in the late 19th century and also after 1958 (up to 3 ma^{-1}), the latter being explicable in terms of recent cliff stabilisation schemes updrift at Barton. These have starved the study area of protective beach sediment, thus exposing the cliff toe to marine erosion and accounting for differences in retreat between the cliff top and toe.

A sewage outfall is notable in intercepting drift and producing a zone of down-drift erosion (Figure 4.6). Its effect was studied by measuring profiles across the cliffs at 50 m intervals downdrift of the Becton Bunny sewage outfall. The photo-grammetric software associated with the analytical plotter was used to reposition each of 326 sample points per profile at the same X and Y coordinates for each epoch. Sequential plots of the measured coordinate heights clearly show the landward migration of the key morphodynamic zones between epochs (Figure 4.7). Retreat is especially rapid close to the outfall, which operates as a transport barrier. The profiles also suggest that the "front" of this zone of retreat is migrating progressively downdrift.

The intensively studied coastal segment incorporated between 6000 and 9000 X, Y, and Z coordinated points per epoch. The interpolated grids were then subtracted from each other to produce contour maps of change between consecutive images (Figure 4.8). Such analyses are especially effective in representing the distribution of change and, by incorporating process knowledge, can also assist in elucidating its causes.

Results can also be determined volumetrically. Analysis based on the retreat of the extensively mapped features revealed erosion of 1.4×10^6 m^3 of sediment from the entire study area between 1967 and 1993. This is obviously a major input to the regional coastal system and has implications for any future management decisions. More detailed volumetric analysis was applied to the intensive site using the mass balance approach in combination with process knowledge. The geological control of slope degradation processes, afforded by the near-horizontal Tertiary stratigraphy, results in strong correlation between morphological zones and specific process mechanisms (e.g. Barton, 1973; Barton and Coles, 1984). Thus, by calculating volumetric changes for specific horizontal slices which correspond to process domains (Figure 4.7) it is possible to quantify process rates much more effectively than hitherto (Table 4.3).

Becton Cliff Top Recession

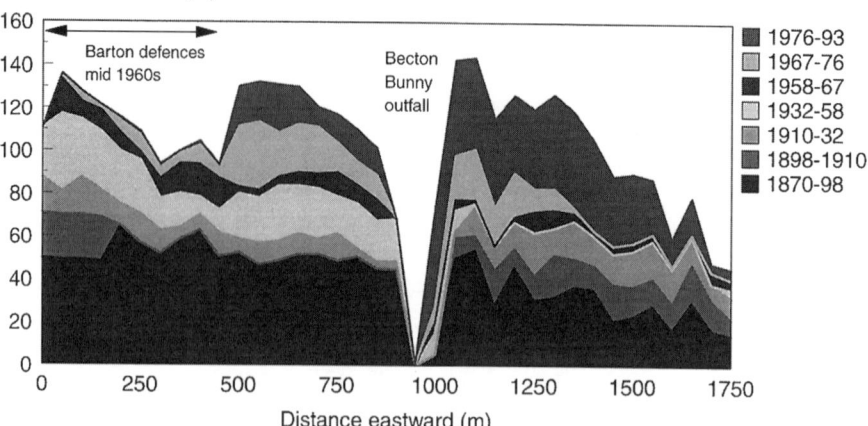

Erosion since 1870 (m)

Becton MHW Recession

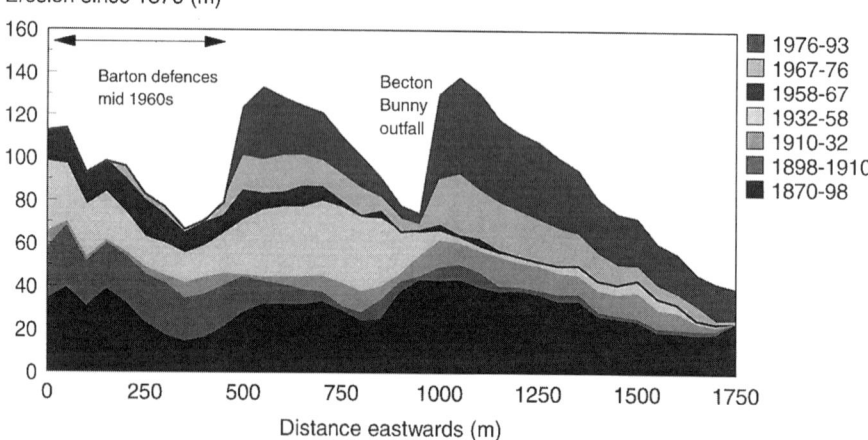

Erosion since 1870 (m)

Figure 4.6 Historical and longshore variations in recession. Note that trends are detected more easily when data are simplified and presented in this manner

Losses have not been uniform, indicating that both the relative and absolute activities of different processes vary through time. Comparison of losses from the lower (marine erosion) and upper (slumping, spalling and bench sliding) cliffs nevertheless reveals a process balance (proportions of loss constant) operating over the full study period. These results accord with the work of Chandler and Brunsden (1995) in which similar steady-state behaviour was identified at the Black Ven cliff system in Dorset. The crucial difference is that the results here are based upon continuity of mass, whilst those from Dorset involved continuity of slope angle.

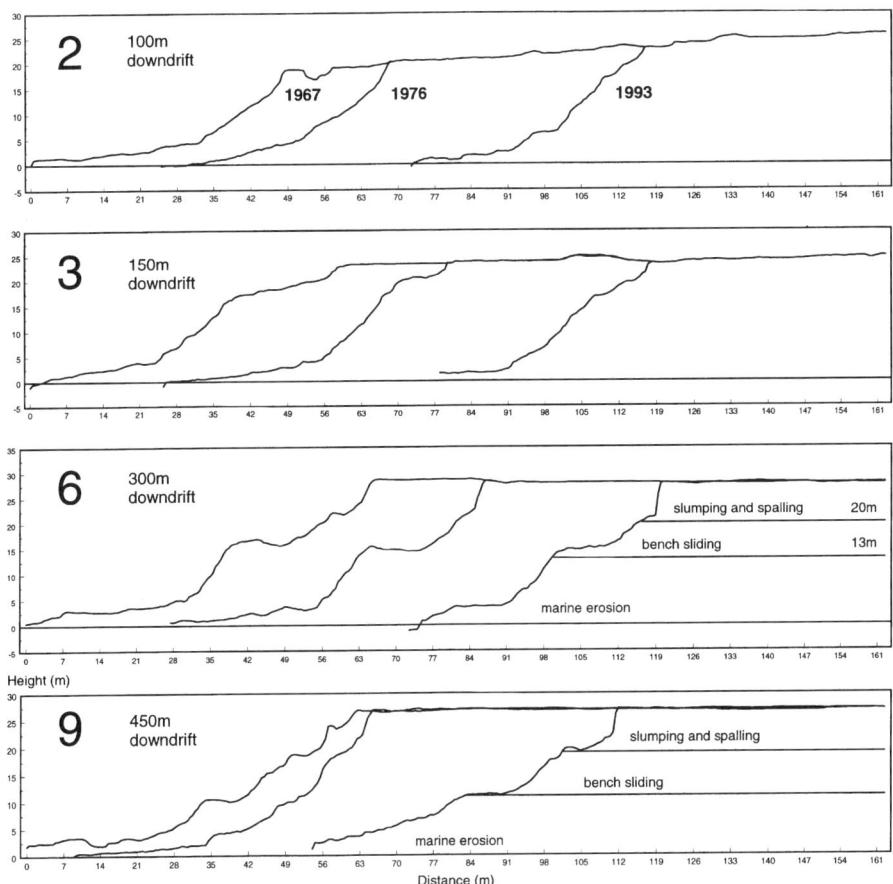

Figure 4.7 Selected cliff profiles plotted at intervals downdrift of the Becton Bunny outfall. Note that a characteristic scarp and bench morphology is retained as retreat proceeds and that specific degradation processes operate within the zones defined

Further research aims to improve resolution by analysing more of the photographic time series, whilst unifying the relationships between continuity of mass and slope angle.

Analysis of Soil Surface Evolution

Another important application of sequences of images is the ability to quantify form evolution and to begin to test the appropriateness of some models of morphological evolution. During the evolution of the soil surface by raindrop impact, the soil structural units are broken down and the released material redistributed and organised across the surface (Farres, 1978; Valentin and Bresson, 1992). The result of these processes is to produce a lowering of soil surface, a

Contour Map of Change

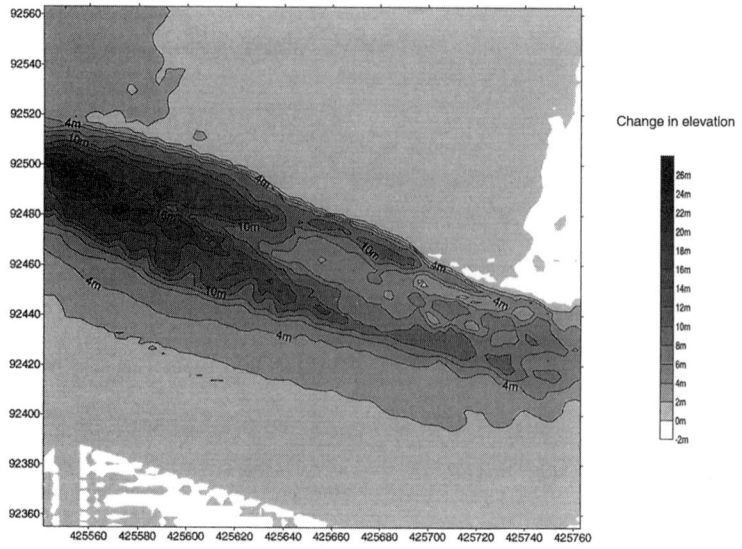

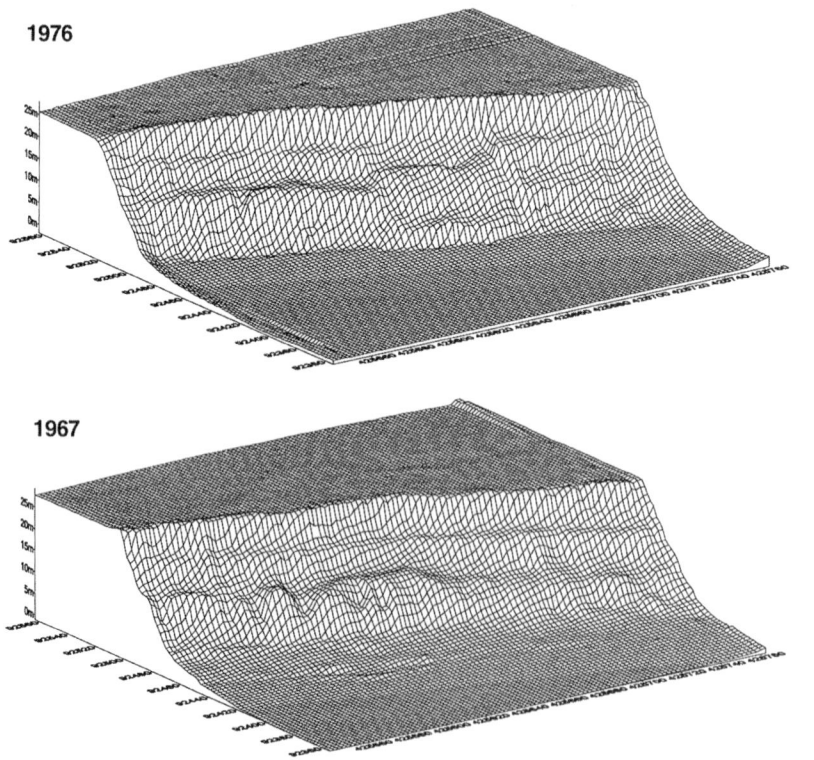

Table 4.3 Erosion volumes expressed according to morphodynamic zones within the intensive study area. Rates can be related to specific processes by reference to Figure 4.7

Contour	1967–1976			1976–1993			Total	
(m OD)	(m^3)	(m^3a^{-1})	(%)	(m^3)	(m^3a^{-1})	(%)	(m^3)	(m^3a^{-1})
>20	43 751	4861	30.42	53 125	3125	23.91	96 886	3726
13–20	27 417	3046	19.06	59 852	3521	26.94	87 269	3357
2–13	49 791	5533	34.62	88 683	5217	39.91	138 474	5326
1–2	22 866	2541	15.90	20 524	1207	9.24	43 390	1669
Total	143 825	15 981	100	222 194	13 070	100	366 019	14 078

Table 4.4 Typical soil surface property changes derived from photogrammetric data. Note that all heights are in millimetres unless otherwise stated and are based upon an arbitrary coordinate system

	Time t_0 initial state	Time t_1 after exposure	Difference $t_1 - t_0$	Description
Mean height	9582.7	9573.1	−9.6	Lowering
Height st dev.	17.3	16.4	0.9	Smoothing
Volume (mm^3)	5.6895E+07	5.04105E+0.7	6 484 500.0	Volume change

reduction in roughness and a compaction of the soil surface through time. These effects can be simply quantified from the data obtained by photogrammetry. Mean height change, variation of height values about this mean and the differences in volume between surfaces give a simple, direct measure of the three basic responses (Table 4.4). Many other statistical devices can be used on the data to characterise the changes, for example variograms (Oliver *et al.*, 1989) or fractals (Bertuzzi *et al.*, 1990). One potentially useful approach is to consider the functional relationship between heights at t_1 to their initial heights at t_0. The coefficients from such analysis can then be interpreted in terms of standard geomorphological models of slope change (Phillips, 1995). Additionally, from such analysis, the potential exists to explore the residuals from the statistical fit in terms of the non-morphological properties of the surface as recognised in the soil surface feature maps already described.

Measuring Bank Erosion

The final application of analytical photogrammetry developed here shows the power of the technique to deal with non-standard photographic formats and the added advantage of a non-invasive measurement technique of an object's surface. The aim

Figure 4.8 DTMs of the intensive study area for 1967 and 1976. The contour plot illustrates the distribution of net change between epochs

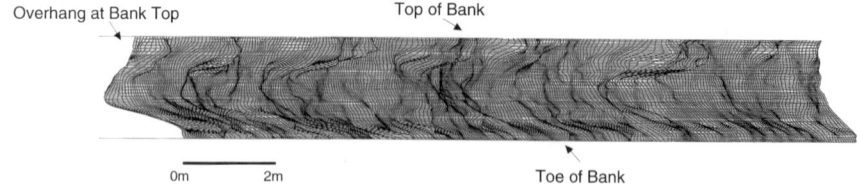

Overhang at Bank Top   Top of Bank

0m   2m   Toe of Bank

Figure 4.9 DTM for a 20 m section of the River Yarty bank

of this study is to determine the relative importance of bank erosion in supplying sediment to the channel, as opposed to sediment supplied from slopes. The study is situated on the River Yarty in Devon and involves determining the amount of sediment supplied to the channel by a 60 m eroding bank face. Comparisons of the site as it was 20 years ago (Hooke, 1977) with recent visits, suggest that the reach in question is extremely active, with bank erosion of several metres occurring in the past two decades.

Traditionally, bank erosion monitoring is achieved by intensive fieldwork and the use of erosion pins and/or surveying. These methods have several limitations: (i) the fieldwork is time-consuming; (ii) they involve interference with the bank face being measured; and (iii) the spatial resolution is limited to the points under investigation. Photogrammetry reduces spatial sampling problems and enables bank erosion to be monitored with minimal contact with the bank face itself.

The use of terrestrial photogrammetry for measuring bank erosion dates back to the 1970s when two studies were carried out using analogue photogrammetry (Painter *et al.*, 1974; Collins and Moon, 1979). However, these two studies suffered from a low level of accuracy caused by the direct measurement of the position and orientation of the camera during photo-acquisition. It is now more efficient to determine these parameters mathematically by using analytical photogrammetry and providing ground control.

In this case study, camera stations were set up along a 60 m baseline on the opposite side of the river, but parallel with the bank face. The camera used was a Pentax 645 with an additional glass plate mounted in the camera to produce the required fiducial marks on the ISO 100 film. The stations were positioned to give a conventional stereoscopic overlap of 60% between photographs in order to generate an adequate base–distance ratio. Control for the stereo-models was provided by survey poles on the top of the bank, in addition to six small white targets for each stereo-pair mounted on the face of the eroding bank. These temporary targets were added with minimal disturbance to the bank face itself. Measurements to all control points were determined using a Geodimeter 412 total-station, from each end of the baseline, and their coordinates established within one single local coordinate system.

Figure 4.9 shows a DTM from one epoch of part of the near-vertical eroding bank face. Once subsequent images are produced of the same bank sites, it becomes a routine matter to calculate volumetric changes between the images and hence a quantitative analysis of the spatial distribution of the bank erosion along the reach of the river. Analysis of these structural changes and temporal distribution may well allow greater insight into the mechanisms of bank erosion.

CONCLUSION

Lane *et al.* (1993, 1994) have clearly documented the potential of analytical photo-grammetry as a tool for geomorphological research. Developed here are some of the practicalities associated with such an approach as seen across a number of applications covering various temporal and spatial scales. In particular, the specific advantages of analytical photogrammetry in terms of accurate, non-invasive data capture have been demonstrated. Equally, the ability to remeasure the elevation at exactly the same location within a stereo-model on a number of occasions permits an indicator of precision to be determined. This is essential to test the effectiveness of the technique and in particular to provide one indicator describing the quality of derived data.

Additionally, the ability to create time series DTMs of the same surface adds a vital component to our understanding of change in form. By calculating volumetric differences between sequences of surfaces, important information can be derived, which can then be incorporated into models of morphological change. Such data are also becoming increasingly necessary for application of numerical simulation models, e.g. of beach or river planform change (Lane *et al.*, 1995). The data obtained from analytical photogrammetry is digital and thus directly transferable into GIS and data analysis software. This facility has great potential not only in terms of visual display of form and statistical description, but in the way in which photogrammetrically derived data can be integrated with alternative spatial data such as maps. Photography is being increasingly used by various authorities, organisations and agencies, and the growth in such data sources enhances the future feasibility of analytical photogrammetry as a tool in environmental monitoring and planning. However, users of such techniques should understand fully the associated advantages and limitations inherent in such an approach.

The technique of analytical photogrammetry is not without its disadvantages, most notably the need for expensive equipment and specifically the need for skilled operators. One of the requirements and benefits, as demonstrated here, is the combination of technical expertise and geomorphological experience. Only when both combine can the full potential of such an approach be fully realised. Another of the problems which arises, as a very consequence of the power of the technique, is the amount of data generated. The manipulation of the stored data can often become a non-trivial problem. One final aspect of the technique which should also be noted is that collecting data from the stereo-models created can be time-consuming, even for a skilled and experienced user. Technological advances in digital photogrammetry are helping to overcome this potential problem (Walker, 1995; Wrobel, 1991) but are likely to exacerbate the data volume issue.

ACKNOWLEDGEMENTS

The authors acknowledge the assistance of Andrew Bradbury (NFDC), Nick Cooper, Bill Duane, the Countryside Council for Wales (CCW), Rod Jones (CCW), Justine Moore and Dr Peter Collier.

REFERENCES

Ackermann, F., 1992. Kinematic GPS control for photogrammetry. *Photogrammetric Record*, **14**, 261–276.

Allamaras, R. R., Burwell, R. E., Larson, W. E. and Holt, R. F., 1966. *Total porosity and random roughness of the interrow zone as influenced by tillage*. US Department of Agriculture Conservation Research Report No. 7, Washington DC.

Barton, M. E., 1973. The degradation of the Barton clay cliffs of Hampshire. *Quarterly Journal of Engineering Geology*, **6**, 423–440.

Barton, M. E. and Coles, B. J., 1984. The characteristics and rates of the various slope degradation processes in the Barton clay cliffs of Hampshire. *Quarterly Journal of Engineering Geology*, **17**, 117–136.

Bertuzzi, P., Rauws, G. and Courault, D., 1990. Testing roughness indices to estimate soil surface roughness changes due to simulated rainfall. *Soil and Tillage Research*, **17**, 87–99.

Bradford, J. M. and Huang, C., 1992. Applications of a laser scanner to quantify soil microtopography. *Soil Science Society of America Journal*, **56**, 14–21.

Brunsden, D., 1974. The degradation of a coastal slope, Dorset, England. In *Progress in Geomorphology* eds E. H. Brown and R. S. Water. Institute of British Geographers Special Publication No. 7, 79–98.

Brunsden, D. and Jones, D. K. C., 1980. Relative time scales and formative events in coastal landslide systems. *Zeitschrift für Geomorphologie N.F. Supplementband*, **34**, 1–19.

Carr, A. P., 1962. Cartographic record and historical accuracy. *Geography*, **47**, 135–145.

Carr, A. P., 1980. The significance of cartographic sources in determining coastal change. In *Timescales in Geomorphology*, eds R. A. Cullingford, D. A. Davidson and J. Lewin, John Wiley & Sons, Chichester, 69–78.

Chandler, J. and Brunsden, D., 1995. Steady state behaviour of the Black Ven mudslide: the application of archival analytical photogrammetry to studies of landform change. *Earth Surface Process and Landforms*, **20**, 255–275.

Chandler, J. H. and Cooper, M. A. R., 1988. Monitoring the development of landslides using archival photography and analytical photogrammetry. *Land and Minerals Surveying*, **6**, 576–584.

Chisholm, N. W. T., 1990. Coastal Air Survey: changing tides of fortune? *Photogrammetric Record*, **13**, 533–560.

Clark, M. J., Ricketts, P. J. and Small, R. J., 1976. Barton does not rule the waves. *Geographical Magazine*, **48**, 580–588.

Collin, R. L. and Chisholm, N. W. T., 1991. Geomorphological photogrammetry. *Photogrammetric Record*, **13**, 845–854.

Collins, S. H. and Moon, G. C., 1979. Stereometric measurement of streambank erosion. *Photogrammetric Engineering and Remote Sensing*, **45**, 183–190.

Combs, J. E., 1980. Planning and executing the photogrammetric project. In *Manual of Photogrammetry*, 4th edition, ed. C. C. Slama. American Society of Photogrammetry, Falls Church, 367–412.

Cornelius, S. C., Sear, D. A., Carver, S. J. and Heywood, D. I., 1994. GPS, GIS and geomorphological fieldwork. *Earth Surface Processes and Landforms*, **19**, 777–787.

Cox, R. C. A., 1992. The benefits of forward motion compensation for aerial survey photography. *Photogrammetric Record*, **14**, 5–17.

El Ashry, M. (ed.), 1978. *Air Photography and Coastal Problems*. Benchmark Papers in Geology No. 38, Dowden, Hutchinson and Ross, Inc., Stroudsburg, Pennsylvania, 425pp.

Farres, P. J., 1978. The role of time and aggregate size in the crusting process. *Earth Surface Processes and Landforms*, **3**, 243–254.

Fryer, J. G., Chandler, J. H. and Cooper, M. A. R., 1994. On the accuracy of heighting from aerial photographs and maps: implications to process modellers. *Earth Surface Processes and Landforms*, **19**, 577–583.

Ghosh, S. K., 1988. *Analytical Photogrammetry*, 2nd edition. Pergamon Press, New York.

Granshaw, S. I., 1980. Bundle adjustment methods in engineering photogrammetry. *Photogrammetric Record*, **10**, 181–207.

Hooke, J. M., 1977. *An analysis of changes in river channel patterns*. Unpublished PhD thesis, University of Essex.

Hooke, J. M. and Kain, R. J. P., 1982. *Historical Change in the Physical Environment: A Guide to Sources and Techniques*. Butterworth, Sevenoaks, 236pp.

Hooke, J. M., Horton, B. P., Moore, J. and Taylor, M. P., 1994. *Upper River Severn (Caersws) Channel Study*. Report to Countryside Council for Wales, University of Portsmouth, 165pp.

Keckler, D., 1995. *Surfer for Windows User's Guide*. Golden Software Inc., Colorado.

Kumler, M. P., 1994. An Intensive comparison of triangulated irregular networks (TINs) and digital elevation models (DEMs). *Cartographica*, **31**(2), Monograph 45.

Lane, S. N., Richards, K. S. and Chandler, J. H., 1993. Developments in photogrammetry; the geomorphological potential. *Progress in Physical Geography*, **17**(3), 306–328.

Lane, S. N., Chandler, J. H. and Richards, K. S., 1994. Developments in monitoring and modelling small-scale river bed topography. *Earth Surface Processes and Landforms*, **19**, 349–368.

Lane, S. N., Richards, K. S. and Chandler, J. H., 1995. Morphological estimation of the time-integrated bed load transport rate. *Water Resources Research*, **31**(3), 761–772.

Li, Z., 1992. Variation of the accuracy of digital terrain models with sampling interval. *Photogrammetric Record*, **14**, 113–128.

Neill, L. E., 1994. Accuracy of heighting from aerial photography. *Photogrammetric Record*, **14**, 917–942.

Oliver, M., Webster, R. and Gerrard, J., 1989. Geostatistics in Physical Geography. Part 1: Theory. *Transactions of the Institute of British Geographers*. N. S. **14**, 259–269.

Painter, R. B., Blyth, K., Mosedale, J. C. and Kelley, M., 1974. The effect of afforestation on erosion processes and sediment yield. In *Effects of man on the interface of the hydrological cycle with the physical environment*, International Association of Hydrological Sciences Publication 113, 62–68.

Petrie, G., 1990. Photogrammetric methods of data acquisition for terrain modelling. In *Terrain Modelling in Surveying and Civil Engineering*, eds G. Petrie and T. J. M. Kennie. Thomas Telford, London, 26–48.

Phillips, J., 1995. Nonlinear dynamics and the evolution of relief. *Geomorphology*, **14**, 57–64.

Press, W. H., Flannery, B. P., Teukolsky, S. A. and Vetterling, W. V., 1986. *Numerical Recipes*. Cambridge University Press, Cambridge.

Shelford, A., Inkpen, R. J. and Payne, D., in press. *Spatial Variability of Weathering on Portland Stone Slabs*. Processes of Urban Stone Decay, Donhead Publishing, London.

Short, T., 1992. The calibration of a 35mm non-metric camera and the investigation of its potential use in photogrammetry. *Photogrammetric Record*, **14**(80), 313–322.

Stevens, D., McKay, W. M. and May, M. R., 1992. Topographic surveying: the Jubilee Line extension survey. *Photogrammetric Record*, **14**, 85–98.

Tait, D. A. (1980) Instrumentation for close range photogrammetry. In *Developments in Close Range Photogrammetry-1*, ed. K. B. Atkinson. Applied Science Publishers, London, 39–61.

Thieler, E. R. and Danforth, W. W., 1994. Historical shoreline mapping (I): Improving techniques and reducing positioning errors. *Journal of Coastal Research*, **19**(3), 549–563.

Trudgill, S. T., 1976. The marine erosion of limestone on Aldabra Atoll, Indian Ocean. *Zeitschrift für Geomorphologie Supplementband*, **26**, 164–200.

Trudgill, S. T., Viles, H. A., Inkpen, R. J. and Cooke, R. U., 1989. Remeasurement of weathering rates, St. Paul's Cathedral, London. *Earth Surface Processes and Landforms*, **14**, 175–196.

Valentin, C. and Bresson, L.-M., 1992. Morphology, genesis and classification of surface crusts in loamy and sandy soils. *Geoderma*, **55**, 225–245.

Walker, A. S., 1995. Analogue, analytical and digital photogrammetric workstations: practical investigations of performance. *Photogrammetric Record*, **15**, 17–25.

Wolf, P. R., 1983. *Elements of Photogrammetry*, 2nd edition. McGraw-Hill, Singapore.

Wrobel, B. P., 1991. The evolution of digital photogrammetry from analytical photogrammetry. *Photogrammetric Record*, **13**, 765–776.

5 Quality, Use and Visualisation in Terrain Modelling

M. J. MCCULLAGH

Department of Geography, University of Nottingham, UK

ABSTRACT

An outline is given of various approaches to terrain modelling, with explanation of the reasons for choosing different methods in different situations. Common types of error found in models are discussed, particularly in relation to inclusion of drainage network information. A test bed example of a low quality model is used to demonstrate these weaknesses. The effects of input data types, scale of model and required accuracy are considered in relation to a general survey of data sources and available modelling systems on different platforms. Terrain models require good visualisation systems if they are to be useful in many geomorphological applications. The standard facilities found in many Geographical Information System modelling packages have been rapidly overtaken in terms of flexibility and photorealism by those increasingly available in true three-dimensional modelling systems. These systems have become cheaper and ubiquitous on a wide variety of platforms, making the creation of detailed landscape views a possibility for all researchers. Not only visualisation is required: animation is also a very necessary component to show dynamic aspects of geomorphological models. Once again, the necessary systems are now present to generate animations reasonably easily, but quality real-time animation remains the province of expensive computer systems.

INTRODUCTION

Terrain modelling should not be considered a technique exclusively related to terrain. The modelling processes used to generate digital terrain or general elevation models are common to a wider user audience concerned with general interpolation methods. Similar modelling techniques can be found in geology, chemistry and even the car industry, where the "terrain" is each car body panel. The accurate modelling of geographical terrain and similar rather ill-behaved surfaces is one of the more difficult problems in surface modelling, as it is very awkward to find simple or even complex mathematical functions which will perform reliably in all situations (Watson, 1992). More recent approaches using fractal methods have different but equally severe problems.

This chapter is not concerned with a search for the holy grail of digital terrain representation, but more with a review of the adequacy of the various mixtures of

Landform Monitoring, Modelling and Analysis. Edited by S. N. Lane, K. S. Richards and J. H. Chandler.
© 1998 John Wiley & Sons Ltd.

interpolation methods which have applicability to varying terrain surface situations (Lam, 1983; Tobler, 1985; McCullagh, 1988; Weibel and Heller, 1991). For example, the methods used to develop models from digitally controlled stereoscopic air photographs are necessarily different from those used to generate terrain from field survey or even from pre-existing paper maps (Weibel and Heller, 1991). This is because: (a) the data collection systems are wildly different; (b) the data quantities available from the source document are effectively unlimited and may be resampled for the air photograph sitting on a stereo-plotter, but cannot be enhanced without further survey for paper map or field survey sources; and, most importantly, (c) the model requirements and accuracies are likely to be very different for different methods.

Once a method appropriate to the problem in hand has been determined, the platform and software must be chosen. The exact selection will depend on many factors, but there is an increasingly large range of systems available on low-end computing machines that, in terms of user friendliness if not always in terms of complete functionality, are satisfactory substitutes for much larger and more expensive systems. The cheaper systems found on PCs have an advantage as well: they usually have support provided to allow the migration of the terrain models they have generated into a plethora of excellent viewing software. This may include full photorealistic images, and even animation systems. The second part of this chapter is concerned with outlining some of the possible pathways beyond terrain modelling and examining their relative merits.

WHAT TYPE OF TERRAIN MODEL?

The type of terrain model used depends partly on the type and distribution of available input data, and partly on the fit required to the data in the final model. In the latter case, where a generalisation of the surface is required rather than a precise fit, the modelling function will tend to be based on a more or less simple mathematical function, probably a low order polynomial. The problem with such a model is that it does not fit the surface at any of the data points, but may have the virtue of representing the trend of the surface well if the data are noisy. This might be the case where there is inherent error in the data values that have been collected, not necessarily in terms of absolute accuracy but perhaps caused by surface roughness. A case in point might be a field survey of a hillside where the general shape of the hill slope is desired rather than every detail of the surface, but where the survey perforce has to measure height at any and all available points, many of which may well be located in local and irrelevant undulations.

Most model users require something more accurate in the sense that values entered as data at specific locations should be exactly reproduced in the final model, and areas between the data points should, generally, be representative of the shape of the surface. This rather begs the question of what one means by good surface shape. This question is rarely answered in precise form, and generally only receives the response that a smooth surface is required, at least where there are no cliffs, V-shaped river valleys, roads, embankments, dams or other surface impedimenta.

Table 5.1 Modelling styles in different disciplines

	Typical data			Typical surface	
	Type	Form	Example	Modelling	Form
Geology	Random	Point and line	Bore and seismic	Triangle and grid	Smooth/fault
	Regular	3D grid	3D seismic	3D grid	Planar
Engineering	Survey	Point and line	Site plan	Triangle	Planar
Photogrammetry	Regular	Point	Air photo	Grid	Resampled
Geography	All	All	Terrain	Triangle and grid	Smooth/cliff

It is interesting to note that different disciplines have different views on how terrain surfaces should look, usually based on the type of data used, its mode of collection, and some received wisdom concerning surface form. Table 5.1 outlines the difference in approach in four disciplines. Geologists commonly use either randomly located data in the form of isolated point patterns such as bore holes, or data organised in linear patterns as a result of seismic data collected along linear paths. In both cases the data are collected without direct reference to the sub-surface horizon because it is not directly observable. The preferred surface generated from such a data set is smoothly rounded, except where definite evidence of faulting exists. The advent of three-dimensional (3D) seismic survey, where a regular grid is present at the data collection stage, has redefined the modelling requirements to such an extent that what now predominates is the need for good enhancement and visualisation tools rather than modelling ones.

At the other end of the spectrum, a civil engineer (e.g. Petrie and Kennie, 1990) may well carry out a precise field survey with carefully positioned data points which are chosen in such a manner as to form an even coverage of the area and to ensure that no high or low spots have been missed. The required form of model is then usually a triangular data structure formed directly from the data points with planar facets interpolated across the triangles. This forms the most error-free model as it coincides with the approach used to obtain the data. It is knowledge of the surface itself which makes the modelling approach dissimilar to that imposed on the geologist.

The photogrammetrist requires a still different modelling approach. The terrain surface is known and sampled from air or satellite photographs, and therefore the surface is potentially exactly known, as in the case of the engineer. But the advantage the photogrammetrist has is that the image can be resampled directly from the photograph in highly variable areas such as breaks of slope (Tempfli, 1986). This is impossible for the geologist, and can only be achieved by a return to the field for the engineer. The process of model creation could therefore be as accurate as the stereo-machine used to extract the height information and of very high spatial resolution. The process works on a regular grid sampling basis, with infilling on a finer grid where local slopes warrant it.

The geographer or geomorphologist uses a great variety of sources and spatial data types. A common form has been the digitised contour, gained from manually drawn maps based on field survey (Petrie and Kennie, 1990). Contours from any source have most undesirable properties in that they (a) represent a discrete set of height values, (b) provide little surface differentiation in areas of low slope, (c) tend to generate excessive amounts of data along the path of the digitised line, but none between, which leads to very poor spatial distribution over the model area, (d) are rarely accompanied by heighted drainage lines to hold valley and ridge shape, and (e) are usually the result of manual generalisation from the original source document and hence not true representations of the terrain surface. The lines can be thinned automatically to reduce data quantity to sensible levels and give a more even spatial distribution depending on the spatial accuracy required of the final model, but it is very difficult to supplement contour data in low gradient areas other than by insertion of interpolated spot height and contour line features. Existing heighted drainage information is available in a very low proportion of available digitised contour sets, partly because it is a tedious and lengthy process to handle manually and, as will be discussed later, automation only partly solves the heighting problem.

Triangular networks are very common as underlying spatial representations of the surface generated, but often they are converted to rectangular grids for visualisation owing to imperatives in the display software used. There is a requirement that all data points be present on the final surface, that it be smoothly curved except where there is a break of slope, and that interpolation between data points should not introduce unexpected bumps and hollows. Most of the effort in terrain modelling lies in achieving these goals by careful choice of data input and diligent editing of the resulting surface. Terrain modelling is very definitely not a "load and go" process, as considerable user interaction is nearly always required to achieve a good result unless the quantities of good data are prodigious.

DATA SOURCES

The wise terrain modeller always looks round for already digitised data – or complete models – before reaching for a paper map and manual digitising table. There are a growing number of sources for digitised data, of varying scale quality and reliability, available by either mail or the Internet, and either free or costly. In all cases, the source of the data should be investigated to try to get some idea of its authenticity and expected accuracy. Some data sets have been deliberately degraded before entering the public domain, and may contain considerable induced inaccuracy.

If one wants a model of part of the USA, the data are usually available at media production cost, or even free, from such sources as the US Geological Survey node of the National Geospatial Data Clearinghouse (on the World Wide Web (WWW) at www.nsdi.usgs.gov), the Eros Data Centre (edcwww.cr.usgs.gov), and many other US government organisations. Ready-built models are available for quadrangle data at various scales, as are contour line and other data sets. A good starting point for a list of those worth searching, for all countries of the world, is held on the Geographical Information System (GIS) server at the University of Edinburgh (www.geo.ed.ac.uk).

In Europe, the situation is complicated by the fact that most governments do not place their topographic survey data in the public domain, but maintain copyright and charge "cost retrieval economic prices" for it, although their citizens have already paid for it through taxes. This means that their WWW pages tend not to hold data but order forms! Charging does have its good points. For instance, the Ordnance Survey (OS) (www.open.gov.uk/ordsurv) has good coverage in terms of both models and digitised data. Special arrangements have been made for education and researchers in terms of access to OS digital data, but so far only for seven very limited areas of the country. These can almost be guaranteed never to coincide with one's research interest. Hopefully this situation may change soon.

Data cost and availability often depend on scale. At scales greater than 1:1 000 000 digitised data become far cheaper and more easily available. The Digital Chart of the World, a vector product originally produced for the Defense Mapping Agency in the USA, is available on a suite of four CD-ROMs for about £200 from Chadwick-Healey. The set contains vector digitising of the Operational Navigation Chart (ONC) 1:1 000 000 map series. It is not very useful for terrain modelling on anything other than a regional scale and then only in areas of reasonable relief, as the contours, digitised in feet, are never closer than 250 ft and have a default interval of 1000 ft. The CDs do contain, nevertheless, a full culture and administrative data set in addition to the contours, and provide a very valuable global resource. Reading the files on the CD is not easy as the format of the data is convoluted. This has enabled resellers to offer the same set imported into other proprietary GIS and mapping systems for over £2000. Raster data are also commonly available at regional scales. Gridded height data from Spot and other newer satellites are also a good source of terrain modelling, provided that one is dealing with areas of several square kilometres rather than small site investigations, one can afford the cost, and one does not mind vertical inaccuracies in the final model of possibly 10 m or more. Positional accuracy may well be worse. At global scales there are a number of 5 min and better altitude raster data sets where altitude (and depth) models are readily available. The World Data Centre run by National Ocean and Atmospheric Administration at National Centre for Atmospheric Research in Boulder, USA (www.ngdc.noaa.gov) can provide these at little more than media cost.

MODELLING METHODS

Users of terrain models have usually preferred rectangular maps. This resulted in early modelling systems taking the line of least resistance by marrying the user desire for squareness to the computer's ability to handle matrix models, producing interpolated rectangular regular grid-based models that could be printed on the primitive line printer output devices then available. The maps became more sophisticated later and were graph plotted in the form of contour maps using pen

Table 5.2 A summary of the advantages and disadvantages of using a regular grid spatial representation compared with a modified Delaunay triangular representation of a terrain data set of N data points

	Rectangular grid	Triangular structure
Size of spatial structure	Approaching N^2 nodes will require interpolation, sometimes more	$2N$ triangles
Speed of calculation	Slow, because of interpolation	Fast, no interpolation
Reproducibility of data point value at data location	Very poor for coarse grids, high accuracy requires a cell size less than half the spacing of the closest pair of data points	Exact
Data density adjustment	None, grid usually over-dense in sparse data areas	Optimally sized triangles are adjusted to data density
Interpolation process	Many options, critical	None required
Break line representation	Quite difficult	Easy
Surface smoothing	Simple and fast fitting to grid cells, but needs careful choice	Irregular triangular shapes require careful, sophisticated and relatively slow fitting methods
Smooth line contouring	Wastes time in low data density areas	Equally slow in all areas
Ease of graphic display	Simple, but large data volumes	More complex structure, but scan conversion fast with small data volumes

plotters. The basic interpolated form was still the grid (Shepard, 1969). When graphics started to appear on terminal screens, the easy correspondence between an interpolated height grid and the screen raster was maintained. The problems inherent in an interpolated grid are shown in Table 5.2.

By the 1980s it became clear that a better form of internal representation of the model was a Delaunay triangulation, which kept all the original data as vertices of the network and generated an optimally equilateral triangular set. The later modification of this Delaunay triangulation, to maintain break of slope lines and other important features of the landscape, enabled the triangular representation in theory to be the closest approximation to the reality of the terrain, under the assumption that the data points themselves were correctly chosen (McCullagh, 1988). At this stage the model created would be very similar to that often required for a civil engineering site investigation. A further step was taken when it became clear that it was necessary to fit some form of curved surface patch to the triangles within the surface representation in order to model terrain accurately when the available data points were not well distributed spatially. The field survey of the civil

Figure 5.1 A simplistic fractal terrain model generated by using the 2D "plasma" fractal in the freeware program Fractint. The landscapes look real until studied closely, when it becomes clear that the drainage pattern is undeveloped and lacks continuity

engineer does not need this curvature as the survey locations are carefully chosen to allow a planar triangular form. The general terrain modeller does require the ability to curve the triangular facets of a model, as data sources such as contours are very poorly distributed in a spatial context. The methods of interpolation that can be used, within either a regular rectangular or an irregular triangular spatial representation, to create grid nodes or to calculate smooth surface approximations are well established (Tobler, 1979, 1985).

More recently, as visualisation processes have become more sophisticated and the demand for more photo realistic views has increased, other methods of creating terrain have become important, particularly the use of fractal theory (Mandelbrot, 1982; Crilly *et al.*, 1993) both to evaluate and characterise terrain (Goodchild and Mark, 1987; Carver, 1994; Polidori, 1995), and to generate convincing views of mountain scenery with software packages such as VistaPro. Terrain has been either entirely fractal generated as a completely fictional landscape, or based on a terrain model modified using fractal techniques to provide a convincing surface roughness. The problem with most fully fractal landscapes is that they do not themselves obey some of the inherent requirements of landscape, such as requiring all drainage to flow downhill or generating connected river valleys. An example of a very simple fractal landscape generated by Fractint (freeware from www.micros.hensa.ac.uk), using its two-dimensional (2D) plasma function is given in Figure 5.1, where a supposed lake area is surrounded by mountains. The drainage problems are very apparent when the mountains are closely investigated. None of the valleys are genuinely interconnected or exhibit a truly fluvial drainage pattern. Fractal modified terrain models overcome most of these disadvantages but do not create a more exact model. The roughness introduced into the landscape may be statistically correct and visually effective but will not be locationally accurate.

MODELLING ACCURACY PROBLEMS

All modelling systems have problems in certain areas in terms of accuracy of representation of the surface. This section illustrates four different problems common to most systems using manual point and line digitised data that result from different data inadequacies. The final part of this section is a discussion as to whether meaningful statistics can be generated to represent the level of error present in a model. The most common problems leading to modelling error are usually caused by: data quantity and distribution in low gradient areas, break lines and surface bounce, incorrect height labels, and the river heighting problem. Figure 5.2a shows a 700 point sparse contour and spot height set, including a few break lines inserted for cliffs and river locations taken from a 50-year old imperial OS map. The average data density is about one data point per 150 m², over an area 4.5 km by 3.5 km. A triangulation was used to represent the spatial structure of the area; smooth patches were fitted to the triangles, which were then mapped onto a regular rectangular grid with a cell size of 25 m by 25 m. The height model generated from this data set can be seen in Figure 5.2b, where dark values indicate low ground, and light values high ground. Many other methods could have been used to create a similar terrain model. It would still be open to investigation to check whether it suffered from any of the following problems.

Data Quantity and Distribution

A comparison of the modern OS 1:50 000 sheet shown in Figure 5.2c and then draped over the area in Figure 5.2d indicates some of the modelling inadequacies. In general, the contours shown in the view should look horizontal if an accurate model has been created. In this view many of them appear to be partly running up and down the hillside, indicating error. Only the major contours were digitised at an interval of 250 ft. This was adequate within limits in the steep areas, but can be seen to be completely insufficient on the valley bottom. It is not necessary when selecting digitised data to maintain a constant contour interval; indeed it is usually foolish to do so. An insufficient number of spot heights were digitised which accounts for the lack of accuracy of the peaks of the hills, where no contour data were present to hold the shape of the hill. The large interval between sampled contours accounts for much of the visible non-horizontality of the displayed contours.

Figure 5.3a is a magnified view of the floor of the valley entrance, and illustrates the difficulty of dealing with areas of low slope where there is insufficient local control in the model. There are three main problems: lack of a properly digitised river to hold the altitude and position of the valley floor, lack of spot heights to indicate local highs and lows, and some doubt about the registration of the river on the OS map itself. The river was represented by very few points in the digitisation in

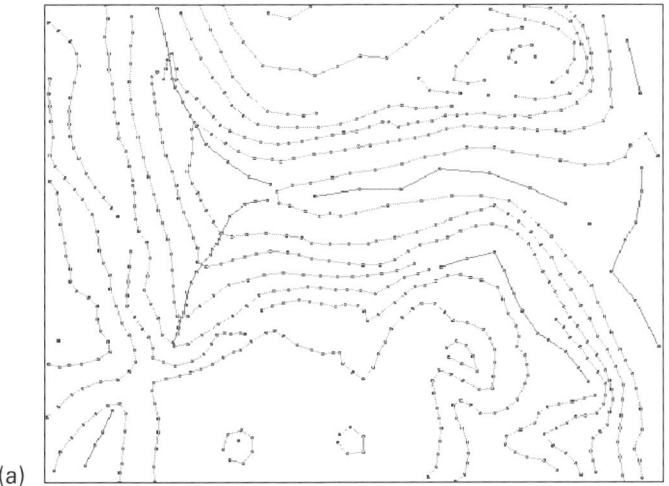

(a)

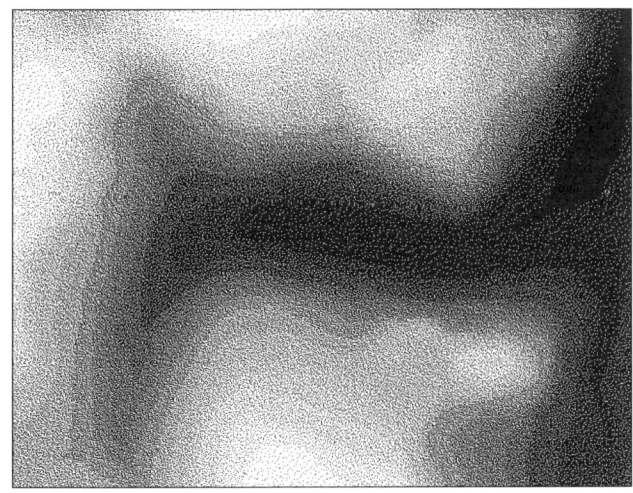

(b)

Figure 5.2   The sequence of four diagrams shows the development of a poor terrain model for a small part of the Cairngorms in Scotland, from digitised contour data to visualisation. (a) Problems of data sampling and density. Digitised points are shown as dots, and logical lines as collections of dots. Dark lines indicate locations where the model ought to include breaks of slope, for instance along rivers and cliff edges. (b) The model generated, with low areas dark, and high altitudes light-coloured. (c) The OS 1:50 000 sheet covering the area, and includes far more altitude detail, some of which could and should have been used to enhance (a). (d) A photorealistic view of the valley with the OS map draped over the landscape to indicate areas of successful and unsuccessful modelling. OS maps reproduced with permission. © Crown Copyright. C4/88-14

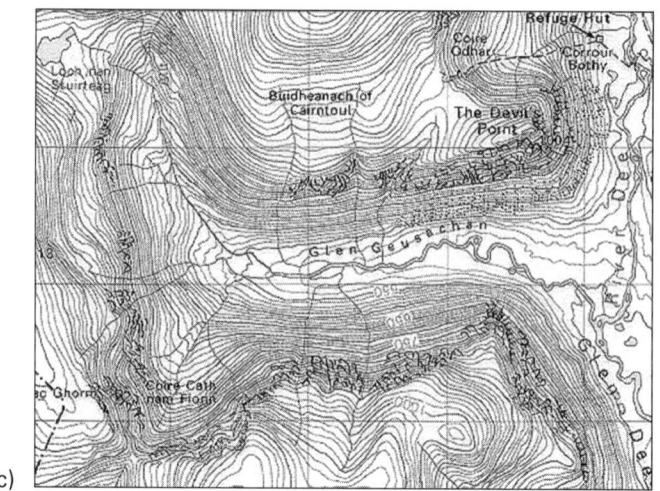

(c)

(d)

Figure 5.2 (continued)

Figure 5.2a, and certainly does not show the meanders seen in Figure 5.2c. This results in the river appearing to "see-saw" down the valley rather than maintaining a low point position. The only solution is a higher digitising resolution.

Spot heights have not been used to hold the shape of the floor of the valley, which is why the contours in the foreground wander aimlessly across the terrain with little accuracy. This is a classic problem in modelling as the insertion of sufficient extra data in the form of spot heights to keep the form of the valley in this

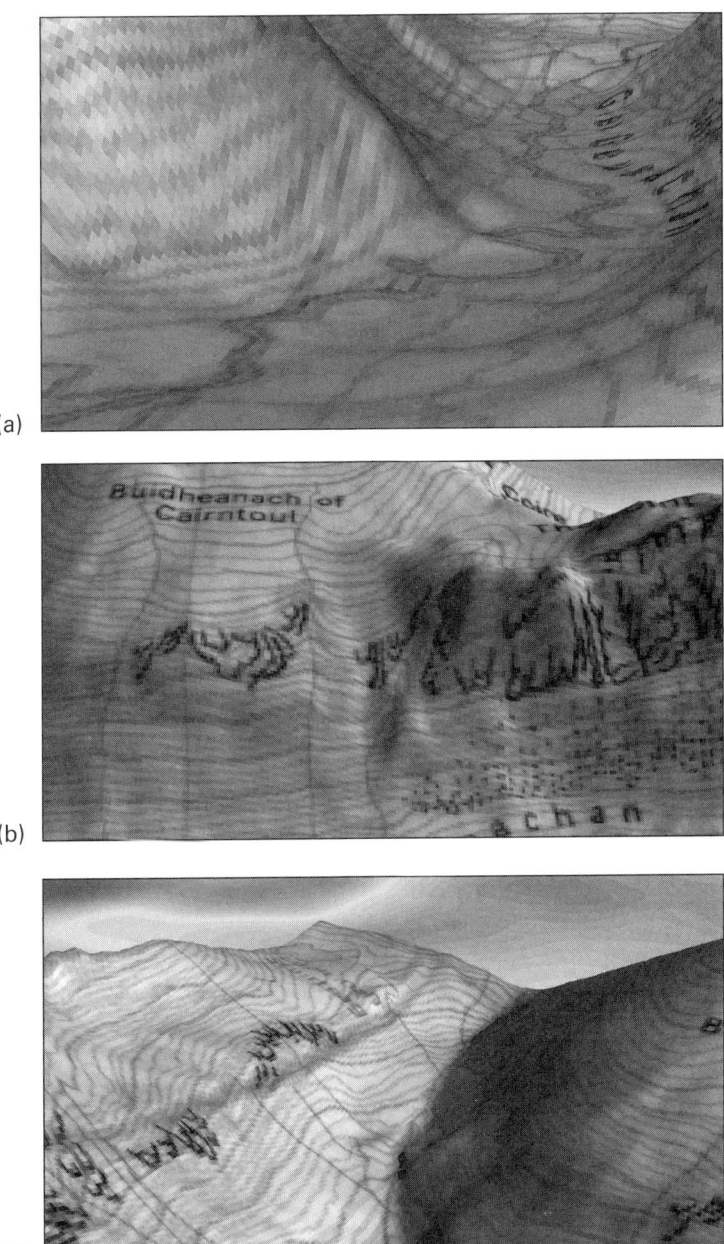

(a)

(b)

(c)

Figure 5.3 Close-up views illustrate problem areas in the DTM generated from Figure 5.2a. (a) Incorrect river placement and low gradient modelling problems; (b) the difficulties encountered with overshoot in the surface fitting function; (c) the "Roman fortification" syndrome resulting from mislabelled heighting of digitised contour lines. All three views demonstrate the need to set a digitising resolution that will suit the desired accuracy of the final product. In this case, although the valley is in the correct place, the detail is rather poor

area would be prohibitive unless digitised stereo air photograph cover were available. The spatial resolution of the model must reflect the purpose to which the model will be put. In this case, a general view of the entire valley would not be expected to include the very high level of detail required to hold the surface in these very flat areas. If the whole area were rather flat extra data would be essential. The contour set and spot heights could not be sufficient even if amplified by inserting all the stream and ridge lines.

The final problem of apparent registration failure is related to the printed map and the correspondence of the river and contours. Close examination of the OS map plan view in Figure 5.2c, and the close-up in Figure 5.3a shows that the river actually touches the contour in many places, while being in the centre of the valley floor further up-river. This is probably due to cartographic knowledge not evident from the map. Air photographs of the area may well show a local small river cliff. Unfortunately the modeller cannot easily do anything about this without further information.

Break Lines and Surface Bounce

The digitised data shown in Figure 5.2a contain very little in the way of break line information. There are a few lines, rather poorly digitised, indicating the river course, and also a ridge line on the south side of the valley inserted to control overall shape at the top of the hill. Figure 5.3b shows the situation along the north wall of the valley where no break line is present. The left (west) side of the north wall has reasonable form, given the limited number of points, but the right (east) side shows the effect of not including a break line where there is a very marked change of slope at the top of the wall. The long "bouncing" parapet is clearly shown by the vertical shifting of the map contours and the very rounded nature of the local "peaks". This is a completely fictitious feature resulting from the numerical function used to fit a curved surface patch through triangles at the top of the mountain (McCullagh, 1981). These triangles have extremely steep local slope estimates resulting in extreme bounce during height interpolation in the central area of the triangle, away from the data at the vertices. The OS map shows cliffs, and the insertion of a break line along the cliff edge would ensure a proper match to the real surface.

Incorrect Heighting

A particular problem often associated with digitised contours used for model building is that of incorrect heighting. This can be difficult to spot when model building unless the system used requires a complete closed set of contours as input. This is often simply not achievable from many data sources, and so incorrect heighting cannot be checked in advance of model building, but only during later visualisation of some sort. The effect is very characteristic and is shown in Figure 5.3c. This view is of the northwest end of the valley and, apart from showing the need for more data, shows a possible "Roman ramp and ditch fortification" structure along the west side. This blemish is highlighted particularly well when the sun is shining from the right direction and casting shadows across the ditch. The cause rests with at least one misheighted line lying between two correct ones.

River Heighting

In the small example being considered in Figures 5.2 and 5.3 there is little difficulty in providing heighted river information. When using digitised data supplied by others it is very unusual to be able to obtain any heighted river, ridge or break line data. Yet this is vital to hold the shape of the bottom of valleys and the top of ridges. Break line data of any sort usually consist only of x,y location and any junction points, because of the difficulty of ascribing a height to every single digitised point during the digitising operation, be it manual line following or automatic scanning. It is possible, however, to create a heighted data set from these lines as long as it is possible to intersect the break line set with the contour set in which it lies. There are a number of ways in which this can be done, but all require the calculation of all possible intersections between contours and break lines – a very lengthy process for a large data set of perhaps many hundred thousand data points. One solution is shown in Figure 5.4, a fragment of a Canadian 1:250 000 topographic map, where the break lines (rivers in this case) have been converted into a quad-tree (Samet, 1990), and then intersected with a quad-tree representing the contours. By using the tree structure as the intersection operator it is possible to increase the speed of the process by about two orders of magnitude. The tree does not create the final intersection location as it has a definite spatial limit, but it does indicate which vectors of the river tree and contour tree need to be tested for intersection.

Once the intersection heights have been determined, interpolation can be used to calculate heights for the string lying between intersections. Problems still abound. For instance, the section of river line "dangling" above or below the first or last intersection must be heighted. In addition, inspection of Figure 5.5 reveals that other data are necessary in order to form a reasonably good interpolation of height along the river line. Junctions rarely lie on a contour intersection but must result in the same height being assigned from any branch of the river system. Similarly, lake heights, often not specified in digitised line data files, must be estimated even where there are numerous streams flowing into the lake, and possibly more than one flowing out, all with different possible estimates of lake height. Once the rivers and break lines have been heighted using this type of approach, they can then be added to the other digitised data to provide full control in the final model. The resulting model will then exhibit at least a consistent drainage pattern and good ridge control. Alternative or even additional approaches are concerned with ensuring drainage continuity in the final gridded model (Hutchinson, 1989).

Error Measurement

It is very difficult to produce good error measures for the accuracy of terrain models of whatever form. Fortunately, few measurements are now made based on comparing the accuracy of reproducibility of the original digitised line work. This used to be a popular method until it was realised that the modelling systems achieved

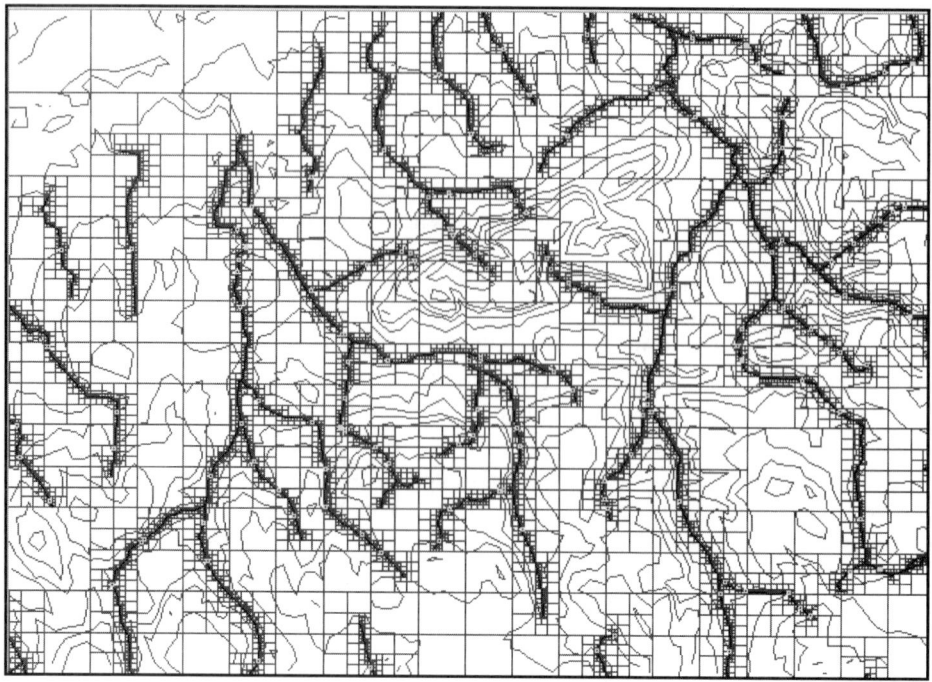

Figure 5.4 Heighting of drainage networks is a difficult problem because of the need to handle large data quantities. This example is taken from an area in Nova Scotia where there were upwards of 100 000 data points in both the river and contour sets. One solution is, as here, to intersect the river (shown in black) as represented by its quad-tree (square grey quads) with the contours set (grey) to give heights at contour crossings (filled white dots at intersections). This solves part of the problem, enabling heights to be interpolated between some of the contour crossing points.

near-perfection along the input lines or where other input data were located, but were often totally unreliable between the original lines. Using a mixture of measures seems more appropriate than simply boiling the accuracy of an entire map down to a single error statistic (Wood and Fisher, 1993). Some measures use the root mean square error of a set of test points chosen randomly over the map, usually fewer than 100 in number. This technique has the usual problems of attempting to use non-spatial statistics to measure a spatial phenomenon (see Wise, Chapter 7). On the other hand, it is extremely simple and rapid to calculate. Another approach is to look at the histogram of altitude and to check it for periodic cusping. This is usually a sign of interpolation failure caused by areas being interpolated flat when they should exhibit more vertical relief, often because of inaccurate estimation of local gradients at data points. Shearer (1990) provides an assessment of some basic methods applied to models generated by a number of different systems. When a regular grid is generated as the final model, the size of grid cell has an important effect on accuracy (Bolstad and Stowe, 1994; Hodgson, 1995).

There may be a problem in determining error in a model owing to a lack of something "correct" for comparison. If the model is being generated from digitised

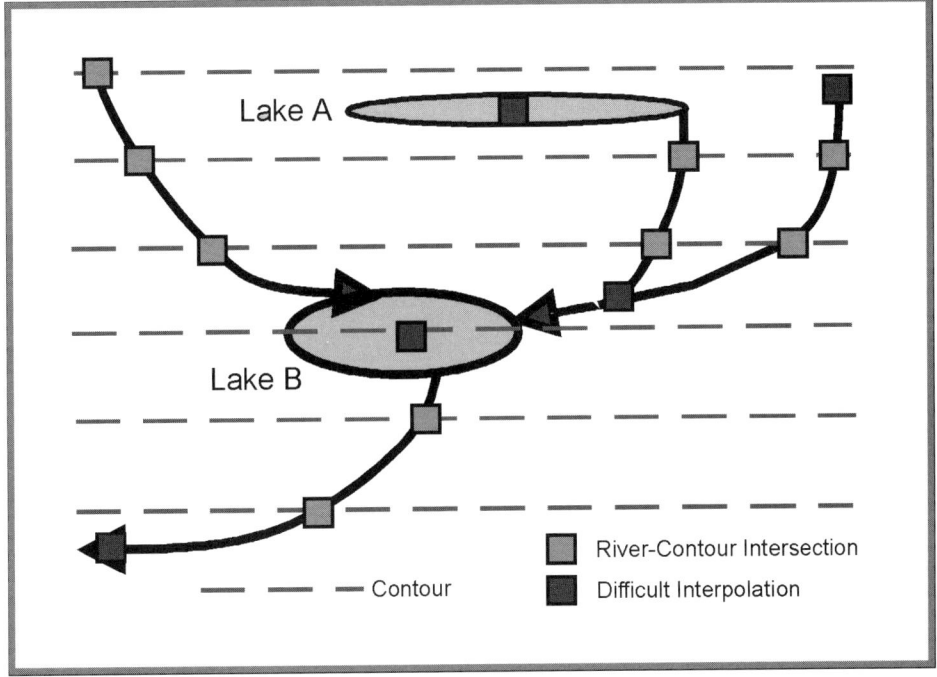

Figure 5.5 Problems abound in heighting river digitising. Simple interpolation between contour crossings will work fairly well, but this diagram illustrates the need for the use of a far more complex interpolation system, in association with knowledge about the drainage network, to achieve successful heighting

line and point data from a manual map, it may be possible to check it partially by taking heights from stereo air photographs, if available. If the model is based on stereo air photographs or a satellite image, the only recourse is to do a field survey. Occasionally it is possible to check the spectral or fractal content of a model (Goodchild, 1980; Culling and Datko, 1987) to see if the right kind of model has been generated. These functions measure the dimensionality of the model rather than the locational accuracy. They are roughly equivalent to assessing the type of landscape generated rather than the detail of particular elevation information.

MODELLING RESOURCES

There are a wide range of modelling systems available on a diverse selection of computer platforms. This discussion merely points out a few of these as examples of their types. Table 5.3 gives an approximate idea of the source, market and platform of typical modelling packages. In some cases they are stand-alone systems, and in others they are part of a much larger and more comprehensive package of GIS or computer-aided design (CAD) facilities.

Table 5.3 Sample modelling systems on a range of platforms (after McCullagh, 1996)

	Model name	Authorship	Address and availability
PC Systems	Panacea	Siren Systems	michael.mccullagh @nottingham.ac.uk
	VistaPro	Virtual Reality Labs Inc.	USA, tel: 001-800-829-8754
	Landscape Explorer	WoolleySoft	100332.2104@compuserve.com, or micros.hensa.ac.uk
PC Systems and work stations	DTM/W/G	Intergraph	www.intergraph.com
	Microstation (InSites/Site-Works)	Bentley	www.bentley.com
	Arc/Info	ESRI	www.esri.com
	Moss	Moss Systems Ltd	Barclays House, 51 Bishopric, Horsham, RH12 1QJ, UK
	DtmCreate	LaserScan	www.lsl.com

PC software is available either for normal purchase, or as shareware. Panacea is an example of the former but is only in the DOS world at the moment. The archive at www.micros.hensa.ac.uk contains several shareware packages of which Landscape Explorer, a Windows-based product, is a good example. Its capabilities are limited in terms of the size of data sets that it can hold, but its facilities for import, export, contour tracing from bit images, and interpolation are really quite sophisticated. As with many PC packages, there is the common (and correct) assumption that different software wheels can be used interchangeably on the modelling wagon via the usual interface formats. Visualisation may be the strong point of one package (such as VistaPro), and modelling of another. Software for the PC is now sufficiently cheap that many systems can be used and put to work on those aspects in which they excel.

The other attractive feature about PC modelling systems is that they tend to have genuinely friendly user interfaces. This does not yet apply fully to systems that are to be found on both PCs and work stations, or on work stations alone. They all tend to be stand-alone packages which are often quite difficult to interface to each other, and cost may also be high. Often the systems can deal with larger data sets, and sometimes with a greater sophistication. Moss is an example of a civil engineering CAD system that grew, and now has a good grip on the highway engineering market.

A number of vendors have approached terrain modelling along the route of GIS and remote sensing. The TIN and GRID modules in Arc/Info are probably the present GIS market leaders, partly because ESRI has such a firm grip on the market through a huge installed user base, and partly because it has quite good modelling facilities. Intergraph and Bentley, which started in CAD and then moved into GIS and remote sensing, have parted company but offer very similar systems, sometimes on rather specialised hardware, but also on PCs. The success of Bentley's port to the Windows NT operating system and Windows 95 will be interesting to watch.

The advantages of the work station over the PC have traditionally been speed and the ability to handle "real" problems. This is still so, but only to a limited extent. Most modellers operating in the Windows environment can handle large data sets of more than a few hundred thousand points with reasonable ease, merely by adding memory. Speeds have risen along with increasing algorithm and graphic interface sophistication.

THE USE OF MODELS

Software designers often fail to see beyond the creation of a model to where it will be used by others in many different areas. The range of applications is vast (see Moore *et al.* (1991) for a review), from ridiculous but very effective backdrops for animated fantasy through to sublime research measurement on damp hill slopes. Scale, source of data, method of construction and need all vary, but have one major requirement in common: they all require some form of visualisation and possibly animation to allow both interpretation and understanding. It has been long-established practice to represent models in the form of contour maps, probably as a matter of reassurance to the user rather than as a useful device. Similarly, isometric and wire frame diagrams were very popular displays, sometimes overlaid with manually added information, and occasionally in stereo (McCullagh and Sampson, 1972).

Fortunately, matters have progressed (McLaren and Kennie, 1989), and computer graphics (Earnshaw and Wiseman, 1992), particularly the photorealism school of graphics (Watt and Watt, 1992), now provide the tools to appreciate modelled terrain as a landscape draped with overlying imagery and possibly with artificial constructs representing the user's research findings or other application. The impact of viewing landscape, correctly proportioned and complete with any vegetative and human culture, and analytical results allows a researcher not only to present findings in papers and talks in an easily assimilable fashion (Hearnshaw and Unwin, 1994), but also provides the key to using the analytical power of the human brain. If the scene is animated providing movement of the viewpoint and also allowing painted constructs positioned on the surface, such as the extent of a flood, to be changed dynamically, much more may be gained in visual and analytic terms. For example, the sun could be moved realistically and insolation characteristics calculated exactly, or the results of fluvial, hydrological, oceanographic or erosional simulations can be displayed in "real" lapse-time. Use can be made of the growing variety of modelling tools. This includes not just the well established constructive solid geometry CAD tools that generate such robotic views of the world, but the newer graphics constructs such as fractals for surface roughness, and procedural animations such as particle systems for droplet flows and analytical systems for wave and other motions.

A reasonable question is "How much does all this cost?" and, for the more mature and wise researcher, a rather more cogent one is "How long does it take to

Table 5.4 Sample visualisation and animation software (after McCullagh, 1996)

Class	Name	Functionality	Cost (£)	Availability
PC: DOS	Idrisi	Drape overlays	100	Clarke University
PC: Windows	VistaPro	Render and animation	85	USA, tel: 001-800-829-8754
PC: Windows	ENVI	Render and fly-through	2000+	support@floating.demon.co.uk
PC Shareware: DOS	PolyRay	Ray trace and animation	25	xander@mitre.org and micros.hensa.ac.uk
PC Shareware: Windows	PoVCAD	CSG modeller	25	72114.2060@compuserve.com
PC: Windows	Animator Studio	CSG modeller and animator	400	AutoDesk Ltd, tel: 01483-303322
PC: Windows	3D-Studio	CSG, ray trace and animation	2000	Autodesk Ltd, tel: 01483-303322
PC: Windows and Windows 95	Dream 3D	Ray trace and animation	300	Corel 6
PC: Windows	Caligari TrueSpace	Ray trace and animation	400	Roderick Manhattan Group
PC Shareware: Windows	GoldWave	Sound capture, player and editor	15	chris3@garfield.cs.mun.ca and micros.hensa.ac.uk
PC Shareware: DOS	DTA	Dave's *.TIGA Animator . . .	25	76546.1321@compuserve and micros.hensa.ac.uk
PC Freeware: Windows	AAPlay	plays *.fli and *.flc animations	0	Autodesk Ltd, tel: 01483-303322
PC Shareware: DOS	CMPEG	Create MPEG movies without sound	0	stefan@lis.e-technik.tu-muenchen.de and micros.hensa.ac.uk
PC Shareware: Windows	VMPEG (lite version)	Plays MPEG movies	0	stefan@lis.e-technik.tu-muenchen.de and micros.hensa.ac.uk
Work Station	Arc/Info	Drape overlays	500+	www.esri.com
Work Station	Performer	Real-time graphies	1000	But you need an SGI machine preferably with a top range Onyx!

create the views and animations?" The answers, for generating basic terrain models, vary from nothing to everything. Table 5.4 may help to put the problem of costs and performance into perspective. Costs can be very low, or even non-existent using freeware or shareware. The products in this category are very good modellers indeed, but have a fairly hostile general user interface in the sense that the control of any scene has frequently to be written in a "C"-like format. This is actually quite straightforward once one gets used to it. Figure 5.2d is a case in point. Twenty lines of PolyRay script generated the basic valley view. PolyRay is only one example of a large number of ray-tracers and animator programs available from the Internet. The www.micros.hensa.ac.uk address has more, and many others are referred to on the various graphics bulletin boards. More sophisticated Windows user interfaces and possibly faster renderers tend to cost money. The 3D-Studio software is very good, but is now threatened by a number of commercial rivals such as Caligari and, most recently, Corel's Dream 3D, newly introduced as part of the Corel 6 package.

ANIMATION

The time required to create single frame, reasonably high resolution photorealistic views on any platform tends to be measured in minutes at least. The time needed to create a fully ray-traced animation sequence of perhaps 30 seconds varies from many days on your desktop PC, to only a few days on a work station, to almost real time on a powerful and purpose-built Silicon Graphics machine. Even on the latter using specialist Performer software, the picture should not be too complicated or the frame rate drops below an acceptable minimum of 15 frames per second (fps). Even the 15 fps limit in real-time animation for a state-of-the-art machine does not include full photorealistic ray-traced standards, but is good enough to fool the eye most of the time on simple scenes. There is a lot of scope for both speeding up the process and cost reduction.

The plethora of digital video formats and sound tracks for recording digital animations is still a major problem, and will continue to be so for some time until standards finally settle. The difficulty is that 30 seconds of animation at 25 fps using low resolution images (320 by 200), which are considered just good enough to show on NTSC (USA standard) television, occupies 144 MB of disk space in uncompressed form. Two problems result: that of transferring the video data sufficiently fast from disk to screen so no Chaplinesque results occur, and storing a sensible quantity of video data (say an hour) in a sensible amount of disk space. One hour might require 20 GB of disk space.

The solution is compression of both sound and vision using either software or hardware approaches. Individual images can be compressed either without loss using GIF, TIF, BMP and other formats, or with an "acceptable" loss using the more complex encoding provided by JPEG (Joint Picture Expert Group) and similar formats. This typically leads to compression factors of 75% or more for full colour images. Video compression can go one stage further by compressing not just within each frame, but by looking at the differences between successive frames, and only encoding areas that change. This can give compression of well over 95% for many video sequences. A number of systems exist, but they tend to be dominated now by the older AutoDesk Animator formats (FLI and FLC), and the newer MPG (Motion Picture Expert Group) format that handles both sound and vision and achieves compression levels about three times higher than FLI/FLC. The DTA software listed in Table 5.4 is very effective at providing all forms of Animator output with excellent colour. The CMPEG program is rather less effective in terms of colour, but generates reasonable MPG sequences. The penalty that must be paid for "small" file sizes, measured in some hundreds of megabytes per hour of animation, is not some "acceptable" loss in picture quality, but also a substantial (though measured only in seconds per frame) software encoding process. The decoding can be performed by software (e.g. VMPEG in Table 5.4), or by hardware boards which are fast, reasonably cheap and widely advertised in the PC magazines. Software decoding on Pentium 100 PCs can reach over 25 fps for "standard" 320 by 200 images. Hardware solutions can not only perform consistently in real time, but often come with encoding chips as well as decoding facilities and hence can be used for MPG movie generation as well as playback.

Animation provides the dynamic medium essential to comprehending complex spatial interaction models. It is difficult to provide examples in this paper, but Figure 5.6 shows one frame from a 30 second sequence developed to show a physical process animation. The terrain model used was the same as in the previous illustrations and the physical process being modelled (without extreme realism) was a hail storm. Hail, falling from the cloud at the top of the picture, changes to water droplets on contact with the terrain surface. The hail falls in a semi-turbulent fashion and lands randomly on the landscape. The droplets obey the laws of gravity before and after their conversion to water droplets. These then roll over the landscape, following the natural drainage channels in the model at a speed controlled by gravity and the local terrain. Ponding occurs frequently in low gradient areas, and waterfalls cascade off the corrie side. This is not a serious simulation of the overland flow of water, but is intended to show how animation can be used to demonstrate the products of simulation in a wide variety of environments. The full movie sequence can be viewed in McCullagh (1996).

The animation and simulation of the physical system use a form of procedural modelling using particles to make its point. During the animation process, particles can be created that effectively have a life of their own. They can be made to obey physical laws of gravity, to interact with each other and with the "scenery". In this case, the particles considered are water droplets, but could for instance have been cars on a motorway. Similar animations could be made of flooding and tidal processes, wave action and erosion. Particle systems and the development of analytical modelling of materials in computer graphics have made possible complex modelling of movement, not just of individual objects such as water droplets but also of connected particles such as skin, cloth or water surfaces that respond to some form of wave-like motion.

CONCLUSION

Many methods of creating models are now available, many suited for a particular problem rather than being a panacea for all types of surface generation. Digital models are now available for most parts of the world, though at varying levels of accuracy, scale and cost. Terrain models are now generated and used routinely in areas varying from hydrology to weather forecasting. Even so, there still seems to

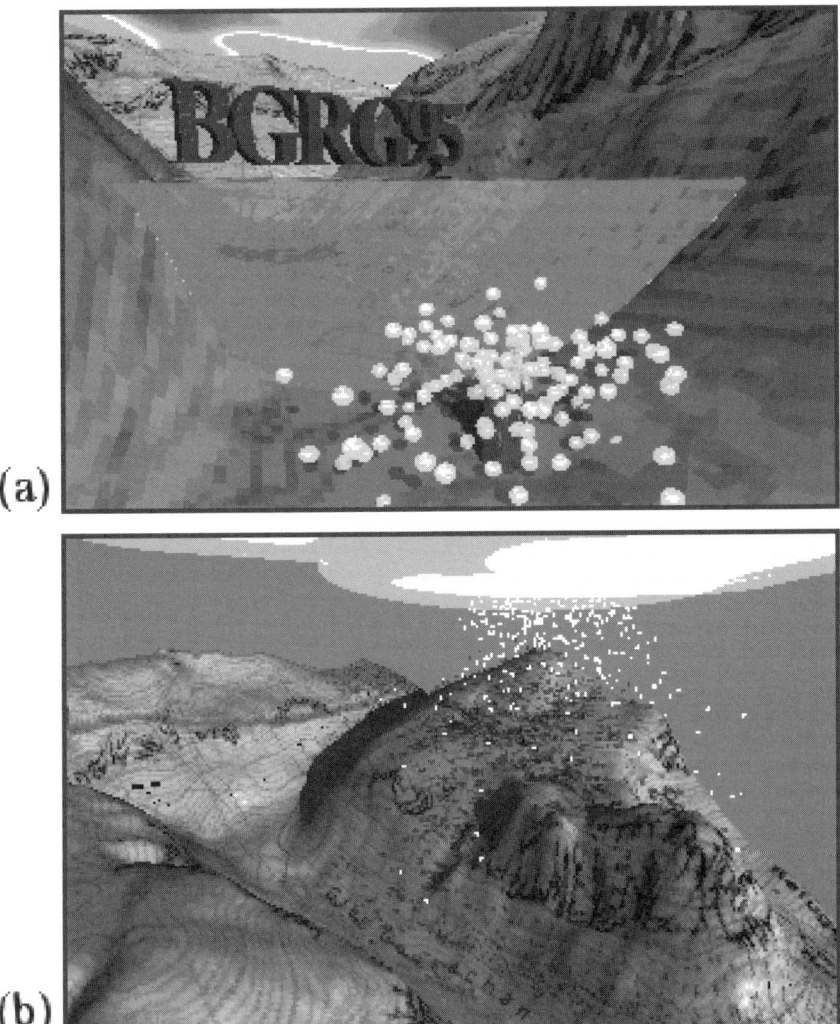

(a)

(b)

Figure 5.6 Visualisation can include animation to show dynamic relationships between variables and landscape. Figure 5.6a shows a dam leak simulation with the BGRG95 logs. Figure 5.6b was a preliminary attempt to demonstrate the use of particle animation by showing the movement of particles (snowdrops) on a hillside in terms of overland flow, in an attempt to demonstrate the way models of landscape systems can be made to come alive and better inform the viewer as to the functioning of the model

be a need for detailed models of specific research sites, and for good specialist models at a regional or world scale. Satellites, digital air photographs and the Global Positioning System are rapidly filling some of the requirements, but owing to cost there is still a high demand for "home-built" models customised for particular purposes. Surprisingly perhaps, the paper map is still one of the best and most easily obtainable data set sources from which to generate a model. An available but coarse 50 m or 25 m spacing grid model will not be very useful for many small valley studies. Unfortunately such models are frequently used, more because they are there than because they are suited to the purpose.

The next step is realistic visualisation of the uses to which the models are put: stages in physical processes, comparison with simulations of those processes, and animation of the visualisations to demonstrate the dynamic nature of many systems. Visualisation is still a challenge, although many of the tools are now in place, and fully (and freely in some cases) available to those who want to use them. Animation is still in its infancy, as evidenced by the rapidly changing computer video and movie standards being proposed. The range of tools available via the Internet and for purchase on a wide range of platforms is increasing daily. The time has come to use them to enable researchers to view the dynamic nature of their modelling systems and thus develop a better understanding of the processes involved.

REFERENCES

Bolstad, B. V., and Stowe, T., 1994. An evaluation of DEM accuracy, elevation, slope and aspect. *Photogrammetric Engineering and Remote Sensing*, **60**, 1327–1332.

Carver, S. J., 1994. Vector to raster conversion error and feature complexity: an empirical study using simulated data. *International Journal of Geographical Information Systems*, **8**, 261–270.

Crilly, A. J., Earnshaw, R. A. and Jones, H. (eds), 1993. *Applications of Fractals and Chaos.* Springer-Verlag: Berlin.

Culling, W. H. and Datko, M., 1987. The fractal geometry of soil covered landscape. *Earth Surface Processes and Landforms*, **12**, 369–385.

Earnshaw, R. A. and Wiseman, N., 1992. *An Introductory Guide to Scientific Visualisation.* Springer-Verlag, 156pp.

Goodchild, M. F., 1980. Fractals and the accuracy of geographical measures. *Mathematical Geology*, **12**, 85–98.

Goodchild, M. F. and Mark, D. M., 1987. The fractal nature of geographic phenomena. *Annals of the Association of American Geographers*, **77**, 265–278.

Hearnshaw, H. M. and Unwin, D. J. (eds), 1994. *Visualisation in Geographical Information Systems.* Wiley, Chichester, 243pp.

Hodgson, M. E., 1995. What cell size does the computed slope/aspect angle represent? *Photogrammetric Engineering and Remote Sensing*, **61**(5), 513–517.

Hutchinson, M. F., 1989. A new procedure for gridding elevation and stream line data with automatic removal of spurious pits. *Journal of Hydrology*, **106**, 211–232.

Lam, N. S., 1983. Spatial interpolation methods: a review. *American Cartographer*, **10**, 129–149.

McCullagh, M. J., 1981. Creation of smooth contours over irregularly distributed data using local surface patches. *Geographical Analysis*, **13**, 51–63.

McCullagh, M. J., 1988. Terrain and surface modelling systems: theory and practice. *Photogrammetric Record*, **12**, 747–779.

McCullagh, M. J., 1996. *Visualisation and Use of Terrain Models in Physical Systems Modelling*, Third International Conference/Workshop on Integrated Geographic Information Systems and Environmental Modelling, Jan 1996, Santa Fe. US National Center for Geographical Information and Analysis (at UCSB), CD-ROM publication, 10pp and video.

McCullagh, M. J. and Sampson, R. J., 1972. User desires and graphic capability in the academic environment. *Cartographic Journal*, **9**, 109–122.

McLaren, R. A. and Kennie, T. J. M., 1989. Visualisation of digital terrain models: techniques and applications. In *Three Dimensional Applications in Geographic Information Systems*, ed. J. Raper, Taylor and Francis, 79–98.

Mandelbrot, B., 1982. *The Fractal Geometry of Nature*, Freeman, Cooper, San Francisco, 668pp.

Moore, I. D., Grayson, R. B. and Ladson, A. R., 1991. Digital terrain modelling: a review of hydrological, geomorphological, and biological applications. In *Terrain Analysis and Distributed Modelling in Hydrology*, eds K. J. Beven and I. D. Moore, Wiley, 7–34.

Petrie, G. and Kennie, T. J. M. (eds), 1990. *Terrain Modelling in Surveying and Civil Engineering*. Whittles Publishing, Latheronwheel, Caithness, 351pp.

Polidori, L., 1995. Fractal-based eevaluation of relief mapping techniques. In *Fractals in Geosciences and Remote Sensing*, eds G. Wilkinson, I. Kanellopoulos and J. Mégier. Joint Research Centre, Report EUR 16092 EN.277-297.

Samet, H., 1990. *The Design and Analysis of Spatial Data Structures*. Addison-Wesley, Reading, Mass, 493pp.

Shearer, J. W., 1990. The accuracy of digital terrain models. In *Terrain Modelling in Surveying and Civil Engineering*, eds G. Petrie and T. J. M. Kennie, Whittles Publishing, 315–336.

Shepard, D., 1969. *A Two-Dimensional Interpolation Function for Computer Mapping of Irregularly Spaced Data*. Harvard Papers in Theoretical Geography 15, Harvard University, 20pp.

Tempfli, H., 1986. *Composite/Progressive Sampling, A Program Package for Computer Supported Collection of DTM Data*. ACSM-ASP Convention, Washington DC, 9pp.

Tobler, W. R., 1979. Lattice tuning. *Geographical Analysis*, **11**, 36–44.

Tobler, W. R., 1985. Smooth multidimensional interpolation. *Geographical Analysis*, **17**, 251–257.

Watson, D. F., 1992. *Contouring: A Guide to the Analysis and Display of Spatial Data*. Pergamon, 321pp.

Watt, A. and Watt, M., 1992. *Advanced Animation and Rendering Techniques: Theory and Practice*. Addison-Wesley, New York, 455pp.

Wcibel, R. and Heller, M., 1991. Digital terrain modelling. In *Geographical Information Systems, Vol 1: Principles*, eds D. J. Maguire, M. F. Goodchild and D. Rhind, Longman, New York, 269–297.

Wood, J. D. and Fisher, P. F., 1993. Assessing interpolation accuracy in elevation models. *IEEE Computer Graphics and Applications*, **13**, 48–56.

6 What Do Terrain Statistics Really Mean?

IAN S. EVANS

Earth Surface Systems Research Group, Department of Geography, University of Durham, UK

ABSTRACT

Digital terrain/elevation models are now widely (and in some countries, freely) available, and it is easy to calculate surface slope and curvature and to interrelate their distributions, even for large data sets. But statistics for a single area, basin or region are difficult to interpret on their own; context for such interpretation is offered here. Drawing on a range of examples, some generalisations about statistics of slope and curvature distributions are offered. The range of results expected with 100 m, 50 m and 30 m grid meshes is identified, and correlations between statistics both between and within areas are interpreted. Three checks on the repeatability of terrain statistics produce reassuring results for altitude and gradient, but large differences in estimates of skewness and kurtosis of profile and especially of plan convexity. A number of geomorphologically significant dimensions of variability are identified, implying the poverty of models (e.g. spectral, fractal) based on two or three parameters.

INTRODUCTION

During the 25 years since the 1970 conference *Spatial Analysis in Geomorphology* (Chorley, 1972), geomorphology has moved away from the study of form for the sake of denudation chronology or for its own sake, and towards the study of process for its own sake. The core of geomorphology is the interaction between form and process (both widely defined), and a plea is made here that this core should be supported on the one hand by analyses of actual land surface form, as it is on the other by field or laboratory observations of process. Geomorphology should attempt explanation of variations in land surface form. Therefore it needs to identify the important terrain statistics and establish how they vary. Explanation may be sought in variations in materials or history, or in the modes of operation of geomorphological processes in different combinations.

To some of us, "terrain analysis" means, especially, quantitative analysis of terrain. This is the line I took in 1970 (Evans, 1972) in anticipation of digital topographic data becoming more widely available, in gridded spot height or digitised contour form. Having found some difficulty in obtaining slope estimates by relating adjacent contours in the computer, as one would by hand, I concentrated on the calculation of local surface derivatives from gridded data. The characterisation

Landform Monitoring, Modelling and Analysis. Edited by S. N. Lane, K. S. Richards and J. H. Chandler.
© 1998 John Wiley & Sons Ltd.

of an area in terms of summary statistics of these derivatives, and Pearson product–moment correlation between the derivatives, was proposed and developed as a system of general geomorphometry in a series of papers, reports and chapters (e.g. Evans, 1980, 1990).

After 25 years, it is clear from publications such as Petrie and Kennie (1990), Pike and Dikau (1995) and this volume, that expectations of increased availability and use of gridded altitude data have been more than fulfilled. Large numbers of digital elevation models (DEMs) are available through the Internet. Such data are available, often at a considerable price, for most industrialised countries. The rate of increase of computer power has exceeded my wildest dreams, if not those of visionaries such as David Bickmore, but we are still hungry for more data, of higher resolution. I maintain that surface derivatives remain basic to the characterisation of land surface form, although they do not alone provide a complete characterisation.

Related variables have been calculated from digital terrain models (DTMs), for example by Franklin (1987), Depraetere (1987) and Eyton (1991). Terrain statistics, especially for gradient and relief, are proving useful in the analysis of landslide occurrence (Pike, 1988b; Ohmori and Sugai, 1995). They are used in terrain classification, by Dikau *et al.* (1995) for New Mexico and by Guzzetti and Reichenbach (1994) for Italy, based on DEMs of 200 and 230 m resolution respectively. No doubt finer meshes will be used in future even for large areas: 30 m or 50 m will capture slopes in all but the most finely dissected areas. So far, such studies have emphasised slope, relief and various indices; there is a need for more systematic use of moments of altitude derivatives, especially curvature.

Breakthroughs have been made in tracing slope or drainage lines through digital topographic data (Douglas, 1986; Moore *et al.*, 1988) in ways which might have strained older computers. Drainage tracing in eight directions is now routine, though not without problems, and is exemplified in a number of other chapters in this volume (e.g. Wise, Chapter 7). It provides indices such as (cumulative) contributing area upslope, distance downslope to a stream, and more complex ones measuring the structure of a drainage network (Moore *et al.*, 1991; Nogami 1995; Richards *et al.*, 1995). These are very important for process and should be used to complement the statistics from derivatives, on which this brief chapter will concentrate.

Here I will consider the following questions. (i) How can we interpret summary statistics and correlations for derivatives, and how do they relate to older (pre-computer) geomorphological indices of the land surface? (ii) What values are reasonable for these statistics, obtained for DTMs and DEMs of several grid meshes, for glacial cirques, and for drumlins? Each specific landform type may be expected to have particular ranges for each summary statistic; for each type, some statistics will be more meaningful, others less so. Reference to these values may aid interpretation in studies where results are calculated for only one or two matrices. (iii) How repeatable are the statistics as the surface is resampled, e.g. with a displaced grid? The accuracy of the data themselves is not considered here, but in a number of other chapters in this volume (e.g. McCullagh, Chapter 5; Wise, Chapter 7). (iv) What does the land surface complexity revealed by these statistics imply for applications of the fractal models which have become popular in recent years?

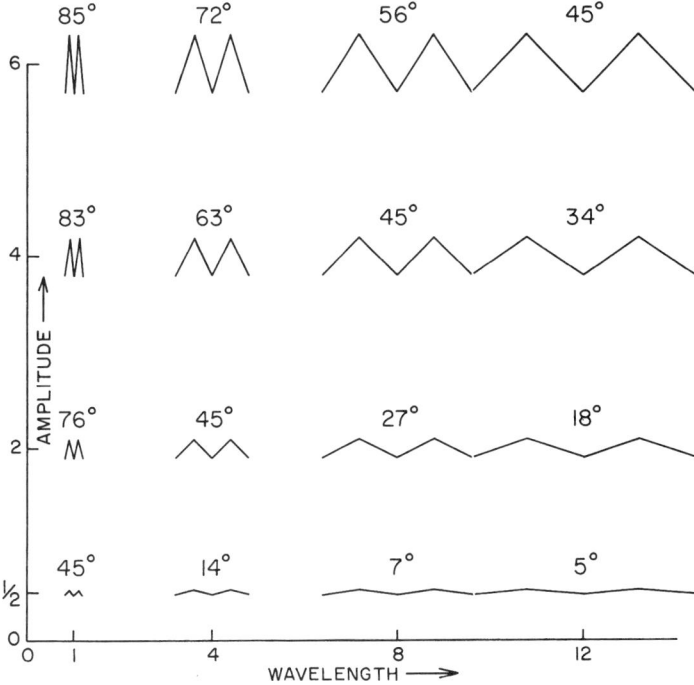

Figure 6.1 Two-parameter model profiles: amplitude versus wavelength

INTERPRETATION OF SUMMARY STATISTICS

This chapter is necessarily mainly empirical, but we begin by considering a simple model of variability along a profile and the implications this has for altitude (and later, gradient) frequency distributions. Figure 6.1 shows a two-dimensional model of statistical variability. If amplitude (relief) is plotted against wavelength (grain), then gradient (held constant for each of these simple profiles) can be represented by a series of lines radiating from the origin, varying from low in the bottom right to high in the top left. Along any of these lines of constant gradient, the topography can be described as "fine" in the bottom left, but increasingly "coarse" to the top or the right. The two-wave profiles shown on the face of the graph are for wavelengths of 1, 4, 8 and 12, and amplitudes of 0.5, 2, 4 and 6 units.

Profiles of equal amplitude will produce the same altitude frequency distribution, but only if they have the same shape. Profile shapes vary, and thus variations in altitude frequency distributions imply two further statistical dimensions for profiles (Figures 6.2a and 6.2b). Massive topography, with a few low areas, gives negative skew, contrasted to positive skew for open topography with occasional high spots. Near-average values are commonly dominant, but topographies of plateaus with broad valley floors give negative kurtosis of the altitude distribution (either

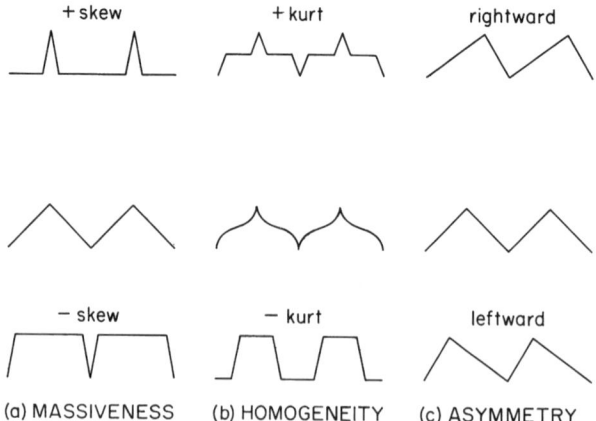

Figure 6.2 Three further types of variation in model profiles

bimodality, or a broad mode of altitude). A further obvious variation in profiles is between leftward and rightward asymmetry (Figuer 6.2c). This makes five dimensions so far for profiles; Posey (1946) took this further, for profiles on engineering surfaces. These properties interact to give further theoretical dimensions for surfaces over two horizontal dimensions. There are sound reasons, then, for expecting the land surface to be complex in terms of its statistical variability. We should not expect one or two "magic numbers" to provide a summary of any accuracy.

This leads us back to the consideration of the altitude surface and its derivatives, slope and curvature. Such considerations are highly relevant to questions regarding terrain as a consequence of tectonics (Fielding *et al.*, 1994) or as a control of sediment yield (Summerfield and Hulton, 1994). Altitude, slope and curvature distributions may be summarised for each tectonic region and for each drainage basin, and used in the prediction of sediment yield along with climatic, land cover and surface materials variables. Analysis of these variables at the global scale requires global DTMs with grid meshes considerably finer than 1 km; these exist now (Fielding *et al.*, 1994), and may shortly become generally available.

In his lecture at Cambridge in 1995, Dikau presented an impressive array of geomorphologists, starting with A. Penck, who had recognised the importance of altitude, slope and curvature for process studies as well as for landform classification. Slope is a vector, composed of slope gradient (or angle) and slope aspect (or azimuth); the rates of change of each give profile and plan convexity, two components of the "curvature", or second derivative vector. The first earth scientist to recognise the importance of these five variables explicitly was perhaps Aandahl (1948). He was followed by Curtis *et al.* (1965), Speight (1968) and Ahnert (1970) before Evans (1972) and Krcho (1973) more formally incorporated the five into geomorphometric systems. At the same time, all five were used in the texts by Young (1972) and Carson and Kirkby (1972).

Table 6.1 Interpretations of summary statistics for frequency distributions of altitude and its derivatives

---

Altitude
Standard deviation: a more stable measure of variation than range = RELIEF
Skewness: whether the mode of altitude is low or high compared to the range = MASSIVENESS (inverse). Relates closely to Hypsometric Integral
Kurtosis: extent of "tails" = HOMOGENEITY

Slope Gradient
Mean = STEEPNESS
Standard deviation: variability of slope = HETEROGENEITY
Skewness: mode low or high = LIMITATION
Kurtosis = MODALITY

Slope Aspect, Azimuth
Analyse modulo 360° (full circle) or modulo 180° (folded over)
Vector strength (modulo 360°) = DIRECTEDNESS
Vector strength (modulo 180°) = ORIENTATION (degree of)
Modulo 90°: for ERROR

Profile Convexity: rate of change of gradient
Mean = NET CONVEXITY IN PROFILE
Standard deviation: surface irregularity in profile = CURVATURE IN PROFILE
Skewness = SHARPNESS OF CONVEXITIES compared to CONCAVITIES
Kurtosis: ERRORS (PROBABLY)

Plan Convexity: rate of change of aspect
Mean = NET CONVEXITY IN PLAN
Standard deviation = CURVATURE IN PLAN
Relates closely to Drainage Density
Skewness = SHARPNESS OF CONVEXITIES compared with CONCAVITIES
Kurtosis: ERRORS

Table 6.1 lists the summary statistics for the frequency distributions of these five, except for mean altitude (which is of climatic and locational rather than morphometric significance). Following some interpretation, a suggested term for the property measured is given in capitals. Interpretation of those for altitude is easy and is discussed above. A graph relating the hypsometric integral (a classic index in drainage basin analysis, related to earlier measures of "dissection" or "aeration"; Strahler, 1952) to skewness of altitude for model distributions was presented in Evans (1972, p. 47). Difficulties increase as we move towards higher derivatives and higher moments.

Moment measures for slope gradient represent several further aspects of surface roughness. Gradient variability, measured by standard deviation rather than range, measures the heterogeneity of topography and would be high, for example, for a well-dissected plateau. Gradient measured in degrees (even more so if expressed as percentage, i.e. tangent) commonly gives positively skewed distributions, except for rather steep topography. This reflects the limitation of gradient to positive values; as the mode approaches zero, a "tail" can extend only in a positive direction. Although, at the other end of the scale, the mode is unlikely to approach the upper limit of 90°, rock mass strength or other slope stability considerations can provide an upper limit and, if this is approached over much of the landscape, negative skew of gradient may result. Kurtosis of gradient will be high if there is a long "tail" on either side, or low (negative as defined here) if there are multiple modes, which is quite common (e.g. a floodplain or a limiting slope).

Gradient at a point is defined as the maximum rate of change of altitude; aspect is the direction (azimuth) of this change. There is little room for disagreement over these concepts, but there are many different ways in which they can be calculated from gridded altitude data. Skidmore (1989) demonstrated the superiority of fitting a local quadratic (or a third-order finite difference) to a 3 × 3 submatrix, as advocated by Evans (1980). Since the data are never exactly accurate, it is an advantage that a local quadratic slightly smooths the surface suggested by the data. Guth (1995) showed that though this "eight neighbours unweighted" method produced a lower mean gradient than methods based on fewer points (such as "steepest adjacent neighbour"), it also produced the best (least spiky) gradient frequency distributions, and was necessary to obtain values of aspect (not just eight classes). The steeper gradients produced by "steepest adjacent neighbour" for peaks, pits, ridges and valleys do not relate to the central point, for which the lesser gradient of the local quadratic is more appropriate. The local quadratic is the only method considered by Skidmore or by Guth which can give curvature at the same point as it gives slope. My position remains that if altitude values are rounded (to the nearest metre is common) or have any error, the smoothing provided by the local quadratic is beneficial, and this definition remains preferable to the others discussed by Skidmore and Guth (including the four-point method used by Eyton (1991)), and to the exact-fit partial cubic of Zevenbergen and Thorne (1987) – all the more so if curvature is of interest.

In general, slope aspect (azimuth) should not be analysed in the same way as the other four variables, but by circular statistics (Fisher, 1993). These will be considered only briefly here. The degree of concentration of vectors on the circle is expressed simply by the vector strength, modulo 360°; this gives the "directedness" of the surface, the two-dimensional version of the asymmetry in model profiles of Figure 6.2c. There may, however, be two opposed modes of slope aspect, e.g. SE and NW, and it is useful to measure the degree of "orientation" of the area or landform by a "folded-over", modulo 180°, vector strength analysis. Of rather more limited value is a second folding-over to give modulo 90° analysis, checking the strength of four opposed modes. This proved to be useful in the assessment of error caused by bilinear interpolation (Evans, 1984).

Surface curvature is more difficult to measure because the second derivative is more sensitive to error in the initial altitude data and to any interpolation techniques used. Also, there are alternative ways of defining components of curvature: compare Zevenbergen and Thorne (1987) with Evans (1980) and Eyton (1991). It seems simplest to me to consider rates of change of gradient and of aspect, giving convexity in profile and in plan respectively, with concavity represented by negative values. This sign convention has been adopted by geomorphologists since they consider change along profiles in a downslope direction, hence concavity is a reduction in gradient (see also Eyton, 1991). These two variables, measured in degrees/(100 m), clearly express two complementary components of surface curvature (Figure 6.3).

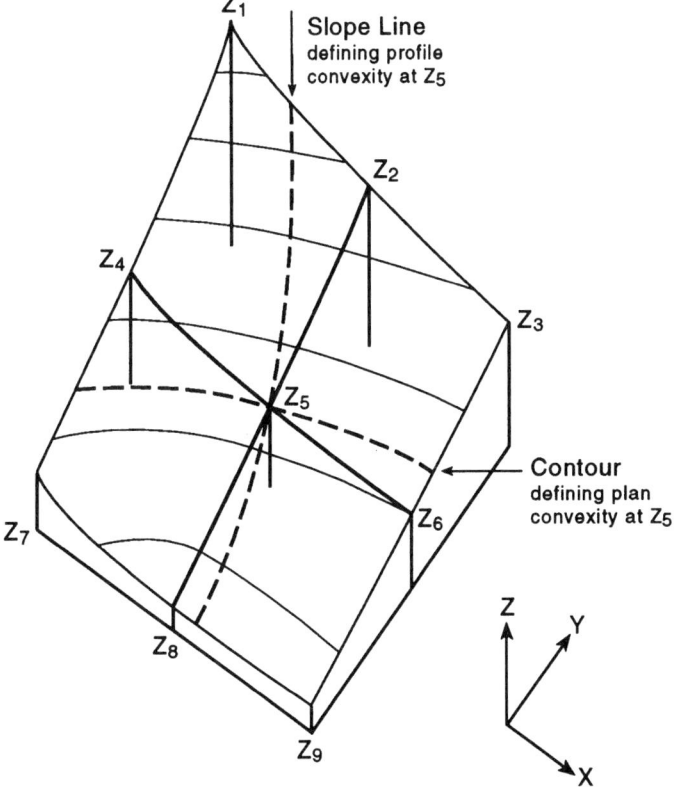

Figure 6.3 Definition of profile and plan convexity

They are measured here by fitting a six-parameter local quadratic to each 3×3 submatrix, using polynomials, and using this equation to estimate rate of change of gradient and of aspect at the central point (Z_5 in Figure 6.3) (Evans, 1979).

Though conceptually distinct, profile and plan curvature provide statistics which are interpreted in similar ways (Table 6.1). Means measure net convexity or concavity in the area considered. For large, arbitrary areas such as rectangles, we would expect the means of curvature statistics to approach zero as convexities and concavities balance out. The standard deviation is the most important convexity statistic, expressing the degree of curvature of the surface. Curvature in plan relates closely to that important though subjective index, drainage density. Gardiner (1990) has reviewed the many different ways in which drainage density may be defined. When the criterion is contour crenulation (sharp indentations in successive contours), the question of "how sharp" is resolved differently by different operators, and inconsistently by the same operator. Standard deviation of plan convexity is an objective measure of contour indentation at the given grid mesh, and further studies are needed of its variation in relation to drainage density.

Skewness reflects the sharpness of convexities compared with concavities. For example, if a surface consists largely of cirques or of landslide scars, small areas of sharp convexity are balanced by larger areas of gentle concavity, giving positive skewness both in profile and in plan. Finally, some high values of kurtosis are found for these second derivatives; although true extreme values are possible, especially in plan for contours wandering over floodplains, the highest values of kurtosis probably reflect poor data quality causing outliers.

The meaning of (Pearson product–moment) correlations should be straightforward. That many of them are low is no problem; it is an indication that we have four nearly independent variables (aspect being excluded). For large areas, positive or negative correlations are feasible in most cases: the more restricted ranges noted below provide information about the topographies considered. Particular tendencies may be expected when specific landforms (or drainage basins) are considered. Correlations are less meaningful where there are extreme values (where skewness or kurtosis is strong) and where relationships are non-linear, as is quite common, so it is very advisable to study scatter plots as well. Triangular scatters, reversals and extreme values require explanation.

RANGE OF VALUES ENCOUNTERED

The comparison of summary statistics is useful for similar landforms, or areas of similar size, analysed at the same grid mesh. Table 6.2 is focused on results for 50 m grid mesh but the "large DTMs" are supplemented by some at 30 m and 100 m mesh. All three grid meshes are in common use: the range of meshes here is narrow in logarithmic terms, but gradient and curvature statistics will vary even across this range. Finer grid meshes should not affect altitude statistics, but they do give higher and more varied gradients, and more varied convexities. The ranges given in rounded terms in Table 6.2 might form an initial point of comparison for researchers with new results (see also O'Neill and Mark (1987) and Klinkenberg (1992)).

The "large DTMs" of Table 6.2, many of which were considered by McClean (1991), are between 5 and 17 km across. A number of DTMs analysed more recently have been added, together with altitude and gradient statistics for 53 10 × 10 km DTMs for the Wessex area of England (Evans, 1984). For altitude, gradient and profile convexity, negative skews occur but positive skews may be more pronounced. As expected, means and standard deviations are highly varied.

Standard deviations of plan convexity are often more than ten times those of profile convexity. Profile convexity always correlates positively with altitude, but the Pearson coefficient ranges from near zero to 0.8. The altitude–gradient correlation can be moderately negative or positive, reflecting "convex-up" or "concave-up" topography respectively. Profile convexity always correlates positively, but weakly, with plan convexity.

Values are also given in Table 6.2 for a small set of drumlins in Cumbria (Evans, 1987), and for an extended data set of cirques, covering nearly half the cirques in

Table 6.2 Range of values encountered. The large DTMs are for rectangles or drainage basins, and have grid meshes of 30, 50 or 100 m. The cirque and drumlin DTMs are at 50 m mesh; they are small and have irregular outlines

| | **Altitude** | | | | | |
| | Mean (m) | | St. dev. (m) | | Skew | |
	min.	max.	min.	max.	min.	max.
Large DTMs (30, 50 or 100 m)	54	881	5	429	−0.54	1.68
75 Cumbrian cirques, 5–95%	314	759	44	109	−0.17	0.94
7 Cumbran drumlins	45	174	4	9	−0.57	0.29

| | **Gradient** | | | | | |
| | Mean (°) | | St. dev. (°) | | Skew | |
	min.	max.	min.	max.	min.	max.
Large DTMs (30, 50 or 100 m)	1.1	33.5	0.3	16.3	−0.37	2.30
75 Cumbrian cirques, 5–95%	19.5	33.9	4.9	13.0	−0.82	0.48
7 Cumbrian drumlins	2.4	5.9	1.4	2.3	−0.40	3.06

| | **Profile Convexity** | | | | | |
| | Mean (°/100 m) | | St. dev. (°/100 m) | | Skew | |
	min.	max.	min.	max.	min.	max.
Large DTMs (30, 50 or 100 m)	–	–	0.9	54.9	−1.90	6.34
75 Cumbrian cirques, 5–95%)	−5.4	4.2	11.1	30.4	0.04	2.22
7 Cumbrian drumlins	−1.1	8.0	3.6	7.8	−0.84	1.57

| | **Plan Convexity** | | | | | |
| | Mean (°/100 m) | | St. dev. (°/100 m) | | Skew | |
	min.	max.	min.	max.	min.	max.
Large DTMs (30, 50 or 100 m)	−1.8	1.9	52	3776	−199	36.2
75 Cumbrian cirques, 5–95%	−26	1	30	102	−3.0	1.9
7 Cumbrian drumlins	16	59	50	130	−0.9	7.9

| | **Correlations** | | | | |
| | Altitude–Gradient | | Altitude–Profile Convexity | | |
	min.	max.		min.	max.
Large DTMs (30, 50 or 100 m)	−0.50	0.72		0.02	0.81
75 Cumbrian cirques, 5–95%	0.14	0.82		0.20	0.71
7 Cumbrian dumlins	−0.52	0.42		0.38	0.75

| | **Correlation** | | **Weighted Vector Strength** | |
| | Profile Convexity– Plan Convexity | | modulo 360° | |
	min.	max.	min.	max.
75 Cumbrian cirques, 5–95%	0.09	0.47	0.39	0.93
7 Cumbrian drumlins	0.09	0.29	0.05	0.89

Cumbria. Both of these are based on small DTMs with 50 m mesh, less than 1.5 km across, generated by manual interpolation (thus avoiding some of the problems discussed by McCullagh in Chapter 5) from 1:10 000 scale maps with photogrammetric contours every 5 m or 10 m respectively. This data set of general geomorphometric statistics for cirques is quite distinct from that of specific geomorphometric indices analysed by Evans and Cox (1995) and Evans (1994). The cirque DTMs were either generated by the author, or edited by him from student-generated DTMs. Because some of the extreme values probably reflect data errors, values for the 5th and 95th percentiles replace minimum and maximum in Table 6.2.

As expected, most statistics for the two specific landform types cover more limited ranges than the larger DTMs considered above, and each is contained within the range for those DTMs. Cirques are high and drumlins low on mean and standard deviation of both altitude and gradient, and on standard deviation of profile convexity; not surprisingly, they have no overlap on these five statistics. Although cirques approach the top of the DTM ranges for both these gradient statistics, this is not so for standard deviation of altitude. For its limited extent, a cirque is varied in altitude, but greater variation is seen in a matrix several kilometres across which covers glacial troughs as well as cirques. With extensive areas of cliff, cirques with limited floors have more negatively skewed gradient distributions than any large DTMs; their skewness of altitude is usually positive, but not always so. Skewness of gradient or altitude is less pronounced for cirques than for larger DTMs.

Cirques are concave landforms delimited by convex rims, while drumlins are the reverse. Depending partly on size (affecting the ratio of peripheral to interior sample points), mean convexity in profile or plan can be negative or positive in either case. Sharp convexities and gentler concavities should give positive skewness for cirque convexity: this materialises for profile but not for plan. The standard deviation and skewness of plan convexity can be very high for DTMs which contain flat areas where contours can be very convoluted. The more reasonable values for cirques and drumlins (after censoring) suggest good data quality, but may owe more to the absence of floodplains. Extreme values of plan convexity can nevertheless be found on flat cirque floors or on the (convex) summits of drumlins.

The altitude–gradient and altitude–profile convexity correlations are always positive for cirques. For drumlins, the more varied altitude–gradient coefficients reflect non-linear relationships (steepest slopes at mid-altitude), but profile convexity always increases strongly with altitude from negative around the edge to positive near the summit. "Directedness" is moderate or high for cirques, but low for drumlins and very low for large DTMs. Reassuringly, drumlin slopes are moderately well oriented.

Results from this set of cirques are portrayed in Figures 6.4 to 6.6, where each cirque name is plotted by the program Stata (Stata Corporation, 1995) with up to eight characters, centred on the plotted point. The standard deviation of altitude correlates +0.63 with the logarithm of cirque area, so large cirques plot high and small cirques low on Figure 6.4. There is a weak correlation (+0.33) between standard deviation of altitude and mean gradient, with well-developed cirques

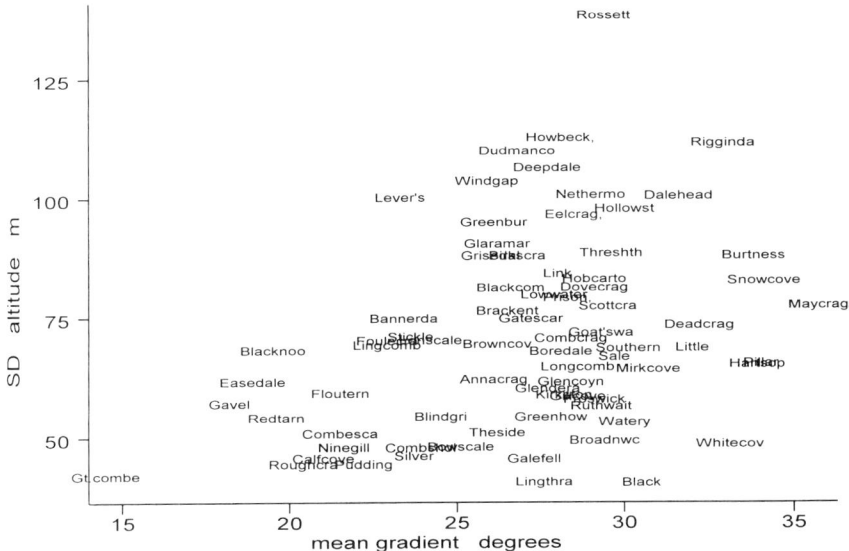

Figure 6.4 Plot of standard deviation of altitude against mean gradient for 75 Cumbrian cirques (identified by first eight characters of name)

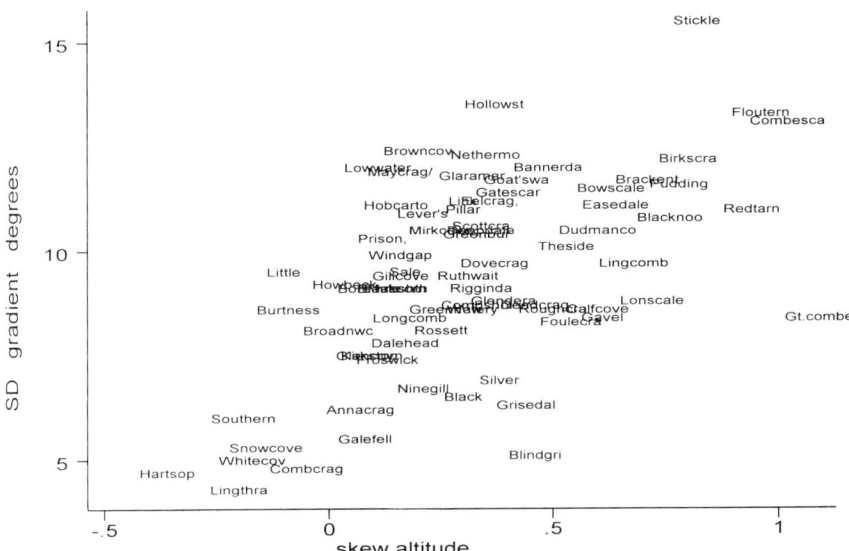

Figure 6.5 Plot of standard deviation of gradient against skewness of altitude for 75 Cumbrian cirques

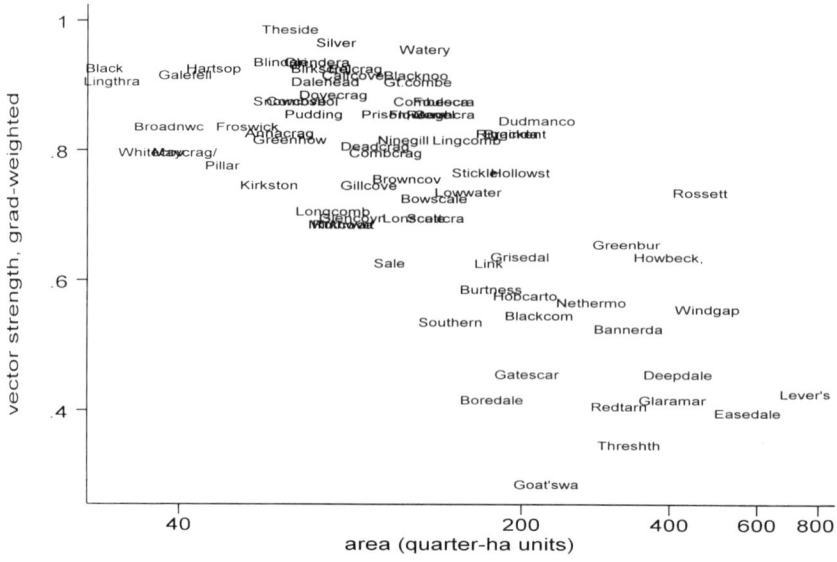

Figure 6.6 Plot of vector strength against area for 75 Cumbrian cirques

plotting above this trend, and poorly developed ones, below. The extreme point on the left of Figure 6.4, Great Combe in Dentdale, is the only one in this data set which is on Carboniferous rocks.

Figure 6.5 arrays the 75 cirques on two different axes. Better-developed cirques tend to have a high standard deviation of gradient, as a steep headwall is contrasted with a gentle floor. An extensive, gently sloping floor leads to positive skew of altitude, so cirques with these plot on the right. Finally, Figure 6.6 shows the negative correlation (−0.61) between vector strength and the logarithm of cirque area. The headwalls of small cirques tend to have a consistent direction, while large cirques have greater plan closure with varied headwall aspects giving low vector strengths. All three of these plots show related variables: many others could be presented in which the two variables are unrelated, yet the position of individual cirques within the broad scatter is of interest.

REPEATABILITY

Are these statistics sufficiently well based to permit meaningful interpretation? Sometimes their value may be greatly reduced by poor data. Even if the data are excellent, it is possible that sampling of the surface may be inadequate to provide reliable estimates of statistics for areas, especially for small DTMs. One check is to start with a DTM which is denser than required, and sample it at coarser mesh, repeating this with displacement of the sample grid relative to the terrain. This is done here in three small experiments, where four displaced incidences of a given grid are taken for the same area.

First there is a very large DTM for the Quillan area of the French Pyrenean foothills (Evans, 1979). This was generated by the Gestalt photomapper, with a mesh of 4.544 m, covering an area 2.508 × 5.235 km with 552 × 1152 grid points. Sampling every fifth row and column gives a 110 × 229 grid with a 22.72 m mesh; every eighth gives a 69 × 144 grid with a 36.35 m mesh. Four such grids with 9.09 m displacement were analysed, for each sampling density. A similar approach was taken to a 25 m mesh DTM of Sale Pot cirque (located above Hawes Water in Cumbria, England); all four 50 m mesh grids contained within this were analysed.

Table 6.3 gives the lowest and highest of the four estimates obtained for each statistic in each of the three experiments. The difference between the four replicates is given both in absolute terms (the range) and, where appropriate, in relation to the mean. The range is used because we are dealing with only four estimates in each case. It should be judged relative to the actual estimates (hence minimum and maximum of the four are given), but the relative range is less meaningful for statistics where both positive and negative values are possible (all skewness, kurtosis and correlation values, and means of profile and plan convexity). Range/mean is given where it is possibly relevant.

Errors in estimates increase for higher moments, for smaller data sets and for near-zero correlations. Replication is very good in both absolute and relative terms for most altitude and gradient statistics (mean, standard deviation and skewness). Kurtosis is not badly estimated except for the gradients of Sale Pot. Standard deviations are estimated reasonably for profile convexity, but are disappointingly variable for plan convexity. Mean profile convexity is estimated well for Quillan (5 and 8) and consistently (negative) for Sale Pot. Mean plan convexity is near-zero for Quillan and clearly negative for Sale Pot. Skewness and kurtosis for profile and plan convexity have very high relative errors, but for profile convexity the ranges of estimates are not too great in absolute terms; both are consistently positive for Quillan and near-zero for Sale Pot.

From these results, the main concerns are for skewness and kurtosis (and even standard deviation) for plan convexity. It would be useful to summarise the frequency distribution of plan convexity with quantile or other robust methods, as well as with moment-based statistics. The additive properties of the latter, however, are an advantage in dealing with convexity distributions.

Estimates of gradient-weighted vector means are repeatable within a few degrees; compared with the 360-degree range of possible values, this is a good performance even for the small Sale Pot data set. Vector strength can vary between 0 and 1, and the observed ranges are low in absolute terms. The two strongest correlations, of altitude with gradient and with profile convexity, are highly reproducible. Error margins for the others are acceptable in absolute terms, except for some near-zero correlations which are too weak to be of interest. Note that for the cirque Sale Pot, the gradient–profile convexity correlation is stronger than for Quillan; it is thus more repeatable, despite the much smaller size of the data set. On the whole, the results for correlations and vector analyses are reassuringly reproducible.

Table 6.3 Repeatability under displacement: range of results for four replicates

Altitude

| | Mean (m) | | | | Standard deviation (m) | | | |
	Min.	Max.	Range	Range/ Mean	Min.	Max.	Range	Range/ Mean
Quillan5	351.7	352.3	0.5	0.1%	74.02	74.14	0.12	0.2%
Quillan8	351.3	352.2	0.9	0.3%	73.80	74.22	0.42	0.6%
Sale Pot	621.5	623.1	1.6	0.3%	65.5	67.1	1.6	2.4%
	Skewness				Kurtosis			
Quillan5	1.074	1.085	0.011	1.0%	0.718	0.752	0.034	4.6%
Quillan8	1.078	1.092	0.014	1.3%	0.728	0.774	0.046	6.1%
Sale Pot	0.167	0.190	0.023	1.3%	−1.232	−1.177	0.055	4.6%

Gradient

| | Mean (°) | | | | Standard deviation (°) | | | |
	Min.	Max.	Range	Range/ Mean	Min.	Max.	Range	Range/ Mean
Quillan5	13.65	13.69	0.04	0.3%	8.59	8.60	0.01	0.1%
Quillan8	12.93	12.98	0.05	0.4%	8.03	8.06	0.03	0.4%
Sale Pot	29.59	29.83	0.24	0.8%	9.28	9.49	0.21	2.2%
	Skewness				Kurtosis			
Quillan5	0.555	0.563	0.008	1.4%	−0.420	−0.400	0.020	4.9%
Quillan8	0.545	0.557	0.012	2.2%	−0.302	−0.325	0.023	7.4%
Sale Pot	−1.094	−1.029	0.065	6.1%	0.324	0.534	0.210	49.8%

Profile Convexity

| | Mean (°/100 m) | | | | Standard deviation (°/100 m) | | | |
	Min.	Max.	Range	Range/ Mean	Min.	Max.	Range	Range/ Mean
Quillan5	−2.393	−2.42	0.027	1.1%	27.47	27.73	0.26	0.9%
Quillan8	−2.239	−2.34	0.103	4.5%	18.25	18.80	0.55	3.0%
Sale Pot	−4.73	−5.58	0.85	16.2%	11.65	12.14	0.49	4.1%
	Skewness				Kurtosis			
Quillan5	0.274	0.395	0.121	35.7%	4.423	5.447	1.024	21.2%
Quillan8	0.482	0.650	0.168	29.8%	4.023	5.341	1.318	28.7%
Sale Pot	−0.230	0.244	0.474	−	−0.369	0.417	0.786	−

Plan Convexity

| | Mean (°/100 m) | | | | Standard deviation (°/100 m) | | | |
	Min.	Max.	Range	Range/ Mean	Min.	Max.	Range	Range/ Mean
Quillan5	−1.04	1.62	2.66	−	344	424	80	20.0%
Quillan8	−2.17	3.89	6.06	−	202	235	33	15.2%
Sale Pot	−25.54	−23.43	2.11	8.5%	43.60	58.86	15.26	30.6%
	Skewness				Kurtosis			
Quillan5	−22.62	11.76	34.39	−	234	1864	1630	−
Quillan8	−5.35	7.10	12.45	−	165	403	238	−
Sale Pot	−4.55	−0.86	3.69	−	2.16	33.29	31.13	−

Table 6.3 (*cont.*)

	Gradient-weighted Vector							
	Mean (°)				Strength			
	Min.	Max.	Range	Range/Mean	Min.	Max.	Range	Range/Mean
Quillan5	075.22	075.75	0.53	–	0.026	0.027	0.001	3.8%
Quillan8	073.33	074.99	1.66	–	0.026	0.028	0.002	7.4%
Sale Pot	134.24	137.20	2.96	–	0.605	0.616	0.011	1.8%

	Correlation							
	Altitude–Gradient				Altitude–Profile Convexity			
	Min.	Max.	Range	Range/Mean	Min.	Max.	Range	Range/Mean
Quillan5	0.409	0.410	0.001	0.2%	0.163	0.168	0.005	3.0%
Quillan8	0.429	0.431	0.002	0.5%	0.223	0.226	0.003	1.3%
Sale Pot	0.678	0.711	0.033	4.8%	0.754	0.762	0.008	1.1%
	Altitude–Plan Convexity				Gradient–Profile Convexity			
Quillan5	0.044	0.053	0.009	19.0%	0.021	0.025	0.004	17.0%
Quillan8	0.074	0.095	0.021	24.8%	0.027	0.032	0.005	16.9%
Sale Pot	0.237	0.258	0.021	8.5%	0.430	0.453	0.023	5.2%
	Gradient–Plan Convexity				Profile Convexity–Plan Convexity			
Quillan5	0.026	0.035	0.009	30.2%	0.111	0.131	0.020	17.0%
Quillan8	0.030	0.076	0.046	91.5%	0.133	0.169	0.036	24.6%
Sale Pot	0.344	0.394	0.050	13.3%	0.094	0.187	0.093	65.5%

				Size of Data Sets
Data	No. of points	Mesh (m)	Displacement (m)	Comments
Quillan5	110 × 229	22.72	9.09	Every fifth row and column of Quillan DTM
Quillan8	69 × 144	36.35	9.09	Every eighth row and column of Quillan DTM
Sale Pot	*c.* 108	50	25	Every second row and column in Sale Pot 25 m DTM

IMPLICATIONS FOR FRACTAL MODELS

A large number of land surface descriptors have been discussed above; they have been shown to be meaningful and (mostly) repeatable, while being based economically on the widely used scientific concepts of taking surface derivatives, moment measures or vector statistics, and correlations. The list is not exhaustive, notably in being "point-based" and lacking variables for position in relation to the slope and drainage network. Naturally there are relationships between the different statistics, if they are correlated over a set of areas as in Figure 6.5, but these are not strong enough to collapse the long list of statistics into a few statistical dimensions.

Table 6.4 Statistical dimensions of the Wessex land surface, for 53 areas, each 10 × 10 km, analysed from 50 m grids (after Evans, 1984)

Property	Statistical descriptor (key variable)
Gradient	Mean gradient
Massiveness	Skewness of altitude
Level (or Bias)	Mean altitude
Profile convexity	Skewness of profile convexity
Orientation	Weighted vector strength (modulo 180°)
Plan convexity	Standard deviation of plan convexity
Altitude–Convexity	Correlation of altitude with profile convexity
(Profile) Variable	Standard deviation of gradient
Directedness	Weighted vector strength (modulo 360°)

Table 6.5 Deficiencies in unifractal models

1. Some landforms are scale-specific
2. Landform shape varies with size
3. Variograms are curved throughout
4. Variograms differ by azimuth
5. Closed pits are less frequent than summits
6. Surface properties vary with relative height
7. Land surfaces are lineated

A number of multivariate studies, such as those of Depraetere (1987) and Pike (1987, 1988a), have shown that more than five independent statistical dimensions are required to represent terrain variability, and possibly as many as 12. Table 6.4 lists the nine main dimensions or properties, some moderately intercorrelated, and the key variables used to measure them, which Evans (1984) concluded were required to represent terrain variability in the lowlands of Wessex (southwest England).

The land surface, then, is statistically complex. Efforts to summarise it with one or two parameters, if not doomed to complete failure, are likely to generalise to such an extent that important and interesting aspects of terrain variability are excluded. This led Evans and McClean (1995) to summarise the numerous deficiencies of unifractal models of the land surface, of the type presented by Mandelbrot (1982). These are listed in Table 6.5; quantitative evidence of the first four deficiencies is presented by Evans and McClean (1995), while the final three are visual contrasts evident to those familiar with real terrain. Application of the inappropriate unifractal model has led to many of the practical difficulties listed in Table 6.6. Multifractal models represent a movement in the right direction (Lavallée *et al.*, 1993), but still exhibit a number of the deficiencies in Table 6.5; a more complex model is needed, possibly a compound one with fractal(?) slopes related to fractal drainage nets.

Table 6.6 Further problems with fractal analysis of the land surface (from Evans and McClean, 1995). D is the fractal dimension

(a) D varies between adjacent **sub-areas**
(b) D varies for different altitude **contours**
(c) D varies between estimation **methods**
(d) D estimates are **grid-mesh** dependent
(e) D varies depending on the **section** of variogram or variance spectrum analysed, and the degree of **convexity** or concavity tolerated
(f) Data sets **large** in linear dimension are required to establish D over several logarithmic decades
(g) Estimates of D in excess of 2.5, suggesting **anti-persistence** of trends, are problematic in view of the known properties of the land surface
(h) The land surface needs at least five, and probably 12, independent **statistical dimensions** to describe its variability from one area to another

CONCLUSIONS

Altitude, gradient, aspect, profile convexity and plan convexity are basic attributes of the land surface at a point (small neighbourhood): as derivatives (zero, first and second) of the land surface, they form a coherent system for its description and analysis, and are easily calculated. For convexity (curvature), alternative definitions have been offered, but rates of change of gradient and of aspect provide the simplest definitions: their wider use will provide more comparability between researchers, facilitating the cumulation and progression of geomorphological knowledge.

Areas should be characterised by moment statistics of these derivatives and correlations between them, rather than by complex indices: moment statistics are efficient, easily interpreted and, for the most part, reproducible summaries of local land surface form. As calculated from gridded DTMs of medium resolution, means and standard deviations have been shown to be repeatable for all derivatives; for altitude and gradient, skewness and kurtosis are also repeatable. There are large differences in estimates of skewness and kurtosis for profile and plan convexity, and these four statistics should be used only with great caution. Though they do have value in locating gross errors, convexity variables probably should not be used for gridded data which are excessively rounded or of low accuracy. As yet we cannot specify what accuracy is required of altitude data to prevent distortion of convexity values.

Intercorrelations between moments and correlations of surface derivatives are strong enough to make many of them redundant: studies at 50 m grid mesh have found the key variables listed in Table 6.4 to be adequate. Further tests of the robustness of these dimensions are desirable. These point-based variables need augmentation by measures based on slope/drainage tracing. Variation can be studied both between and within landform types, each of which has a characteristic range of values. Comparisons should be for equal grid meshes, for similar methods of data generation and for sets which do not vary too much in area.

Although the same statistical techniques can be applied at any grid mesh, the meaning of the results may vary. Nogami (1995) has emphasised the need for a mesh of 50 m or finer, at least in Japan. At scales much broader than 100 m, many erosional valleys are lost and the results will be of less relevance to exogenetic processes, though still significant in relation to endogenetic processes. At scales much finer than 25 m, different processes may be dominant. Thus for macro-relief and micro-relief, lists of statistical dimensions and key variables may differ from those in Table 6.4.

With the exception of fractal analyses and drainage tracing, the amount of research on terrain characterisation remains limited. Urgent needs for future work include relating the statistics discussed here to positional variables such as those from drainage tracing, and mapping variation in the statistics so that they can be related more closely to perceived landform types such as tablelands, basin and range, etc. Further studies of the effect of grid mesh (e.g. Guth, 1995) should be made for the DTM scales which are now widely available, with close attention to the mode of data generation.

These statistics demonstrate the statistically multidimensional character of the land surface, and thus show up the relative poverty of fractal and spectral models. A broader basis against which terrain statistics can be compared would encourage wider use of, and familiarity with, the basic properties of land surface form. Geomorphology should not lose sight of the need to account for quantitative variations in landforms.

ACKNOWLEDGEMENTS

I am grateful to Colin McClean for work on fractal models and analyses, to Margaret Young for computing the Quillan results, to Martin Robson for generating the Sale Pot DTM, and to all those who contributed data sets. Comments by Nick Cox and two referees led to improvements in the manuscript.

REFERENCES

Aandahl, A. R., 1948. The characterisation of slope positions and their influence on the total N content of a few virgin soils in Western Iowa. *Soil Science Society of America, Proceedings*, **13**, 449–454.

Ahnert, F., 1970. An approach towards a descriptive classification of slopes. *Zeitschrift für Geomorphologie N.F. Supplement Band*, **9**, 70–84.

Carson, M. A. & Kirkby, M. J., 1972. *Hillslope Form and Process*. Cambridge University Press, Cambridge, 475pp.

Chorley, R. J. (ed.), 1972. *Spatial Analysis in Geomorphology*. Methuen, London, 393pp.

Curtis, L. F., Doornkamp, J. C. and Gregory, K. J., 1965. The description of relief in field studies of soils. *Journal of Soil Science*, **16**, 16–30.

Depraetere, C., 1987. *Classification automatique interrégionale à partir de MNT issus de la BDZ*. Institut Géographique National, Paris, ANGEO/DELI/SCME Rapport 4.

Dikau, R., Brabb, E. E., Mark, R. K. and Pike, R. J., 1995. Morphometric landform analysis of New Mexico. *Zeitschrift für Geomorphologie N.F. SupplementBand*, **101**, 109–126.

Douglas, D. H., 1986. Experiments to locate ridges and channels to create a new type of Digital Elevation Model. *Cartographica*, **23**, 29–61.

Evans, I. S., 1972. General geomorphometry, derivatives of altitude, and descriptive statistics. In *Spatial Analysis in Geomorphology*, ed. R. J. Chorley. Methuen, London, 17–90.

Evans, I. S., 1979. *An Integrated System of Terrain Analysis and Slope Mapping*. Final Report on Grant DA-ERO-591-73-G0040, Department of Geography, University of Durham, 192pp.

Evans, I. S., 1980. An integrated system of terrain analysis and slope mapping. *Zeitschrift für Geomorphologie N.F. SupplementBand*, **36**, 274–295.

Evans, I. S., 1984. Correlation structures and factor analysis in the investigation of data dimensionality: statistical properties of the Wessex land surface, England. *Proceedings, International Symposium on Spatial Data Handling '84*. Geogr. Inst., Universitat Zürich-Irchel, Zürich, Switzerland, vol. 1, 98–116.

Evans, I. S., 1987. A new approach to drumlin morphometry. In *Drumlin Symposium*, eds J. Menzies and J. Rose. A.A. Balkema, Rotterdam, 119–130.

Evans, I. S., 1990. General geomorphometry. In *Geomorphological Techniques*, 2nd edition, eds A. Goudie *et al*. Unwin Hyman, London, 44–56.

Evans, I. S., 1994. Lithological and structural effects on forms of glacial erosion; cirques and lake basins. In *Rock Weathering and Landform Evolution*, eds D. A. Robinson and R. B. G. Williams, J. Wiley, Chichester: 455–472.

Evans, I. S. and Cox, N. J., 1995. The form of glacial cirques in the English Lake District, Cumbria. *Zeitschrift für Geomorphologie N.F.*, **39**, 175–202.

Evans, I. S. and McClean, C. J., 1995. The land surface is not unifractal: variograms, cirque scale and allometry. *Zeitschrift für Geomorphologie N.F. SupplementBand*, **101**, 127–147.

Eyton, J. R., 1991. Rate-of-change maps. *Cartography and GIS*, **18**, 87–103.

Fielding, E., Isacks, B., Barazangi, M. and Duncan, C., 1994. How flat is Tibet? *Geology*, **22**, 163–167.

Fisher, N. I., 1993. *Statistical Analysis of Circular Data*. Cambridge University Press, Cambridge, 277pp.

Franklin, S. E., 1987. Geomorphometric processing of digital elevation models. *Computers and Geosciences*, **13**, 603–609.

Gardiner, V., 1990. Drainage basin morphometry. In *Geomorphological Techniques*, 2nd edition, eds A. Goudie *et al*. Unwin Hyman, London, 71–81.

Guth, P. L., 1995. Slope and aspect calculations on gridded DEMs. *Zeitschrift für Geomorphologie N.F. SupplementBand*, **101**, 31–52.

Guzzetti, F. and Reichenbach, P., 1994. Toward a definition of topographic divisions for Italy. *Geomorphology* (Elsevier), **11**, 57–74.

Klinkenberg, B., 1992. Fractals and morphometric measures: is there a relationship? *Geomorphology* (Elsevier), **5**, 5–20.

Krcho, J., 1973. Morphometric analysis of relief on the basis of geometric aspects of field theory. *Acta Geographica Universitatis Comenianae, Geographico-physica*, Nr. 1, Slovak Pedagogical Publishers, Bratislava, 7–233.

Lavallée, D., Lovejoy, S., Schertzer, D. and Ladoy, P., 1993. Nonlinear variability of landscape topography: multifractal analysis and simulation. In *Fractals in Geography*, eds N. S. N. Lam and L. De Cola, Prentice-Hall, Englewood Cliffs, NJ, 158–192.

Mandelbrot, B. B., 1982. *The Fractal Geometry of Nature*. W.H. Freeman, New York, 468pp.

McClean, C. J., 1991. *The Scale-free and Scale-bound Properties of Land Surfaces: Fractal Analysis and Specific Geomorphometry from Digital Terrain Models*. PhD thesis, University of Durham, Department of Geography, 308pp.

Moore, I. D., Grayson, R. B. and Ladson, A. R., 1991. Digital terrain modelling: a review of hydrological, geomorphological and biological applications. *Hydrological Processes*, **5**, 3–30.

Moore, I. D., O'Loughlin, E. M. and Burch, G. J., 1988. A contour-based topographic model for hydrological and ecological applications. *Earth Surface Processes and Landforms*, **13**, 305–320.

Nogami, M., 1995. Geomorphometric measures for digital elevation models. *Zeitschrift für Geomorphologie N.F. SupplementBand*, **101**, 53–67.

Ohmori, H. and Sugai, T., 1995. Toward geomorphometric models for estimating landslide dynamics and forecasting landslide occurrence in Japanese mountains. *Zeitschrift für Geomorphologie N.F. SupplementBand*, **101**, 149–164.

O'Neill, M. P. and Mark, D. M., 1987. On the frequency distribution of land slope. *Earth Surface Processes & Landforms*, **12**, 127–136.

Petrie, G. and Kennie, T. J. M. (eds), 1990. *Terrain Modelling in Surveying and Civil Engineering*. Whittles, Caithness.

Pike, R. J., 1987. Information content of planetary terrain: varied effectiveness of parameters for the Earth. *Lunar and Planetary Science*, **18**, 778–781.

Pike, R. J., 1988a. Toward geometric signatures for geographic information systems. *International Geographic Systems Symposium, Proceedings* III. NASA, Arlington, 15–26.

Pike, R. J., 1988b. The geometric signature: quantifying landslide-terrain types from DEMs. *Mathematical Geology*, **20**, 491–511.

Pike, R. J. and Dikau, R. (eds), 1995. Geomorphometry. *Zeitschrift für Geomorphologie N.F. SupplementBand*, **101**.

Posey, C. J., 1946. Measurement of surface roughness. *Mechanical Engineering*, **68**, 305–306, 338.

Richards, K., Arnold, N., Lane, S., Chandra, S., El-hames, A., Mattikalli, N. and Chandler, J. H., 1995. Numerical landscapes: static, kinematic and dynamic process-form relations. *Zeitschrift für Geomorphologie N.F. SupplementBand*, **101**, 201–220.

Skidmore, A. K., 1989. A comparison of techniques for calculating gradient and aspect from a gridded DEM. *International Journal of Geographical Information Systems*, **3**, 323–334.

Speight, J. G., 1968. Parametric description of land form. In *Land Evaluation*, ed. G. A. Stewart, Macmillan, Melbourne, 239–250.

Stata Corporation, 1995. *Stata Statistical Software: Release 4.0*. Stata Corporation, College Station, Texas.

Strahler, A. N., 1952. Hypsometric (area-altitude) analysis of erosional topography. *Geological Society of America, Bulletin*, **63**, 1117–1141.

Summerfield, M. A. and Hulton, N. J., 1994. Natural controls of fluvial denudation rates in major world drainage basins. *Journal of Geophysical Research, B*, **99**, 13 871–13 883.

Young, A., 1972. *Slopes*. Oliver and Boyd, Edinburgh, 288pp.

Zevenbergen, L. W. and Thorne, C. R., 1987. Quantitative analysis of land surface topography. *Earth Surface Processes and Landforms*, **12**, 47–56.

7 The Effect of GIS Interpolation Errors on the Use of Digital Elevation Models in Geomorphology

STEVE M. WISE

Department of Geography, University of Sheffield, UK

ABSTRACT

Digital elevation models (DEMs) are being increasingly used in many areas of earth science but little work has been done on the effect of errors in the DEM on the results produced. Most DEM creation methods produce characteristic patterns of errors in the interpolated elevation values, and while the overall level of accuracy of the elevation values may be acceptable, it is known that the calculation of surface derivatives, such as slope and aspect, are sensitive to interpolation artefacts. In this study, four of the DEM creation methods commonly available to UK researchers were used to produce DEMs from a set of digitised contours. The root mean square error (RMSE) of elevation was similar in all cases (1.2–1.8 m) even though visualisation techniques indicated clear artefacts in many of the DEMs. A more sensitive error assessment method was developed by measuring slope aspect directly from the contours, and comparing this with estimates of aspect derived from the DEMs. The RMSE values ranged from 8.6° to 55.7°, suggesting that this is a more useful measure of DEM quality. Analysis of the spatial pattern of errors, and of the results of various hydrological analyses, showed that differences between the results arose because of complex interactions between the interpolation algorithms and the algorithms used in calculating surface derivatives and flow directions.

INTRODUCTION

Digital elevation models (DEM) are becoming an increasingly popular tool in many forms of environmental research including geomorphology, hydrology (Moore *et al.*, 1991; Dikau, 1993) and environmental modelling (Goodchild *et al.*, 1993). It is not hard to see why this should be, since the DEM is a computer representation of one of the fundamental objects of study in all these disciplines: the earth's surface. Given the increasing importance of DEMs, it is important that the reliability of the results produced using them is fully understood. Much of the work reported in the literature appears to assume that a DEM is an accurate representation of the terrain surface, and that any result derived from it, such as slope angle, slope aspect or

Landform Monitoring, Modelling and Analysis. Edited by S. N. Lane, K. S. Richards and J. H. Chandler.
© 1998 John Wiley & Sons Ltd.

catchment area, is essentially correct. However, this is far from being the case. Skidmore (1989), for example has shown that the different algorithms used to calculate gradient and aspect in different Geographical Information System (GIS) packages can produce quite different results from the same DEM. A similar effect was reported by Srinivasan and Engel (1991), who also showed that different algorithms could result in quite different predictions for soil erosion rates.

The commonest form of DEM is the gridded model, where elevation is estimated for each point on a regular grid. Other DEM types are also used, such as the triangulated irregular network (TIN) (Peucker et al., 1978) and contour-based models (Moore, 1988; Moore et al., 1988; Li and Stuart, 1994). Although these model structures are closely related to the form of the terrain rather then being dictated by computer architecture (Mark, 1979), their computational complexity has restricted the number of commercial implementations and hence restricted their use by the general research community. This paper is therefore concerned with gridded DEMs and, for simplicity, the term DEM will be used to refer to them.

Almost all sources of elevation data (contours, field survey, photogrammetry) produce irregularly spaced measurements from which a regular grid of heights must be estimated by interpolation (Weibel and Heller, 1991). A wide range of interpolation methods have been used, and interest has naturally focused on the accuracy of the estimated elevations (Brown and Bara, 1994; Clarke et al., 1982; Hannah, 1981; Ley, 1986; Li, 1994; Monckton, 1994; Robinson, 1994; Wood and Fisher, 1993) and, to a lesser extent, on the effect of errors on the results produced using the DEM (Carter, 1992; Lee et al., 1992). Accuracy is normally measured by comparing elevations in the DEM with "true" values of elevation on the terrain, where the true values can come from spot heights (Monckton, 1994), manual interpolation of contours (Clarke et al., 1982), independent field measurement or more accurate sources of data (e.g. larger scale maps).

The error is normally expressed as the root mean square error (RMSE):

$$\text{RMSE} = \sqrt{\frac{\sum\limits_{i=1}^{n} d_i^2}{n}} \qquad (7.1)$$

where $d_i = Z_{est} - Z_{obs}$. It is also possible to calculate the mean error (ME), which measures whether the DEM is systematically under- or over-estimating elevation:

$$\text{ME} = \frac{\sum\limits_{i=1}^{n} d_i^2}{n} \qquad (7.2)$$

However, many applications of DEMs do not use the elevation values directly, but use the DEM to estimate derivatives of the surface such as gradient, aspect and curvature (Evans, 1990), or more complex measures such as estimates of erosion potential or the topographic index of TOPMODEL (Moore et al., 1991). There are three main sources of error in the estimation of surface derivatives:

A	B	C
D	E	F
G	H	I

Figure 7.1   Labelling of cells in the 3 × 3 window used for DEM calculations

1. errors in the original elevation measurements;
2. errors in the interpolation of elevation values on the DEM grid;
3. errors due to spatial sampling effects.

It is easy to demonstrate that quite small errors in the elevation values (i.e. errors in category 1 and 2 above) can potentially lead to large errors in estimated derivatives. Most derivative estimation techniques are based on processing a 3 × 3 window of pixels around the pixel in question as shown in Figure 7.1.

The simplest method of estimating slope in the Y direction is to measure the difference in elevation between B and H, and divide by the distance between them:

$$\tan y = \frac{dz}{dy} = \frac{Z_B - Z_H}{2d} \tag{7.3}$$

where y is the slope in the Y direction Z_B is the elevation at pixel B, Z_H is the elevation at pixel H and d is the pixel spacing.

The Ordnance Survey 1:10 000 scale DEM data has a reported accuracy of ±1 m, with a horizontal grid spacing of 10 m. If each elevation value could be in error by 0.5 m, this would mean that estimates of the tangent of the slope could vary by 5% (a total error range of 1 m over a horizontal distance of 20 m). At low slope angles this translates to a variation of 5° in the estimate of slope, at higher angles this will reduce until for 45° slopes the possible error range is 3°.

An alternative to equation 7.3 is often used for calculating slope because it is less sensitive to errors in the elevation data (Horn, 1981):

$$\tan y = \frac{(Z_A + 2Z_B + Z_C) - (Z_G + 2Z_H + Z_I)}{8d} \tag{7.4}$$

with notation as in equation (7.3). This is effectively a form of weighted smoothing, and has been shown by Skidmore (1989) to give more accurate estimates of slope.

The third source of error in the derivatives arises from the fact that the DEM is a discrete sample from a continuous surface. Even if the elevations in the grid were error-free, the only gradients which could be measured exactly are in the eight cardinal directions – all others must be estimated in some way. What is normally of interest is the maximum gradient at a point and the direction of this maximum (i.e.

the terrain aspect), and two methods are commonly used to estimate these from the gridded elevations.

The first method is to calculate the gradient in the X and Y directions, using equation (7.3) or (7.4) above, and then estimate maximum gradient (G) and aspect (A) as follows:

$$\tan (G) = \sqrt{\tan^2 x + \tan^2 y} \tag{7.5}$$

$$\tan (A) = \frac{-\tan y}{\tan x} \tag{7.6}$$

The second method is to fit a surface to the nine elevation values and then calculate the derivatives directly from this (Evans, 1980; Zevenbergen and Thorne, 1987).

In practice the two methods are often equivalent, since there may be no degrees of freedom in fitting the surface to the data points, and so the parameters of the surface equation can be calculated directly from the nine elevation values (this is true of the method of Zevenbergen and Thorne, 1987). However, as Skidmore (1989) has shown, the estimated values produced by different algorithms can be markedly different. These differences will then manifest themselves in differences in the results of subsequent analyses. For example, Srinivasan and Engel (1991) have looked at the effect of differing slope estimation algorithms on the estimation of soil erosion potential, and Fisher (1993) has shown that viewsheds calculated using different algorithms can differ by as much as 50%.

The aim of this paper is to investigate the effects of some of these errors on the results of DEM analysis. Of the three sources of error, errors in the original elevation data were not considered, since researchers are often forced to use existing data sources and therefore have no control over this source of error. The focus is therefore on the effect of different interpolation methods and different algorithms on the results of the analyses. Given the wide range of elevation data sources, interpolation methods, DEM data models and types of analyses, this is potentially a very broad topic, and the scope was narrowed in the following ways.

(i)   Only gridded DEMs were considered since these are widely used and the software for using them is more widely available than for TINs or contour-based models.
(ii) Only interpolation from contours was considered. This is partly because contours are known to pose quite severe problems for interpolation, and partly because this is the most common source of elevation data for researchers who are forced to create their own DEMs.
(iii) Only commonly available software was used (the GIS packages ARC/INFO and IDRISI.) Other systems may well be able to produce results superior to these two, but will not be so widely available.

From a gridded DEM it is possible to calculate all the main surface characteristics needed for both general geomorphometric work and more specific analyses (Evans, 1972; Moore *et al.*, 1991). For the purpose of this initial study only aspect was considered, because this can be measured directly from the contour map to provide

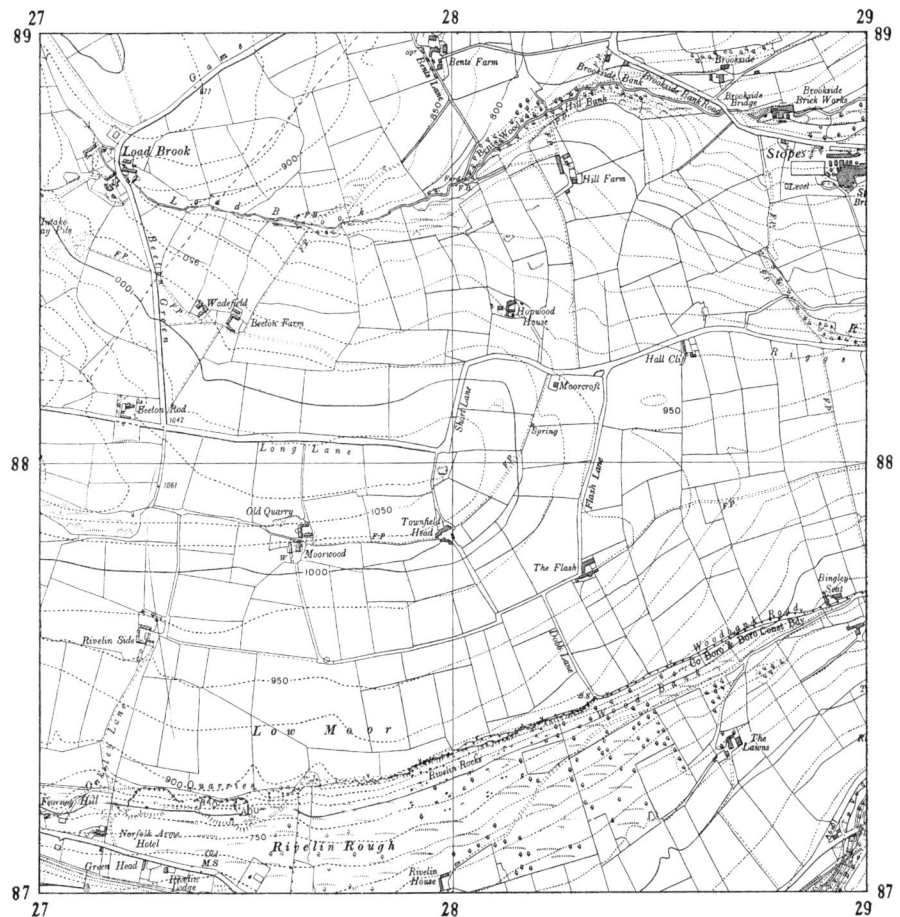

Figure 7.2 Study area: 2 × 2 km extract from Ordnance Survey 1:10 560 map. © Crown copyright. Note: all maps cover the same area as shown in this figure unless otherwise stated

a check on the accuracy of the DEM-derived values, and because a number of useful hydrological applications, such as extraction of the drainage network and identification of catchment areas, are based solely upon this.

STUDY AREA

The study location was a 2 km square area of the Peak District on the western edge of Sheffield as shown in Figure 7.2. This was chosen because a morphological map of the area already existed (Waters, 1958) and it was the intention to try and use this information to improve the production of the DEM by identifying important breaks of slope.

The area is underlain by rocks of Upper Carboniferous age, with Millstone Grit sandstones forming the ridge between two eastward draining streams, the Load Brook to the north and the Rivelin to the south. The ridge has a relatively broad crest, terminated on the south by an escarpment of resistant sandstone (marked Rivelin Rocks on Figure 7.2) within the Millstone Grit series. The slope below this has been affected by periglacial activity during the Devensian, resulting in a 0.5–1 m deposit of head material overlain by blocks which have fallen from the scarp. This material has also been subject to slumping and rotational slides, resulting in a very uneven topography. The northern slope of the ridge, without a sandstone escarpment, has shallower slopes and a smoother topography.

Producing Gridded DEMs from Contours

Contours are a common source of elevation data, but unfortunately particular problems arise when attempting to interpolate a regular grid from them. As Mark (1986) and Band (1993) point out, many authors have assumed that digitised contours can simply be treated as a set of point samples of elevation and have fed them into one of the many general purpose interpolation procedures such as distance weighted average or trend surface fitting (Burrough, 1986). Most interpolation methods base the interpolated value at a point on values from nearby data points. In the case of contours, many of these will have the same data value and so the resulting DEM will contain flat areas where the interpolated values are identical.

Contours are not simply random samples of elevation from a surface, but are a model of the surface containing far more information than just the elevation (Robinson, 1994). Any individual contour can be used to find the aspect of the surface, which will be normal to the contour at all points. The distance between contours in the direction of steepest slope can be used to estimate gradient. The pattern of contours is indicative of the shape of the surface, including such important features as ridges, valleys and passes.

In interpolating height at a point between contours, a map reader will use this extra information by interpolating along the line of steepest slope between contours, and possibly allowing for any marked slope convexity or concavity as indicated by the general contour spacing in the area. Similarly, any automated procedure should ideally try to exploit the extra information contained in contours, and specialised algorithms have been developed to tackle this problem (Weibel and Heller, 1991).

Using the two GIS packages, a DEM was produced using the following four methods.

> (i) IDRISI INTERCON. This starts with a raster grid containing any estimates of elevation which are available in the rasterised version of the contours plus any spot heights. To estimate the value at each of the other pixels, the program produces a series of profiles running in each of the four cardinal orientations (N–S, E–W, NE–SW, NW–SE) including one along each of the edges of the area, for which purpose the heights at the corners of the area must be supplied. At each cell of unknown height, the height is interpolated from each profile plus the slope of the profile at that point. The height assigned to the cell is that interpolated by the profile with the greatest slope (Eastman, 1992).

(ii)   ARC/INFO–TOPOGRID. This is based on the work of Hutchinson (1989) and
       has several powerful features: (a) elevation is interpolated by identifying the lines
       of steepest descent between contours and then interpolating along these lines,
       rather than along the fixed orientations defined within INTERCON; (b)
       information on drainage lines can be supplied and is used to ensure that the
       DEM does not contain pits (i.e. small areas which are entirely surrounded by
       higher ground and which are often produced by interpolation and rounding
       errors); and (c) the algorithm automatically identifies points of maximum
       curvature along the contours, usually ridge or drainage lines, which are used to
       ensure that the DEM is hydrologically correct.

(iii) ARC/INFO–TIN from contours. One way to produce a gridded DEM is first to
       produce a TIN and then derive a grid, since elevation can be estimated at any
       point across the TIN. Until recently this represented the only way to produce a
       gridded DEM from contour data in ARC/INFO, but despite the recent
       introduction of TOPOGRID the method is still of interest. The crudest approach
       is simply to treat the digitised contour as a set of point samples of height which
       will then be triangulated. This approach suffers the same sort of problems as the
       point interpolation approach, since no allowance has been made for the fact that
       the points are not independent samples of elevation. A common problem is that
       three adjacent points on the same contour will, following triangulation, generate
       a triangle with zero slope and undefined aspect (Robinson, 1994). This can also
       happen with points on either side of a ridge, producing a flat ledge on the ridge.
       More sophisticated software has been developed which can cope with these
       problems (Christensen, 1987), but this is not so widely available as ARC/INFO.
       Within ARC/INFO the problems must be resolved by adding extra data points,
       or by thinning the number of points representing the contours.

(iv)   ARC/INFO–TIN from feature-specific points. The real advantage of the TIN
       model is that it is designed to represent the main features of the surface (Peucker
       et al., 1978; Mark, 1979). The triangulation should be designed so that all the
       main ridge and drainage lines are represented as triangle edges, and all the peaks
       and passes are vertices. The normal way to attempt to achieve this is to input all
       the important breaklines as well as the contours, and force them to be
       incorporated into the triangulation. This does not totally solve the problem, since
       it will still result in many small triangles; these are due to the way in which the
       contours have been digitised rather than representing any significant portion of
       the terrain surface.

       Therefore an alternative approach was tried here, which has been suggested
       by a number of authors (Mark, 1986; Heil and Brych, 1978; Eklundh and
       Mårtensson, 1995). In this method, the contour map is sampled along the key
       structural elements of the landscape to produce a series of points which are then
       triangulated. This is a more time-consuming approach, but it does allow the user
       a greater degree of control over the appearance of the DEM, and theoretically
       should produce a DEM which is a good representation of the terrain surface.

Method

The contours from the 1:10 560 map were digitised using ARCEDIT. The heights
at the corners of the map were estimated by eye, and a separate coverage was
produced containing simply these four points (this was required for INTERCON
and also ensured that the TIN methods did not cut off the corners, leading to grid
points with unknown height).

For simplicity, the four DEMs will be referred to by the shortened names shown
in Table 7.1.

Table 7.1 Name used to refer to DEMs produced by each interpolation method. A suffix of D (e.g. ICD) indicates a depressionless DEM, i.e. after smoothing and the removal of pits

Method of creation	Name
INTERCON	IC
TOPOGRID	TG
TIN from contours	TC
TIN from feature-specific points	TF

IC was produced by exporting the digitised contours from ARC/INFO to IDRISI where they were input to the INTERCON command; the corner heights are also supplied to this command by the user. The IC DEM was then transferred back to ARC/INFO where all subsequent analysis was done.

TG was produced using the TOPOGRID command in ARC/INFO which read the digitised contours and corner points.

To produce TC, the contours and corner heights were first input to the ARC/INFO CREATETIN command to produce a TIN (Figure 7.3a). A grid was then interpolated from this TIN using the TINLATTICE command with the quintic interpolation option. This fits a polynomial through the corners of the triangle in which any given grid point lies in such a way that interpolated values do not change sharply along triangle edges (ESRI, 1995).

To create TF, the contours were not used in full but were sampled along the main ridge and drainage lines which were first identified by eye from the map (Figure 7.2) and then digitised. The contours and structure lines were then overlaid using the ARC/INFO IDENTITY command which calculates the intersections between the two sets of lines. The resulting line coverage was generalised using the Douglas–Peucker algorithm (Douglas and Peucker, 1973) so that all detail along the contours was removed, resulting in a coverage where the only points are the intersections between the structural lines and the contours. This set of points was then used to construct a TIN (Figure 7.3b). In comparison with the TIN from the contours, the feature-specific TIN has fewer, larger triangles and fewer flat triangles (shown shaded in both diagrams). The remaining flat triangles could only be resolved by adding additional height information, but this was not done because the intention was to compare the results of interpolating from the same information by a number of different methods. Finally, TF was created using TINLATTICE in the same way as was done for TC.

All grids were created with 10 m pixels with height estimated at the centre of the pixels, producing a 200 × 200 DEM.

The normally accepted method of reducing interpolation artefacts is to smooth the DEM by running a low pass filter over it, replacing each pixel with the average of itself and its eight neighbours. More sophisticated techniques have been developed (e.g. Hannah, 1981) which attempt to identify the parts of the DEM that

(a)   (b)

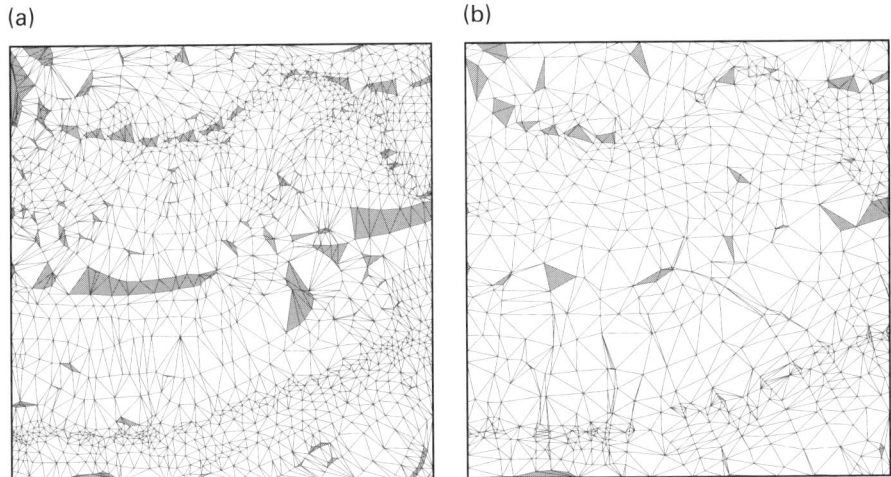

Figure 7.3 Triangulated irregular networks. Flat triangles are shaded. (a) TIN produced from digitised contours. (b) TIN produced from feature-specific points along contours

are most affected and make appropriate corrections based on the nature of the problem, but these are not available in the packages used here.

It is also well known that gridded DEMs contain pits or sinks: pixels which are lower than all their neighbours and which will then affect any subsequent hydrological analysis. All the smoothed DEMs were also checked for these. The worst case was TF, with eight sinks amounting to 16 pixels (out of a total of 40 000).

Any modification of the DEM will obviously alter the elevation values. Brown and Bara (1994) found that smoothing a photogrammetrically produced DEM with a 3×3 filter removed the banding artefacts but reduced the mean elevation values by 2.9 m. Larger filter windows had an even more marked effect. However, smoothing is likely to remove some of the more obvious errors in the calculation of slope, aspect and curvature and so all the DEMs produced in this study were smoothed twice and had all sinks removed (using the ARC/INFO FILL command) to see what effect this would have on the accuracy of results produced using them. This produced four further DEMs (ICD, TGD, TCD and TFD).

Accuracy of DEM Elevations

The 1:10 000 map of the study area contains 32 spot heights, with their elevations determined by a mixture of ground and air survey methods. These points were distributed across the western end of the ridge and along the road running along the ridge top. The height at each of the spot height locations was extracted from each DEM and these values were used to assess the accuracy of the elevation values (Table 7.2).

Table 7.2 Elevation accuracy statistics. All were based on 32 spot heights from the Ordnance Survey 1:10 000 map of the study area

	DEM error (m)	Minimum error (m)	Maximum (m)	Mean error (m)	RMSE
Original	IC	−2.1	2.0	0.00	1.7
	TG	−2.0	2.1	0.28	1.8
	TC	−1.7	2.5	0.16	1.6
	TF	−1.8	2.4	0.06	1.3
Smoothed	ICD	−1.8	2.3	0.03	1.5
	TGD	−1.8	2.1	0.22	1.7
	TCD	−1.8	2.6	0.13	1.6
	TFD	−2.0	2.4	0.00	1.2

The RMSE values are very similar, and are all within the normally acceptable limits for DEM accuracy. The results from the DEMs were compared by performing a series of paired sample t-tests on the error values (the d_i values in the notation of equation (7.1)). IC and TG were found to have errors which differed at the 10% significance level (t = −1.82, p = 0.078). All other comparisons showed no significant difference in the error values.

The smoothing process has improved the RMSE of all DEMs except TC, but paired t-tests on the error values showed that these differences are not statistically significant. This is in contrast to the results of Brown and Bara (1994) who reported an increase in RMSE after smoothing.

The main conclusion would seem to be that in terms of elevation accuracy, all DEMs are very similar. However, these figures are only based on 32 data points, a 0.08% sample, so it is perhaps not surprising that the test is not very discriminatory. Graphical plots reveal that there are quite substantial differences between the DEMs.

Figure 7.4 shows contour plots produced from three of the DEMs. The poor quality of the INTERCON DEM (IC) is clear from the many spikes on the contour lines. Smoothing removes these and produces a DEM which gives similar contours to those generated from TG.

Figure 7.5 shows the elevation histograms for the DEMs, both original and smoothed. The marked spikes in the cases of IC and TG occur at the heights of the original contours, and show that these methods are producing too many pixels with values which are the same as the original contours. The two TIN methods produce generally smoother histograms, but again there are spikes, in this case resulting from flat triangles.

The lower set of histograms is for the smoothed DEMs. The smoothing technique appears to be quite effective at dealing with the oversampling of contour heights (IC and TG) but it is not so effective at dealing with the flat triangles of the TIN, since these will cover a much larger area and only pixels along the edge of the flat area will be changed by the filtering process. One solution to this problem is to remove the flat triangles by providing extra height information. An alternative would be to develop a specific post-processing algorithm which could identify flat areas and

Figure 7.4 Contour maps produced from select DEMs: (a) INTERCON; (b) INTERCON after smoothing; (c) TOPOGRID

interpolate new elevation values based on the nature of the surface in the surrounding areas.

Even more revealing are maps based on derivatives of the surface, rather than on elevation alone. Figure 7.6 shows maps of the aspect values produced by the same two DEMs, IC and TG. The aspect values were calculated using the ASPECT function in the ARC/INFO GRID module which uses equations (7.4) and (7.6). The DEM produced by INTERCON clearly shows the effect of the profiles used in its construction with systematic artefacts oriented in the cardinal directions. These artefacts are not produced by any of the other methods, as illustrated by the example of TG (Figure 7.6b). In the case of TG, however, there is the suggestion of some small artefacts around the location of some of the points digitised along the contours. These are particularly noticeable on the southern flank of the main ridge.

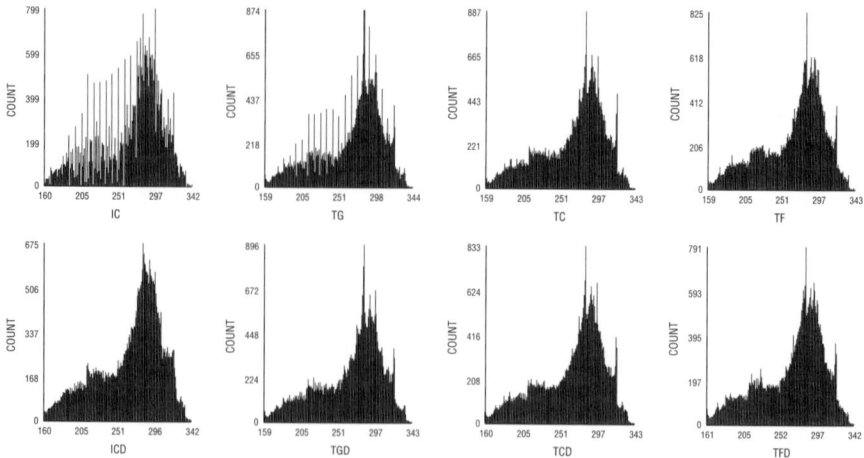

Figure 7.5 Histogram of elevation values for all DEMs. See Table 7.1 for meaning of DEM identifiers. Upper row shows results for original DEMs, lower row shows results after smoothing and removal of pits

Comparison of Figures 7.6a and 7.6b reveals a difference in the success of the INTERCON and TOPOGRID algorithms in picking out the main watersheds. The watershed along the main east–west ridge is well distinguished in the case of TG, with a well defined change from north- to south-facing slopes. Compare this with IC where, especially at the eastern end a very strange pattern of slope aspects is predicted. Similarly, the small portion of the Rivelin stream in the southeast corner is picked up by TG, whereas IC shows parts of the northern bank of the stream draining away from the stream!

Figures 7.6c and 7.6d show detailed views of the same portion of Figures 7.6a and 7.6b respectively, near the western end of the main ridge. There is a small drainage channel which runs eastwards from the hill top and then turns northeast (the line of this channel has been inferred from the contour crenulations). The aspect values should differ for slopes on the left and right hand banks of this channel, as shown for both DEMs in the lower part of the slope, and for TG in the upper part. On the upper slopes, however, INTERCON produces an aspect pattern which ignores the existence of the channel, and again will predict water from the left hand side flowing away from the channel in a northeasterly direction. The difference arises because INTERCON does not take account of the slope direction information contained in the contours, so that it bases its estimate of the steepest slope on the shortest distance between the contours; in the case of the area discussed, this shortest distance is in a northeasterly direction, although the contours clearly show a southeasterly facing slope just north of the channel.

These diagrams reveal that there are important differences between the DEMs which are not revealed by the standard measurement of DEM quality, the RMSE of elevation. What is needed is a method of assessing the accuracy of the surface derivatives, since it has been seen that these are sensitive to errors in the DEM, and

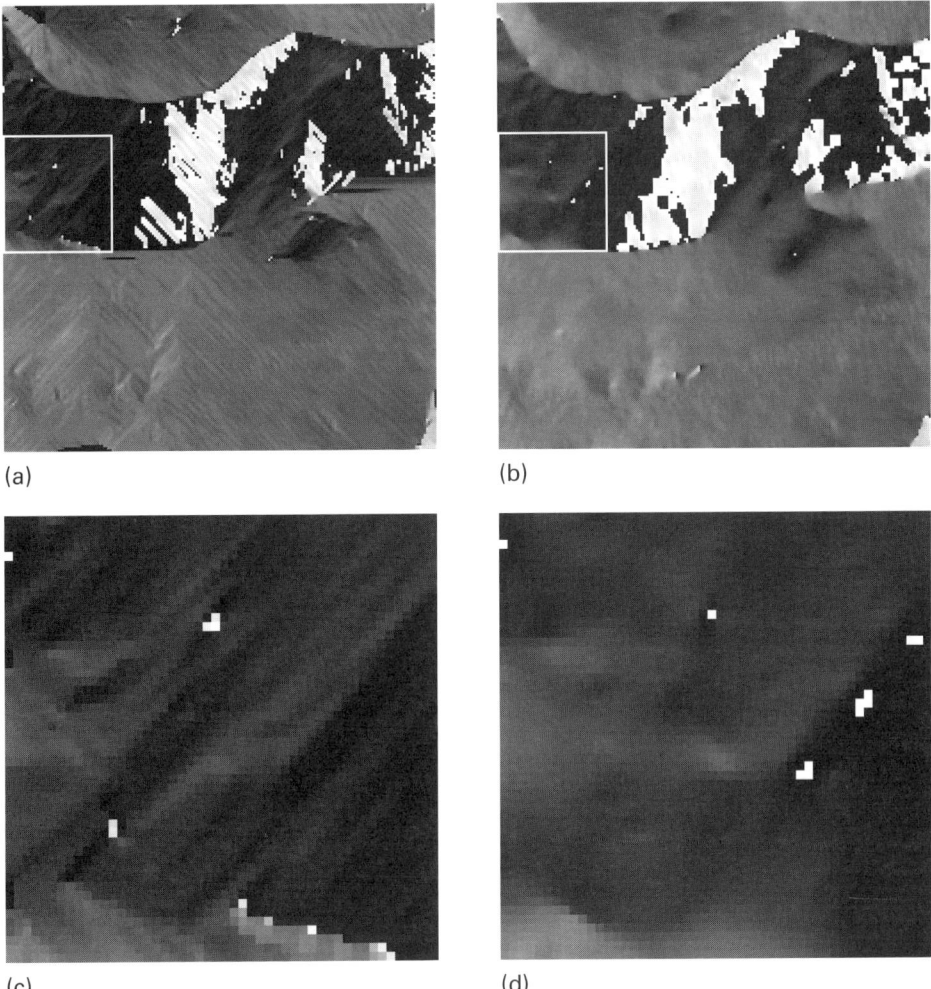

(a)

(b)

(c)

(d)

Figure 7.6 Maps of aspect values for selected DEMs. In all four maps, the grey-scale runs from white at 0° clockwise to black at 359°. (a) INTERCON; (b) TOPOGRID; (c) and (d) show detail of INTERCON and TOPOGRID respectively, together with contours and inferred line of drainage on hillslope

this is covered next with the development of a method for measuring the RMSE of aspect.

Accuracy of Aspect Values

Measures of accuracy always raise the vexed question of what the true value is, since no observation or measurement can be made without error. In the case of derivatives of natural slopes, there is the added complication that these derivatives cannot be defined independently of scale. The terrain surface displays many of the characteristics of a fractal surface and, although not a true fractal in the mathematical or statistical sense, the terrain surface does possess the property that it is rough at all scales, from the whole landscape down to almost arbitrarily small scales. Therefore at no point does the surface become smooth enough for mathematical derivatives to be unambiguously defined.

This means that measurements made from larger-scale maps or from fieldwork are not directly comparable with values derived from a 1:10 560 map as in this case. The most accurate source of information on the surface derivatives is therefore the contour map itself since, as described above, contours represent not only elevation but indirectly the gradient, aspect and curvature of the terrain.

It would not normally be acceptable to use the same data source for both producing a set of measurements and testing their accuracy, but the DEM interpolation is based upon the elevation of contours, whereas the accuracy test will be based on their aspect. The contours probably do not reflect the true slope aspect since they will have been generalised in the production of the map. Really what is being tested is how well the DEMs manage to represent the aspect information held on the contour map, and not how well they represent the true, but unknown, aspect.

Calculating Aspect from Contours

Skidmore (1989) estimated aspect from contours manually in his study, but it is also relatively easy to do the calculations from the digitised contours themselves. Given two points along a digitised contour, the perpendicular to the line connecting them will represent a close approximation to the slope aspect at a point midway between them. This assumes that the points have been digitised to form a reasonably accurate representation of the path of the contour, which is normally the case (it is this density of digitising which leads to some of the problems when digitised contours are used for interpolation). The perpendicular to the line could define two possible slope aspects, at 180° to each other. To determine which is the true one it is necessary to know which direction is upslope. This was done by ensuring that contours were always digitised with high ground to the right. This is relatively easy to do in the ARCEDIT module of ARC/INFO which can show the direction of a line by means of an arrow and which provides a FLIP command to reverse this direction.

A FORTRAN program was written which calculated the aspect for each contour segment and the position of the centre of the segment. These data were read back into ARC/INFO as a point coverage and rasterised to produce a grid of correct aspect values. The point coverage was also used to select the equivalent pixels from the existing grids of aspect derived from each DEM.

In a few cases, because of the close spacing of the points along the digitised contours, more than one sample point fell in the same raster pixel, giving two or more calculations of aspect to compare with the estimate from the grid itself. The solution adopted was to filter out these multiple samples by modifying the

FORTRAN program so that it calculated aspect only from those contour segments which were at least 20 m long. As well as making it unlikely that two points fell in the same 10 m pixel, this also ensured that aspect was being measured over a similar distance from the contours and from the DEMs.

The digitised contours had 1728 line segments, 301 of which were less than 20 m long, leaving a sample of 1427 points at which a measurement of aspect was available. This represents a 3.6% sample of all the pixels in the DEMs, spatially distributed throughout the area. As a further check on the representativeness of the sample, aspect was calculated using the TG DEM and then the pixels representing the sample locations extracted. The frequency distributions of the aspect values for all 40 000 pixels of the grid and for the 1427 sample pixels were compared using the Kolmogorov-Smirnov goodness-of-fit test. The D statistic was 0.03 (critical value at $p = 0.01$ is 0.04) which is evidence that the sample of 1427 values is representative from the full population of 40 000 values.

Calculating Aspect from the DEMs

There are a number of methods for calculating aspect from a gridded DEM which can produce quite different results (Skidmore, 1989). Here, the two which are available in ARC/INFO were used, to see how the different algorithms would interact with the different types of interpolation artefact in the DEMs:

(i) ASPECT: an ARC/INFO GRID function which uses the method described in equations (4) and (6).
(ii) CURVATURE: also an ARC/INFO GRID function which uses the method of Zevenbergen and Thorne (1987) to fit a partial quartic equation to the nine data points in the 3×3 window, from which slope, aspect and curvature can all be calculated. The equation is designed to go through all nine points exactly, which means that the various slope derivatives can all be calculated directly from the nine elevation values. In the case of aspect, the only difference from the previous method is that slope in the X and Y directions is calculated according to equation (3), i.e. solely from the pixels which are orthogonal neighbours. This means that the method is potentially more sensitive to errors in the elevation values.

Figure 7.7 shows histograms of the aspect values calculated from the contours (labelled C) and from the original DEMs using both the ASPECT and CURVATURE algorithms. The graph for C shows a bimodal distribution, with peaks around 180° and 0/360° representing the main south- and north-facing slopes of the ridge. Two interesting features are the spikiness of the histogram, which may relate to the irregularity of the terrain, and a marked peak at 180°, for which there is no obvious explanation. The histograms for IC and TG show spikes at 45°, 135° and 180° as in C, but with far higher frequencies. When pixels with these values are plotted from the full DEM it is clear that many of those in IC relate to the artefacts shown in Figure 7.6, but those in TG show no obvious spatial pattern.

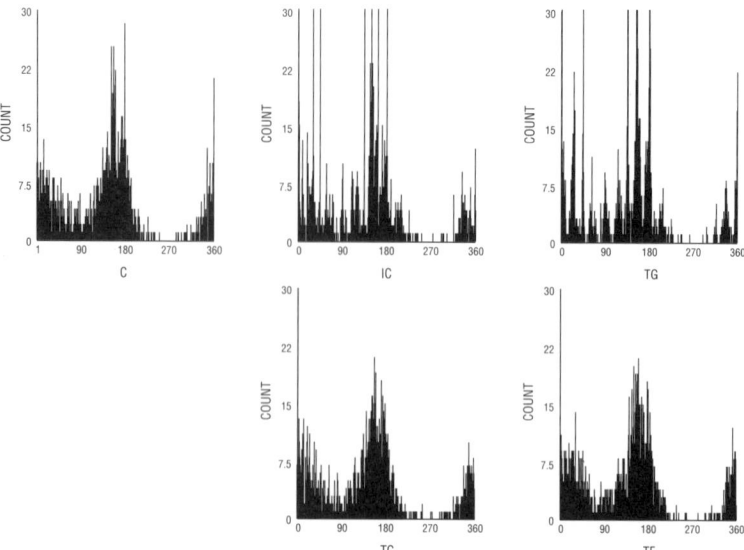

Figure 7.7 Histograms of aspect at 1427 sample points. C show values measured directly from the contours. The other histograms are for values at the same sample locations from each of the four DEMs, calculated using the ARC/INFO ASPECT command

Both IC and TG have a peak at 31–32°, which is not present in C. These are the values which would be produced if pixels B and C in the 3 × 3 window (Figure 7.1) were lower than the other six. Again, pixels with these values show clear spatial patterns in IC but not in TG.

The distributions from the TIN-based DEMs (TF and TC) are the closest to C, with no tendency to exaggerate any of the modal values.

Figure 7.8 shows histograms produced from the same original DEM (IC) and illustrates the effects of smoothing and the difference between the two aspect algorithms. The benefits of smoothing are marked in this case, with a large reduction in the size of the spikes. It is also clear that the CURVATURE algorithm does not perform as well as that used by ASPECT. The difference between the two algorithms is reduced by the smoothing, but even here the CURVATURE algorithm produces a large mode at 270°, despite the fact that there are very few west-facing slopes in the study area.

It is also possible to compare the aspect calculations with those from the contours point-by-point to see the degree of discrepancy, and to look at the spatial pattern of errors. One complication is the fact that the number system is circular, i.e. 359° and 1° are in fact only 2° apart. Therefore the difference was calculated as follows:

(i)   calculate difference as True − Estimated; and then
(ii) if this is greater than 180 (or less than −180) then calculate 360 − difference, and reverse the sign.

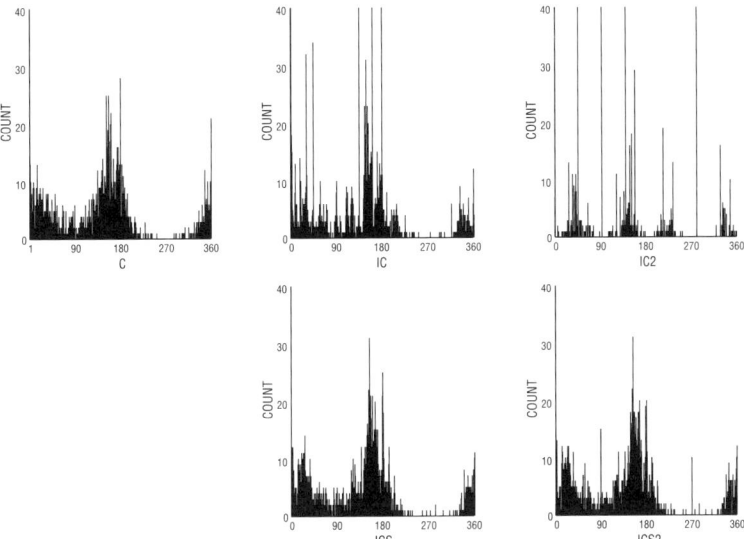

Figure 7.8 Histogram of aspect values from the same DEM (INTERCON) illustrating the effects of smoothing and different aspect algorithms. Top left histogram shows data derived from contours. In group of four, top row is for original DEM, lower row for smoothed, left hand column is using the ASPECT algorithm, right hand column the CURVATURE algorithm

This will produce a signed difference indicating whether the estimated value is clockwise (positive difference) or anticlockwise (negative difference) with relation to the true value. It seems unlikely that there would be a systematic bias to the errors in this respect and indeed it was found that all but one of the mean differences in aspect were very close to zero.

The error statistics are summarised in Table 7.3, from which a number of conclusions can be drawn

   (i)   The extremely poor performance of the combination of the INTERCON DEM and the CURVATURE algorithm (Zevenbergen and Thorne, 1987) is confirmed with a 56° RMSE. What makes this particularly disturbing is that IDRISI uses this exact algorithm to calculate aspect and therefore it is imperative that anyone using this package examines their results with great care.

  (ii)   The CURVATURE algorithm consistently produces poorer results than the ASPECT algorithm, with higher RMSE values from equivalent DEMs.

 (iii)   Smoothing is again beneficial in terms of the RMSE figures expect for TOPOGRID with the ASPECT algorithm. In the case of the INTERCON/ CURVATURE combination, this simple smoothing reduces the RMSE figure from 56° to 15°, which is actually better than the result for TF.

 (iv)   Overall, the TOPOGRID DEM performs best with the lowest RMSE figures.

  (v)   The simple TIN-based method (TC) actually does better than that based on feature-specific points (TF). The main improvement in the case of the TF TIN was the removal of many of the flat triangles (Figure 7.3) but maps of the aspect

Table 7.3 Aspect accuracy statistics. All were based on a sample of 1427 points for which aspect was measured directly from the digitised contours. The significance of the r values between slope and the error is indicated as follows: * 5% level, ** 1% level, *** 0.5% level. Critical values for r when $n = 1427$ were estimated using the Fisher z-transformation as described in Neave (1979, p. 35)

Interpolation method	Aspect calculation algorithm	DEM	Min. error (degrees)	Max, error (degrees)	RMSE (degrees)	Error/slope correlation (r values)
INTERCON	ASPECT	IC	−165	127	12.4	−0.07***
		ICD	−131	59	10.8	−0.09***
	CURVATURE	IC	−177	150	55.7	0.09
		ICD	−117	97	15.4	−0.06*
TOPOGRID	ASPECT	TG	−56	31	8.6	−0.06*
		TGD	−72	45	9.4	−0.11***
	CURVATURE	TG	−50	93	11.3	0.07
		TGD	−71	45	9.3	−0.11***
TIN – contours	ASPECT	TC	−170	92	11.5	−0.07***
		TCD	−163	52	11.4	−0.09***
	CURVATURE	TC	−174	161	13.1	−0.08***
		TCD	−164	56	11.4	−0.09***
TIN – FS points	ASPECT	TF	−180	122	19.1	−0.14***
		TFD	−158	122	17.4	−0.14***
	CURVATURE	TF	−170	118	20.7	−0.15***
		TFD	−158	120	17.4	−0.15***

errors showed that these were not in fact related to the flat triangles. (Aspect on the flat surfaces is undefined.) The poorer performance is therefore presumably related to the fact that the TF TIN is a more simplified model of the terrain, with fewer triangles representing the slopes, although further work will be needed to explore this properly.

(vi) All methods produce maximum and minimum errors which are very large, indicating that in some circumstances any of the methods can produce aspect values which are almost 180° from what they should be.

It might be expected that elevation errors would have a greater effect on aspect calculations in areas of low gradient. Examination of the areas around some of the extreme error values revealed cases where elevation differences of less than 0.2 m caused aspect errors of 60°. However, it was also clear that there were cases where the errors were being caused by systematic errors such as those illustrated in Figure 7.6c, and these could occur on quite steep slopes. As a check on the importance of gradient, the maximum slope at each sample point was calculated from the TG DEM (which appears to be the most accurate on the evidence presented here), and correlated with the aspect error (ignoring the sign). The correlation coefficients are shown in Table 7.3, together with their level of significance in a one-tailed test. Almost all the correlation coefficients are negative, as expected, and all but two are statistically less than zero. However, even the largest (for TF) only implies that gradient is explaining 2% of the variation in the aspect errors, so although gradient may be a factor, it is not a major one.

Analysis of Flow Direction

It was decided to see whether the differences found in the accuracy of the aspect calculations would manifest themselves in further analysis based on aspect. Many hydrological applications of DEMs are based upon identifying the direction water will take over the surface, which essentially depends upon the direction of the slope (Band, 1993).

Once the direction of flow from each pixel has been estimated, it is possible to identify which pixels should contribute water to any given point on the DEM and hence the catchment area for that point can be derived. The total number of pixels contributing water to a point can also be calculated. Those pixels which receive flow from more than a certain number of pixels can then be categorised as channels (Moore *et al.*, 1991; Jenson and Domingue, 1988).

The ARC/INFO FLOWDIRECTION function uses the algorithm of Jenson and Domingue (1988) to calculate the flow direction upon which other analyses are based. This uses the direction of steepest descent from the central pixel, i.e. the single lowest pixel from the eight surrounding ones. This is actually one of the aspect algorithms tested by Skidmore (1989), but he found that it performed very poorly compared with other methods. It is also possible to estimate the flow directions by reclassifying the aspect values, which would then make it possible to compare the results from the DEMs with those derived from the contours. Unfortunately, flow directions derived in this way cannot then be used in the channel identification algorithm because they can generate flow paths which loop back on themselves and so cause the flow accumulation algorithm to hang.

Flow directions were calculated for all eight DEMs, using three methods: (i) reclassification of aspect values derived from ASPECT; (ii) reclassification of aspect values derived from CURVATURE; and (iii) using the FLOWDIRECTION function in ARC/INFO GRID.

The "true" flow directions were calculated by reclassifying the aspect values derived from the contours, although, as explained above, this is not directly comparable with the values derived using the FLOWDIRECTION algorithm.

The assignments to the flow direction categories were then compared with those found by reclassifying the true aspect values by calculating a χ^2 statistic, with the results shown in Table 7.4. These results largely mirror the results of the aspect comparisons, with smoothing having a beneficial effect in all but two cases and TOPOGRID being the most reliable interpolation method overall.

In many ways, aspatial statistics like this can be potentially misleading because the derivation of drainage channels or catchment areas will be very sensitive to the location of errors in assigning flow directions. Figure 7.9a shows the direction of flow assigned in a portion of the IC DEM near the stream which flows across the northern half of the area. As Figure 7.9b shows, changing the flow direction of two of the pixels would have a dramatic effect on the area which is predicted to contribute water to the outflow pixel, which in turn would affect the location of those pixels which receive enough water to be defined as channels.

Figure 7.10 shows the result of deriving the channel network by calculating the accumulated flow to each pixel, and then selecting all those pixels with a value

Table 7.4 Flow direction accuracy. Each figure is the χ^2 statistic calculated by comparing the number of pixels in each flow direction category with the number assigned by reclassifying the contour-derived aspect values. Figures less than 12 indicate no significant difference between the true and DEM-derived values at the 1% level

Interpolation method	DEM	ASPECT	CURVATURE	FLOWDIRECTION
INTERCON	IC	46.1	14535.4	249.0
	ICD	12.0	38.6	11.2
TOPOGRID	TG	40.4	22.9	51.6
	TGD	5.1	4.4	8.4
TIN – contours	TC	6.0	12.1	91.1
	TCD	1.7	2.1	9.2
TIN – FS points	TF	10.4	8.9	51.1
	TFD	12.4	13.4	19.7

greater than 100 (the grids were also trimmed around the edges to remove obvious edge effects such as channels flowing along the edge of the grid). The general pattern of drainage channels predicted is actually quite similar, although there are clearly differences of detail. All the DEMs predict a tributary channel joining the Load Brook near the northern edge of the area, but the predicted length of this channel is different in every case. Similarly, the pattern of drainage on the southern flank of the main ridge is broadly similar but differs in detail. For example, only two of the DEMs (IC and TC) succeed in identifying the Rivelin River which just appears in the southeastern corner of the area. In the case of TG, the flow which is derived from the large hillslope immediately to the northwest of this river is fragmented into several parallel channels, and in the case of TF it is diverted where a single channel suddenly turns left through 45° just before reaching what would be the main river valley. What is particularly interesting about these cases is that the TG DEM is the one which has consistently produced the most accurate predictions of aspect and flow direction, and the TF DEM is one in which the river channel would have been explicitly supplied to the interpolation process as a point at which a change of gradient occurred.

The threshold which is used to define the channels will clearly determine just how detailed a drainage network is produced. In Figure 7.11, a value of 1000 has been used which means that only the major drainage channels have been identified. Several significant differences remain, and one in particular is evidence of another source of problems.

All four DEMs predict a small channel near the western end of the south-facing ridge slope. In the case of TG and TC, there is another similar channel just to the east, but this is missing from IC and TF. In both cases, this is because the flow direction algorithm routes all the flow from a pixel in just one direction which leads to the prediction of a large number of sub-parallel flowlines, as is clear from Figure 7.10. In order for a hillslope channel to be predicted, some of these parallel flowlines must converge. Whether or not this happens may depend on quite small variations in predicted flow directions on the slope. If a DEM contains parallel

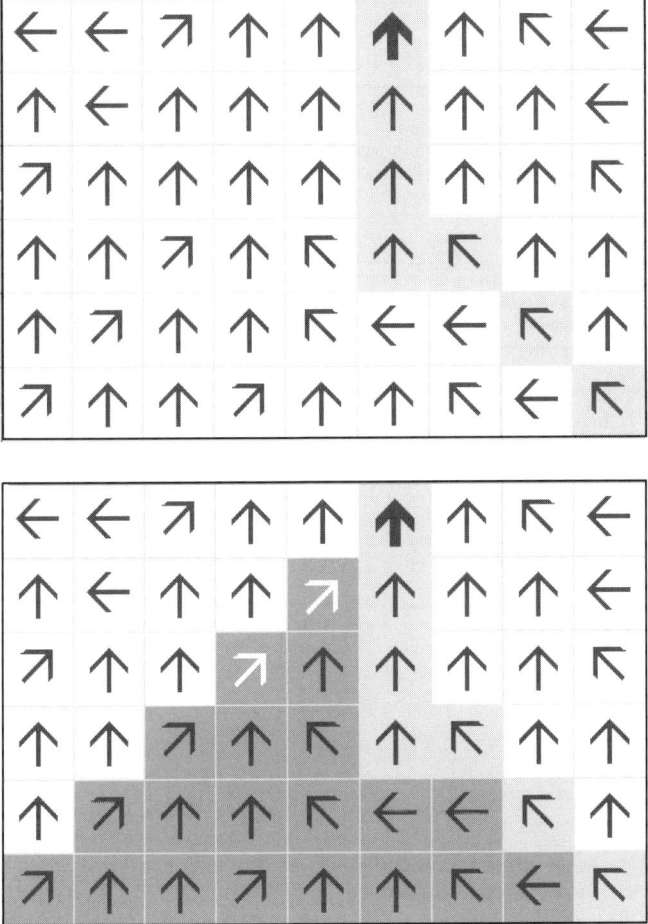

Figure 7.9 Effect of flow direction errors on the prediction of hillslope hydrology. (a) Flow directions for part of the IC DEM, and (shaded) those pixels which contribute flow to the "outlet" (identified with a bold arrow). (b) The effect of changes of 45° in the flow direction of two pixels (identified with white arrows) leading to a drainage area predicted at 27 pixels rather than six

artefacts, as is the case with IC, this is likely to exacerbate the problem and this is in fact what has happened in the case of IC. It might be expected that the TIN-based methods would not suffer from this problem, but it is also the reason for the missing channel in TF. On the upper slopes, the flow pattern which is predicted funnels flow towards the line of this drainage channel, but the two sets of flows each meet a different set of south-facing slope pixels, and therefore flow in parallel to the slope foot, rather than coalescing into a single channel.

This problem with the single flow direction algorithm is well known, and more sophisticated algorithms have been developed which route flow into all the down-slope pixels (Moore *et al.*, 1991; Quinn *et al.*, 1991). However, these are not yet

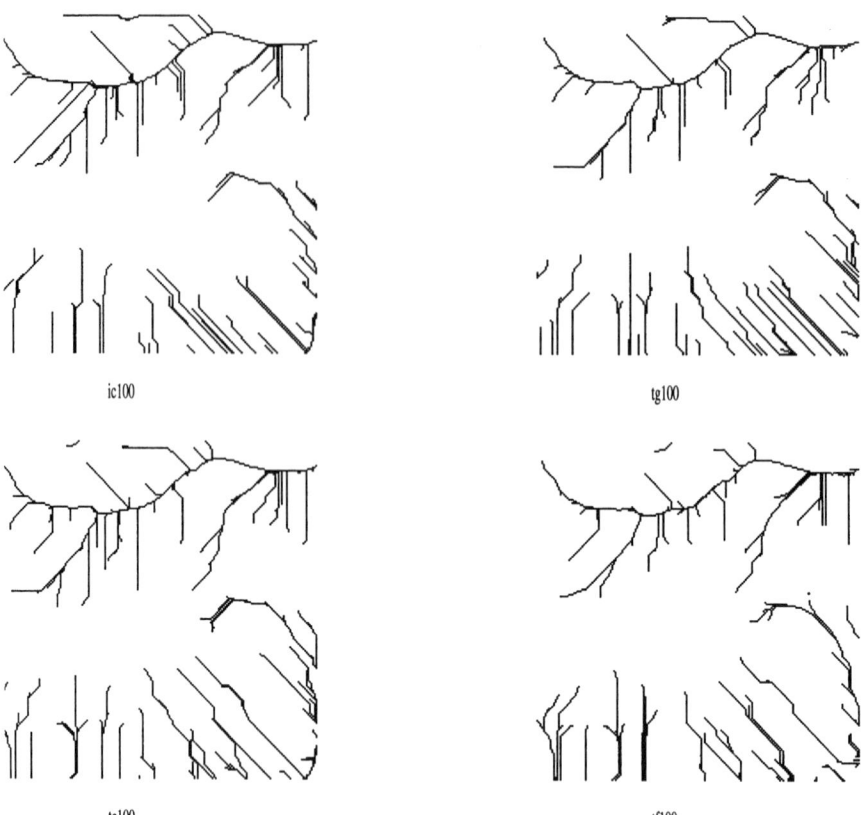

ic100

tg100

tc100

tf100

Figure 7.10 Drainage channels predicted on the basis of selecting pixels which receive flow from more than 100 pixels. The tendency for the flow routing algorithm to produce sub-parallel channels on hillslopes is particularly clear in the southeastern corner of the map

available in standard packages, although it is possible to program ARC/INFO to use this algorithm using its macro language (C. Stocks, personal communication).

CONCLUSIONS

This study has only begun to explore what is a potentially important topic, namely the reliability of results produced using digital elevation models in common GIS packages. However, from this initial study, based on one small DEM, a few tentative conclusions may be drawn.

It is clear that the RMSE of elevation values is an insensitive measure of DEM quality. All the DEMs produced in this study had very similar RMSE values, and all were within the range normally considered acceptable. However, these relatively small errors in elevation hid systematic variations, which became obvious as soon as

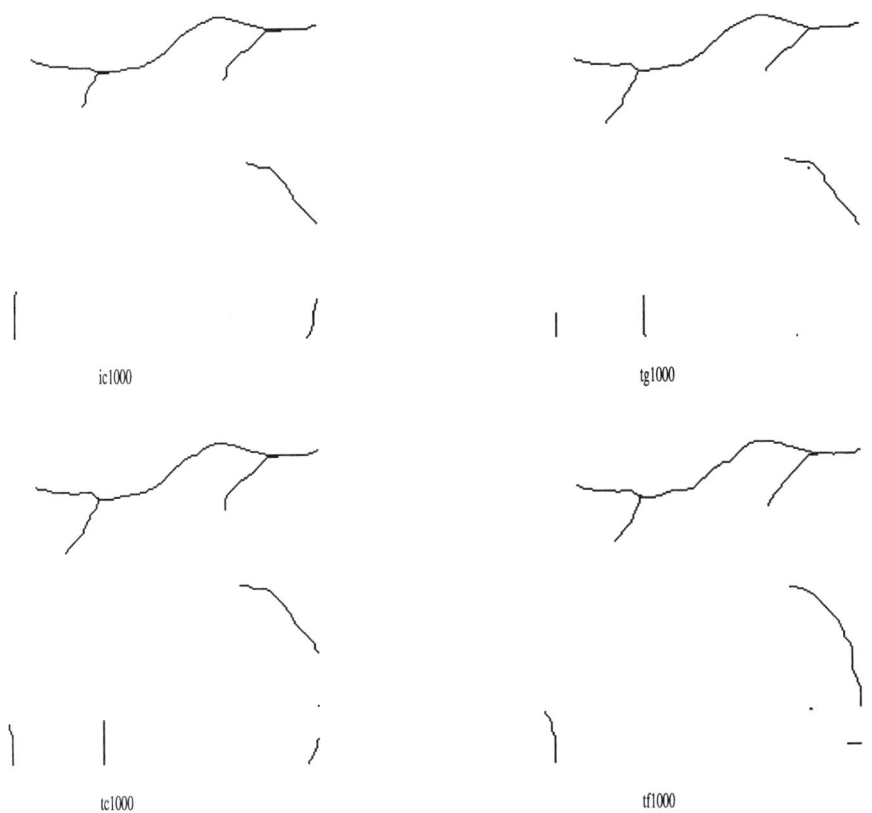

Figure 7.11 Drainage channels predicted on the basis of selecting pixels which receive flow from more than 1000 pixels

surface derivatives were calculated and displayed. By measuring the true value of aspect from the original contours, it was possible to calculate the RMSE of aspect and this proved to be a more sensitive indicator of the presence of artefacts in the DEM. A very similar exercise could be carried out for the calculation of slope angles. The programming necessary to do this from the contours would be more complex because the calculations depend on being able to find the intersection between the perpendicular to the contour and its upslope and downslope neighbours. However, the calculation of aspect is relatively easy, and it is suggested that further work be done to explore the use of this as a measure of DEM quality.

It is difficult to assess the effect of these artefacts on the results of the drainage extraction based on the evidence presented in this paper. Certainly at the local scale, when basing predictions of channels on relatively small supporting areas, large differences appeared in the results produced from the different DEMs. This suggests that applications which require an estimate of the area draining to each pixel on a DEM could be extremely sensitive to the nature and quality of the DEM. Conversely, when the main drainage channels were extracted, the results from the different DEMs were comparable, suggesting that at some scale the effect of interpolation errors may be reduced. Further work will be needed to test this suggestion and, if it seems to be a general result, to define the scale at which the impact of DEM quality reduces in significance.

ACKOWLEDGEMENTS

I would like to thank Dr I. S. Evans and one anonymous referee whose comments on the original version of this paper were most helpful. Figure 7.2 is based upon the Provisional edition of sheet SK28NE of the Ordnance Survey 1:10 560 maps, and all the other maps are based upon contours digitised from this map. All are published with the permission of the Controller of Her Majesty's Stationery Office and remain crown copyright.

REFERENCES

Band, L. E., 1993. Extraction of channel networks and topographic parameters from digital elevation data. In *Channel Network Hydrology*, eds K. Beven and M. J. Kirkby, Chichester, Wiley, 13–42.

Brown, D. G. and Bara, T. J., 1994. Recognition and reduction of systematic error in elevation and derivative surfaces from 7.5 minute DEMs. *Photogrammatic Engineering and Remote Sensing*, **60**(2), 189–194.

Burrough, P. A., 1986. *Priciples of Geographical Information Systems for Land Resources Assessment*. Oxford University Press, Oxford.

Carter, J. R., 1992. The effect of data precision on the calculation of slope and aspect using gridded DEMs. *Cartographica*, **29**(1), 22–34.

Christensen, A. H., 1987. Fitting a triangulation to contour lines. In *Proceedings of AutoCarto 8*, ASPRS and ACSM, Falls Church, Virginia, 57–67.

Clarke, A. L., Gruen, A. and Loon, J. C., 1982. The application of contour data for generating high fidelity grid digital elevation models. *Proceedings of Auto-Carto, 5*, ASPRS, Falls Church, Virginia, 213–222.

Dikau, R., 1993. Geographical Information Systems as tools in geomorphology. *Zeitschrift für Geomorphologie NF SupplementBand*, **92**, 231–239.

Douglas, D. H. and Peucker, T. K., 1973. Algorithms for the reduction of the number of points required to represent a digitised line or its caricature. *The Canadian Cartographer*, **10**, 112–122.

Eastman, J. R., 1992. *IDRISI Version 4.0: Technical Reference*. Clark University, Worcester, MA.

Eklundh, L. and Mårtensson, U., 1995. Rapid generation of digital elevation models from topographic maps. *International Journal of Geographical Information Systems*, **9**(3), 329–340.

ESRI, 1995. *ARC/INFO Online Documentation*. Environmental Systems Research Incorporated, Redlands, CA.

Evans, I. S., 1972. General geomorphometry, derivatives of altitude, and descriptive statistics. In *Spatial Analysis in Geomorphology*, ed. R. J. Chorley, Methuen, London, 17–90.

Evans, I. S., 1980. An integrated system for terrain analysis and slope mapping. *Zeitschrift für Geomorphologie*, **36**, 274–295.

Evans, I. S., 1990. General geomorphometry. In *Geomorphological Techniques*, ed. A. J. Goudie, London, Unwin Hyman, 44–56.

Fisher, P. F., 1993. Algorithm and implementation uncertainty in viewshed analysis. *International Journal of Geographical Information Systems*, **7**(4), 331–347.

Goodchild, M. F., Parks, B. O. and Steyaert, L. T., 1993. *Environmental Modeling with GIS.* Oxford University Press, Oxford.

Hannah, M. J., 1981. Error detection and correction in digital terrain models. *Photogrammetric Engineering and Remote Sensing*, **47**(1), 63–69.

Heil, R. J. and Brych, S. M., 1978. An approach for consistent topographic representation of varying terrain. *Proceedings of DTM Symposium*, ASPRS, Falls Church, VA, 397–411.

Horn, B. K. P., 1981. Hill shading and the reflectance map. *Proceedings of the IEEE*, **69**(1), 14–47.

Hutchinson, M. F., 1989. A new procedure for gridding elevation and stream line data with automatic removal of spurious pits. *Journal of Hydrology*, **106**, 211–232.

Jenson, S. K. and Domingue, J. O., 1988. Extracting topographic structure from digital elevation data for geographic information system analysis. *Photogrammetric Engineering and Remote Sensing*, **54**(11), 1593–1600.

Lee, J., Snyder, P. K. and Fisher, P. F., 1992. Modeling the effect of data errors on feature extraction from digital elevation models. *Photogrammetric Engineering and Remote Sensing*, **58**(10), 1461–1467.

Ley R. G., 1986. Accuracy assessment of digital terrain models. In *Proceedings of AutoCarto London*, *1*, Royal Institute of Chartered Surveyors, London, 455–464.

Li, Y. and Stuart, N., 1994. Construction and analysis of contour-based digital terrain models for hydrological applications. *Proceedings of GISRUK '94*, 274–279. Department of Geography, University of Leicester..

Li, Z., 1994. A comparative study of the accuracy of digitial terrain models (DTMs) based on various data models. *ISPRS Journal of Photogrammetric Remote Sensing*, **49**(1), 2–11.

Mark, D. M., 1979. Phenomenon-based data structuring and digital terrain modelling. *Geo-Processing*, **1**, 27–36.

Mark, D. M., 1986. Knowledge-based approaches for contour-to-grid interpolation on desert pediments and similar surfaces of low relief. *Proceedings of 2nd International Symposium on Spatial Data Handling*, 225–234.

Monckton, C. G., 1994. An investigation into the spatial structure of error in digital elevation data. In *Innovations in GIS 1*, ed M. F. Worboys, London, Taylor and Francis, 201–211.

Moore, I. D., 1988. A contour-based terrain analysis program for the environmental sciences (TAPES). *Transactions of American Geophysical Union*, **69**, 345.

Moore, I. D., O'Loughlin, E. M. and Burch, G., 1988. A contour-based topographic model for hydrological and ecological applications. *Earth Surface Processes and Landforms*, **13**, 305–320.

Moore, I. D., Grayson, R. B. and Ladson, A. R., 1991. Digital terrain modelling: a review of hydrological, geomorphological and biological applications. *Hydrological Processes*, **5**(1), 3–30.

Neave, H. R., 1979. *Elementary Statistical Tables.* George Allen and Unwin, London.

Peucker, T. K., Fowler, R. J., Little, J. J. and Mark, D. M., 1978. The triangulated irregular network. *Proceedings of DTM Symposium*, ASPRS, Falls Church, VA, 516–540.

Quinn, P., Beven, K., Chevallier, P. and Plancher, O., 1991. The prediction of hillslope flow paths for distributed hydrological modelling using digitial terrain models. *Hydrological Processes*, **5**, 59–79.

Robinson, G. J., 1994. The accuracy of digital elevation models derived from digitised contour data. *Photogrammetric Record*, **14**(83), 805–814.

Skidmore, A. K., 1989. A comparison of techniques for calculating gradient and aspect from a gridded digital elevation model. *International Journal of Geographical Information Systems*, **3**(4), 323–334.

Srinivasan, R. and Engel, B. A., 1991. Effect of slope prediction methods on slope and erosion methods. *Applied Engineering in Agriculture*, **7**(6), 779–783.

Waters, R. S., 1958. Morphological mapping. *Geography*, **43**, 10–17.

Weibel, R. and Heller, M., 1991. Digital terrain modelling. In *Geographical Information*

Systems, eds D. J. Maguire, M. F. Goodchild and D. W. Rhind, Longman, Harlow, 269–297.

Wood, J. D. and Fisher, P. F., 1993. Assessing interpolation accuracy in elevation models. *IEEE Computer Graphics and Applications*, **13**(2), 48–56.

Zevenbergen, L. W. and Thorne, C. R., 1987. Quantitative analysis of land surface topography. *Earth Surface Processes and Landforms*, **12**, 47–56.

8 Landform and Lineament Mapping using Radar Remote Sensing

COOMAREN P. VENCATASAWMY[1], CHRIS D. CLARK[1] and R. J. MARTIN[2]

[1] Sheffield Centre for Earth Observation Science, University of Sheffield, UK
[2] School of Mathematics and Statistics, University of Sheffield, UK

ABSTRACT

Radar imagery, because of its sensing geometry and use of microwave radiation, provides a different view of the earth and is theoretically superior to optical imagery (Landsat etc.) for mapping landforms and lineaments. There are, however, a number of difficulties in using synthetic aperture radar (SAR) images. They cannot be handled using the same methods as optical imagery because of image distortions and the process by which the images are formed. Special tools for processing and enhancement are required. This paper reports on work exploring different techniques with the aim of finding the best approaches for dealing with SAR data for geomorphological applications. Methods for geocoding and terrain correction are described. The emphasis is on producing high quality images for photo-interpretation, rather than automatic feature detection algorithms which are not yet able to match the performance of the visual approach.

INTRODUCTION

The recent growth of interest in large scale geomorphology and geology has been partly prompted by, and partly assisted by, the widespread availability of satellite imagery. Such images typically cover areas of 100 by 100 km and greater, and thus have the capacity to reveal hitherto unknown large scale structures, landforms or patterns that are only faintly discernible or even invisible when viewed from ground level (e.g. Clark, 1990). Remote sensing has thus become another technological tool that can assist in geomorphological and geological investigations. New discoveries can be made by utilising this large scale view and field evidence can be maximised by using imagery to extend mapping into unvisited locations or between widely located point data. At the simplest level, satellite images can be acquired as photographic prints and treated in a similar manner to aerial photographs. The advantage in ease of use is weighed against the underutilisation of the spatial and spectral resolution of the data, and the fact that the analysis must necessarily be by

Landform Monitoring, Modelling and Analysis. Edited by S. N. Lane, K. S. Richards and J. H. Chandler.
© 1998 John Wiley & Sons Ltd.

manual means (i.e. photo-interpretation). Processing of digital data, on the other hand, can fully utilise the data's resolution and some tasks can be performed semi-automatically. The cost here is in the level of specialist remote sensing knowledge and equipment required and time required for analysis. In many cases the digital method is prohibitive and this probably explains why the use of remote sensing in geomorphology has been fairly limited. Nevertheless, remote sensing activity using optical sensors such as Landsat and SPOT is now widespread in geology and increasing in geomorphology. The techniques for analysis of such data are well known and widely taught, so it is becoming increasingly possible for geomorphologists to undertake research using remotely sensed data.

Technological developments led to the launch of the ERS-1 satellite in 1991 which acquires, *inter alia*, radar imagery known as SAR (synthetic aperture radar). This imagery is fundamentally different to the optical type of data collected by Landsat and SPOT, requiring different processing techniques. Knowledge of what can be gained from radar is still in its infancy, as are the techniques required to utilise it. However, on theoretical grounds, and on the basis of an increasing number of investigations, it is apparent that radar data should be of great potential for geomorphological and geological research. This is particularly so for the detection and mapping of landforms and lineaments, which is the subject of this paper. The principal advantages of SAR for geomorphology include the recording of microwave wavelength energy which provides different information to that from visible light or infrared radiation, and the oblique imaging geometry which helps to enhance and highlight topography.

The use of radar imagery in geomorphological and geological research will not be extensively reviewed here. Rather, the first aim is to provide some explanation of this form of data, and the second is to provide guidance, based upon our ongoing research, on appropriate methods of processing and analysing such data for geomorphological purposes. Our main test area is a 10 by 20 km block covering the area around Kendal, Sedbergh, Baugh Fell, Whernside and the Howgill Fells in the Lake District, England. Four ERS-1 SAR scenes, winter and summer Landsat thematic mapper (TM) images, a digital elevation model, field data and published maps are being utilised. Figure 8.1 illustrates the difference in landform detectability afforded by different types: a terrain corrected SAR image, and a summer and winter Landsat image. It is apparent that drumlins and geological structure are best displayed on the winter TM, closely followed by the SAR image, with the summer TM image being fairly useless.

Whilst the oblique viewing geometry and use of active microwave radiation provide the basis for geomorphological investigation, significant difficulties arise. If the topography is greater than gently undulating then considerable distortions due to relief occur. Moreover, image speckle reduces the visibility and detectability of landforms. Fortunately these problems can be partly overcome. Without a basic understanding of the peculiarities of SAR data it would be easy to make mis-interpretations from imagery and so the next section gives a brief description of how SAR images are formed. Different types of images are described and sources of further information are outlined. The focus of attention is on SAR images collected from space-borne rather than airborne platforms. This is because images of the

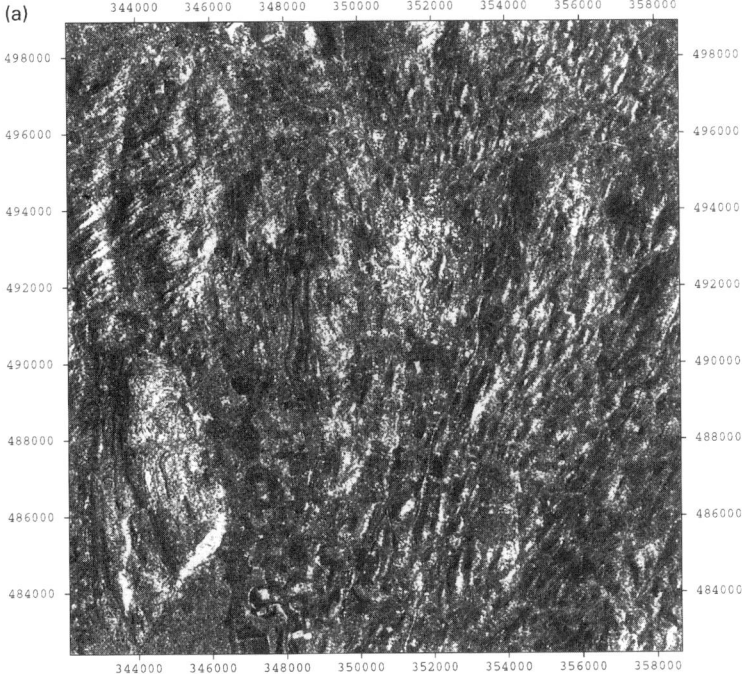

Figure 8.1 Comparison of the visibility of drumlins and geological lineaments on space-borne images. (a) Terrain corrected ERS-1 SAR image acquired during the descending path of the satellite, with a look direction towards the NW; (b) Landsat TM image (band 5) acquired in summer (high solar elevation); (c) Landsat TM image (band 5) acquired in winter (low solar elevation). A large field of drumlins (around grid reference 354000, 490000) and geological structure (around 344000, 496000) in Silurian greywackes and mudstones can clearly be seen on both the SAR and winter TM images, but not on the summer TM image. The area is in northern England with the town of Kendal at 352000, 492000. The images are corrected to the Ordnance Survey grid, presented with units in metres

former now exist in archive for most of the earth's land surface whilst airborne cover is limited. Some applications of SAR images for mapping geological lineaments and glacial landforms are reviewed. The tools for handling SAR images are described and advice is provided, based on our own experimentation, of alternative approaches. The problems of choosing, filtering and geocoding SAR images are addressed. A method of data reduction for visual interpretation is also proposed.

FUNDAMENTALS OF RADAR REMOTE SENSING

A researcher wishing to use radar imagery for a particular geographical area of investigation is likely to use data from the ERS satellites because these represent the largest source of archive radar images for which close to complete global coverage

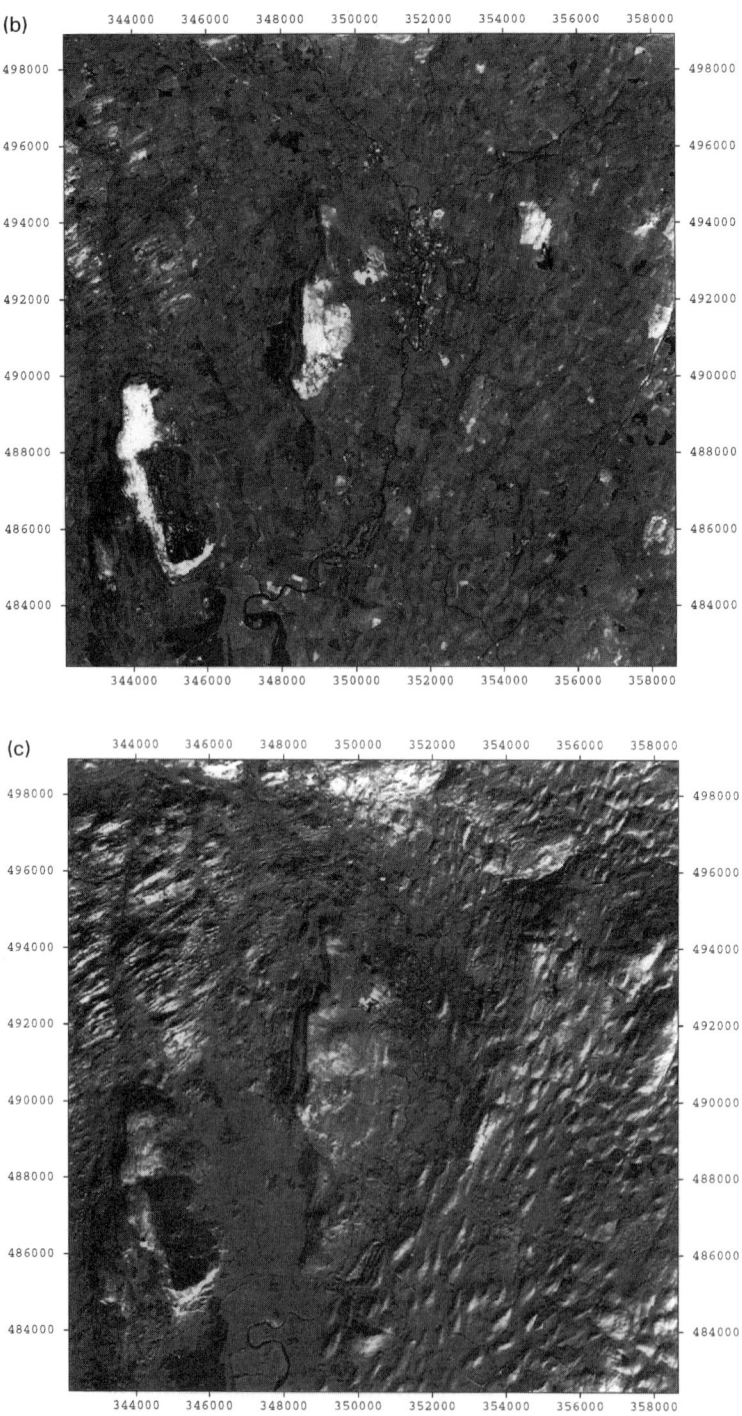

Figure 8.1 *(continued)*

is available. For this reason we focus our attention on the processing and use of ERS data, although it should be noted that many other space-borne and airborne systems have acquired data but on a much more limited basis. The specifications of other space-borne SAR, as well as the World Wide Web addresses of home pages where more information can be obtained, are listed in the Appendix. The most promising space-borne missions are ERS-2 and its tandem operation with ERS-1, and Radarsat, whose radar beam can be shaped and steered to alter the incidence angle and resolution. At the time of writing Radarsat images are not yet available, but if the system continues to operate there will soon be widespread coverage. Way and Smith (1991) describe a wide range of SAR systems used in remote sensing.

A SAR sensor is an active monostatic instrument that sends out a pulse of microwave energy and records the returned echo. Two entities are recorded: the time taken for the echo to return (this is a measure of the distance of the target from the sensor and is called the slant range) and the nature of the echo itself. The echo carries two parcels of information: the Doppler shift from the original carrier wave, and the strength of the echo (the amplitude or its squared value, known as the intensity). The strength of the echo depends on factors such as slope angle, moisture (or more specifically, the surface dielectric properties) and surface roughness. The Doppler shift (change in frequency) occurs because of the motion of the sensor relative to the object and is used to refine the azimuth (along-track) resolution. The ground range (cross-track) resolution of an ERS-1 raw image is 22.1 m at the far range (away from sensor) and 28.9 m at the near range (close to sensor). The azimuth resolution is 4.45 m.

The main differences between SAR images and optical images are the following.

- *Microwave sensing*: the signal is measured using microwaves (i.e. wavelengths of 1 to 50 cm) which interact with the ground surface in fundamentally different ways to optical wavelengths. Different information is thus recorded in the signal (see above).
- *Oblique viewing geometry*: as the electromagnetic waves are emitted from and recorded using the same platform, it is essential that the look direction of the sensor is oblique to the ground surface rather than close to the vertical.
- *Day/night and all-weather imaging*: as the sensor emits its own radiation it does not need to rely on illumination from the sun, and the microwaves travel freely through cloud. This eliminates the logistical problem with passive sensors of finding cloud-free images, a difficulty which often greatly hinders the utility of Landsat and SPOT images.

The more complex viewing geometry creates numerous problems in the way images are produced and the quality of information they contain, particularly in high relief terrain. It has been shown, for example, that in some hilly areas it is not even possible to identify forested from non-forested areas due to the overriding terrain signal (Beaudoin *et al.*, 1994). For our purposes, however, the main strength of SAR is considered to be its sensitivity to slope angle. The effect of the sensitivity of backscatter to topography is analogous to viewing a landscape with a low sun angle: quite subtle changes in topography can be detected. For our test sites we

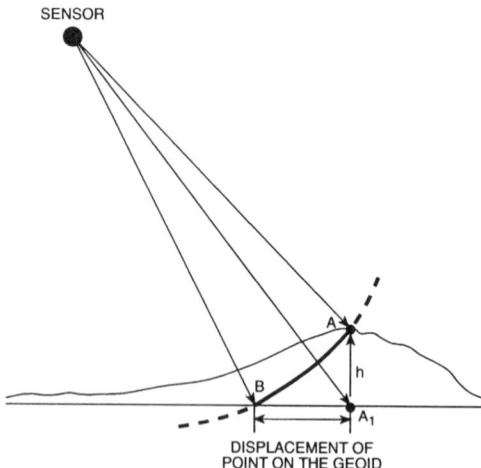

Figure 8.2 Geometric distortion due to relief. As the points A and B are the same distance from the sensor this means that A becomes mapped onto B in the image rather than to its true location at A_1. This relief displacement increases with increasing height, h. After Curlander and McDonough (1991)

have found that the slopes constitute most of the information content of the SAR data, and that the vegetation cover is subordinate to this. Unfortunately, this sensitivity to slope can create problems, as a number of terrain distortions are introduced into the image. One major local distortion is that since SAR measures range it is prone to errors in target height. For example, in Figure 8.2, A and B are at the same distance from the sensor, that is they have the same slant range and thus considered to have the same cross-track displacement. Fortunately this type of distortion can be removed using terrain correction (see below). Foreshortening and layover are two other distortions that arise from topographical variation, and these are practically impossible to remove. Foreshortening (Figure 8.3a) occurs when the slope of the local terrain is less than the incidence angle. This results in features close to the sensor (near range) appearing compressed compared to those further away from the sensor (far range). The layover condition (Figure 8.3b) occurs in steep terrain where the slope is greater than the incidence angle. For ground areas sloping towards the sensor, the effective incidence angle becomes smaller, thus increasing the cross-track pixel spacing, and the reverse occurs for slopes away from the radar. In relatively high relief areas a layover condition may exist, such that the top of a mountain appears nearer than the base, resulting in a severely distorted image. Additionally, echo signals from multiple target locations will arrive at the SAR antenna simultaneously, resulting in a high response. Finally, "shadowing" (Figure 8.3c) occurs when objects in the near range prevent the pulse from continuing any further and a dark region with no information about the terrain whatsoever is thus recorded in the imagery. Changes in the backscatter in the shadows are only due to noise in the sensor and so cannot be

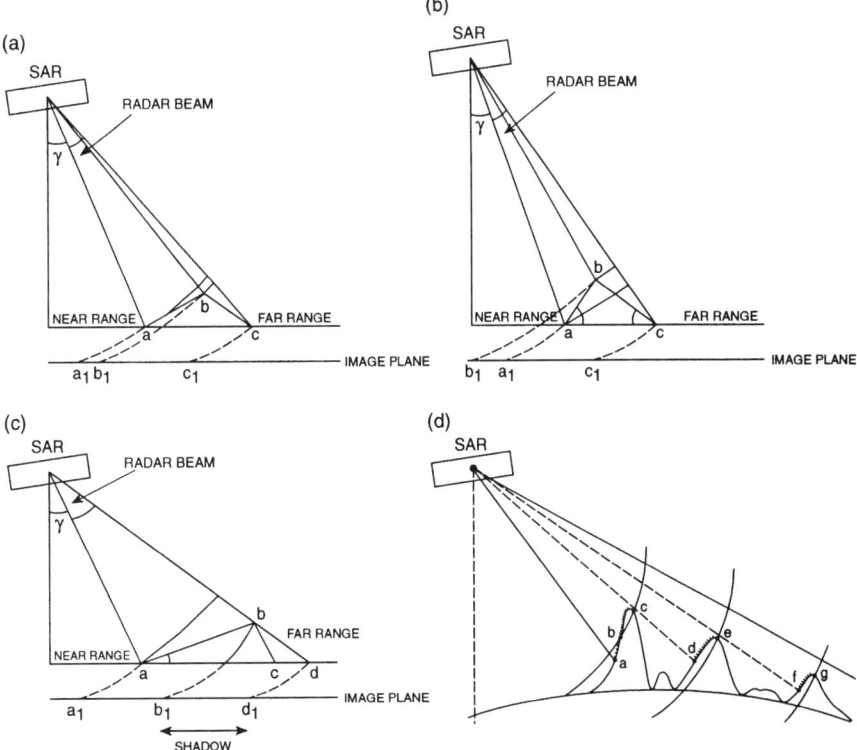

Figure 8.3 Geometric distortions arising from the effects of relief, but which are extremely hard to correct. (a) Foreshortening occurs when the slope of the local terrain is less than the incidence angle (γ). The facing slope a–b becomes compressed to a_1–b_1. (b) Layover occurs in steep terrain where the slope angle is greater than the incidence angle. As the top of a mountain, b, is closer to the sensor than the base, a, this causes it to lay over in the image, i.e. an incorrect positioning of b_1 relative to a_1. (c) As the illumination is from an oblique source, areas of rugged terrain frequently suffer from radar shadow. No information is thus recorded from the region b_1–d_1. (d) Over a region with varying topography a combination of distortions may arise. Points b and c will suffer from layover and will be positioned incorrectly in relation to a. Information from the region between c and d will be lost due to radar shadow. Foreshortening occurs at slope facet d–e. Further radar shadow occurs at e–f, and foreshortening at f and g. After Curlander and McDonough (1991)

analysed to extract information. In some cases, a complicated mixture of layover and shadowing (Figure 8.3d) occurs, distorting the image further (Kropatsch and Strobl, 1990).

Another factor that arises from the oblique viewing geometry is that the view must be in a certain direction: the "look direction". SAR imagery from a satellite such as ERS-1 can be acquired from the ascending or descending part of its orbit, i.e. South–North or North–South. The descending pass records daytime imagery. While the satellite is on its ascending part of the orbit on the other side of the world it records data during the night. The time of day makes little difference to lineament detection, but the orientation of the orbit is of relevance as this determines the look direction, and is thus of great importance. The look direction has the effect of introducing a bias in the detection of lineaments and landforms. Those that are oriented perpendicular to the look direction will be most easily detected and those parallel may not appear at all. Isiorho (1984) found, by plotting the directions of detected lineaments as a rose diagram, that this was the case. Graham and Grant (1991) found that lineaments oriented within ±20° of a line perpendicular to the look direction are enhanced. Plate I shows a composite of two ERS-1 images (see plate section at the end of Chapter 11). One image (rendered yellow) was acquired during the descending path of the satellite with a look direction of 290° from north, whereas the second image (blue) was acquired during its ascending path with a 70° look direction. Interpretation of the colours in this figure reveals the differences in lineament information content between the images. Linear features that are highlighted by just yellow or blue are only detected effectively by one of the images, whereas yellow and blue features are detected in both. It is usual for SAR images of different look directions to appear very different even though they are of the same area. A consequence of the look direction bias is that interpretation of the spatial distribution of lineaments and distribution of orientation must be executed with caution. For example, it is not valid to infer absence of a feature based upon image evidence alone.

Another problem that SAR images suffer from is speckle, which has the effect of reducing the resolution of an image and blurring small features within it. Further details about speckle and speckle reduction strategies are given below.

Most of the material presented in this paper is based on the use of ERS-1 SAR images in PRI (Precision Image) format. The PRI format is a three-look amplitude image for which major distortions due to the sensor geometry have been corrected, and it is the most widely used and recommended format. A three-look image is one formed by averaging three independent frames of the same scene (see below). Its spatial resolution is 25 m by 25 m because the original image has been processed and resampled, although the data are provided as 12.5 m square pixels. Pixel values are not in any real units and so if quantitative measurements are required (for example, in forest monitoring) it is essential to calibrate the data to decibels (dB). This is not necessary here as the calibration does not affect visual perception.

It is important to note that SAR data can be collected by a variety of means, such as with different incidence angles, frequencies and polarisation. ERS-1 uses C-band waves which have a wavelength (λ) of 5.6 cm and the polarisation of the waves is VV. Polarisation can be horizontal (H) or vertical (V), and can be either transmitted or received. A vertically polarised wave is one in which the electric field is constrained to a vertical plane only. A sensor is said to have a VV polarisation if it sends out vertically polarised waves and records only vertically polarised waves. The incidence angle on a flat surface varies from 20.1° at the near range to 25.9° at the far range and is 23° at its centre. Unlike Radarsat and SIR-B (see Appendix), ERS-1 has a "fixed" incidence angle.

There are many further aspects relating to the acquisition and processing of SAR data; details may be found in Curlander and McDonough (1991), Leberl (1990), Schreier (1993), Quegan (1995a, b), Glasbey and Horgan (1995).

SOME APPLICATIONS OF SAR IN MAPPING LINEAMENTS AND LANDFORMS

The measured signal (also called backscatter) is dependent upon moisture, roughness and slope of the viewed terrain element, and these factors largely determine the most appropriate geomorphological applications. A large field of interest lies in the ability of SAR to detect changes in soil moisture (e.g. Normand *et al.,* 1994). Recent work demonstrates that it is possible to gain estimates of soil moisture at the catchment scale, given certain conditions. This is of great use for hydrological modelling, flood prediction and agricultural applications. In sparsely vegetated regions, the sensitivity to roughness has the potential to be used for mapping aspects of surficial geology (Millington *et al.,* 1995). Many authors have reported success in mapping lineaments and glacial landforms using radar images (spaceborne or airborne) supported by auxiliary information such as Landsat imagery, aerial photographs or fieldwork. The high dependence of backscatter on slope angle makes SAR well suited for landform and lineament mapping, which is the aspect of interest here. This section reviews some of the papers describing the methods used and the results reported.

Isiorho (1984) used real aperture airborne radar (X Band, $\lambda = 3.1$ cm) with dual look and depression angle imaging, Landsat imagery and fieldwork to map geological features in Nigeria. The spatial resolution of the radar image was 30 m. The methods used were a slight modification of routine photo-interpretation involving the delineation of lithologic units and structure using drainage pattern, vegetation, tone, texture and shape, and then correlating outcrops from one location to another. Stereoscopic viewing under different lighting conditions (transmitted and reflected light) and monoscopic viewing were used. Folds that were previously mapped were discriminated on the radar imagery and several probable fault lines, which had not been previously mapped, were also identified. Linear features were also evident and small, well defined pegmatitic dykes were found. It was concluded that all known stratigraphic units in the study area were visible in the radar images, but some characteristics could be traced across stratigraphic boundaries and needed to be checked against field surveys and published or unpublished data.

The use of radar imagery for lineament mapping has now become more widespread. Airborne SAR images were used by Roy *et al.* (1993) to map lineaments of an area in Quebec in order to understand better the tectonic setting; this is of significance because of the earthquake that recently occurred in the region and the continuing seismic risk. For a tectonically active region in Taiwan, airborne SAR has been used in conjunction with SPOT imagery and a digital elevation model to produce maps of neotectonics which are to be used in a hazard assessment (Deffontaines *et al.,* 1994). Guo *et al.* (1993) used radar imagery acquired from NASA's space shuttles (SIR-A and SIR-B) and airborne SAR for lithological discrimination, tectonic mapping and mineral exploration in China.

The combination of radar and optical data has been explored by many, as their synergistic use has been shown to aid lineament detection (e.g. Daily *et al.,* 1979; Welch and Ehlers, 1988; Harris *et al.,* 1990; Toutin *et al.,* 1992; Yésou *et al.,* 1993; De Séve *et al.,* 1996). These authors have geometrically registered and combined images from, for example, Seasat, SIR-B, ERS-1 and airborne SAR, with optical/ infrared imagery such as Landsat, SPOT and MOS. Such geometric correction of SAR data is not easy, and yet it is critical if SAR is to be merged with other data sources. De Séve *et al.* (1996) evaluated two techniques of geometric correction for the merging of ERS-1 and Landsat TM data, and quantified how the residual errors affect measurements of lineament length and orientation.

Ford (1984) identified and mapped geomorphological and glaciological features on Seasat radar (L Band, $\lambda = 23.5$ cm) images of scenes that are located in the drumlin drift belt, Ireland, and in the Alaska Range, Alaska. The SAR images had a ground resolution of 25 m and were enlarged to photographic prints at a scale of 1:250 000 for landform mapping. Corresponding Landsat multispectral scanning system (MSS) images were used to compare the clarity of features mapped from the SAR images. The landforms of interest were mostly drumlins of different sizes. Faults could also be delineated from the images, while lateral moraines were found to be individually distinguishable. It was found that regional distributions of drumlins and associated streamlined features are only just visible on the MSS images but are clear enough on the SAR to be able to be mapped in detail.

Graham and Grant (1991) used C-band airborne SAR images to interpret the geomorphological features of central Newfoundland, Canada. The SAR data collected had two look directions at right angles to each other to ensure that all lineament orientations were imaged equally well. Spatial resolution was 20 m, and images were taken in autumn to minimise the background surface roughness responses due to variable vegetation height. Image prints were enlarged to match the scale of the bedrock and Quaternary compilations and then previewed as a mosaic, which should be interpreted with the shadows facing the observer. This makes the images appear more natural and the features clearer to see. Lineaments were broadly differentiated as glacial or structural on the basis of shape, length, height and con- tinuity. Structural lineaments were subdivided by linearity into curvilinear features (folds and foliation) and rectilinear features (faults and shear zones, joints and fractures). Areal patterns of tone and texture, representing variations in bedrock roughness, till texture and surface moisture, were also noted. Glacial lineaments (drumlins, drumlinoids, crag-and-tail hills, fluting) were found to be the most promi- nent and more common features on the radar images. Various structural features (folds and foliation, faults and fractures) were also recognised. Brief descriptions of how these features manifest themselves on radar images were also given. Radar captured virtually all ice-flow features, even those as small as a few metres, resolving outlines, patterns and scale variation of morainal topography more clearly than air photographs, especially in vegetated areas. The detection of faults had varied success, with new features being mapped and some mapped features not being visible.

The work of Clark (1990, 1993, 1994) and Boulton and Clark (1990) on large scale glacial geomorphology using Landsat data has now been extended to include ERS-1 radar imagery. Over 80 SAR images of the Ungava region of Arctic Canada have been photo-interpretatively mapped for glacial lineaments in order to aid ice- flow reconstruction for this sector of the Laurentide Ice Sheet (Knight and Clark, 1995; Clark *et al.*, 1995, 1996). A field expedition was made to the Larch River area, Northern Quebec, in order to validate ice-flow interpretations made from the

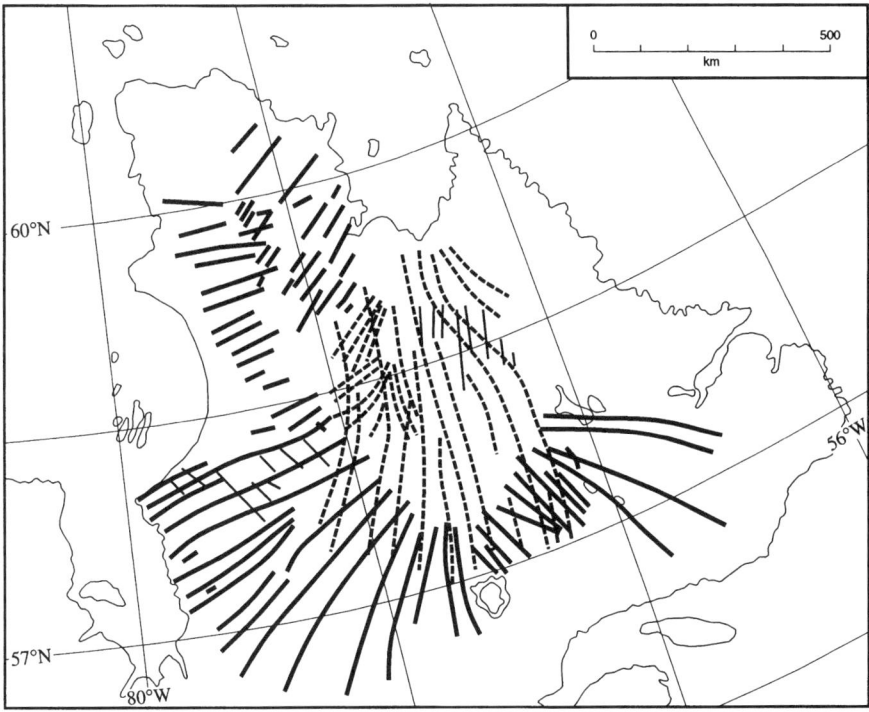

Figure 8.4 Example of ice-flow patterns in eastern Canada summarised from over 120 000 photo-interpreted lineaments mapped from 80 ERS-1 SAR scenes. These patterns record ice-flow dynamics of the last ice sheet to have covered Canada (Laurentide Ice Sheet), and can be used to reconstruct its behaviour (Knight and Clark, 1995; Clark *et al.*, 1996)

imagery. The combined use of extensive striae mapping and aerial photograph interpretation confirmed the radar interpretations and added further detail. Figure 8.4 displays some of the ice-flow patterns that have been revealed using the SAR images. Glacial lineament mapping has also been achieved for Ireland and parts of England, which will form the basis of reconstruction of the last ice sheet (cf. Mitchell and Clark, 1994). Individual glacial lineaments mapped from SAR scenes covering Ireland are illustrated in Figure 8.5. Clark and Knight (1994) note the extra level of drumlin information that can be extracted from ERS-1 SAR imagery compared to Landsat scenes.

TECHNIQUES FOR HANDLING SAR IMAGES

The main objective of this paper is to establish and provide guidance for potential users on the most appropriate methods for visually extracting landform and lineament information from radar images. A longer research aim is to use automated image segmentation methods and pattern recognition to derive such information.

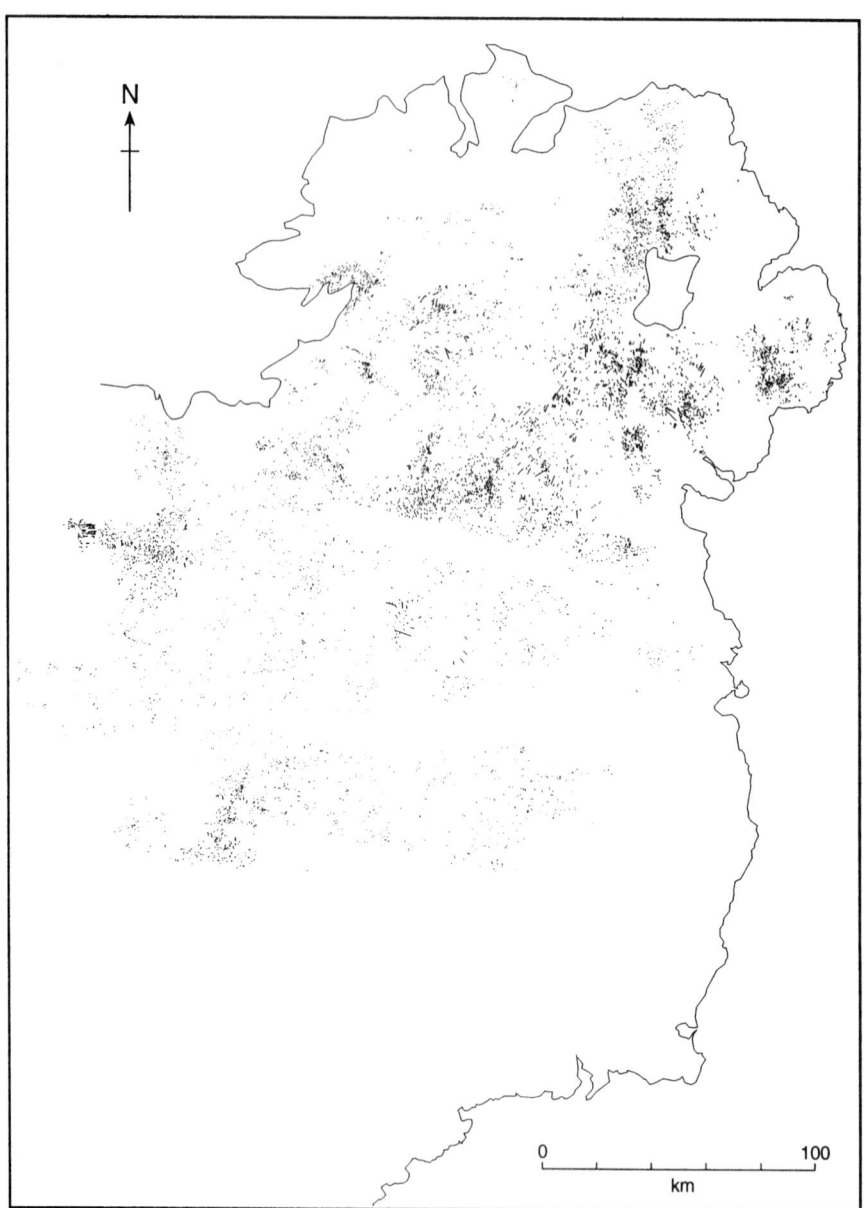

Figure 8.5 Individual glacial lineaments mapped from eight ERS-1 SAR scenes covering most of Ireland. These are being used to reconstruct the dynamics of the former ice sheet

To our knowledge, no algorithms published so far can detect the subtle kind of lineaments which geomorphologists and geologists are interested in.

In this paper, therefore, the focus is on the simpler and hitherto unanswered question: given a SAR image, how should it be treated to optimise the visual expression of landforms and lineaments for photo-interpretation? This is a much harder question to answer than might be expected because the tools, techniques and understanding developed for optical remote sensing rarely apply. Whilst techniques for processing Landsat-type images have been widely disseminated and incorporated into image analysis software, the situation for radar data is that commercially available software has only recently incorporated radar image processing routines. The knowledge of how best to use these techniques is still largely unreported or unknown. The remote sensing literature has a wealth of techniques that have been specifically developed for analysing SAR data, but these have only rarely been evaluated for particular applications. We aim to evaluate published radar techniques, if necessary develop hybrids, and then make both visual and statistical tests of their performance. From this, it may be possible to recommend a range of approaches that minimise the current difficulties which arise when using SAR images.

We discuss the data requirements necessary for the efficient mapping of a given area, and address the problem of speckle and how to choose an appropriate speckle reduction filter. Geocoding and terrain correction of SAR images are described, and methods for contrast enhancement and data reduction are proposed.

Decision-making Regarding Image Acquisition

It would be highly desirable for a geomorphologist or geologist interested in mapping the landforms and lineaments of an area to have access to: (i) an ascending pass SAR scene; (ii) a descending pass SAR scene; (iii) a winter Landsat TM image; and (iv) a high resolution digital elevation model (DEM).

If only one SAR scene of an area of interest is to be acquired, it should be either of the ascending or descending scenes which has a look direction perpendicular to the likely linear structures that may be present, if this is known. The preferred option is to use imagery obtained during both passes, as this permits the maximum amount of information to be detected. Ideally, these look directions would be orthogonal to each other, but the orbit track does not permit this and the angular difference is fixed for any given latitude. For instance, the angular difference for our test site is 40°.

If a further look direction is required, recourse can be made to imagery obtained in winter using an optical sensor: an appropriate source is Landsat TM. This has a spatial resolution of 30 m, which is similar to the ERS-1 SAR. As an example, Greenbaum *et al.* (1993) describe the acquisition and use of winter Landsat TM images for Britain. The acquisition of data in the winter months is crucial, because Landsat records its data at a fixed time of day for any point on the earth. In winter, solar elevation is at its lowest, and hence is best for enhancing subtle topographic features. There is, however, considerable difficulty in acquiring such data, as the chances of there being cloud-free scenes are often very low. Indeed, for most parts of the world it is extremely unlikely that cloud-free winter TM data exist in the archive. Fortunately, the amount of information that can be extracted from a single SAR image, assisted with auxiliary information (e.g. geological maps, air photos, fieldwork), is sufficient in most cases.

If the field area of interest contains any appreciable topographic relief, it is desirable to obtain a DEM of comparable spatial resolution to the imagery. This permits much of the geometric distortion, outlined above, to be removed, and allows the image to be transformed into a geometry that fits a chosen map projection. This is essential if the precise location of lineaments is required, but can be avoided when trends and spatial distributions are the main focus. This relaxation is fortunate because the availability of high resolution DEMs is very poor and mostly restricted to parts of Europe and North America. For our work on English and Irish glacial lineaments it was possible to perform full terrain correction, but for the Canadian Arctic suitable DEMs were not available.

Distortions that arise from the viewing geometry of the sensor in relation to topography frequently make SAR images (and, to a lesser extent, winter TM) unusable for any mapping purposes in high relief terrain. Whilst it is hard to define limits on when data become unusable, our experience suggests that only in areas with low to medium relief terrain can SAR images be sensibly used. On an ERS-1 SAR image of the English Lake District (relief of the order of 800 m), for example, it is hard to recognise any of the major mountains, and approximately 20% of the land surface is hidden by radar shadow or rendered useless by foreshortening.

Having obtained SAR imagery, the next stage is to execute a number of processing steps. Upon loading a SAR scene in an image analysis system, it is most unlikely that the image will be in such a form as to be useful for visual interpretation. The following processing stages are usually required: (i) speckle reduction; (ii) geometric and/or terrain correction; (iii) contrast enhancement; and (iv) data volume reduction. The order in which these steps are executed is important for the final result. It would be wrong, for example, to perform data reduction by resampling prior to speckle reduction, as this would irretrievably embed the structure of the speckle within the data. The sequence described above has been found to be most appropriate for visual interpretation.

Speckle Reduction

Speckle can be considered to be noise which reduces the visibility of features in an image. Its effect is similar in appearance to a television picture with poor reception, and has the infuriating property that the closer you look at an image the harder it is to distinguish features clearly. It appears in SAR images because of the coherent processing of the returned signal. For example, consider a resolution cell as consisting of a number of scatterers which reflect the incident wave as small wavelets with different phases and amplitudes. At the sensor, these wavelets are added together vectorially (Figure 8.6) and so can produce either constructive interference (a high response) or destructive interference (a low response). This fluctuation in backscatter is referred to as speckle, and although speckle is usually referred to as noise, it does contain information about a resolution cell which may be useful for certain applications, for example to determine the roughness of a resolution cell or

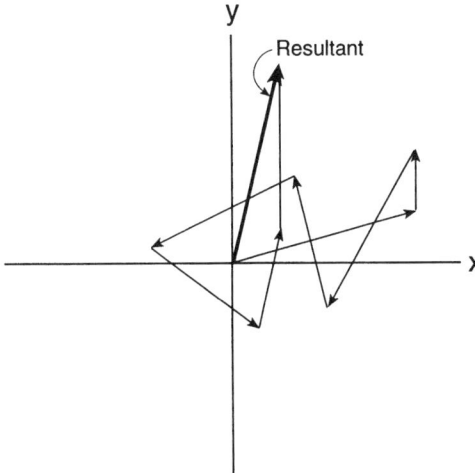

Figure 8.6 Vector addition of wavelets. The magnitude and orientation of waveforms arriving at the sensor from a single pixel can be represented on a phasor diagram (arbitrary Cartesian representation of vectors). All wavelets (represented by vectors) are added together as magnitude and phase to produce the resultant vector, which is the final pixel value. As the magnitude and the phase are random, this causes adjacent pixels to have widely varying values producing the appearance of speckle

in interferometry (Dainty, 1984). Speckle occurs in all types of coherent imagery such as acoustic imagery and laser illuminated imagery, and so its statistical properties have been well studied (Goodman, 1984).

As speckle blurs small features in SAR images, one may wish to remove it while preserving features such as edges. Korwar and Pierce (1981) showed that the probability of detecting a small feature in an image decreases drastically with speckle. A number of speckle reduction routines exist; the most common use averaging of independent frames in the processing procedures (Porcello, 1976). This reduces the speckle at the expense of resolution. ERS-1 SAR PRI images are three-look images where the variance due to speckle has been reduced by a factor of one-third, with the spatial resolution being trebled. A number of speckle reduction algorithms can then be applied to the processed image. These can be classified into two groups: adaptive and non-adaptive filters. The two most common non-adaptive filters are the mean and median filter. These replace the middle pixel value in an m by m window by the mean or median of all the pixel values in that window. The mean filter has a smoothing effect and tends to blur edges and destroy texture information, as well as reducing speckle. The median filter removes high and low pixel values and so blurs some features of interest. The main strength of non-adaptive filters is that they are not based on any model of image and speckle formation (see below), and can be applied to almost any image or at any stage in the manipulation process. For example, it could still be reasonable to apply the filter to a georeferenced image, whereas the application of model-based filters would be inappropriate. A further advantage is that these filters can be executed with very simple algorithms that are available in many image analysis packages.

Adaptive filters also use an *m* by *m* window, but use local statistics gathered from the neighbourhood to guide the modification of the central pixel value. Common adaptive filters are the Nagao filter (Nagao and Matsuyama, 1979), the Lee filter (Lee, 1981), the K-average filter (Davis and Rosenfeld, 1978) and the Sigma-Lee filter (Lee, 1983). The Nagao filter looks for the most homogeneous neighbourhood around each pixel in an image, and then replaces the value of that pixel by the average in the selected neighbourhood area. It uses wedge-shaped neighbourhoods in order to avoid edges and uses the local variance as a measure of homogeneity. The Lee filter uses local statistics and local gradient information to detect edges and their orientation, and redefines the neighbourhood accordingly to avoid the edge. The value of the central pixel is then replaced by the average of the pixels in this new neighbourhood. In the modified version of the K-average filter proposed by Rao *et al.* (1995), the number of nearest neighbours used in calculating the mean to replace the central pixel is determined by the local variance. With the Sigma-Lee filter, the central pixel value is replaced by the average of neighbouring pixel values which are within two standard deviations of the central pixel value. The filter is based on the assumption that the pixel values have a Gaussian distribution. Its mean is taken to have the same value as the central pixel and its standard deviation is taken to be the product of the mean and the standard deviation of the speckle. Note that the standard deviation of the speckle can be estimated theoretically and depends on the number of looks: it is 0.3 for a three-look amplitude image. Martin and Turner (1993) propose a refinement to the Sigma-Lee filter by using a weighted mean. The weights are calculated as the probability of occurrence of a given pixel value using the Gaussian distribution. The first three filters are not based on any model of image formation, except that the variance is a suitable statistic for detecting changes in pixel values. Thus the same observations about versatility of the filters as in the previous paragraph apply. The Sigma-Lee filter, although based on a multiplicative model for the speckle and a particular distribution for the pixel values, could be used for any image, and even for varying standard deviations of speckle.

Filters developed using models of image and speckle formation are the Frost (Frost *et al.*, 1989) and the Kuan (Kuan *et al.*, 1987) filters. They express the imaging mechanisms mathematically to estimate a "true" pixel value. This is achieved at large computational cost and a number of approximations have to be performed to successfully implement the methods. The MAP filter (Lopes *et al.*, 1990) is a modification of the Kuan filter. It is based on the assumption that the distributions for the observed intensity (or amplitude) and the texture are known. The Maximum *a posteriori*, (MAP), estimate of the "true" pixel value (without speckle) is then calculated using Bayes' theorem. Based on an overall statistical model for the observed values and the texture, an estimate of the true pixel value, given the local statistics (calculated from the image), is obtained. The filter is adaptive. The coefficient of variation (the ratio of the standard deviation to the mean) is used to enhance the MAP filter by detecting edges and corner reflectors in the image. Effective reduction of speckle may involve a number of passes of filters. There has been a recommendation, for example, for a pass of Sigma-Lee (3 by 3), followed by Sigma-Lee (5 by 5) followed by Nagao (7 by 7) (e.g. Erdas, 1994).

When faced with the task of filtering an image, three questions arise: (i) which filter should be used? (ii) what is the most appropriate size for m, the dimension of the window; and (iii) how many successive passes should be made, and should these be by different filter types and/or window sizes? Unfortunately these questions remain to be answered on a rigorous basis. A number of authors (Durand *et al.*, 1987; Paudyal and Aschbacher, 1993; Mascarenhas *et al.*, 1991) have compared speckle filters. However, the methods used were either qualitative (based on visual assessment) or were not specific to photo-interpretation. Furthermore, the most appropriate method may be image- and application-dependent. In experimenting with speckle reduction filters, two main problem arise: (i) the large number of possible permutations; and (ii) uncertainty as to how to test the performance of the results in a quantitative manner. Added to this is the processing time involved, which can be of the order of 1–3 h for a full scene with a current UNIX workstation (Sun Sparc 20). It is sometimes worth asking whether you need to filter your image at all. All the filters mentioned above are derived under certain assumptions which are sometimes violated. The estimate of the true pixel value is only an approximate one, and features such as roads, texture and even features of interest, such as moraines or geological structure, may become blurred. In some cases this might be a high price to pay for a slight increase in the clarity of the image. All of the filters mentioned above use neighbourhood information to predict the value of the central pixel. Intuitively one might think that the central pixel value would be best modelled using its eight nearest neighbours, even for homogeneous areas. The spatial autocorrelation of the pixel values drops from 0.6 for the first neighbours to almost zero for the second set of neighbours in the direction of flight (azimuth) and in homogeneous regions.

While experimenting with different speckle reduction filters it was found that increasing the processing window of the filter may produce artefacts in the output. The Sigma-Lee filter, for example, creates noise in the form of dark spots in the smoothed images. The noise can reach a high level even for a 7 by 7 window. The analysis is still at its preliminary stage and the effect of noise on visibility has yet to be determined. As this noise has not been reported elsewhere, it is suspected that its occurrence may be image-dependent, but this effect has been noticed in many of our images. The effectiveness of the different filters using a 3 by 3 window has been examined, and it was found that the MAP filter produced a clear and sharp image. However, when the output images from the different filter types were examined it was found that the pixel values varied little between applications of different filters, with a difference of less than 10 at most pixels. This small difference occurred even in non-homogeneous regions where landforms were present, and suggests that for a single pass of a 3 by 3 window, the choice of filter type makes little difference to the outcome. The MAP filter, however, reduces large pixel values more than the others do. This may be the reason why the output image appears sharper, in that the output is better contrast stretched. A more elaborate investigation supported by statistical analyses has yet to be performed. We tentatively propose the MAP filter using a 3 by 3 window as the best speckle reduction filter, although it may require up to three passes to adequately reduce the speckle. The second choice would be the Sigma-Lee filter, still with a 3 by 3 window. This may require just two passes, but may produce noise.

Geocoding of SAR Images

Satellite images do not have the same geometry as maps and must therefore be modified if they are to be used to produce maps of even reasonable accuracy, or if they are to be used in conjunction with other sources of data, or in a Geographical Information System (GIS). The internal geometry of an image may need to be warped to remove effects that arise from various distortions, and the overall image rotated to the north to be able to fit a map projection. The final product is a resampled image whose pixels have known coordinates relating to a chosen map projection. There are three approaches to geocoding: (i) mathematical transformation based on a model that describes the satellite's orbit and a model that describes the shape of the earth; (ii) as in (i), but with a small number of ground control points (GCPs: features that can be located in the image and on a map) used to constrain a number of errors due to uncertainties in the parameters (e.g. sensor position) which are not precisely known; and (iii) many GCPs are located and a polynomial transformation is derived and used to warp the image to the map geometry.

The third option is widely used, particularly for Landsat-type data. This is because many optical sensors form images by viewing the earth's surface vertically beneath the satellite and so simple angular measurements away from the nadir position represent distance. However, for radar, the sensor measures distance between an object on the earth and the sensor using time lag in the return of a pulse of energy. This results in SAR data having a non-linear geometry expressed in range (direction perpendicular to orbit track) and Doppler axis. This non-linearity, coupled with the extreme sensitivity of SAR to topographic relief, makes the images hard to correct to a map geometry. The use of a polynomial transformation, for example, simply does not work (De Séve *et al.*, 1996) owing to the accumulation of systematic errors arising from both non-linearity and relief distortion. These errors often result in displacements of features by a few hundred metres, and this displacement is neither uniform nor linear across the image. In spite of these problems, this is the recommended procedure if locational accuracy is not of great concern. It is adequate for most geomorphological and geological applications, where the distribution and orientation of features is of most importance rather than their precise location. If the area of interest has medium to high relief, much of the imagery will be uninterpretable due to terrain distortions and so this level of correction is likely to be insufficient.

The most rigorous and potentially accurate approach to SAR geocoding involves the development of a model-based system which expresses every aspect of the image geometry mathematically. The model should take into consideration the exact position of the sensor in its orbit, the geometry of the sensor (an appreciation of how the sensor collects information from the ground and the geometry it uses to store the raw data) and a model for the earth to account for its curvature. Then, using a limited number of GCPs, the parameters of the model can be estimated and any errors due to the assumed position of the sensor can be assessed. It is also recommended to take account of variation in topographic relief which is termed "terrain correction". This uses the change of height above the smooth geoid (the earth model) and is necessary to both rectify relief distortion and modify the overall geometry so that it conforms to the chosen map projection. An image corrected by this method is easier to interpret. Terrain correction requires the use of a DEM at a similar grid resolution to the SAR data. This presents a problem beyond the expense of acquiring such data, in that many geomorphological and geological investigations may be exploratory in nature, and located in remote and poorly mapped areas. For these cases it is unlikely that the area has been mapped to the required accuracy, and even more unlikely that a DEM is available. In some circumstances it may be possible to use maps to produce a DEM by means of digitising the contours and using a spatial interpolation algorithm, but this method is time-consuming and many artefacts may arise from the interpolation (see McCullagh, Chapter 5, and Wise, Chapter 7). Developments in digital photo-grammetry now allow DEMs to be generated automatically using either scanned aerial photography or stereo satellite imagery such as SPOT. This development will undoubtedly improve both the quality and availability of DEM data suitable for SAR terrain correction.

Whilst there have been a number of terrain correction algorithms published, they have only recently become more readily available as commercial software, and many image analysis systems still do not incorporate them.

To geometrically correct our images two different methods have been used:

(i)   a model designed by Curlander (1984), Curlander *et al.* (1987) and Kwok *et al.* (1987) that solves three equations – the range equation, the Doppler equation and the equation for the earth's model – to determine the cartographic coordinates of a given pixel; processing was executed on a version of this model adapted by A. Sowter at the National Remote Sensing Centre, England

(ii)   a model based upon rigorous photogrammetric principles and developed by Toutin and Carbonneau (1992), Toutin *et al.* (1992) and Toutin (1994); pro-cessing was executed using PCI software.

In this paper we shall describe the use of the photogrammetric approach using PCI software. The main disadvantage of this software over the first method is that it requires more GCPs (about 10) and it also requires the elevation above sea level in addition to Eastings and Northings. The accuracy of the geocoded image can be of the order of two pixels and is restricted by the accuracy of the positioning of the GCPs. The following steps were necessary. The first involved locating 15 GCPs on both the image and on 1:50 000 Ordnance Survey map sheets. The map coordinates of the GCPs were measured and the elevation estimated using adjacent contour data. The GCPs consisted of points on railway lines (which provided the clearest features on the image), bends on rivers, tips of lakes (when these could be clearly identified) and motorways (sometimes very difficult to follow). We avoided GCPs from forest corners and the coast as these seem unreliable. Then, using 1:25 000 sheets covering the GCP areas, the coordinates were refined and elevations checked. The parameters of the model were estimated and a root mean square error (RMSE) produced for each and all control points. GCPs with a high RMSE contribute most to the overall error. These were checked against the map and deleted if necessary, ensuring that at least 10 GCPs remained. The parameters of the model were again estimated until the RMSE values produced were sufficiently small compared to the size of a resolution cell. This process of GCP selection is widely adopted but can introduce a bias to the estimate of true mispositioning error. Some tests on the corrected image showed that the "true" RMSEs were in fact much higher than originally estimated. To avoid this bias, a number of check points could be used to assess the accuracy of the fitted model. However, some difficulties remained: how to collect the check points with sufficient accuracy and how to assess them. These problems have already been partly solved in Statistics using methods such as cross-validating or jack-knifing (Efron and Tibshirani, 1993). Cross-validating is a very simple method that consists of splitting the data (GCPs in our case) into a number of roughly equal sized parts. The model is then fitted to all but one of the parts and the prediction error (RMSE in this case) is estimated for the omitted part. This is repeated until each part has been omitted once and the prediction errors are combined to give an improved estimate of the error. The jack-knife is similar to cross-validating but leaves out one observation at a time. These methods have yet to be applied in any of the models used for geocoding. Note additionally that the accuracy of the geometric correction cannot be refined using more GCPs (Cheng and Toutin, 1995); this can only be achieved by choosing more reliable GCPs.

Two methods were used to assess the accuracy of the geocoded image. More GCPs were located (they are easier to find once the image has been terrain corrected) and used to provide a better estimate of the RMSE. In most cases the "true" RMSE was five to six times that given by the model. Railway lines were digitised from the maps and overlaid on the corrected image (Figure 8.7) to provide a visual as well as quantitative assessment of mispositioning error.

One of the processing steps used during geocoding is resampling. Since pixels need repositioning, interpolation between pixel values in the original image is required. The ideal resampling function is the sinc function, but this is not practicable because an infinite number of pixels would be needed. There are three approximate methods commonly used for resampling: nearest neighbour, bilinear and cubic convolution. Nearest neighbour can create step patterns in structures that are horizontal or vertical in the original image; it is, however, the least computationally intensive. Bilinear resampling yields good results although it tends to smooth the image. Cubic convolution was finally adopted; this provides the greatest sharpness since it approximates more closely to the sinc function. It typically takes about three days to execute a good terrain correction of a SAR image, and the final accuracy is of the order of 30 m.

Contrast Enhancement

ERS-1 SAR amplitude images (PRI images) are supplied in 16-bit format. In other words, the amplitude of the echo is recorded on an integer scale of 0 to 65 535. Most display devices use the pixel values to specify the brightness with which the pixel is illuminated on a computer screen. The range of possible data values and their distribution usually results in a very poor display of the image, often with only a few grey levels utilised. This problem is exacerbated by the limitation of only 256 grey levels by most display devices and/or the software. The task is therefore to

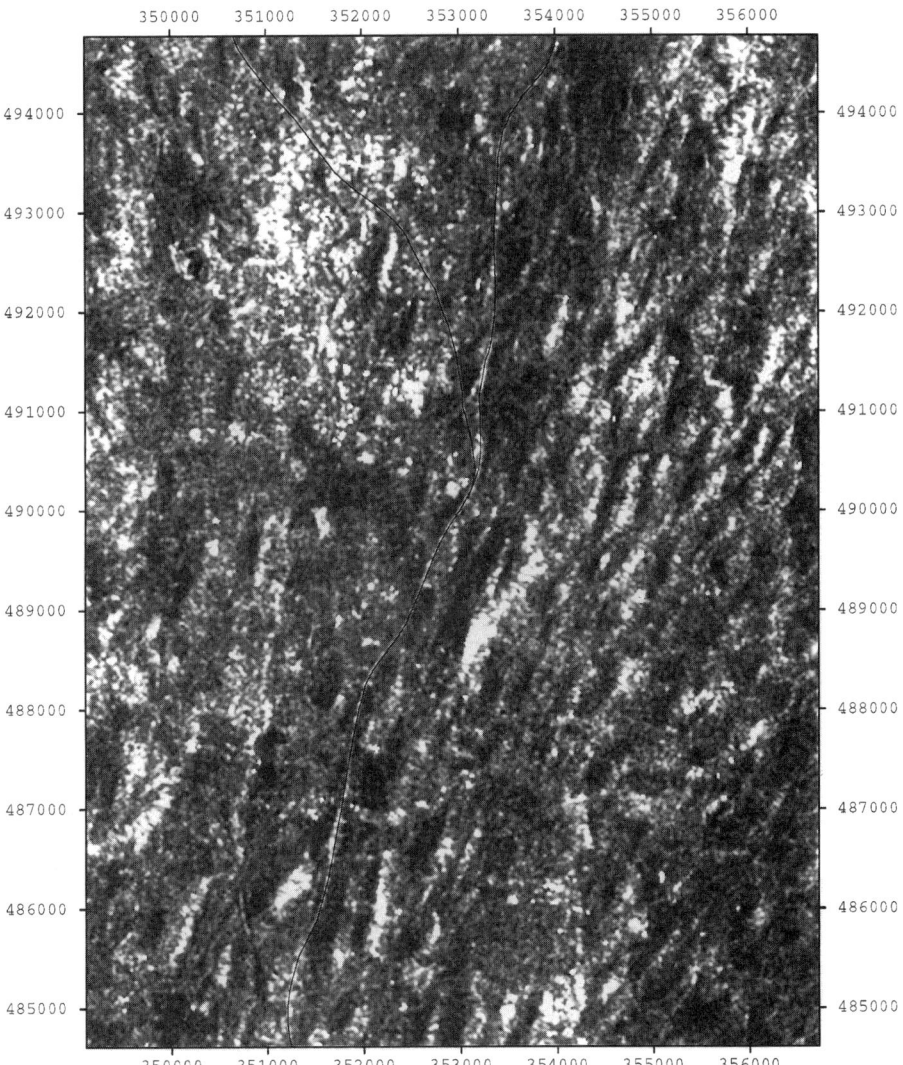

Figure 8.7 A final independent check of geocoding accuracy over the whole image can be made by overlaying a digitised railway network, which provides a better assessment of mispositioning error than by RMSE estimates which are implicitly biased. An appreciation of the high accuracy can be seen by the good fit of the digitised railway to its representation in the image

transform the data so that they can be represented on the screen using the maximum number of grey levels, which should result in an image with good contrast and brightness.

A typical image histogram resembles a positively skewed distribution with a very long tail, and is extremely hard to contrast stretch by the usual methods employed for Landsat data, which is often assumed to approximate to a Gaussian distribution. It is the very large tail extending into high values that presents the main problem. Histograms show that in most cases 99% of the measured amplitude values fall within a small range (1–1500), and it is only the remaining 1% that occupy the wider range of 1500 to 65 535. Pixel values in this higher range are usually corner reflectors or slope facets with an effective incidence angle close to zero. There are a number of methods of transforming the data for the purposes of display, the most popular being a square root transformation. The square root of each pixel value is taken, which has the effect of reducing the high values more than lower values, and often produces the desired result for image display. Experimentation has revealed that the best procedure involves truncating the upper tail of the histogram by replacing all pixel values above a certain threshold with the threshold value. If the threshold is low enough, this method drastically decreases the range of pixel values, and in this experimental work a value of 1500 was usually adopted for the threshold value. An improved value can be found by examining the histogram and selecting a value where the histogram appears to begin its long tail. A valuable benefit of this truncation is that it reduces the visual effects of speckle. A logarithm transformation has also been proposed (e.g. Glasbey and Horgan, 1995). This decreases the range of the pixel values, and at the same time transforms the distribution of the data towards Normality. However, we have observed that when the pixel values vary greatly across an image, the logarithmic transformation is not ideal, making the image less contrast stretched and uniformly whiter.

Data Volume Reduction

Following from the simple processing step discussed above, the data now have a much smaller range (e.g. 0–1500). However, as most display and software systems usually expect 8-bit data (values between 0 and 255), it is necessary to reduce the data further. Reducing the dynamic range of the data from about 1500 to 256 clearly involves a loss of information. This is not a problem for visual interpretation of images, as 256 grey levels are more than the human eye can distinguish. It is also highly advantageous to reduce the data to 8-bit as there is a considerable saving in the size of the data file. An ERS-1 SAR image is usually about 132 Mb in size and reduction to 8-bit format halves this to 66 Mb.

There are a number of methods for rescaling the data to 8-bit, such as linearly or exponentially, or by square root or logarithmic transformations. A number of rescaling methods have been tried but the simple linear rescaling method was found to be adequate after using the method described above for contrast enhancement.

Although ERS-1 SAR PRI images are supplied with a pixel size of 12.5 m, this is not the nominal spatial resolution of the sensor which is 25 m. Once the speckle reduction and rescaling steps have been performed, it is a good idea to resample the image to your chosen resolution, usually 25 m. This is achieved by the values in a 2 by 2 window of four pixels being averaged to a single 25 m pixel value. This step also has the effect of reducing apparent speckle in the image and reduces the data volume by a factor of four. The SAR image is now only about 16.5 Mb in size, about 12% of its original volume. This is important as it drastically reduces the disk size that is required, and is particularly useful if many SAR scenes are being used as a mosaic. It also significantly reduces the time required for further manipulation of the image, and the time it takes to display an image.

PHOTO-INTERPRETATION AND MAPPING TECHNIQUES

Photo-interpretation is a skill that developed with the advent of aerial photographs and is now used with a wide range of images. It consists of visually recognising and mapping features from an image using the interpreter's experience of its appearance in terms of tone, texture, pattern, scale, shadow and context. Whilst photo-interpretation skills are easily transferred from aerial photographs to optical satellite imagery, this is not so because of the geometric distortions and speckle. Due to the recording of microwave energy it is often hard to find reference features which help to navigate across an image. Roads, rivers and lakes, for example, are often hard to locate or are invisible, whereas railway lines (because of the dielectric properties of metal) are easy to detect. It can take a long time to develop the experience necessary to recognise the landforms or lineaments of interest, especially if the more subtle and previously unmapped features are the main concern. It is wise to start in a well known area with ancillary aids such as field maps and aerial photographs.

The most important aspect of SAR photo-interpretation, particularly if the features of interest have a topographic expression, is to be aware of the look direction. This is critical as it permits visualisation of the image in terms of topography, i.e. brighter tones representing slopes facing the look direction and *vice versa*. To the inexperienced, it is common for some features to appear sometimes as hollows or depressions, thus inverting the topography. This is a simple optical illusion which can be rectified by turning the image upside down. Topographic perception is also improved if shadows face towards the bottom right corner of a computer screen. Whilst we have described and recommended speckle reduction, as this usually improves image clarity and makes it easier to detect features, experienced SAR photo-interpreters often find that this is not necessary. This may be because they have become so accustomed to speckle that it ceases to distract them.

Photo-interpretative mapping from digital satellite data can be achieved by two methods. One approach is to produce hard copy prints on a high quality printer and then to trace features of interest. If the printer is not of sufficiently high quality this is not ideal, as most printing techniques are good at representing homogeneous regions but are poor at portraying subtle changes in tone or fine structures. A better approach involves on-screen digitising, in which lineaments are represented by a sequence of simple lines with nodes at each end. This method has the advantage of being able to use the high quality display of the computer monitor and permits mapping at a multitude of scales, by zooming in or out when required. It also means that the interpretations are captured digitally, which is preferable if the information is to be used in a GIS. Digital lineament maps can then be overlaid or combined with ancillary data, such as structural geology, drift cover or field maps, or can be exported to packages for performing data representation such as rose diagrams, and statistical analysis. Clark and Wilson (1994) report on software (available by ftp) file transfer protocol which uses digitised lineament data to produce outputs such as line length, orientation, parallel and perpendicular spacing, nearest neighbour distances and an assessment of how the spatial distribution of lineaments compares to a purely random pattern.

CONCLUSIONS AND RECOMMENDATIONS

It has been shown that SAR images, used either alone or in conjunction with optical imagery, are of great use for mapping landforms and lineaments. The tools required to extract the necessary information automatically have yet to be fully developed and existing methods need to be assessed on a rigorous basis. It is still the case that the best method for extracting lineament information is to process the image to improve lineament visibility and then to use photo-interpretation. A number of speckle reduction filters were examined and the MAP filter is considered to be the most appropriate. Simple methods for enhancing the image and decreasing the data volume have also been proposed. Different methods of geocoding and terrain correction were examined and based on experimentation using different techniques, a processing route to enhance images for photo-interpretation is now outlined (Figure 8.8).

The next phase of this work is to apply statistical performance tests to better understand and evaluate alternative processing techniques, and to explore the performance of segmentation and linear feature detection algorithms. We are currently assessing a number of semi-automatic methods using segmentation, for example the simulated annealing method of White (1991). Interferometry (Zebker *et al.*, 1994) also appears to be a promising method for extracting the structures of interest in the future. As Radarsat is capable of providing images at a variety of spatial resolutions and incidence angles, it is important that the user knows which will be of most use in their work. For a range of test sites, we aim to evaluate the sensitivity of lineament detection to incidence angle in order to ascertain the most appropriate image acquisition parameters for extracting linear geomorphological information.

ACKNOWLEDGEMENTS

We would like to thank the European Space Agency for supplying the SAR images under their pilot project scheme (PP2-UK8), and the University of Sheffield for providing a scholarship to C.P.V. We would also like to thank NERC for the TM imagery used, and Andy Sowter of NRSC for allowing us to use the geocoding software. We would also like to thank the referees and editors for their helpful suggestions for improving this paper.

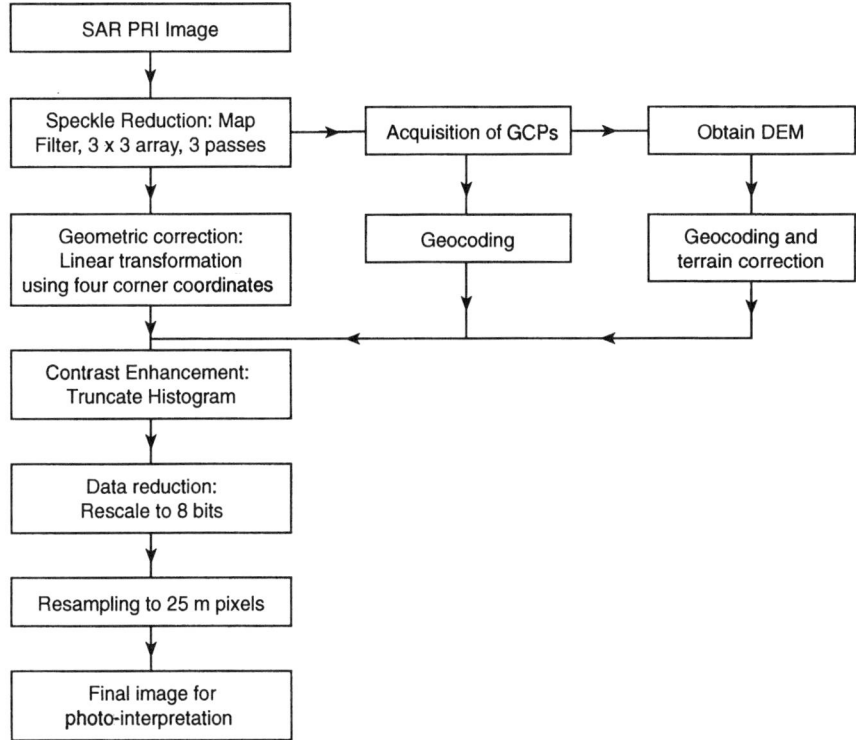

Figure 8.8 A summary of the recommended processing steps for producing images for photo-interpretation. The alternative routes depend on the degree of sophistication of the geometric correction required, with the right hand route being the best

REFERENCES

Beaudoin, A., Deshayes, M., Piet, L., Stussi, N. and Le Toan, T., 1994. Retrieval and analysis of temperate forest backscatter signatures from multitemporal ERS-1 data acquired over hilly terrain. *Proceedings of the first ERS-1 pilot project workshop*, Toledo, Spain, 22–24 June 1994, ESA publication, sp-365, 283–289.

Boulton, G. S. and Clark, C. D., 1990. A highly mobile Laurentide Ice Sheet revealed by satellite images of glacial lineations. *Nature*, **346**, 813–817.

Cheng, P. and Toutin, T., 1995. High accuracy data fusion of airphoto and satellite images. *ACSM/ASPRS Annual Convention & Exposition Technical Paper*, 27 February– 2 March 1995, 2, 453 Charlotte, NC, USA.

Clark, C. D., 1990. Remote sensing scales related to the frequency of natural variation; an example from palaeo-ice flow in Canada. *IEEE Transactions on Geoscience and Remote Sensing*, **GE-28**, 503–508.

Clark, C. D., 1993. Mega-scale glacial lineations and cross-cutting ice-flow landforms. *Earth Surface Processes and Landforms*, **18**, 1–29.

Clark, C. D., 1994. Large scale ice-moulded landforms and their glaciological significance. *Sedimentary Geology*, **91**, 253–268.

Clark, C. D. and Knight, J. K., 1994. ERS SAR data as a tool for landform and lineament identification in previously glaciated areas. *Proceedings of the first ERS-1 pilot project workshop*, Toledo, Spain, 22–24 June 1994, ESA publication, sp-365, 211–218.

Clark, C. D. and Wilson, C., 1994. Spatial analysis of lineaments. *Computers and Geoscience*, **20**, 1237–1258.

Clark, C. D., Knight J. K. and Gray, J. T., 1995. Reconstructing ice dynamic scenarios of the Laurentide Ice Sheet in the Ungava and southern Baffin Island sectors: scenario evaluation and implications. Abstract in proceedings of *INQUA XIV International Congress*, 1995, Berlin, 50.

Clark, C. D., Knight J. K. and Gray, J. T., 1996. Reconstructing ice dynamic scenarios of the Laurentide Ice Sheet. *26th International Arctic Workshop, Arctic and Alpine Environments, Past and Present*, program with abstracts. Institute of Arctic and Alpine Research, University of Colorado, Boulder, 18.

Curlander, J. C., 1984. Utilisation of spaceborne SAR data for mapping. *IEEE Transactions on Geoscience and Remote Sensing*, **GE-22**, 106–112.

Curlander, J. C. and McDonough, R. N., 1991. *Synthetic Aperture Radar – Systems and Signal Processing*. Wiley, New York, Chapter 1, 1–26.

Curlander, J. C., Kwok, R. and Pang, S. S., 1987. A post-processing system for automated rectification and registration of spaceborne SAR imagery. *International Journal of Remote Sensing*, **8**, 621–638.

Daily, M. I., Farr, T., Elachi, C. and Schaber, G., 1979. Geologic interpretation from composited radar and Landsat imagery. *Photogrammetric Engineering and Remote Sensing*, **45**, 1109–1116.

Dainty, J. C., 1984. Recent developments. *Laser Speckle and Related Phenomena*, ed. J. C. Dainty, Springer-Verlag, Berlin, Chapter 8.

Davis, L. S. and Rosenfeld, A., 1978. Noise cleaning by iterated local averaging. *IEEE Transactions on Systems, Man and Cybernetics*, **SMC-8**, 705–711.

Deffontaines, B., Lee, J. C., Angelier, J., Carvalho, J. and Rudant, J. P., 1994. New geomorphic data on the active Taiwan orogen – a multisource approach. *Journal of Geophysical Research – Solid Earth*, **99**, 20 243–20 266.

De Séve, D., Toutin, T. and Desjardins, R., 1996. Evaluation de deux méthodes de correction géométrique d'images Landsat TM et ERS-1 SAR dans une etude de linéaments géologiques. *International Journal of Remote Sensing*, **17**, 131–142.

Durand, J. M., Gimonet, B. J. and Perbos, J. R., 1987. SAR data filtering for classification. *IEEE Transactions on Geoscience and Remote Sensing*, **GE-25**, 629–637.

Efron, B. and Tibshirani, R. J. 1993. *An Introduction to the Bootstrap*, Chapman and Hall, London, Chapter 17, 239–241.

Erdas, 1994. *Field Guide of Erdas Imagine Image Processing Software*. ERDAS, Inc., Atlanta, USA.

Ford, J. P., 1984. Mapping of glacial landforms from SEASAT radar images. *Quarternary Research*, **22**, 314–327.

Frost, V. S., Perry, M. S., Dellwig, L. F. and Holtzmann, J. C., 1989. A model for radar images and its application to adaptive digital filtering of multiplicative noise. *IEEE Transactions on Pattern Analysis and Machine Intelligence*, PAMI-4, 157–166.

Glasbey, C. A. and Horgan G. W., 1995. *Image Analysis for the Biological Sciences*. Wiley, Chichester, 72–73.

Goodman, J. W., 1984. Statistical properties of laser speckle patterns. In *Laser Speckle and Related Phenomena*, ed. J. C. Dainty, Springer-Verlag, Berlin, Chapter 2.

Graham, D. F. and Grant, D. R., 1991. A test of airborne, side looking synthetic aperture radar in Central Newfoundland for geological reconnaissance. *Canadian Journal of Earth Science*, **28**, 257–265.

Greenbaum, D., Marsh, S. and Tragheim, D., 1993. Landsat inputs to UK geological mapping. *Geoscientist*, **3**, 4–6.

Guo, H. D., Zhang, Y. H., Shao, Y., Dong, P. L. and Wang, C., 1993. Geological analysis using Shuttle Imaging Radar and airborne SAR in China. *Advances in Space Research*, **13**, 79–82.

Harris, J., Murray, R. and Hirose, T., 1990. IHS transform for the integration of radar imagery with other remotely sensed data. *Photogrammetric Engineering and Remote Sensing*, **56**, 1631–1641.

Isiorho, S. A., 1984. Radar geology of the Shelleng-Numan area of Nigeria. An evaluation. *International Journal of Remote Sensing*, **5**, 519–531.

Knight, J. K. and Clark, C. D., 1995. Reconstructing ice dynamic scenarios of the Laurentide Ice Sheet in the Ungava and southern Baffin Island sectors: scenario building. Abstract in proceedings of *INQUA XIV International Congress*, 1995, Berlin, 138.

Korwar, V. N. and Pierce, J. R., 1981. Detection of gratings and small features in speckle imagery. *Applied Optics*, **20**, 312–319.

Kropatsch, W. G. and Strobl, D., 1990. The generation of SAR layover and shadow maps from digital elevation models. *IEEE Transactions on Geoscience and Remote Sensing*, **GE-28**, 98–107.

Kuan, D. T., Sawchuk, A. A., Strand, T. C., and Chaval, P. C., 1987. Adaptive restoration of images with speckle. *IEEE Transactions on Acoustics, Speech and Signal Processing*, **ASSP-35**, 373–383.

Kwok, R., Curlander, J. C. and Pang, S. S., 1987. Rectification of terrain induced distortions in radar imagery. *Photogrammetric Engineering and Remote Sensing*, **53**, 507–513.

Leberl, F. W., 1990. *Radargrammetric Image Processing*. Artech House, Boston.

Lee, J. S., 1981. Refined filtering of image noise using local statistics. *Computer Graphics and Image Processing*, **15**, 380–389.

Lee, J. S., 1983. A simple smoothing algorithm for Synthetic Aperture Radar images. *IEEE Transactions on Systems, Man and Cybernetics*, **SMC-13**, 85–89.

Lopes, A., Nezry, E., Touzi, R. and Laur, H. (1990). Maximum a posteriori speckle filtering and first order texture models in SAR images. *IGARSS'90*, vol. III, Washington, 2409–2412. Institute of electrical and electronic engineering Inc., Maryland, USA.

Martin, F. J. and Turner, R. W., 1993. SAR speckle reduction by weighted filtering. *International Journal of Remote Sensing*, **14**, 1759–1774.

Mascarenhas, N. D. A., Ono, S. E., Fernandes, D., Kux, H. J. H., 1991. A comparative study of speckle reduction filters and their application for classification performance improvement. Presented at the *24th International Symposium on Remote Sensing of the Environment*, Rio de Janeiro, Brazil, 27–31 May 1991.

Millington, A. C., White, K., Drake, N. A., Wadge, G. and Archer D. J., 1995. Remote sensing of geomorphological processes and surficial material geochemistry in drylands. In *Advances in Environmental Remote Sensing*, eds F. M. Danson and S. E. Plummer, Wiley, Chichester, 105–122.

Mitchell, W. A. and Clark, C. D., 1994. The last ice sheet in Cumbria. In *The Quaternary of Cumbria: Field Guide*, eds J. Boardman and J. Walden, Quaternary Research Association, Oxford, 4–14.

Nagao, M. and Matsuyama, T., 1979. Edge preserving smoothing. *Computer Graphics and Image Processing*, **9**, 394–407.

Normand, M., Chkir, N., Cognard, A., Imberti, M., Loumagne, C., Ottle, C., Vidal, A. and Vidal-Madjar, D., 1994. Estimation of surface soil moisture from ERS-1 SAR data for hydrological modelling purposes. *Proceedings of the first ERS-1 pilot project workshop*, Toledo, Spain, 22–24 June 1994, 97–102.

Paudyal, D. R. and Aschbacher, J., 1993. Evaluation and performance tests of selected SAR speckle filters. Presented at the *International Symposium "Operationalization of Remote Sensing"*, ITC, Enschede, The Netherlands, 19–23 April 1993.

Porcello, L. J., 1976. Speckle reduction in synthetic-aperture radars. *Journal of the Optical Society of America*, **66**, 1305–1311.

Quegan, S., 1995a. An introduction to SAR theory, statistical properties of ERS-1 SAR data and scattering mechanisms. Unpublished Report.

Quegan, S., 1995b. Recent advances in understanding SAR imagery. In *Advances in Environmental Remote Sensing*, eds F. M. Danson and S. E. Plummer, Wiley, Chichester, 89–104.

Rao, P. V. N., Vidyadhar, M. S. R. R., Rao, T. CH. M. and Venkaratnam, L., 1995. An adaptive filter for speckle suppression in synthetic aperture radar images. *International Journal of Remote Sensing*, **16**, 877–889.

Roy, D. W., Schmitt, L., Woussen, G. and DuBerger, R., 1993. Lineaments from airborne SAR images and the 1988 Saguenay earthquake, Quebec, Canada. *Photogrammetric Engineering and Remote Sensing*, **59**, 1299–1305.

Schreier G. (1993). *SAR Geocoding: Data and Systems*. Wichmann, Germany.

Toutin, T., 1994. Multisource data integration with an integrated and unified geometric modelling. *Presented at EARSEL workshop*, Sweden, June 94.

Toutin, T. and Carbonneau, Y., 1992. MOS and SEASAT image geometric corrections. *IEEE Transactions on Geoscience and Remote Sensing*, **GE-30**, 603–609.

Toutin, T., Carbonneau, Y. and St-Laurent, L., 1992. An integrated method to rectify airborne radar imagery using DEM. *Photogrammetric Engineering and Remote Sensing*, **58**, 417–422.

Way, J. and Smith, E. A., 1991. The evolution of Synthetic Aperture Radar systems and their progression to the EOS SAR. *IEEE Transactions on Geoscience and Remote Sensing*, **GE-29**, 962–985.

Welch, R. and Ehlers, M., 1988. Cartographic feature extraction with integrated SIR-B and Landsat TM images. *International Journal of Remote Sensing*, **9**, 873–889.

White, R. G., 1991. *Simulated annealing applied to discrete region segmentation of SAR images*. Technical Report 4498, Radar Signals Research Establishment (RSRE), Malvern, UK.

Yésou, H., Besnus, Y., Rolet, J., Pion, J. C. and Aing, A., 1993. Merging Seasat and Spot imagery for the study of geological structures in a temperate agricultural region. *Remote Sensing of Environment*, **43**, 265–279.

Zebker, H. A., Werner, C. L., Rosen, P. A. and Hensley, S., 1994. Accuracy of topographic maps derived from ERS-1 interferometric radar. *IEEE Transactions on Geoscience and Remote Sensing*, **GE-32**, 823–836.

APPENDIX

Further Information about SAR Systems and how to Acquire Imagery

This appendix lists some of the features of well-known space-borne SAR systems. The World Wide Web (WWW) address of each system is provided for anyone seeking further information. These WWW pages contain information such as a wide range of images, technical information about the system and its coverage, and explain how to acquire the images. A very informative WWW page that provides intensive information about SAR systems in general is the Remote Sensing Society SAR Special Group. A number of airborne SAR systems exist; two well known ones are AirSar of the Jet Propulsion Laboratory (JPL) and the Dutch system Pharus. Information about the use of SAR for planetary exploration such as the Magellan explorer is also available. The addresses of these home pages are given below. Some of the agencies which manage the space-borne systems and the orbits of the systems are also provided. Table 8.1 gives a summary of the specifications of these systems.

Table 8.1 Specifications of some well known space-borne SAR systems

Name of system	Date launched	Inclination of orbit (deg.)	Altitude (km)	Swath width (km)	Antenna dimensions (m)	Wavelength (cm)	Incidence angle at centre (deg.)	Polarisation	Nominal Resolution (m)	Repeat cycle (days)
ERS-1	July 1991	98.5	782–785	102.5	10 by 1	5.6	23	VV	25	35
ERS-2	April 1995	98.5	782–785	102.5	10 by 1	5.6	23	VV	25	35
Radarsat	November 1995	98.6	793–821	50–500	15 by 1.5	5.6	20–50	HH	25[e]	24
JERS-1	February 1992	97.7	568	75	12 by 2.5	23.5	38	HH	18	44
SIR-A[a]	November 1981	38.0	259	50	2.16 by 9.4	23.5	50	HH	40	—
SIR-B	October 1984	57.0	224 (varied)	20–60	2.16 by 10.7	23.5	15–60	HH	25[e]	—
SIR-C and	Ongoing	57.0	225 (varies)	15–90	12 by 3.7	23.5, 5.8	17–63	HH,HV,VV	30[e]	—
X-SAR[c]	Ongoing	57.0	225 (varies)	15–40	12 by 0.4	3.1	17–63	VV	30[e]	—
Seasat[d]	June 1978	108	805	100	2.16 by 10.7	23.5	23	HH	25	—
Almaz	March 1991	73	300	20–45	15 by 1.5	10	30–60	HH	15[e]	4

[a] Launched from Space Shuttle. Mission lasted 2.5 days
[b] Launched from Space Shuttle. Mission lasted 8.3 days
[c] Launched from Space Shuttle. Mission ongoing
[d] Launched from Space Shuttle. Mission lasted 100 days
[e] Resolution varies as radar beam can be shaped and steered

Remote Sensing Society:	http://axp10.iend.wau.nl/sar/sig/rad_sig.htm
AirSar:	http://www-airsar.jpl.nasa.gov
Pharus:	http://TUDEDV.ET.TUDELFT.NL/www/ttt/rs/pharus/pharus_ home .html
Planetary Exploration:	http://www.jpl.nasa.gov/mip/planet.html

1. ERS-1

Agency:	European Space Agency (ESA)
Orbit:	Near circular, polar and sun-synchronous
WWW address:	http://gds.esrin.esa.it/ERS1.1

2. ERS-2 (same specifications for SAR as ERS-1)

Orbit:	same orbital plane as ERS-1
WWW address:	http://sloth.esrin.esa.it:80/specers2.htm

3. Radarsat

Agency:	Canadian Space Agency
Orbit:	sun-synchronous
WWW address:	http://radarsat.espace.gc.ca/welcome.html

4. JERS-1

Agency:	NASDA (Japan)
Orbit:	sun-synchronous
WWW address:	http://hdsn.eoc.nasda.go.jp/guide/guide/satellite/satdata/jer-s_e.html

5. SIR-A

Agency:	NASA (USA)
Orbit:	various, mounted on Space Shuttle missions
WWW address:	http://www.jpl.nasa.gov/mip/sira.html

6. SIR-B

Agency:	NASA (USA)
Orbit:	various, mounted on Space Shuttle missions
WWW address:	http://www.jpl.nasa.gov/mip/sirb.html

7. SIR-C/X-SAR (four channels)

Agency:	NASA (USA)
Orbit:	various, mounted on Space Shuttle missions
WWW address:	http://southport.jpl.nasa.gov/desc/SIRCdesc.html

8. Seasat

Agency:	NASA (USA)
WWW address:	http://www.jpl.nasa.gov/mip/seasat.html

9. Almaz

Agency:	USSR
Orbit:	non-sun-synchronous
WWW address:	http://gds.esrin.esa.it/15C81B2C/Ceuri.almaz

9 Image Analysis of Aerial Photography to Quantify the Effect of Gold Placer Mining on Channel Morphology, Interior Alaska

DAVID J. GILVEAR[1], TERTIA M. WATERS[1] and
ALEXANDER M. MILNER[2]
[1] Department of Environmental Science, University of Stirling, UK
[2] Department of Geography, University of Birmingham, UK

ABSTRACT

Placer mining for gold in Interior Alaska causes large-scale changes in channel and floodplain morphology, but little is known about the precise geomorphological effects or time scales of recovery following mining. In this study, image analysis of aerial photography was used to detect changes in channel morphology prior, during and following gold placer mining on Faith Creek. The analysis demonstrated that, prior to mining, the study reach resembled an anastomosing gravel-bed river containing a range of in-stream features including pools, riffles, and lateral and mid-channel gravel bars. Unmined McManus Creek, which joins Faith Creek and forms the larger Chatanika River, likewise was shown to be similar to an anastomosing stream and maintains a range of bar forms and morphological features. During mining, Faith Creek was diverted to a channelised reach with a lack of morphological diversity. Once mining ceased, a channel avulsion subsequently occurred and Faith Creek abandoned its channelised course. Image analysis revealed that four years after the avulsion, a new single-thread channel with a range of bed-forms had formed, but with few deep pools. This suggests that on Faith Creek geomorphological recovery following mining, and particularly the formation of well developed pool–riffle sequences, involves a number of large flood events, likely to require more than 10 years. Moreover, the study suggests that the channel is likely to remain incised and single thread as a result of having cut into the post-mining spoil at an elevation higher than the pre-mining valley floor. More generally, the study demonstrates that for low turbidity (< 20 nephlometric turbidity units) gravel-bed rivers, with water depths generally less than 1 m, image analysis applied to aerial photography can provide high resolution spatial data on channel morphology. Such potential information could be beneficial to a range of fluvial studies.

INTRODUCTION

In recent years, a number of new technologies have become available to geomorphologists for mapping and representing landforms. These include computerised digital

Landform Monitoring, Modelling and Analysis. Edited by S. N. Lane, K. S. Richards and J. H. Chandler.
© 1998 John Wiley & Sons Ltd.

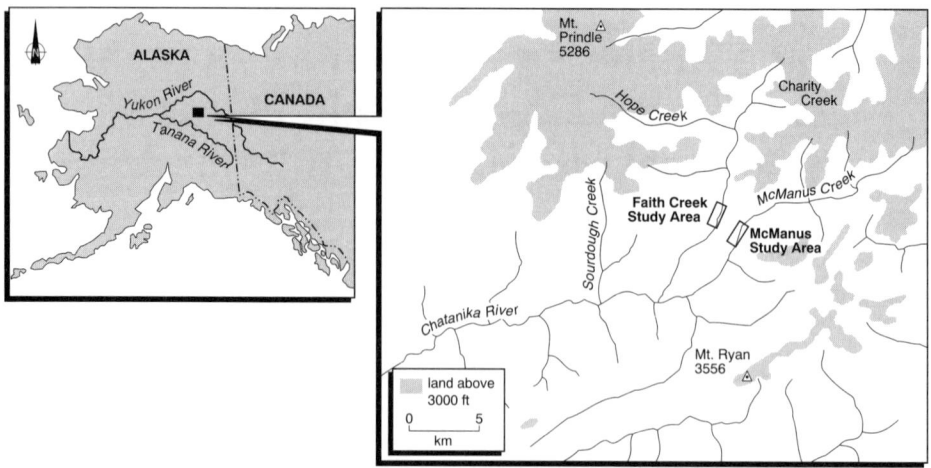

Figure 9.1 The McManus Creek and Faith Creek study areas

terrain models (DTMs), photogrammetry, Global Positioning Systems (GPS), satellite and airborne multispectral imagery and new dating techniques (Cornelius *et al.*, 1994). In some cases they supersede previous techniques while in others they complement existing approaches. Only recently, however, has the contribution of image analysis to remotely sensed data of streams and rivers been fully recognised by fluvial geomorphologists (Milton *et al.*, 1995; Muller *et al.*, 1993). For example, Phillip *et al.* (1989) applied linear contrast stretching and band ratioing to Landsat data of the Ganga River, Bangladesh, to date channel changes between 1975 and 1986 and reconstruct old palaeochannel patterns. Similarly, Gurnell *et al.* (1994) scanned aerial photographs, geometrically corrected and registered them to a base map and digitised bank lines to detect planform change on the River Dee, England. Recent studies have also illustrated the potential of applying image analysis to aerial photographs and multispectral imagery to map channel bathymetry in non-turbid gravel-bed rivers (Winterbottom, 1996; Acornley *et al.*, 1995; Hardy *et al.*, 1994). Image analysis of remotely sensed data should be seen as an "enabling" technology (Milton *et al.*, 1995). It does not seek to replace traditional field-based investigations but to complement them by allowing the results and interpretations of such studies to be extended across wider areas, from the cross-section to the reach and thence to the channel segment scale.

This paper applies image analysis techniques developed by Winterbottom and Gilvear (submitted) to conventional aerial photography in order to provide detailed information on channel morphology. The technique is used to complement field investigations of the geomorphic effects of gold placer mining on impacted streams in the Circle Mining District of Alaska (Figure 9.1). The primary aim of the paper is to suggest that image analysis is a useful tool for geomorphological research, although the authors recognise that further work is required to confirm its wider applicability and to test accuracy on a range of river types. A secondary aim of the paper is to provide an insight into the geomorphic effects of placer mining. Few studies have examined the geomorphological effects of placer mining, except in areas where 19th century mining introduced exceptional amounts of sediment into stream systems prior to environmental protection legislation (e.g. James, 1991; Knighton, 1989; Graf, 1979). The objectives of the geomorphic investigations are to assess time scales of geomorphic and ecological recovery of streams following mining and to provide recommendations for river restoration and ecological recovery. Although studies have examined the effects of placer mining in Alaska on water quality (Bjerklie and LaPerriere, 1985) and invertebrates (Wagener and LaPerriere, 1985), little is known about geomorphological impacts.

BACKGROUND

Gold Placer Mining in Alaska and its Impact on Stream Geomorphology

Placer mining is a type of opencast mining that targets alluvial deposits called placers, and most of Alaska's gold is found in such deposits. Placers are usually located close to the interface between the valley fill and bedrock, and mining typically involves disruption to the stream channel, active floodplain and old terraces adjacent to the valley sides.

In a typical modern Alaskan placer mining operation, heavy machinery is used to scrape away the vegetation mat and organic layers (the overburden) to reveal the underlying alluvial deposits. Meanwhile, the stream is often diverted to a channelised reach at the edge of the valley floor, or onto a low-level terrace, and kept there during mining by the bull-dozing of spoil banks. All alluvial deposits containing gold are removed and transferred to an on-site processing plant where screening, sorting and sluicing is undertaken. Sluicing waters were traditionally discharged directly to the adjacent water course, but the state of Alaska has set turbidity standards for the dilution of placer mining effluent. The receiving stream cannot now exceed 5 NTU (nephlometric turbidity units) above natural conditions and so settling ponds are employed.

Traditionally, after mining, the valley floor was left covered with conical heaps of tailings upon which little vegetation colonised. In this situation the post-mining river is usually found flowing around the tailings and is not able to maintain a natural planform, except where the stream has sufficient power to erode the tailings. The United States Department of the Interior's Bureau of Land Management now requires the mined spoil to be levelled in order to facilitate recovery of the natural floodplain ecosystem. A problem arises because the levelled volume is typically one-third greater than the excavated volume owing to lack of compaction, and so it is impossible to re-create the original valley floor morphology (Figure 9.2A). Three options are available: the whole valley floor can be levelled to an overall higher elevation (Figure 9.2B); the original elevation of the active floodplain can be maintained, but excess material is piled up against the valley sides (Figure 9.2C); finally, a V-shaped section can be created with no level valley floor (Figure 9.2D). Each option has inherent problems including unnecessary disturbance of the valley sides, an unnatural floodplain elevation and/or a lack of valley floor in which a

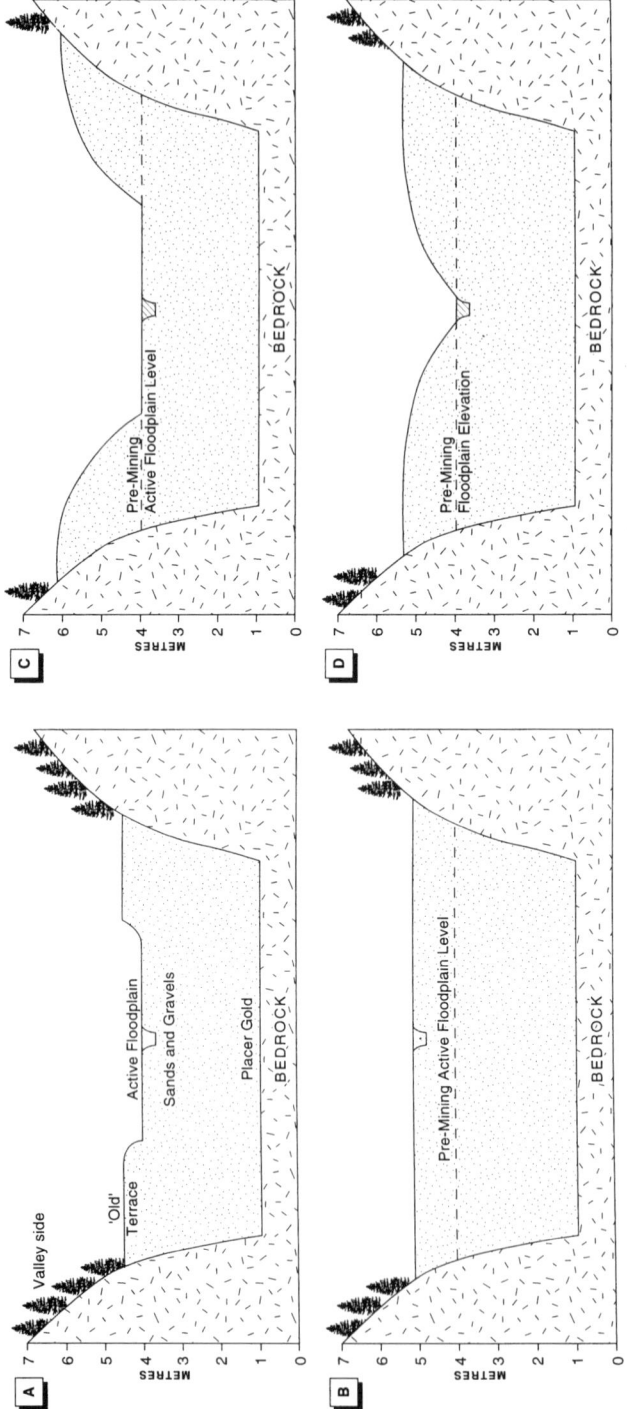

Figure 9.2 Idealised cross-sections showing options for spoil reclamation following mining. (A) Pre-mining; (B) levelling at a higher elevation; (C) creation of a flat valley bottom at a pre-mining floodplain elevation; (D) creation of a V-shape

stream can "carve" a natural morphology. In relation to the stream or river itself a number of post-mining practices are apparent: the stream may be left within its artificial channel, only returning to a more natural course if it erodes the banks of spoil constraining its position and adopts a new course on the levelled spoil; the river water may be diverted back onto the levelled spoil and left to find its own course; finally, the spoil may be contoured and an attempt made to "engineer" a course for the river by creating a channel of appropriate dimensions in which the stream or river can flow (Jackson and Van Haveren, 1984; Karle and Densmore, 1994).

The Theoretical Basis to Applying Image Analysis to Aerial Photography for Mapping Channel Morphology

Mapping of fluvial features using remotely sensed data relies on the presence of measurable differences in electromagnetic radiation between geomorphic features. At the most basic level, water, green vegetation and bare substrate exhibit different spectral response curves and these components may be thought of as the "spectral end-members" of the fluvial environment.

With regard to in-channel morphology, one requires a knowledge of the reflectance from exposed sediments and the effect of inundation by water on substrate reflectance. The reflection of sunlight from unvegetated gravels in the visible spectrum is usually high and allows them to be easily differentiated from water and vegetation. With respect to reflection from the stream water itself, Bouger's law states that, in an adsorbing medium, the intensity of radiation decreases exponentially with distance. Applying this law to aerial photography of river channels, the level of light reflected from the bottom of a water body is not linearly related to water depth, and this is illustrated by field spectra collected from a gravel substrate of the River Tay, Scotland (Gilvear and Winterbottom, submitted; Figure 9.3a). These data firstly show marked differences in spectral response obtained from the substrate with different depths of inundation, and secondly that 10 cm changes in water depth have decreased effects on the difference in spectral response in deeper water. This suggests that spatial variability in reflectance from stream channel bed material may potentially be useful in remotely detecting channel bathymetry in very clear or shallow waters. In deep or turbid water, however, differences in reflectance levels will generally become undetectable between different water depths and such imagery would be unsuitable for bathymetric mapping. To overcome the non-linear response problem, Lyzenga (1981) proposed an algorithm which can be applied to observed reflectances so that they become a linear function of depth:

$$R = \ln(L_1 - L_w) \qquad (9.1)$$

where R is the variable that is linearly related to water depth, L_1 is the observed brightness and L_w is the deepest water reflectance level.

The reflection of sunlight from a water body, however, is related not only to water depth but to water surface backscatter, water turbidity and stream bed reflectance (e.g. sediment type and algal covering). Thus, in Figure 9.3b, the data collected from algae-covered channel substrates on the River Tay, Scotland, have

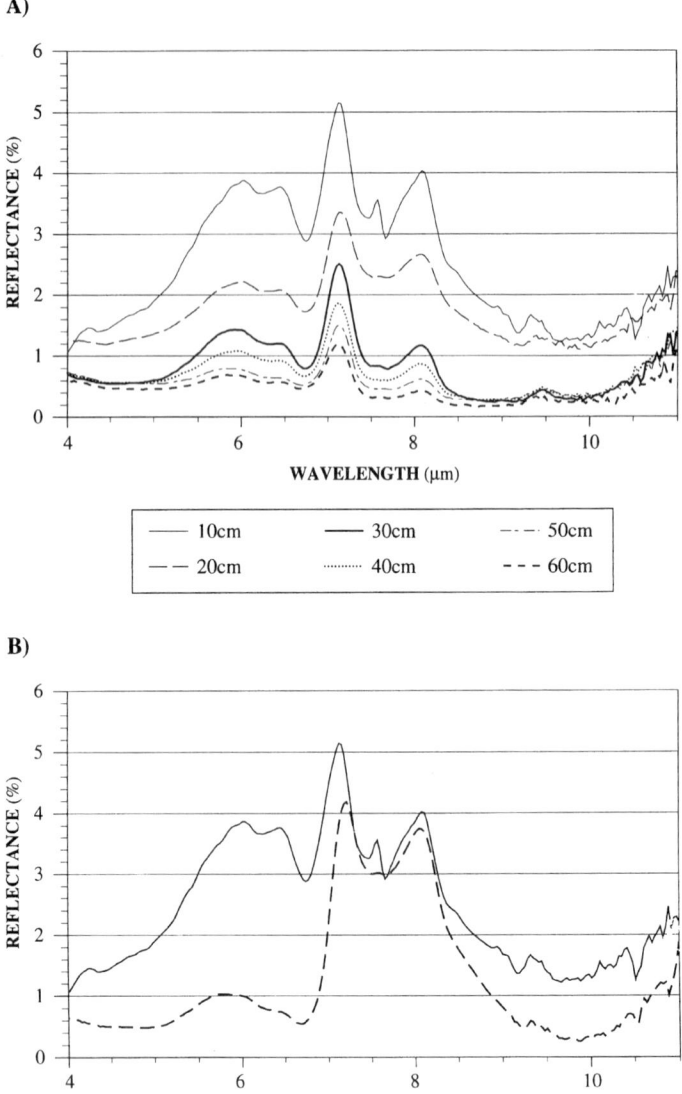

Figure 9.3 Spectral response curves from gravel substrates on the River Tay, Scotland (Winterbottom, 1996): (A) for clean gravels with differing water depths; (B) for "clean" and algae covered gravel substrates with 20 cm of inundation

high absorption in the 0.6–0.7 μm part of the spectrum and obscure the effect of increasing water depth (Winterbottom and Gilvear, submitted). Between 0.7 and 0.8 μm however, although the algal covering causes some amount of absorption, it is still possible to discriminate between different water depths; thus a correlation coefficient of 0.71 was established between measured water depths and the digital number of bands 5, 6, 8 and 3 (visible and near-infrared) of airborne thematic mapper (Daedalus 1268) data. This allowed channel bathymetry to be mapped over a reach of the River Tay comprising pools, riffles and a range of other bedforms. Similarly, Winterbottom and Gilvear (submitted), for the same reach, showed that water depths were well correlated with digital numbers obtained from panchromatic black and white aerial photographs of the reach. However, riffle data had a lower correlation coefficient than pool areas. It was presumed that the broken water over riffles gave abnormally high reflectance values because of increased water surface backscatter. Panchromatic black and white film records visible light in the 0.4–0.7 μm range.

   Thus, in order to use the reflectance from a water body for mapping channel morphology, the effects of differences in surface backscatter, substrate characteristics and turbidity need to be minimal or the effects of spatial variability should be taken into account. In the case of mapping in-stream geomorphological features, spatial variability in surface backscatter and substrate character, in part, define different in-stream morphological units. Thus, high levels of backscatter are a function of water with high surface roughness (rapids, cascades and riffles). Elsewhere, except in the presence of very low sun angles, backscatter normally only affects isolated pixels. It can then be partially compensated for by filtering the image; the value of each pixel becomes the average of itself and its eight nearest neighbours. Similarly, differences in substrate characteristics on gravel-bed rivers usually relate to morphological features. Thus, gravel-bed substrates generally have a high reflectance of white light; any variability tends to be that of finer, darker sediments in pools, highlighting further the spectral differences between shallow and deep water and pools and riffles. Extensive mats of benthic algae, however, can causes differences in spectral response of the substrate, independent of bed morphology, and are a potential source of error. Uniform mixing of water usually precludes local-scale variability in turbidity but, when comparing images from different dates, temporal variability may be a problem.

   Within the context of this paper, the image analysis of remotely sensed data provides a retrospective view of channel morphology before and during placer mining and high spatial resolution data on post-channel morphology which could not be obtained by field survey.

THE STUDY AREA

The Circle Mining District of Alaska between 65 and 66°N (Figure 9.1) has been actively mined for placer gold since the beginning of the last century, and most streams in the area have been mined within the last 100 years. Despite this long history of mining, better methods of working and the current price of gold have maintained and increased both the level and extent of mining, with few streams currently not being mined. Moreover, old workings are being remined with improved techniques for gold recovery. Mining therefore continues to represent a major disturbance on most streams within the region, and its geomorphological effects and time scales of recovery following cessation of mining need to be understood.

In order to investigate how mining affects stream geomorphology and how stream channels "recover" following mining, image analysis techniques have been applied to aerial photography of two stream reaches; firstly to a 2 km reach of Faith Creek mined during much of the 1980s; and secondly to a 850 m reach of McManus Creek, which has never been mined and lies within an almost pristine catchment (Figure 9.1). The second reach will provide the study with a natural control. Both reaches are approximately 16.5 km from the stream source, but at this point McManus Creek is appreciably smaller (8–10 m) than Faith Creek (15–20 m). Both streams naturally resemble anastomosing gravel-bed rivers but during the summer the low flow only occupies a "parent" channel (Riley, 1975). Each anabranch resembles either a meandering or wandering gravel-bed river and the anastomosed nature of the river probably results from the effects of aufeis on channel location during the spring thaw and ice break-up (see below). The Faith Creek study reach was not directly affected by mining in 1966 (when aerial imagery was first available), although mines were located further upstream. The reach was mined during the early 1980s and, by 1989, had been reclaimed by levelling the mine tailings and leaving the channel to recover naturally. Thus, in relation to the post-mining management practices outlined earlier, levelling of the valley floor (Figure 9.2B) was undertaken, with the stream being left within a straight channel on a high-level terrace confined by bull-dozed spoil. Areas upstream are still being mined.

Mean annual precipitation in the area is 260 mm with 60–70% occurring as rain between June and September. A mean annual temperature of –9°C is responsible for discontinuous permafrost in the region and aufeis in the valley floors. Aufeis is formed by hydrostatic pressure which forces groundwater, in liquid phase, upwards through cracks in the overlying ice whereupon it freezes as it spills onto the ice surface, causing a further increase in ice thickness. This can be an important factor in causing channel change upon spring break-up (Scrimigeour *et al.*, 1994).

METHODS

Image Analysis and Ground Truthing

Image analysis initially involves converting the image into digital format. The image is captured as a series of small squares in grid format (512 × 512 or 1024 × 1024) called pixels, with each pixel being assigned a numerical value between 0 (black) and 255 (white) according to its grey-tone (this number of grey tones is much greater than the human eye can differentiate). Aerial photographs of McManus Creek in 1966 and Faith Creek in 1966 (pre-mining), 1989 (immediately following

Table 9.1 Comparison of the scales of the photographs for the three dates and pixel resolution after digitization (1024 × 1024)

Date	Scale	Type	Pixel size (m)	Pixel area (m^2)
McManus Creek				
1966	1:6000	Black and white	2.7	7.28
Faith Creek				
1966	1:6000	Black and white	2.7	7.28
1989	1:3340	Colour	1.42	2.02
1993	1:2500	Colour	1.33	1.77

reclamation) and 1993 were digitised using a Hamamatsu camera under even illumination. Table 9.1 shows the ground resolution of each pixel for the aerial photography scales available for the study reach. Direct comparison of individual photographs from different dates was not possible owing to scale differences and different photograph edge locations. The digital images were cut and combined to allow direct comparison of the same river reaches and comparison of image statistics. The digital imagery data were analysed using RChips image analysis software and a Laserscan Geographical Information System (GIS).

All digital images were initially shade corrected, filtered and the land area masked out so that pixel values representing only the stream could be analysed (Winterbottom, 1996). Shade correction is necessary when comparing aerial photographs of different dates because of differences in illumination and photographic processing. Shade correction relies on converting digital numbers for an area with a constant reflectance (e.g. a roof) on photographs of different dates to the same value. A linear contrast stretch (a procedure which forces all 255 grey-tone values to be utilised even if no pure white or black grey-tones are present on the original image) was applied, the maximum tonal range defining the stretch limits. The effect of linear contrast stretching for one image of Faith Creek is shown in Figure 9.4; the unstretched data presented in (A) thus utilise 146 of the total grey-tone values available while the stretched image (B) has a minimum grey-tone value of 1 and a maximum of 256. The effect of this reassignment of values, in the context of water depth mapping, is that it allows small changes in water depth to be identified. A logarithmic transformation using the Lyzenga algorithm (Lyzenga, 1981) was then applied to the numerical data comprising the digital image. The images were then density sliced (sub-divided into a number of grey-tone ranges) to 15 water depth classes and statistical analysis of the number of pixels in each class undertaken.

In order to compare channel planform changes between dates to a high degree of accuracy, the aerial photographs also had to be rectified; this compensates for distortion caused by aircraft pitch and roll and corrects for scale differences between photographs. Manual rectification of each aerial photograph to others of different dates was undertaken using a stereo facet plotter; resulting maps were then digitised into a Laserscan GIS where the overlay facility was used to examine channel planform changes. Systematic errors associated with inexact rectification were estimated to be less than 5 m which compares with maximum channel displacement

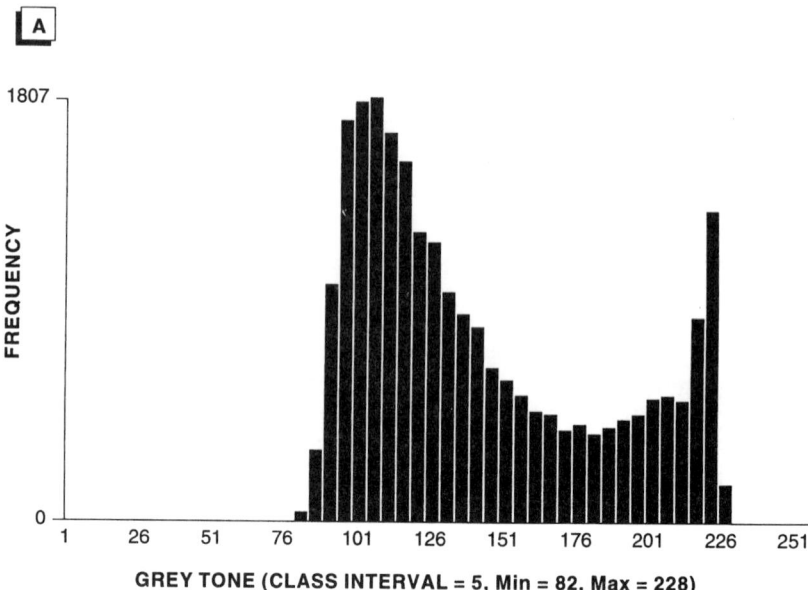

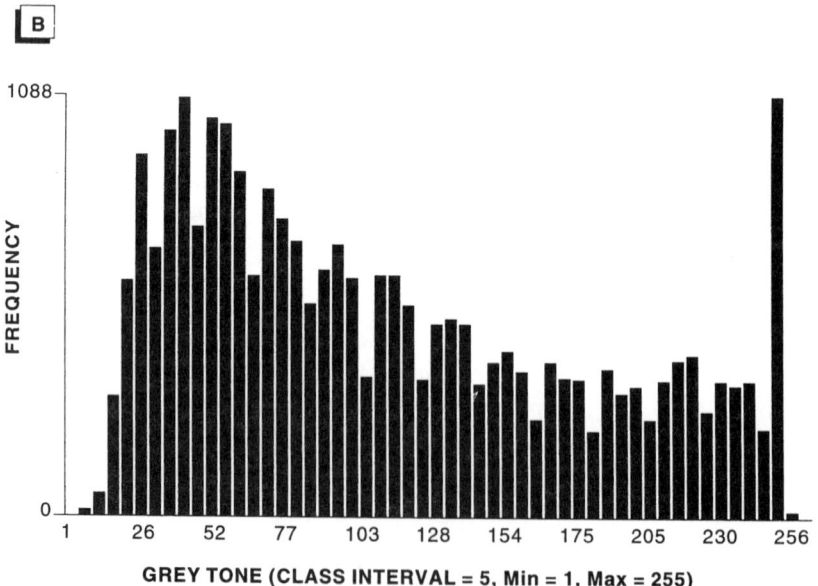

Figure 9.4 (A) The tonal range of the raw digitised image of Faith Creek in 1966 and (B) after a contrast stretch has been applied

exceeding 100 m. Geomorphological mapping of the study reach, particularly noting the location of pools, riffles and areas of turbulent water and variable substrate types, was undertaken also. Computerised rectification was not undertaken owing to the absence of sufficient ground control points (i.e. fixed points identifiable as being at exactly the same position on photographs of different dates) or detailed maps of the study area.

Cross-sections of Faith Creek and the valley floor were surveyed in July 1993, with the ends of each transect identified for subsequent location on the aerial photograph by large white painted boulders (50–100 cm).

RESULTS

Channel Planform

The aerial photography of McManus Creek in 1966 showed it to be an anastomosing river occupying much of the valley floor, with the "parent" channel (Riley, 1975) towards the northern edge of the floodplain (see Figure 9.9, below). Suitable aerial photography of McManus Creek (i.e. not unevenly illuminated due to clouds) for image analysis after 1966 either was not available or at the time of the photography the valley floor was covered in aufeis. However, it is likely that the McManus Creek study reach will be flown in the near future, so allowing natural channel dynamics of rivers within the area to be accurately assessed.

The 1966 aerial photographs of Faith Creek showed it to be an anastomosing river with a "parent" channel. Indeed, its planform was very similar to that of McManus Creek. The main channel occupied the western fringe of the floodplain. By overlaying the three rectified images of Faith Creek and digitising the channel outline, channel planform and changes resulting from the mining were determined (Figure 9.5). In 1989 the stream remained in the channel created during mining on a low-level terrace on the eastern side of the valley (Figure 9.6). The planform geometry was one of a straight, uniform-width channel. In 1993 the stream was on the western edge of the valley floor in a wandering gravel-bed river planform, apparently due to an avulsion out of the channelised reach, and a number of mid-channel bars were apparent. Field visits in 1994 and 1995 suggested that there had been little change in channel position or planform since 1993.

Channel Bed Morphology

The channel bed morphology was determined by image analysis of photographs for McManus Creek in 1966 and Faith Creek in 1966, 1989 and 1993 (Figures 9.7, 9.8 and 9.9). The photographic scale and pixel size after digitising are given in Table 9.1. The pixel values contained within the study reach were assigned either to one of the 14 water depth classes (estimated as being 5 cm class intervals) or to the exposed substratum gravel class, and were analysed to determine the number of pixels in each class, their mean value, standard deviation, skewness and kurtosis. The 5 cm class interval estimate was obtained by comparing the distribution of the

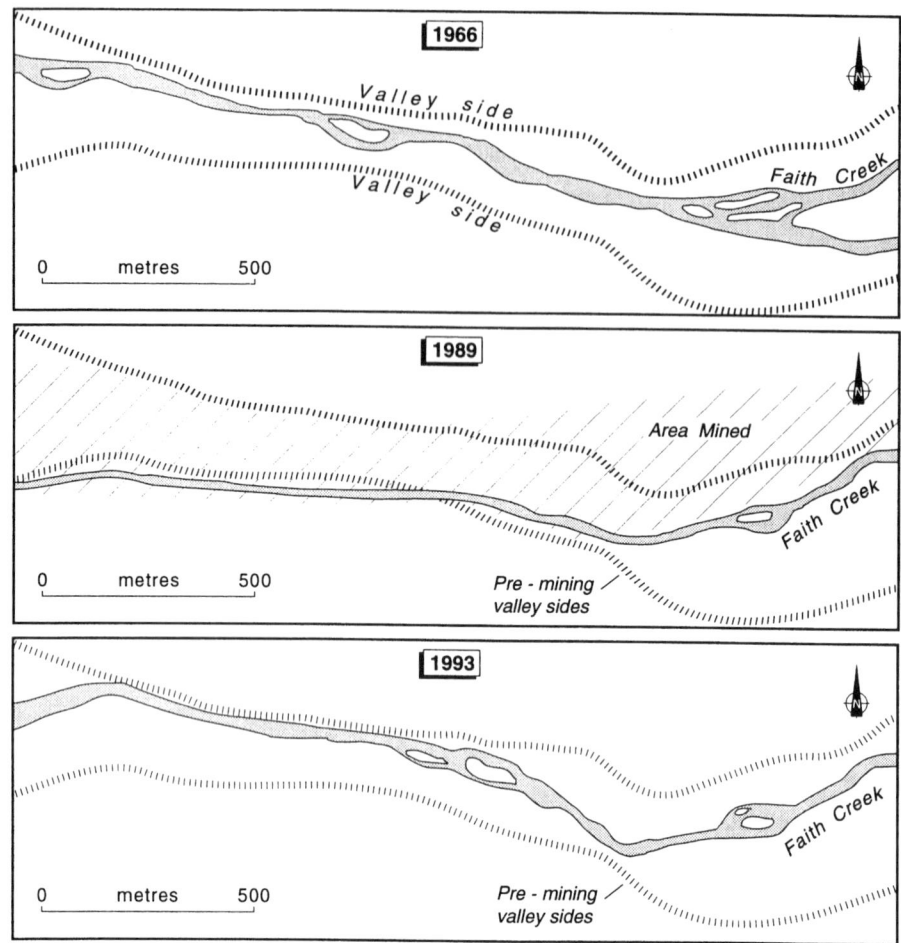

Figure 9.5   Position and planform of Faith Creek in 1966, 1989 and 1993

14 classes in the 1993 photographs with field measurements of water depth collected in the summer of 1993 under low flows. It was not possible to account for stage changes between the date of the photography (14 September 1993) and the ground survey (20 August 1993), but discharge changes during 1993 summer baseflows were small. No rainfall fell between the two dates or in the week previous to the ground survey, and miners on Faith Creek stated that under these conditions stream water levels remain almost constant (< 5 cm). Turbidity in 1966 would probably have been close to natural background levels owing to the limited extent of mining upstream and, in 1993, were measured to be within 5 NTU of background. It was assumed therefore that the image analysis results for these three years would not have been significantly affected by this variable or by a difference in water turbidity between Faith Creek and McManus Creek. In addition, by 1989 effluent discharges

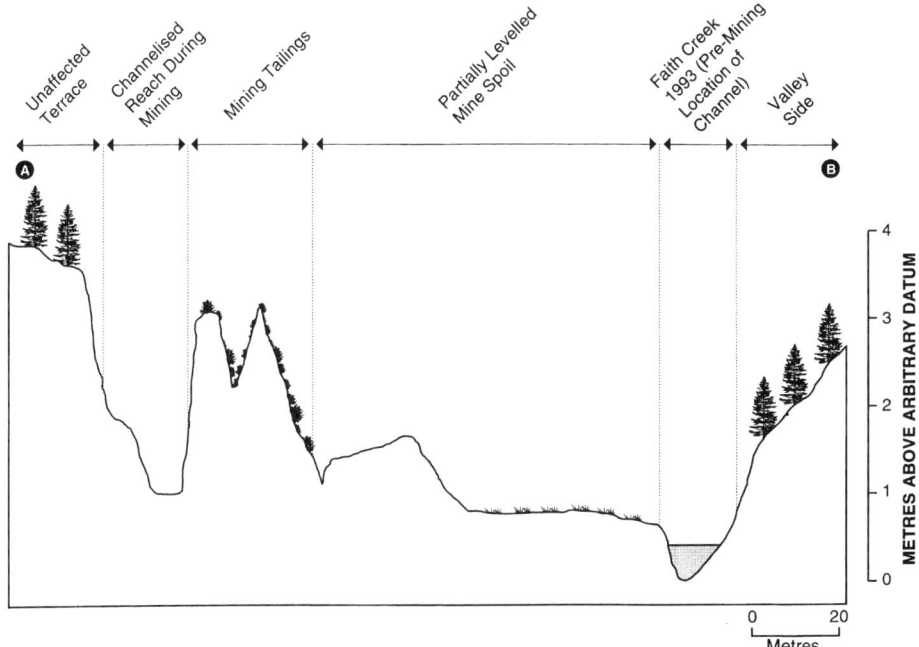

Figure 9.6   A transect across the valley floodplain as surveyed in August 1993 depicting the general topography after the cessation of mining, the position and form of the 1989 channelised reach and the location of the 1993 channel

from settling ponds should legally have been within 5 NTU of background. Channel form and changes in morphology were assessed from indices representing the distribution of estimated water depths and areas of exposed substratum (Figure 9.7; Table 9.2).

In 1966 on McManus Creek, the transformed reflectance distribution (Figure 9.7A) demonstrated a slightly skewed distribution with a large number of pixels classified as moderate water depth (25–50 cm) or deep water (estimated > 50 cm). Thirty-six per cent of pixels were classified as being in the deepest five water classes, with 54% in the following five, the modal value being class 4 (Table 9.2). The standard deviation also indicates a high degree of spread in the range of water depths. This is shown by the summary statistics, which indicate a positive skewness and kurtosis (Table 9.2). Examination of the bed morphology, as depicted by the image analysis, indicated that pools, deep water thalwegs, and a range of mid-channel and lateral gravel bar forms were all present (Figure 9.9).

In 1966 on Faith Creek, the transformed reflectance distribution (Figure 9.7B) demonstrated a significantly skewed distribution with a large number of pixels classified as moderate water depth (estimated 25–50 cm) or deep water (estimated > 50 cm). Forty-six per cent of pixels were classified as being in the deepest five water classes; the modal value was class 4 (Table 9.2). The standard deviation also indicates a high degree of spread in the range of water depths. This is reflected by

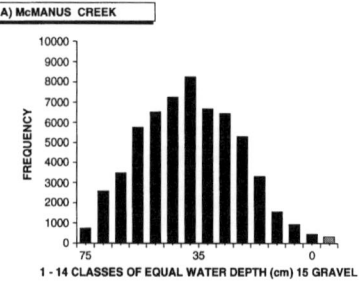

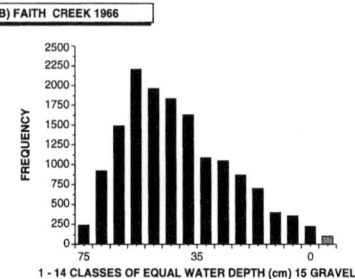

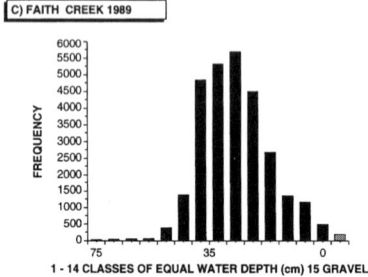

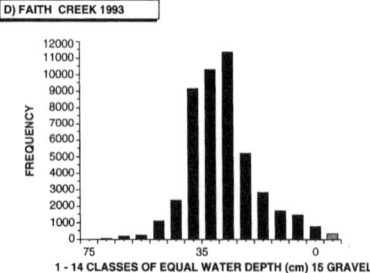

Figure 9.7 Histograms of the transformed image pixel values classified into 15 water depth and exposed gravel classes. (A) McManus Creek; (B) Faith Creek 1966; (C) 1989; (D) 1993

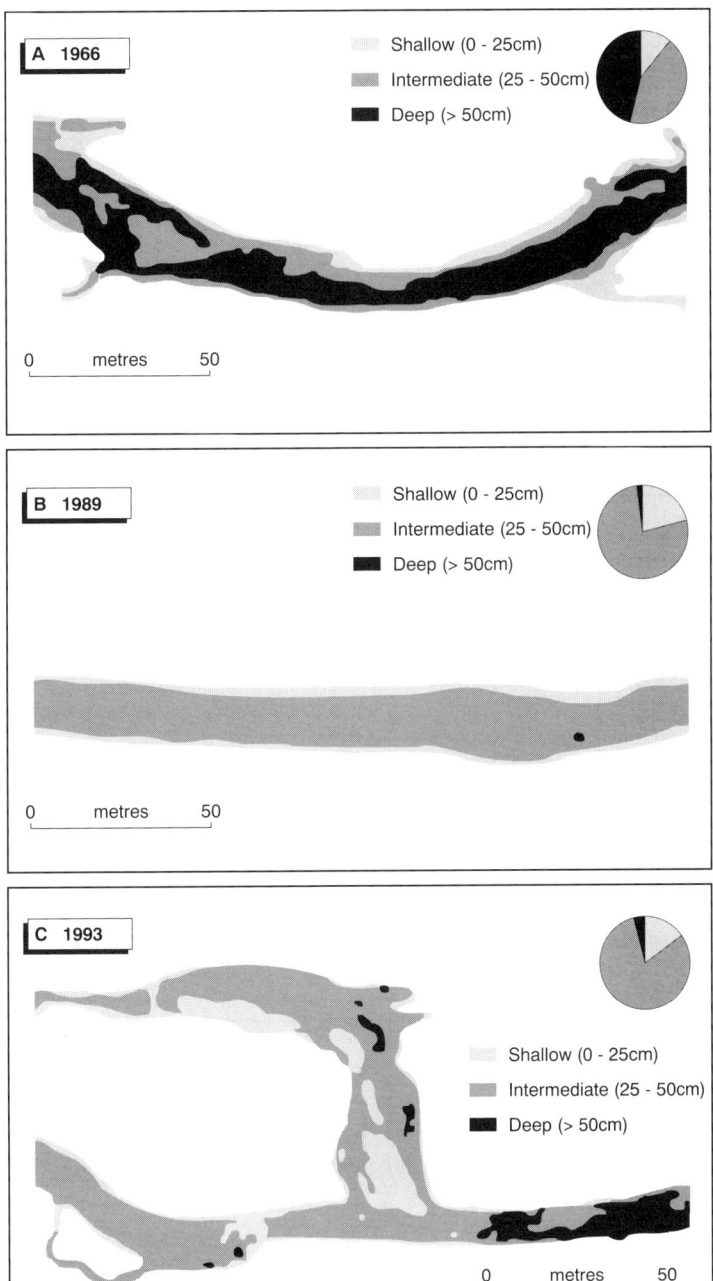

Figure 9.8 Bathymetric maps of one short section of Faith Creek and percentage area within each geomorphic unit for the whole reach, as determined from the image analysis for (A) 1966, (B) 1989, and (C) 1993

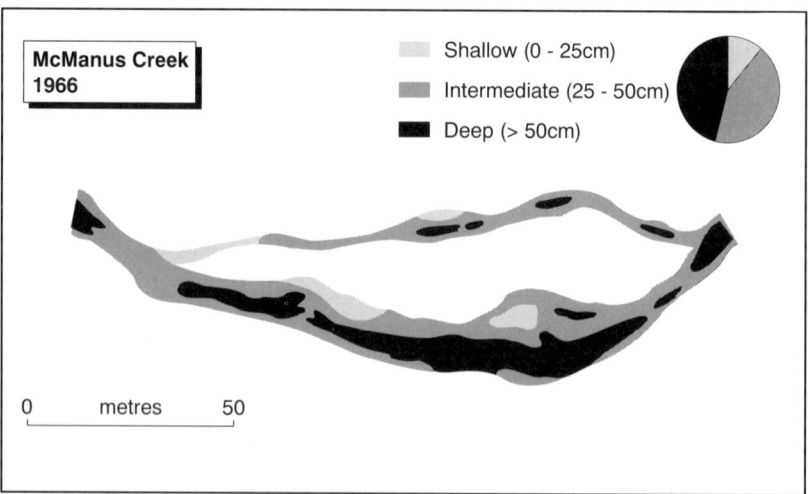

Figure 9.9 Bathymetric map of one short section of McManus Creek and percentage area within each geomorphic unit for the whole reach as determined from the image analysis

Table 9.2 Summary statistics of transformed pixel value distribution for McManus Creek and Faith Creek in 1966, 1989 and 1993

	McManus Creek		Faith Creek	
	1966	1966	1989	1993
Number of pixels	12 933	14 858	28 108	4758
Mean	6.949	6.352	9.010	8.557
Modal class	7	4	9	9
Standard deviation	2.830	3.068	2.008	2.009
Skewness	+0.156	+0.593	−0.361	−0.479
Kurtosis	−0.389	+2.721	3.339	+4.046

the summary statistics, indicating a positive skewness and kurtosis (Table 9.2). Scrutiny of the bed morphology, as depicted by the image analysis, indicated that pools, deep water thalwegs, and a range of medial, lateral and diagonal gravel bar forms existed (Figure 9.8b).

In 1989 on Faith Creek the transformed reflectance distribution (Figure 9.7C) was very different from that in 1966 and from those of McManus Creek. The distribution was negatively skewed with only 2% of values in the deepest five water depth classes; the modal value was class 9. A lower standard deviation was also observed, indicating a lower degree of spread in the range of water depths. The summary statistics show a moderate negative skewness and high kurtosis, reflecting the narrow range of water depth classes (Table 9.2). Figure 9.7C shows the channel at this time to be straight with very few pools or bar forms. The field survey in 1993 indicated that this channel bed was 3 to 4 m above the general level of the valley floor (Figure 9.6).

The 1993 Faith Creek image statistics are similar to those of 1989, reflecting few deep water areas (Figure 9.7D); the modal value was class 9. The summary statistics show a moderate negative skewness and high kurtosis. Despite development of a slightly sinuous thalweg and mid-channel bars (Figure 9.7C), very few deep water areas exist, although there had been a slight increase in moderate water depths at the expense of shallow areas (i.e. movement from classes 10 and 11 to 7, 8, 9). It is apparent that, between 1989 and 1993, some bed morphological forms had developed in the new channel but not enough time had elapsed for the development of deep water areas or marked pool–riffle sequences. The only exception was where a pool had developed due to flow convergence below two confluent channels either side of a medial bar (Figure 9.8C). The lack of pool development may be due, in part, to bed armouring (modal particle size based on Wolman counts at five locations generally varied between 90 and 128 mm in the mined reach compared with 65–90 mm on a comparable site on McManus Creek), presumably caused by incision of the channel and a steeper channel slope. The coarse nature of the channel bed material on Faith Creek was in marked contrast to the grain size of the valley floor deposits outwith areas sorted by streamflows, confirming that particle sorting and bed armouring had occurred within the Faith Creek study reach.

A simple tripartite geomorphological classification was devised based on apportioning the 15 transformed pixel value classes into three types, each incorporating five depth groupings: Type 1 encompasses relatively deep pools (> 50 cm), Type 2 includes runs and glides (25–50 cm) and Type 3 riffles and exposed gravels (< 25 cm). Within the study reach, few, if any, areas exceeded 75 cm depth at the time of ground survey. Exposed unvegetated gravels were not included as a separate geomorphic unit because, when analysing changes for different dates, slightly different water levels could be critical to the classification of the extent of exposed gravels in a shallow gravel-bed stream.

In 1966 in Faith Creek, Type 1 occupied 46% of the channel area (Figure 9.8A) but by 1989 had decreased to 2% of the stream area, increasing to 4% of the Faith Creek channel area by 1993; this slight increase over the four year period was the result of the abandonment of the channelised reach and the creation of a wandering gravel-bed river. By 1993 the channel had not yet developed large pools except in the case of a pool downstream of an anabranch confluence.

A marked change in the area of stream channel in the Faith Creek study reach (sum of the number of pixels multiplied by pixel area) between 1966, 1989 and 1993 is evident (Table 9.3). In 1966 the area was 108 227 m^2 but in 1989 had been reduced by 48% to 56 804 m^2. Over the same stream length McManus Creek had an estimated area of 37 940 m^2. In 1993, although the area of channel on Faith Creek had increased to 84 583 m^2, it was still less than the pre-mining value. The contrast is pronounced when examining the area of available Type 1 morphology. In 1966, 49 284 m^2 of deep water areas existed (45% of channel area). In the channelised reach in 1989 the area was 1115 m^2 and by 1993 the area had increased to 3099 m^2, still only 6% of the pre-disturbance value.

Table 9.3 Changes in area (m²) of shallow, medium and deep water on Faith Creek between 1966 and 1993

Water depth (cm)	1966 (m²)		1989 (m²)		1989 area as % of pre-mining value	1993 (m²)		1993 area as % of pre-mining value
> 50 cm	49 284	(45)	1115	(2)	2	3099	(4)	6
25–50 cm	46 516	(42)	43 883	(77)	94	68 571	(81)	147
< 25 cm	12 427	(13)	11 806	(21)	95	12 883	(15)	103
Total	108 227	(100)	56 804	(100)	52	84 553	(100)	78

Figures in parentheses represent the percentage of channel bed in each water depth class.

DISCUSSION

Geomorphic Effects of Gold Placer Mining on Alaskan Rivers

This pilot study, using image analysis techniques, has demonstrated that gold placer mining causes a decline of in-stream morphological diversity in directly affected reaches. Following the cessation of mining, the image analysis further illustrates that significantly longer than four years is required for a new stream channel to develop pre-disturbance pool-riffle sequences and in-stream morphological forms. The exact time, however, will obviously depend upon the magnitude and frequency of flood events following mining and the rate of sediment supply. Complementary field observations (repeat surveying of 1993 cross-sections in 1994 and 1995) show that Faith Creek is slowly incising into the reclaimed valley floor and maintaining a stable single-thread channel, unlike its pre-mining counterpart. This is most likely due to the river cutting through the levelled spoil which is left at a higher elevation than the pre-mining valley floor (Figure 9.6) to achieve a bed profile commensurate with upstream and downstream controls. Examination of natural stream channel sections on McManus Creek and Faith Creek in the Circle Mining District illustrate that the natural rivers typically also have relatively stable channels with limited signs of bank erosion. Often, however, a number of former channels exist, with the river resembling an anastomosing channel but with one "parent" channel. It is presumed that the occurrence of a number of channels relates to aufeis formation. On break-up, aufeis can control the location of meltwaters and create new channels and bank erosion in certain localities. Thus, although aufeis can accelerate the recovery of floodplain morphological diversity and channel change, the streams are not naturally highly dynamic and recovery times are likely to be greater than tens of years.

Both US Federal and State governments are beginning to require reclamation of mined sites. The US Bureau of Land Management now requires that reclamation include restoring the stream bed to a condition that will provide for recovery of fish and wildlife habitat and stream channel stability (US Department of the Interior, 1988). A knowledge of the nature of the stream bed and morphology prior to mining is thus required, as well as methods of monitoring geomorphological recovery. It appears from this study that in order to achieve a channel form similar to the anastomosing channel prior to mining, a wide valley floor level approximating its pre-mining elevation is required, and thus levelling is best performed according to the approach indicated in Figure 9.2C. In this case the river will be on the same gradient as those undisturbed reaches upstream and downstream. The stream is unlikely to incise into the valley floor and will be free to migrate and develop anabranches. In the case of the management practice illustrated in Figure 9.2B, channel incision is most likely, resulting in a relatively stable single-thread channel with a coarse substrate, and, given the higher elevation of the rest of the valley floor, there is little scope for anabranch channels to form. With respect to the final management option (Figure 9.2D), incision is not likely to occur but channel migration or the development of anabranches is not possible without large amounts of eroded sediment entering the stream system, possibly causing downstream aggradation and/or substrate siltation.

Image Analysis Techniques Applied to Aerial Photography of Rivers

The application of image analysis of aerial photography for mapping and quantifying in-stream habitat types has been shown with reference to an Alaskan case study. The techniques appear to be an ideal method for gaining initial information on channel form and monitoring the effects of activities such as placer mining on the physical attributes of river systems. Although there are errors associated with the technique, this case study demonstrated that it could be a useful "tool" to geomorphologists. Although there will inevitably be a range of factors causing inaccuracies, these may be more than offset by the increased spatial resolution obtainable. Factors causing inaccuracies may include bankside shading, variable substratum reflectance, changing water level, high turbidity, spatial resolution and atmospheric distortion. Sometimes these will be significant and make the technique inappropriate, while elsewhere the factors may be insignificant or at least can be compensated for. For example, shadow cast by riparian vegetation will reduce reflectance levels compared to fully sunlit areas of the same depth. Partial compensation can be gained by digitising around the area affected by shade and adding the average grey-scale level caused by the shading. Spatially variable substrates result in scatter in the reflectance level/water depth relationship but, for habitat mapping, the difference will emphasise and facilitate the differentiation between morphological features. The technique is only applicable to either non-turbid river systems (< 5 NTU) or very shallow turbid river systems (< 25–50 NTU). Water level fluctuations will also invalidate direct comparison between aerial imagery of different dates, although these can be accounted for where water level changes are known. Finally, the spatial resolution of the photography is important. For larger river systems the spatial resolution will always adequately describe morphological variability, even using small-scale photography (Table 9.1), but on smaller streams large-scale photographs (e.g. 1:2500) may be required.

Notwithstanding the limitations of the technique as outlined, the study has demonstrated the potential for the application of image analysis to aerial photographs to provide high spatial resolution information on in-stream morphology.

The ability to gain such information is likely to advance understanding in areas such as bar growth in braided rivers, the dynamics of pool–riffle development and the relationship between bank erosion and avulsions and channel bed topography. Retrospective image analysis on aerial photographs from different years can also document historical changes in in-stream channel morphology. Where ground-truth information is available to validate and quantify the image analysis data, the technique is enhanced, providing more accurate and quantitative information for fluvial geomorphologists. The data also lend themselves, to geostatistics and other statistical analysis. For example, in this study, a low standard deviation and high kurtosis indicated a fairly uniform channel bed whereas a high standard deviation and low kurtosis describe a varied bed topography with a range of bar forms or a pronounced pool–riffle sequence. Simple indices representing different channel morphologies might, therefore, be feasible.

Image analysis, whether applied to aerial photography or perhaps more importantly to multispectral imagery, is therefore a new and important "tool" for fluvial geomorphologists. On large river systems image analysis applied to satellite data sensing can provide important geomorphological information. However, image analysis applied to airborne remote sensing data appears to be most promising, being relevant to all but the smallest headwater streams and those obscured by tree canopies. Spatial resolutions as small as 1 m are possible and new hyperspectral sensors (e.g. Compact Airborne Spectrographic Imager (CASI)) provide very high spectral resolution, allowing different geomorphological and sedimentological features to be detected and mapped at a range of scales. Moreover, repeat surveys can be used to analyse channel change over temporal scales from days to decades. Thus, although the full capabilities of image analysis applied to remotely sensed data have yet to be fulfilled in the mapping and detection of riverine landscape change, it promises to be a rich tool and avenue for research in the forthcoming decade.

CONCLUSION

By applying image analysis to aerial photography of Faith Creek, Alaska, before, during and after placer mining for gold, and to the adjacent undisturbed McManus Creek, the geomorphic effects of placer mining were established. During mining, a straight channel with limited morphological diversity was apparent and four years after the cessation of mining, although the river had adopted a new planform, there was an absence of deep water areas or anabranch channels. This is in contrast to the unmined McManus Creek and Faith Creek prior to mining, where these features were identified as abundant. The geomorphic findings suggest that full post-mining geomorphic recovery requires decades rather than years and that post-mining levelling of the valley floor should strive to produce a valley floor elevation similar to that which existed prior to mining if a return to a more natural channel morphology is seen as important.

The ability of image analysis techniques to examine retrospectively, and concomitantly with field investigations, channel morphology from aerial photographs and to provide high spatial resolution data suggests that it could usefully be applied to a range of fluvial studies complementing traditional methods of field investigation. The greater use of image analysis applied to remotely sensed data in geomorphological investigations of rivers is thus advocated.

ACKNOWLEDGEMENTS

We wish to thank Phillip North for insights into placer mining reclamation and Bill Jamieson for his cartographic skills. Funding for the geomorphological investigations of the effects of gold placer mining have been provided by the US Environmental Protection Agency and NATO. Field assistance in 1995 by Sharon Bruce is also appreciated.

REFERENCES

Acornley, R. M., Cutler, M. E. J., Milton, E. J. and Sear, D. A., 1995. Detection and mapping of salmonid spawning habitat in chalk streams using airborne remote sensing, *Remote Sensing in Action, Proceedings of the 21st Annual Conference of the Remote Sensing Society*, Southampton, 267–274.

Bjerklie, D. M. and La Perriere, J. D., 1985. Gold mining effects on stream hydrology and water quality, Circle Quadrangle, Alaska, *Water Resources Bulletin*, **21**, 235–243.

Cornelius, S. C., Sear, D. A., Carver, S. J. and Heywood, D. I., 1994. GPS, GIS and geomorphological fieldwork, *Earth Surface Processes and Landforms*, **19**, 777–787.

Graf, W. L., 1979. Mining and channel response. *Annals of the Association of American Geographers*, **69**, 262–275.

Gurnell, A. M., Downward, S. R. and Jones, R., 1994. Channel planfrom change on the River Dee meanders, 1876–1992. *Regulated Rivers: Research and Management*, **9**, 35–43.

Hardy, T. B. Anderson, P. C., Neale, C. M. U. and Stevens, D. K., 1994. Applications of multi-spectral videography for the delineation of riverine depths and mesoscale hydraulic features. In *Effects of Human-Induced Changes on Hydrologic Systems*, eds R. A. Marston and V. R. Hasfurther, American Water Resources Association Technical Publication Series TPS-94-3, Maryland, 445-454.

Jackson, W. L. and Van Haveren, B. P., 1984. Design for a stable channel in coarse alluvium for riparian zone restoration. *Water Resources Bulletin*, **20**, 695–703.

James, L. A., 1991. Incision and morphologic evolution of an aluvial channel recovering from hydraulic mining sediment. *Geological Society of American Bulletin*, **103**, 723–736.

Karle, K. F. and Densmore, R. V., 1994. Stream and floodplain restoration in a riparian ecosystem disturbed by placer mining. *Ecological Engineering*, **3**, 121–133.

Knighton, A. D., 1989. River adjustment to changes in sediment load: the effects of tin mining on the Ringarooma River, Tasmania, 1875–1984. *Earth Surface Processes and Landforms*, **14**, 333–359.

Lyzenga, D. R., 1981. Remote sensing of bottom reflectance and water attenuation parameters in shallow water using aircraft and LANDSAT data. *International Journal of Remote Sensing*, **2**, 71–82.

Milton, E., Gilvear, D. J. and Hooper, I., 1995. Investigating change in fluvial systems using remotely sensed data. In *Changing Rivers*, eds A. Gurnell and G. E. Petts, Wiley, Chichester, England, 277–301.

Muller, E., Decamps, H. and Donson, M. K., 1993. Contribution of space remote sensing to river studies. *Freshwater Biology*, **29**, 301–312.

Philip, G., Gupta, R.P and Bhattacharya, A., 1989. Channel migration studies in the middle

Ganga basin, India, using remote sensing data. *International Journal of Remote Sensing*, **10**, 1141–1149.

Riley, S. J., 1975. Some differences between distributing and braiding channels. *Journal of Hydrology*, **14**, 1–8.

Scrimigeour, G. J., Prowse, T. D., Culp, J. M. and Chambers, P. A., 1994. Ecological effects of river ice break-up: a review and perspective. *Freshwater Biology*, **32**, 261–275.

US Department of the Interior, Bureau of Land Management, 1988. *Birch Creek Placer Mining Final Cumulative Environmental Impact Statement*, Bureau of Land Management, Anchorage, Alaska, 14pp.

Wagener, S. M. and LaPerriere, J. D. 1985. Effects of placer mining on the invertebrate communities of Interior Alaska streams. *Freshwater Biology*, **4**, 208–214.

Winterbottom, S. J., 1996. *An analysis of channel changes on the River Tay, Scotland using remotely sensed and GIS techniques.* Unpublished PhD. Thesis, University of Stirling.

Winterbottom, S. J. and Gilvear, D. J. (submitted). Quantification of channel morphology on gravel-bed rivers using aerial photography and mutli-spectral imagery, Regulated Rivers, Research and Management.

Section 2

APPLICATION

10 Terrain-based Approaches to Environmental Resource Evaluation

J. P. WILSON[1] and J. C. GALLANT[2]

[1] Department of Earth Sciences, Montana State University, USA
[2] Centre for Resource and Environmental Studies, The Australian National University, Canberra, Australia

ABSTRACT

The Terrain Analysis Programs for the Environmental Sciences (TAPES) is a set of computer programs for calculating spatially distributed hydrological, geomorphological and ecological variables in natural landscapes. TAPES was originally developed in 1992 and was subsequently modified and added to by various contributors. TAPES utilises the UNIX operating and X window systems and is available in contour- and grid-based systems. Here we review the grid-based versions of these programs, which are most compatible with Geographical Information Systems and remote sensing, and provide an illustrative application of each program for a 76 ha catchment within the Montana Agricultural Experiment Station Red Bluff Research Ranch. Topographic information for the grid-based programs is provided by digital elevation models (DEMs), and all variables are computed for each element of the DEM grid. The programs included are TAPES-G, DYNWET, EROS, and SRAD. TAPES-G calculates slope, aspect, upslope contributing area, curvature (profile, plan and tangential), flow direction and flow path length. A depressionless DEM can optionally be created and the user may select any one of four alternative methods for estimating upslope contributing areas. DYNWET calculates topographic wetness indices based on either a steady-state or quasi-dynamic sub-surface flow assumption. EROS calculates soil loss and erosion potential. SRAD calculates short- and long-wave radiation fluxes, energy balances and indices. Stream networks, sub-catchments and the spatial distribution of individual attributes can be plotted interactively or copied to ARC/INFO for further analysis and display.

INTRODUCTION

Governments, businesses and private citizens make thousands of decisions each day that affect the quality of the natural environment and the use of rapidly depleting natural resources (McAllister, 1982). Surveys or inventories of selected environmental resources are required periodically to (i) inventory the existing resource base, (ii) monitor the impacts of regional and global environmental change, (iii) identify and quantify key environmental problems, (iv) evaluate alternative solutions, and

Landform Monitoring, Modelling and Analysis. Edited by S. N. Lane, K. S. Richards and J. H. Chandler.
© 1998 John Wiley & Sons Ltd.

(v) monitor the results of implementing specific management actions. The increasing availability and advances in mathematical representations of physical processes and Geographical Information Systems (GIS) will alter the ways in which these surveys are conducted in the future (see National Resource Council 1986 and Scott *et al.* 1993 for reviews of surveys incorporating mathematical models and GIS, respectively).

Mathematical models integrate existing knowledge into a logical framework of rules and relationships (Moore and Gallant, 1991) and can be used to (i) improve our understanding of environmental systems, that is, as a tool for hypothesis testing, and (ii) provide a predictive tool for management (Beven, 1989; Grayson *et al.*, 1992b). Many environmental models require spatially distributed inputs because solutions to environmental problems such as soil erosion and non-point source pollution involve changes in the management of landscapes at the hillslope or catchment scale (Moore *et al.*, 1993c). The lack of input data at the required spatial resolution and difficulty of handling multiple model inputs that vary in different ways across landscapes (space) have emerged as the greatest impediments to the successful application of models in environmental surveys.

GIS offers new opportunities for the integration, analysis and display of spatially distributed biophysical and socio-economic data (Goodchild *et al.*, 1993, 1996). Maidment (1996) advocates using GIS as a common data and analysis framework for environmental models, so long as the different data structures, functions and methods for inputting and outputting spatial information in specific GIS and environmental models can be reconciled. Such a framework would need to incorporate tools for extrapolating spatial information to representative and environmentally diverse landscapes to help with the management of land and water resources. The paucity of these types of tools in many commercial GIS combined with the continued popularity of using GIS as a cartographic tool to map essentially two-dimensional phenomena, such as land use, vegetation and soil map unit boundaries, have frustrated recent attempts to effectively integrate GIS databases and environmental models (Moore *et al.*, 1993c; Wilson, 1996).

Following the proposal of Moore and Hutchinson (1991), we advocate using an index approach to characterise the spatial variability of specific processes; this is based on simplified representations of the underlying physics of the processes but includes the key factors that modulate system behaviour (i.e. topography). With this approach we sacrifice some physical sophistication to allow improved estimates of spatial patterns in landscapes (Moore *et al.*, 1993c). The method is able to handle variations in the availability of possible input data and the spatial resolution of those data. Care must be taken in developing these techniques because simplifying assumptions can increase rather than resolve computational complexity. This possibility looms large for systems with many variables and for models with many simplifications where several variables of the original model participate in the simplifying assumptions (Denning, 1990; Moore and Hutchinson, 1991).

Topography plays an important role in the hydrological response of a catchment to rainfall and has a major impact on the hydrological, geomorphological and biological processes that are active in landscapes (Smith and Bretherton, 1972; Speight, 1980; Dikau, 1989). The increasing availability of digital elevation models

(DEMs) and advent of computerised terrain analysis tools has made it possible to quantify quickly the topographic attributes at a large number of points in a landscape (Moore *et al.*, 1991). These attributes can be divided into primary and secondary (or compound) attributes. Primary attributes are calculated directly from a DEM and include variables such as elevation, slope, aspect, curvature (profile, plan and tangential), flow path length, and specific catchment area. Compound attributes involve combinations of two or more primary attributes and can be used to characterise the spatial variability of specific processes occurring in the landscape, such as the steady-state topographic wetness index (Beven and Kirkby, 1979; Quinn *et al.*, 1995) used to predict soil moisture deficit, the $A_s(\tan \beta)^2$ index (Montgomery and Dietrich, 1989, 1992; Montgomery and Foufoula-Georgiov, 1993) used to predict channel initiation, and the erosion index proposed by Moore and Wilson (1992, 1994).

This chapter describes (i) the TAPES (Terrain Analysis Programs for the Environmental Sciences) suite of integrated C and FORTRAN 77 programs and (ii) several topographic attributes, to demonstrate how these tools might be integrated with GIS and other models to provide more robust and effective research and management tools for simulating the functioning of environmental systems and understanding their behaviour under altered conditions.

TERRAIN ANALYSIS PROGRAMS FOR THE ENVIRONMENTAL SCIENCES

The TAPES programs were designed to analyse spatially distributed hydrological, geomorphological and biological processes in topographically complex landscapes (Moore, 1992). They utilise the UNIX operating and X window systems and include tools for drawing graphic displays and for statistically analysing and fitting frequency distributions to the results. The grid-based versions of these programs are discussed here because most existing GIS and remote sensing software packages are based on pixel or cellular data structures and these topographic indices are likely to provide the primary geographic data for applications integrating GIS and environmental modelling during the next 5–10 years.

TAPES-G is a grid-based method of terrain analysis that calculates slope, aspect, principal drainage direction, curvature (profile, plan and tangential), flow path length and specific catchment area at each node in a square-grid DEM (Gallant and Wilson, 1996). These attributes are useful in characterising a wide range of hydrological, erosional, geomorphological and ecological processes occurring in natural landscapes (Table 10.1). The advantages of removing enclosed depressions (pits) from a DEM are reviewed, for example, in Jenson and Dominque (1988), who proposed that pits be filled to the minimum elevation required for drainage. TAPES-G optionally creates a depressionless DEM using the method of Jenson and Dominque (1988). Most of the topographic attributes are calculated with a computationally efficient second-order, central finite-difference scheme centred on the interior node of a moving 3 × 3 window on a square-grid DEM. Slope can be calculated from the directional derivatives or a simpler approximate approach (*D8*

Table 10.1 Primary topographic attributes that can be computed by terrain analysis from DEM data (after Moore *et al.*, 1991, 1993c)

Attribute	Definition	Significance
Altitude	Elevation	Climate, vegetation, potential energy
Upslope height	Mean height of upslope area	Potential energy
Aspect	Slope azimuth	Solar insolation, evapotranspiration, flora and fauna distribution and abundance
Slope	Gradient	Overland and sub-surface flow velocity and runoff rate, precipitation, vegetation, geomorphology, soil water content, land capability class
Upslope class	Mean slope of upslope area	Runoff velocity
Dispersal slope	Mean slope of dispersal area	Rate of soil drainage
Catchment slope	Average slope over the catchment	Time of concentration
Upslope area	Catchment area above a short length of contour	Runoff volume, steady-state runoff
Dispersal area	Area downslope from a short length of catchment	Soil drainage rate
Catchment area	Area draining to catchment outlet	Runoff volume
Specific catchment area	Upslope area per unit width of contour	Runoff volume, steady-state runoff rate, soil characteristics, soil water content, geomorphology
Flow path length	Maximum distance of water flow to a point in the catchment	Erosion rates, sediment yield, time of concentration
Upslope length	Mean length of flow paths to a point in the catchment	Flow acceleration, erosion rates
Dispersal length	Distance from a point in the catchment to the outlet	Impedance of soil drainage
Catchment length	Distance from highest point to outlet	Overland flow attenuation
Profile curvature	Slope profile curvature	Flow acceleration, erosion/deposition rate geomorphology
Plain curvature	Contour curvature	Converging/diverging flow, soil water content, soil characteristics

approach) which calculates the gradient as the steepest slope in one of the four cardinal or four diagonal directions (i.e. from the central node to one of the eight nearest neighbour nodes).

Upslope contributing areas can be estimated using either the classical *D8* algorithm (O'Callaghan and Mark, 1984), the quasi-random *Rho8* algorithm (Fairfield and Leymarie, 1991), the dispersive *FD8/FRho8* algorithm based on Freeman (1991) and Quinn *et al.* (1991), or the stream-tube based *DEMON* algorithm (Costa-Cabral and Burges, 1994). The *D8* algorithm routes flow from each cell to one adjacent cell in the direction of steepest descent as measured by:

$$ s_i = \frac{|z_9 - z_i|}{\lambda \phi_i} \tag{10.1} $$

where s_i is the slope, z_9 is the elevation of the central cell in the 3×3 window, z_i is the elevation of a neighbouring cell ($i = 1,8$), λ is the grid cell size and ϕ_i is 1 for cardinal directions and $\sqrt{2}$ for diagonal directions. This method can only represent parallel or converging flow, so divergent areas in the upper parts of the landscape are not well represented. The use of eight flow directions also results in a poor representation of flow networks where the surface is not oriented in exactly one of those eight directions.

The *Rho8* algorithm seeks to avoid this second problem by introducing a stochastic element into the flow direction calculations which results in accurate average flow direction over large areas. This is achieved by using $\phi_i = 1/(2 - r)$ for the diagonal directions, where r is a uniformly distributed random number between 0 and 1. This algorithm produces more realistic-looking flow networks without dealing with the non-divergent property of the *D8* algorithm, and it also distorts the frequency distribution of contributing area further from the true distribution (Moore *et al.*, 1993b).

The *FD8/FRho8* method represents the divergence of flow in convex topography (hills and ridges) by dispersing flow to all downslope cells, using slope as a weighting factor:

$$ w_i = \left(\frac{(z_9 - z_i)}{\lambda \phi_i} \right)^v \qquad \text{if } z_i < z_9 \tag{10.2} $$

$$ w_i = 0 \qquad \text{if } z_i \geq z_9 \tag{10.3} $$

The w_i values are scaled so that their sum equals one. Freeman (1991) found that an exponent $v = 1.1$ gave the closest match to the expected contributing area on conical surfaces, and that the algorithm should switch to a single flow direction method in convergent areas to avoid dispersion of flow in (presumably) channelled valleys where the contributing area exceeds a user-specified threshold. The *FD8* version of this algorithm uses the *D8* method in valleys while the *FRho8* version uses the *Rho8* method. Holmgren (1994) suggested that the exponent should be much larger than Freeman suggested, probably in the range 6 to 8, and Quinn *et al.*

(1995) have taken up this suggestion and refined the method to include a variable exponent that changes from 1 (full dispersion) at hilltops to a large value (effectively no dispersion) in valleys.

The *DEMON* algorithm routes flow using stream tubes constructed downstream of each cell. Because of this construction it is not limited to cell-to-cell routing as the previous methods are, and it directly models divergence and convergence of flow in response to the shape of the surface without having to resort to flow dispersion. The *FRho8* and *DEMON* methods result in similar frequency distributions of contributing area (Moore *et al.*, 1993b).

Stream networks, sub-catchments and the spatial distribution of individual attributes can be plotted interactively or copied to the ARC/INFO GIS for further analysis and display (Gallant and Wilson, 1996). The topographic attributes calculated by TAPES-G are also required by several other terrain analysis programs to calculate a series of water balance and erosion/deposition indices (discussed below). Alternatively, selected topographic attributes may also be used to define basic relief units for geomorphological and pedological mapping (Figure 10.1). This approach provides a systematic methodology for the derivation of complex geomorphological units that can be used to develop stratified sampling frames for field data collection, because many soil and other important properties are strongly influenced by the geomorphological position of a site (Moore *et al.*, 1993a, c, d).

DYNWET calculates a spatially distributed topographic wetness index based on either a steady-state or quasi-dynamic sub-surface flow assumption using the terrain attributes computed by TAPES-G. The topographic wetness index characterises the location of zones of surface saturation (source areas for partial area runoff) and the spatial distribution of soil water (i.e. the soil water content overlying a shallow impermeable or semi-impermeable layer) in a catchment (Beven and Kirkby, 1979; O'Loughlin, 1986; Moore *et al.*, 1988). This index is a measure of the hydrologic condition of a catchment and is important in describing the distribution and abundance of vegetation, erosion and deposition, and soil properties in natural landscapes (e.g., Moore *et al.*, 1993a, c; Wilson *et al.*, 1994).

The simplest form of the steady-state topographic wetness index is:

$$\chi_i = \ln \left[\frac{A_{si}}{\tan \beta_i} \right] \qquad (10.4)$$

where A_{si} is the specific catchment area (i.e. catchment area draining across a unit width of contour) and β_i is the slope. This steady-state index predicts that points lower in a catchment, and particularly those points near the outlets of the main channels, are the wettest points on the catchment, and the soil water content decreases as the flow lines are retraced upslope to the catchment divide. This pattern will only occur in semi-arid and arid environments (in the unlikely situation) where recharge to a perched water table occurs at a constant rate for the length of time required for every point in the catchment to reach sub-surface drainage equilibrium (Kirkby and Chorley, 1967; Barling *et al.*, 1994).

The quasi-dynamic topographic wetness index is a new index that recognises that the sub-surface flow regime in a catchment rarely, if ever, reaches this steady-state

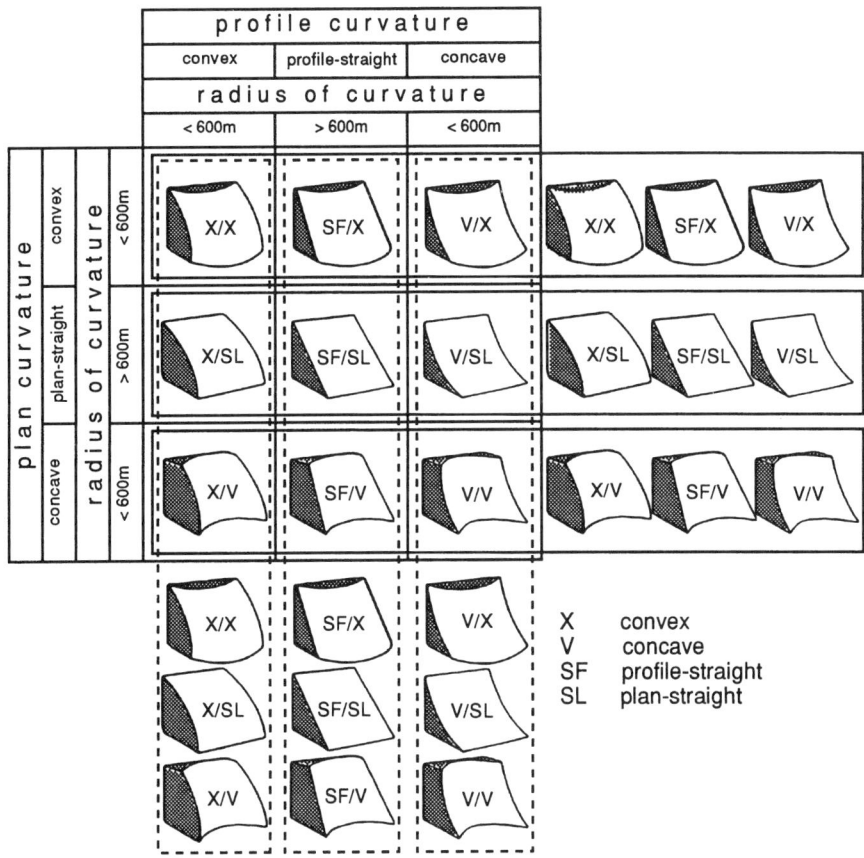

Figure 10.1 Classification of landform elements by plan and profile curvature for the determination of geomorphological relief units (adapted from Dikau, 1989) Concave, straight and convex plan curvatures indicate convergent, parallel and divergent flows, respectively

condition (Moore, 1992; Barling *et al.*, 1994). DYNWET calculates (i) the time for sub-surface flow to travel the length of each cell from selected topographic (slope, upslope drainage area, flow path) and hydrologic (saturated hydraulic conductivity, drainable porosity) attributes; and (ii) the effective upslope contributing area per unit width of contour for each cell limited by the user-specified drainage time (Iida, 1984; Barling *et al.*, 1994). This quasi-dynamic version of DYNWET requires the user to input a drainage time (which typically would be the time between significant rainfall or snowmelt events), uniform or spatially varying soil properties and evapotranspiration (using the short-wave radiation ratio calculated by SRAD discussed below), and the name of the file containing TAPES-G attributes. The quasi-dynamic version assumes that soil water storage shares a linear relationship with the magnitude of the dynamic topographic wetness index. Barling *et al.* (1994) calculated static and quasi-dynamic indices for a catchment near Wagga Wagga in

Australia and found that only the quasi-dynamic index correctly predicted that the topographic hollows (and not the drainage channels themselves) determined the hydrologic response of the catchment.

EROS calculates the spatial distribution of soil loss and erosion and deposition potential in a catchment (Wilson and Gallant, 1996a). The method is based on characterising the sediment transport capacity of overland flow using a stream power approach (Moore and Wilson, 1992, 1994). The program requires the elevation, slope, flow direction and drainage area attributes generated by TAPES-G, and optionally a soil boundary file which contains polygons specifying the boundaries of soil mapping units and a weighting factor for each polygon. The choice of weighting scheme depends on the runoff generation mechanism that is used. The simplest option assumes that overland flow is generated uniformly across the entire catchment ($\mu = \mu_i = 1$). The second option assumes Hortonian overland flow in which μ_i is spatially variable and a function of the infiltration capacity of the soil. The final option assumes saturation overland flow in which overland flow only occurs in zones of saturation defined by some user-specified critical value of the steady-state or quasi-dynamic topographic wetness index. EROS calculates two erosion indices (i) a sediment transport index that is similar to the length–slope factor in the Revised Universal Soil Loss Equation (RUSLE; Renard et al., 1991, 1993) but is applicable to three-dimensional terrain; and (ii) the change in sediment transport capacity across a grid cell. The second index provides a possible measure of erosion and deposition potential in a catchment (Moore and Burch, 1986; Wilson, 1996).

SRAD provides an approximate method of estimating the spatial distribution of (i) direct, diffuse, reflected and global short-wave radiation; (ii) incoming (atmospheric), outgoing (surface), and net long-wave radiation; (iii) net radiation and the short-wave radiation ratio (i.e. the ratio of global short-wave radiation on a sloping surface to that on a horizontal surface) in complex topography (Moore, 1992). The program calculates potential solar radiation as a function of location, slope, aspect, topographic shading and time of year, and then adjusts this estimate using monthly estimates of cloudiness, sunshine and atmospheric transmissivity. Temperature is extrapolated across the surface using a method based on Running et al. (1987), Hungerford et al. (1989), Running (1991), and Running and Thornton (1996). This method corrects for elevation via a lapse rate, for slope-aspect effects via a short-wave solar radiation ratio, and for vegetation effects via a leaf area index (LAI) (Moore et al., 1993c). Radiation and temperature estimates can be produced for periods ranging from one day to one year.

SRAD requires a square-grid DEM (similar to TAPES-G), climate parameter file and optionally a vegetation file as inputs (Wilson and Gallant, 1996b). The climate file specifies several radiation, temperature and land cover parameters. Many of the radiation parameters (i.e. circumsolar coefficient, cloudiness parameter, sunshine fraction, atmospheric transmittance) can be estimated from a nearby climate station with daily sunshine, cloud cover and solar radiation measurements. Some of the temperature inputs (i.e. mean monthly air and surface temperatures) may be obtained directly from records at a climate station located nearby or, preferably, in the area covered by the DEM. Other inputs (i.e. minimum, daylight average and maximum temperature lapse rates) must be computed from records at numerous nearby stations or estimated from published data. The land cover parameters, which include monthly albedo and one or more sets of monthly LAI values (i.e. one for each vegetation type), are usually estimated from published data. A square-grid vegetation file matching the dimensions of the DEM and specifying the vegetation type at each grid cell is needed if multiple LAI profiles are included in the climate parameter file.

COMPUTED TOPOGRAPHIC ATTRIBUTES: COTTONWOOD CREEK CATCHMENT

The upper third of the Cottonwood Creek catchment (76 ha) used for this study is located on the Montana Agricultural Experiment Station Red Bluff Research Ranch south of Norris, Montana. Slopes are moderate to steep and elevations range from 1652 m at the catchment outlet to 1969 m at the southeast corner of the catchment. The main channel is spring fed and runs year round. It is flanked by small, intermittent seeps which feed water laterally into its channel and several ephemeral tributaries (Jersey, 1993). The soils are deep, well drained, and formed in colluvium and material derived from gneiss, schist and granite. Most of the soils have loamy or sandy loam surface textures (Boast and Shelito, 1989). The climate is semi-arid (25–50 cm annual precipitation) and the vegetation is strongly correlated with landscape position and aspect. Grasses cover about 60% of the upper catchment and dominate south-facing slopes and ridge tops. Maple, aspen, willow and snowberry cover about 10% of the catchment and occupy north-facing slopes and lower stream bottoms. Sagebrush and conifers dominate the remainder of the study area (Jersey, 1993).

A DEM was prepared for the study area by digitising 6 m (20 ft) contours and blue lines on 1:24 000-scale United States Geological Survey (USGS) topographic map quadrangles in ARC/INFO and converting these data to a regular 20 m grid with ANUDEM (Jersey, 1993). ANUDEM takes irregular point data or contour data and creates square-grid DEMs. The program automatically removes spurious pits within user-defined tolerances, calculates stream lines and ridge lines from points of locally maximum curvature on contour lines, and (most importantly) incorporates a drainage enforcement algorithm to maintain fidelity with a catchment's drainage network (Hutchinson, 1989).

A depressionless DEM was created and the standard set of primary topographic attributes noted earlier was calculated with TAPES-G. Upslope contributing areas were computed using the *FRho8* and *Rho8* algorithms (described in the previous section). The maximum cross-grading area (threshold) used by TAPES-G to distinguish upland and channel areas (cells) was set at 8000 m^2 (20 cells) based on the proportion of grid cells crossed by "blue lines" on a nearby 1:24 000-scale USGS map quadrangle. This approach tends to generate many more first-order tributaries than are present on the original topographic maps.

The slope and upslope drainage area maps reproduced in Figures 10.2 and 10.3 were prepared from ARC/INFO grids and show the major geomorphological and

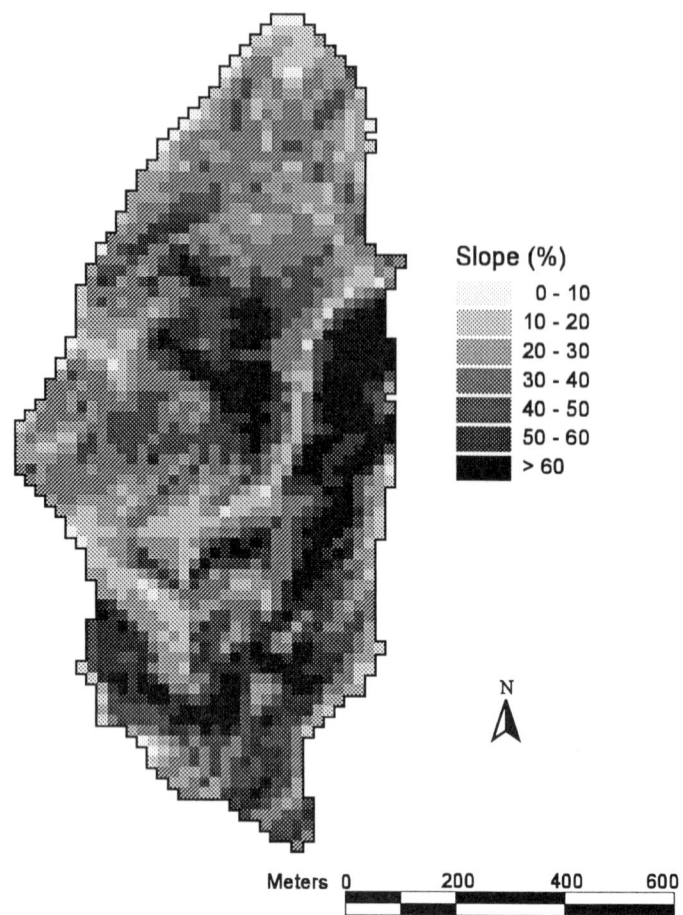

Figure 10.2   Distribution of slope in Cottonwood Creek catchment

hydrologic features of the upper portion of the Cottonwood Creek catchment. These maps (and those which follow) portray the original 20 m by 20 m cells with north at the top of the page. Figure 10.2 shows that the main channel is flanked by steep slopes to the south and east and the western portion of the catchment is split into a series of sub-catchments by a series of north- and south-facing slopes. The cells with uplsope drainage areas exceeding 8000 m^2 show a channel system starting with a series of tributaries near the southern, western and northern catchment boundaries that drain to a main channel that flows in a predominantly northeasterly direction to the catchment outlet on the eastern boundary of the catchment (Figure 10.3).

The slope and upslope drainage area maps also indicate some potential problems that result from the choice of a square-grid DEM and/or the 20 m by 20 m grid resolution. The small size of Cottonwood Creek and its tributaries (< 1 m wide in

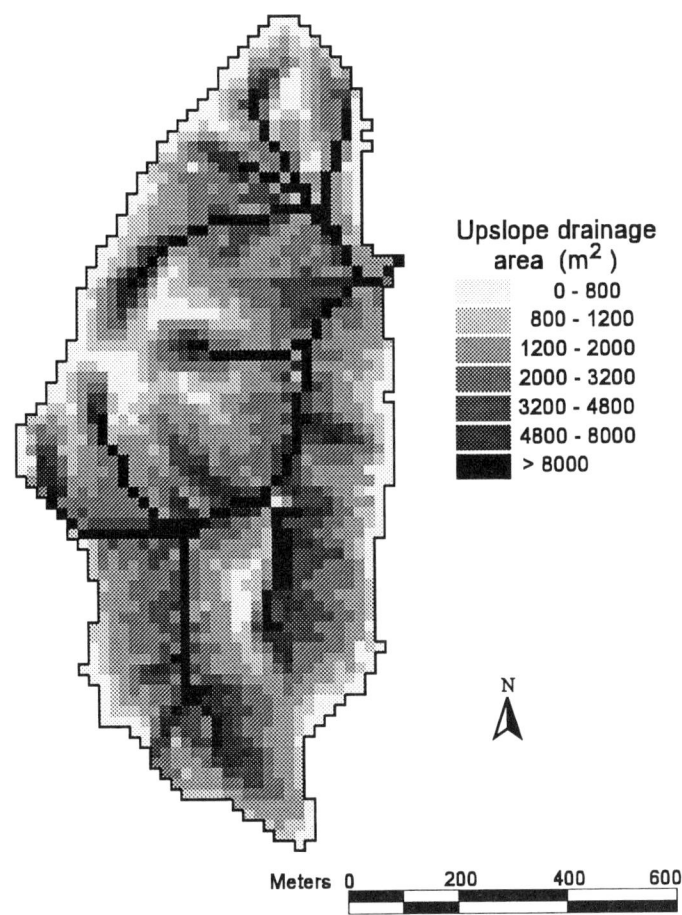

Upslope drainage
area (m²)

	0 - 800
	800 - 1200
	1200 - 2000
	2000 - 3200
	3200 - 4800
	4800 - 8000
	> 8000

N

Meters 0   200   400   600

Figure 10.3 Distribution of upslope drainage area in Cottonwood Creek catchment computed with the *FRho8* method in upland areas and *Rho8* method in channel areas (i.e. cells with upslope contributing areas >8000 m²)

the upper catchment used for this study) relative to the grid resolution accounts for the absence of light shaded cells (i.e. cells with upslope drainage areas <800 m²) along parts of the southwestern catchment boundary and the presence of dark-shaded cells (i.e. channel cells with upslope drainage areas >8000 m²) arranged in linear patterns in Figure 10.3. These results distort the distribution of upslope contributing area and therefore specific catchment area, and they might have been avoided by choosing a finer grid resolution or a hybrid grid–vector representation of topography.

The mean annual daily minimum temperature (Figure 10.4) and short-wave solar radiation (Figure 10.5) attributes were computed in SRAD from the same square-grid DEM used by TAPES-G and a climate parameter file prepared by the authors.

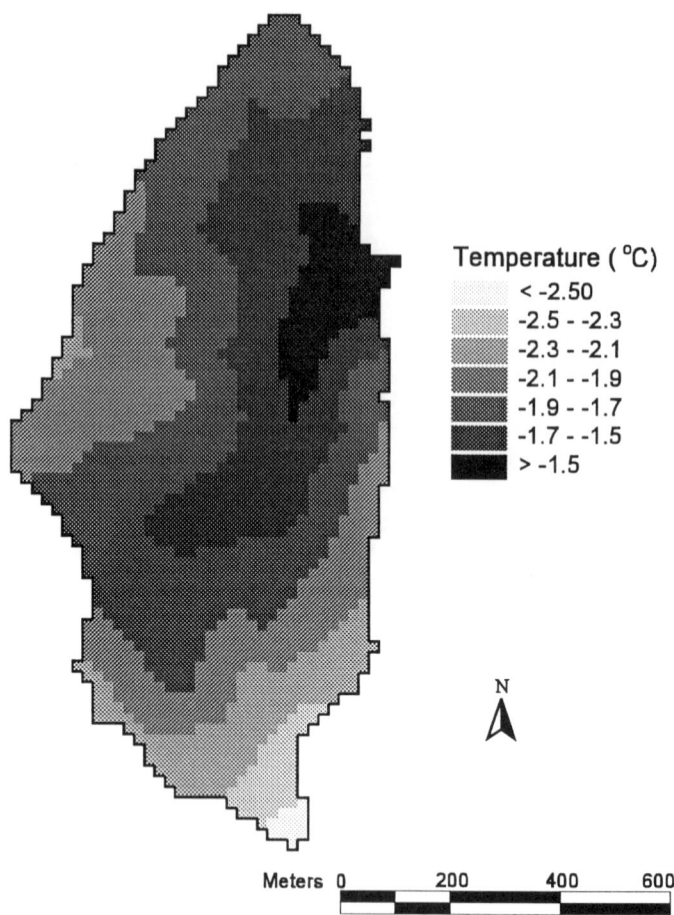

Figure 10.4 Distribution of mean annual minimum daily temperature in Cottonwood Creek catchment

Thirty-year monthly records from the Great Falls and Bozeman W6 Experiment Farm climate stations were used to estimate the radiation and temperature parameters, respectively. The Great Falls station is located 220 km north of the study area and is the only primary solar radiation station in Montana. These station records were used to estimate mean monthly values of the circumsolar coefficient (i.e. the fraction of the diffuse radiation originating near the solar disk that is subject to topographic effects), a cloudiness parameter, sunshine fraction and atmospheric transmissivity. The Bozeman W6 Experiment Farm station is located 22 km east of the study area and was chosen because it is the closest station with long-term air and surface (soil) temperature measurements (Wilson and Gallant, 1996b). Mean monthly air and surface temperatures were obtained directly from station records. The lapse rates required by SRAD were estimated from weighted

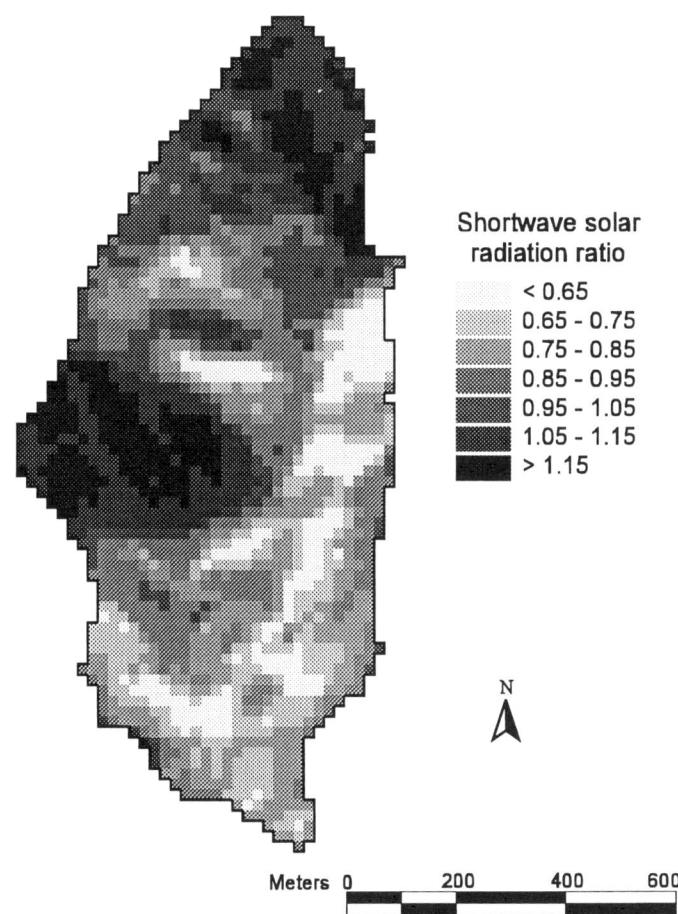

Figure 10.5 Distribution of ratio of global short-wave radiation on a sloping surface to that on a horizontal surface in Cottonwood Creek catchment

regressions of elevation and temperature for pairs of stations within a 200 km radius of the Bozeman W6 station using a similar method to that proposed by Running and Thornton (1996). The monthly albedo and LAI values for Great Falls were obtained from Iqbal (1983) and Barbour and Billings (1988), respectively.

The SRAD results show how the simulated climate in the Cottonwood Creek catchment is modified by topography. For example, mean annual daily minimum air temperature decreases with increasing altitude (Figure 10.4), and high and low values of the short-wave solar radiation ratio indicate south- and north-facing slopes, respectively, and show even more clearly the major topographic features noted earlier (Figure 10.5).

The slope and specific catchment area attributes reproduced in Figures 10.2 and 10.3 were used by DYNWET to calculate a steady-state topographic wetness index

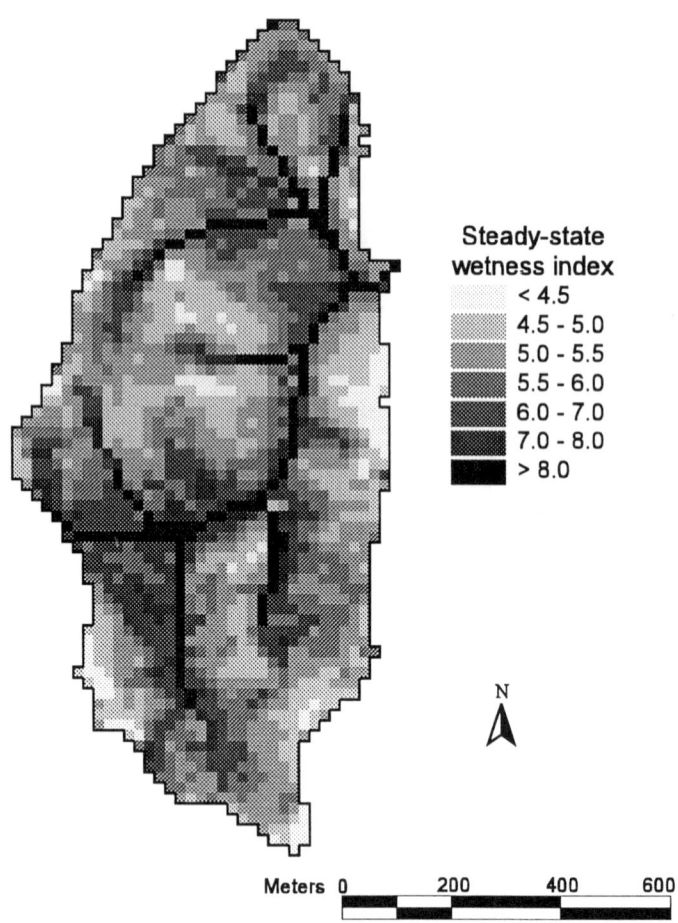

Figure 10.6 Distribution of steady-state topographic wetness index in Cottonwood Creek catchment

(Figure 10.6). High values of this steady-state index ($i > 8.0$) indicate the location of the channel system because the steady-state index predicts that these areas are the wettest points in the catchment. However, the spatial pattern exhibited by the lower values of this index is probably not indicative of the distribution of soil water, because field data collected by Aspie (1989) indicate that the soil water distribution seldom, if ever, reaches the steady-state condition assumed by this index.

A quasi-dynamic topographic wetness index was also calculated with DYNWET assuming a uniform soil depth of 1.3 m, a saturated hydraulic conductivity of 44 mm hr^{-1}, a drainable porosity of 0.24, and a drainage time of 10 days (Figure 10.7). The mean weighted soil depth was estimated from the spatial extent of the different soil mapping units and published soil series descriptions (Boast and Shelito, 1989). The mean weighted saturated hydraulic conductivity was estimated

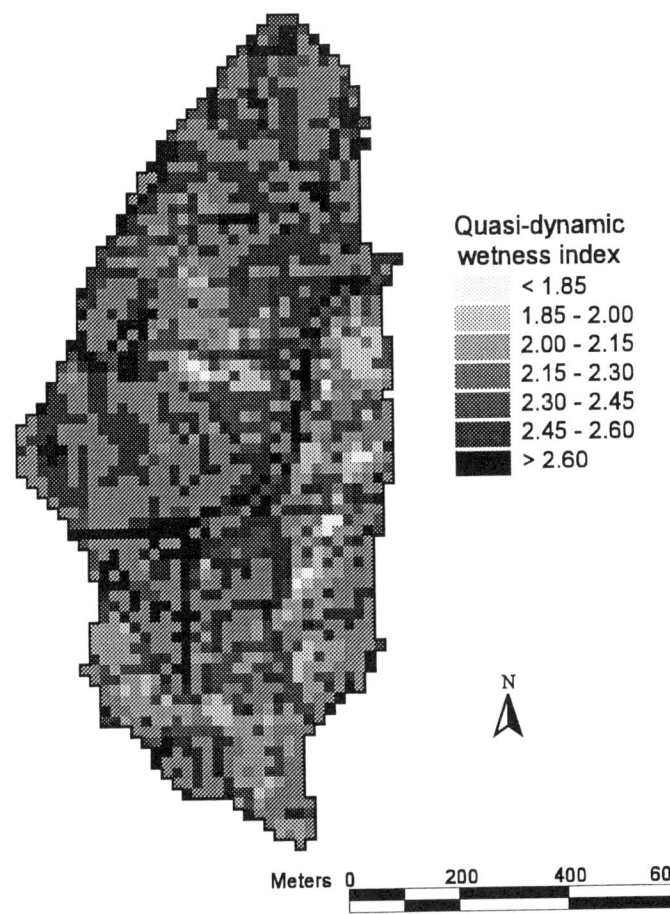

Figure 10.7 Distribution of quasi-dynamic topographic wetness index in Cotton-wood Creek catchment

from soil texture data reported by Boast and Shelito (1989) (weighted by spatial extent and depth) and some published data reporting typical saturated hydraulic conductivities by soil texture class (Rawls and Brakensiek, 1989). The mean weighted drainable porosity was estimated in a similar fashion with the help of published data reporting typical values of this parameter by soil texture class (Ratliff *et al.*, 1983). The spatial distribution of soil water computed with this index is much more dispersed than that of the steady-state index (cf. Figures 10.6 and 10.7). High values of the quasi-dynamic topographic wetness index are scattered throughout the catchment in localised areas with high effective specific catchment areas and relatively gentle slopes. Work is now underway to compare this pattern with the short-wave solar radiation ratio (Figure 10.5) and distribution of vegetation (Jersey, 1993).

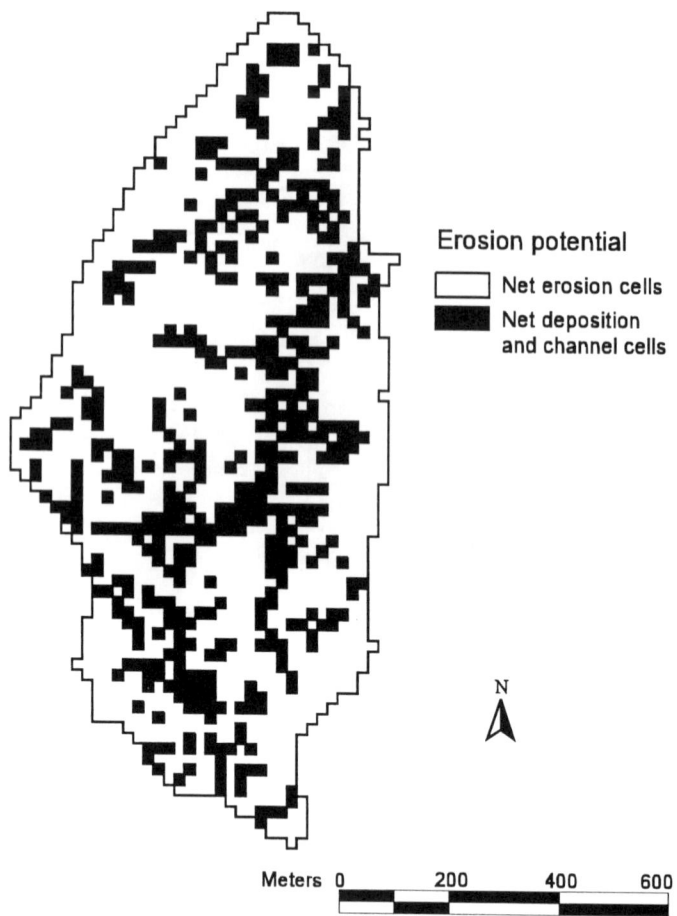

Figure 10.8 Distribution of predicted areas of net erosion and net deposition in Cottonwood Creek catchment based on the change in sediment transport capacity index

The final two topographic attributes reproduced in Figures 10.8 and 10.9 were calculated in EROS assuming that overland flow is generated uniformly across the entire catchment. The first map predicts those areas experiencing net erosion and net deposition based on the change in sediment transport capacity as water is routed across grid cells (Figure 10.8). This type of map is needed because the original and revised versions of the Universal Soil Loss Equation do not distinguish between those areas experiencing net erosion and net deposition, and the model should only be applied to those land areas experiencing net erosion (Wischmeier and Smith, 1965, 1978; Renard *et al.*, 1991, 1993). The final map shows the sediment transport capacity index which is similar to the length–slope factor in RUSLE but incorporates the effects of three-dimensional terrain (Figure 10.9). RUSLE does not

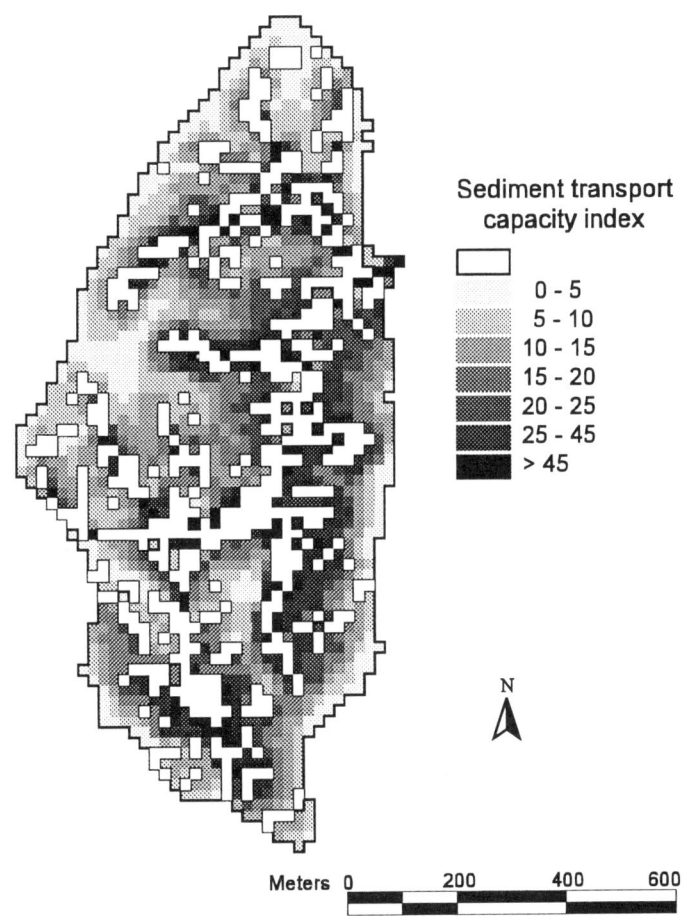

Figure 10.9 Distribution of sediment transport capacity index in Cottonwood Creek catchment

apply to channels nor locations of net deposition. Channel and net deposition cells are shown as white in Figure 10.9 and the sediment transport capacity index was not computed for these cells. Spatially variable soil loss could be estimated from the sediment transport capacity index values reproduced in Figure 10.9 and a series of grids representing the other RUSLE model inputs (Wilson, 1996).

CONCLUSIONS

The computed topographic attributes indicate how generic, knowledge-based modelling techniques that are transportable across environments can be used to examine and possibly resolve complex resource and environmental issues. Knowledge-based methods are useful because they encourage (i) the development of stratified sample designs, (ii) increased efficiency in data collection, (iii) hypothesis testing, and (iv) the calibration of physical system parameters from observations and prior knowledge. The outputs (i.e. computed terrain attributes) can be used by themselves and/or with other types of information to evaluate alternative land uses and/or the susceptibility of landscapes to specific environmental hazards (e.g. accelerated soil erosion, nitrogen and pesticide contamination of ground water). However, the terrain attributes and programs described here are by no means the last word in terrain analysis. There are a number of ways in which these terrain analysis methods might be extended and improved. Four examples are fine (sub-grid) scale features, alternative process models, incorporation of non-topographic spatial variables, and improvements in estimation algorithms.

In some cases, important topographic or hydrologic features cannot be properly represented by a square-grid DEM. Stream channels, for example, are usually too fine for their shape to be captured by the DEM and their contributing area is associated with the whole cell rather than the narrow channel within the cell (see Figure 10.3 for an example of this problem). This result distorts the distribution of contributing area and therefore specific catchment area. One solution to this problem, as proposed by Quinn et al. (1995), is to consider channels separately and not propagate the contributing area of the channel to downstream grid cells. This solution involves a hybrid grid–vector representation of topography. A similar approach was used (for similar reasons) by Grayson et al. (1992a) for modelling channel flow in a contour-based DEM.

Although it is not always apparent to users of terrain analysis, the indices described above are simplified process models and are not applicable in all situations. The topographic wetness index, for example, is based on the assumption that soil hydraulic conductivity decreases exponentially with depth so that sub-surface flow is confined to a shallow layer. If this is not the case, the steady-state and quasi-dynamic topographic wetness indices will be poor predictors of the spatial distribution of soil water, and an alternative index could be developed to better represent the topographic effect on water distribution, perhaps based on ground water potential expressed as a simple elevation difference above a local mean or minimum.

Topographic indices account for the component of the spatial variability of processes that is due to topographic effects. Other spatially variable factors are usually involved, such as soil hydraulic properties and vegetation in the case of soil water. In some instances, the spatial variations in these other attributes are themselves linked to the topographic indices. The spatial variability of soil properties is one case where significant links have been established (e.g. Moore et al., 1993a; Wilson et al., 1994). There are other properties, however, where explicit incorporation of the spatial variation of other important components of process models would substantially improve the predictive accuracy of topographic indices, particularly when working at a broad landscape scale as opposed to the small catchment scale. Surficial geology and, in some cases, climate are likely candidates for inclusion.

Finally, the algorithms for estimating topographic attributes are still being developed. The most difficult attribute to estimate is specific catchment area (and the closely related attribute, upslope drainage or contributing area). The earliest methods for computing contributing area (the *D8* algorithm proposed by O'Callaghan and Mark (1984) and popularised in commercial GIS such as ARC/INFO) did not allow for divergence of flow and gave poor estimates in upland areas. The more recent flow direction methods of Freeman (1991) and Quinn *et al.* (1991) and the *DEMON* method of Costa-Cabral and Burges (1994) both allow divergence in upslope areas. However, the dispersive (*FD8/FRho8*) methods have been criticised (e.g. Holmgren, 1994) and new methods giving improved performance might still be developed.

ACKNOWLEDGEMENTS

This study was funded in part by a grant to the first author from the M. J. Murdock Charitable Trust. The authors thank Ed DeYoung, Lou Glassy, Bob Snyder and Jonathan Wheatley for their assistance with terrain analysis runs and preparation of the grey-scale maps, and two anonymous referees who provided several useful suggestions which improved the manuscript. The final draft of this paper was completed while the first author was a Visiting Professor of Geography and Planning at the University of Southern California.

REFERENCES

Aspie, J. M., 1989. *Influence of groundwater on streambank soil moisture content, storm runoff production and sediment transport in a semi-arid watershed, southwest Montana.* Unpublished MS Thesis, Department of Earth Sciences, Montana State University.

Barbour, M. G. and Billings, W. D., 1988. *North American Terrestrial Vegetation.* Cambridge University Press, Cambridge.

Barling, R. D., Moore, I. D. and Grayson, R. B., 1994. A quasi-dynamic wetness index for characterizing the spatial distribution of zones of surface saturation and soil water content. *Water Resources Research,* **30**, 1029–1044.

Beven, K. J., 1989. Changing ideas in hydrology: The case of physically-based models. *Journal of Hydrology*, **105**, 157–172.

Beven, K. J. and Kirkby, M. J., 1979. A physically-based variable contributing area model of basin hydrology. *Hydrological Sciences Bulletin*, **24**, 43–69.

Boast, R. R. and Shelito, R. G., 1989. *Soil Survey of Madison County Area, Montana.* US Department of Agriculture, Soil Conservation Service, Washington, DC.

Costa-Cabral, M. C. and Burges, S. J., 1994. Digital elevation model networks (*DEMON*): A model of flow over hillslopes for computation of contributing and dispersal areas. *Water Resources Research*, **30**, 1681–1692.

Denning, P. J., 1990. Modelling reality. *American Scientist*, **76**, 495–498.

Dikau, R., 1989. The application of a digital relief model to landform analysis in geomorphology. In *Three Dimensional Applications in Geographic Information Systems*, ed. J. Raper, Taylor and Francis, London, 51–77.

Fairfield, J. and Leymarie, P., 1991. Drainage networks from grid digital elevation models. *Water Resources Research*, **27**, 709–717, 2809.

Freeman, T. G., 1991. Calculating catchment area with divergent flow based on a regular grid. *Computers and Geosciences*, **17**, 413–422.

Gallant, J. C. and Wilson, J. P., 1996. TAPES-G: A grid-based terrain analysis program for the environmental sciences. *Computers and Geosciences*, **22** (in press).

Goodchild, M. F., Parks, B. O., and Steyaert, L. T. (eds), 1993. *Environmental Modeling with GIS*. Oxford University Press, New York.

Goodchild, M. F., Steyaert, L. T., Parks, B. O. *et al.* (eds), 1996. *GIS and Environmental Modeling: Progress and Research Issues.* GIS World Inc, Ft. Collins.

Grayson, R. D., Moore, I. D. and McMahon, T. A., 1992a. Physically-based hydrologic modeling: I. A terrain-based model for investigative purposes. *Water Resources Research,* **28**, 2639–2658.

Grayson, R. D., Moore, I. D., and McMahon, T. A., 1992b. Physically-based hydrologic modeling: II. Is the concept realistic? *Water Resources Research,* **28**, 2659–2666.

Holmgren, P., 1994. Multiple flow direction algorithms for runoff modeling in grid-based elevation models: an empirical evaluation. *Hydrological Processes,* **8**, 327–334.

Hungerford, R. D., Nemani, R. R., Running, S. W. and Coughlan, J. C., 1989. *MTCLIM: A mountain microclimate simulation model.* US Forest Service Research Paper No. 414, Ogden.

Hutchinson, M. F., 1989. A new procedure for gridding elevation and stream line data with automatic removal of spurious pits. *Journal of Hydrology,* **106**, 211–232.

Iida, T., 1984. A hydrologic method of estimation of topographic effect on saturated throughflow. *Transactions of the Japanese Geophysical Union,* **5**, 1–12.

Iqbal, M., 1983. *An Introduction to Solar Radiation.* Academic Press, Toronto.

Jenson, S. K. and Dominque, J. O., 1988. Extracting topographic structure from digital elevation data for geographic information system analysis. *Photogrammetric Engineering and Remote Sensing,* **54**, 1593–1600.

Jersey, J. K., 1993. *Assessing vegetation patterns and hydrologic characteristics of a semi-arid environment using a geographic information system and terrain-based models.* Unpublished MS Thesis, Department of Plant, Soil and Environmental Science, Montana State University.

Kirkby, M. J. and Chorley, R. J., 1967. Throughflow, overland flow, and erosion. *Bulletin of the International Association of Scientific Hydrology,* **12**, 5–12.

Maidment, D. R., 1996. Environmental modeling within GIS. pp. 315-323 in Goodchild, M. F., Steyaert, L. T., Parks, B. O. *et al.* (eds), *GIS and Environmental Modeling: Progress and Research Issues,* eds M. F. Goodchild, L. T. Steyaert, B. O. Parks *et al.,* GIS World Inc, Ft. Collins.

McAllister, D. M., 1982. *Evaluation in Environmental Planning: Assessing Environmental, Social, Economic and Political Trade-offs.* Prentice-Hall, Englewood Cliffs.

Montgomery, D. R. and Dietrich, W. E., 1989. Source areas, drainage density, and channel initiation. *Water Resources Research,* **25**, 1907–1918.

Montgomery, D. R. and Dietrich, W. E., 1992. Channel initiation and the problem of landscape scale. *Science,* **255**, 826–830.

Montgomery, D. R. and Foufoula-Georgiov, E., 1993. Channel network source representation using digital elevation models. *Water Resources Research,* **29**, 3925–3934.

Moore, I. D., 1992. TAPES: Terrain Analysis Programs for the Environmental Sciences. *Agricultural Systems and Information Technology,* **4**, 37–39.

Moore, I. D. and Burch, G. J., 1986. Physical basis of the length-slope factor in the Universal Soil Loss Equation. *Soil Science Society of America Journal,* **50**, 1294–1298.

Moore, I. D. and Gallant, J. C., 1991. Overview of hydrologic and water quality modeling. In *Modeling the Fate of Chemicals in the Environment,* ed. I. D. Moore, Centre for Resource and Environmental Studies, Australian National University, Canberra, 1–8.

Moore, I. D. and Hutchinson, M. F., 1991. Spatial extension of hydrological modeling. In *Proceedings of the International Hydrology and Water Resources Symposium.* Australian Institute of Engineers National Conference Publication No. 91–22, Canberra, 803–808.

Moore, I. D. and Wilson, J. P., 1992. Length–slope factors for the Revised Universal Soil Loss Equation: simplified method of estimation. *Journal of Soil and Water Conservation,* **47**, 423–428.

Moore, I. D. and Wilson, J. P., 1994. Reply to "Comment on Length–slope factors for the Revised Universal Soil Loss Equation: Simplified method of estimation" by George R. Foster. *Journal of Soil and Water Conservation,* **49**, 174–180.

Moore, I. D., Burch, G. J. and MacKenzie, D. H., 1988. Topographic effects on the distribution of surface soil water and the location of ephemeral gullies. *Transactions of the American Society of Agricultural Engineers*, **31**, 1098–1107.

Moore, I. D., Grayson, R. B. and Ladson, A. R., 1991. Digital terrain modeling: a review of hydrological, geomorphological, and biological applications. *Hydrological Processes*, **5**, 3–30.

Moore, I. D., Gessler, P. E., Nielsen, G. A. and Peterson, G. A., 1993a. Soil attribute prediction using terrain analysis. *Soil Science Society of America Journal*, **57**, 443–452.

Moore, I. D., Lewis, A. and Gallant, J. C., 1993b. Terrain attributes: estimation methods and scale effects. In *Modelling Change in Environmental Systems*, eds A. J. Jakeman, M. B. Beck and M. McAleer, John Wiley and Sons, New York, 189–214.

Moore, I. D., Norton, T. W. and Williams, J. E., 1993c. Modeling environmental heterogeneity in forested landscapes. *Journal of Hydrology*, **150**, 717–747.

Moore, I. D., Turner, A. K., Wilson, J. P., Jenson, S. K. and Band, L. E., 1993d. GIS and land surface–subsurface modeling. In *Environmental Modeling with GIS*, eds M. F. Goodchild, B. O. Parks and L. T. Steyaert, Oxford University Press, New York, 197–230.

National Research Council, 1986. *Assessing the Natural Resources Inventory* (two volumes). National Academy Press, Washington, DC.

O'Callaghan, J. F. and Mark, D. M., 1984. The extraction of drainage networks from digital elevation data. *Computer Vision, Graphics and Image Processing*, **28**, 323–344.

O'Loughlin, E. M., 1986. Prediction of surface saturation zones on natural catchments by topographic analysis. *Water Resources Research*, **22**, 794–804.

Quinn, P., Beven, K. J., Chevallier, P. and Planchon, O., 1991. The prediction of hillslope flow paths for distributed hydrological modelling using digital terrain models. *Hydrological Processes*, **5**, 59–79.

Quinn, P., Beven, K. J. and Lamb, R., 1995. The ln(a/tan) index: how to calculate it and how to use it within the TOPMODEL framework. *Hydrological Processes*, **9**, 161–182.

Ratliff, L. F., Ritchie, J. T. and Cassel, D. K., 1983. Field-measured limits of soil water availability as related to laboratory-measured properties. *Soil Science Society of America Journal*, **47**, 770–775.

Rawls, W. J. and Brakensiek, D. L., 1989. Estimation of soil water retention and hydraulic properties. In *Unsaturated Flow in Hydrologic Modeling: Theory and Practice*, ed. H. J. Morel-Seytoux, Kluwer Academic Publishing, Amsterdam, 275–300.

Renard, K. G., Foster, G. R., Weesies, G. A. and Porter, J. P., 1991. RUSLE: Revised Universal Soil Loss Equation. *Journal of Soil and Water Conservation*, **41**, 30–33.

Renard, K. G., Foster, G. R., Weesies, G. A., McCool, D. K. and Yoder, D. C., 1993. *Predicting Soil Erosion by Water: A Guide to Conservation Planning with the Revised Universal Soil Loss Equation (RUSLE)*. US Department of Agriculture, Agriculture Handbook No.703, Washington, DC.

Running, S. W., 1991. Computer simulation of regional evapotranspiration by integrating landscape biophysical attributes with satellite data. In *Land Surface Evaporation: Measurement and Parameterization*, eds T. J. Schmugge and J. Andre, Springer-Verlag, London, 359–369.

Running, S. W. and Thornton, P. E. 1996, Generating daily surfaces of temperature and precipitation over complex topography. In *GIS and Environmental Modeling: Progress and Research Issues*, eds M. F. Goodchild, L. T. Steyaert, B. O. Parks *et al.*, GIS World Inc, Ft. Collins, 93–98.

Running, S. W., Nemani, R. R. and Hungerford, R. D., 1987. Extrapolation of synoptic meteorological data to mountainous terrain and its use for simulating forest evapo-transpiration and photosynthesis. *Canadian Journal of Forest Research*, **17**, 472–483.

Scott, J. M., Davis, F. W., Csuti, B. *et al.*, 1993. GAP analysis: A geographic approach to protection of biological diversity. *Wildlife Monographs*, **43**, 1–41.

Smith, T. R. and Bretherton, F. P., 1972. Stability and the conservation of mass in drainage basin evolution. *Water Resources Research*, **8**, 1506–1529.

Speight, J. G., 1980. The role of topography in controlling throughflow generation: a discussion. *Earth Surface Processes*, **5**, 187–191.

Wilson, J. P., 1996. Spatial models of land use systems and soil erosion: the role of GIS. In *GIS and Spatial Models: New Potential for New Models?*, eds M. Wegener and A. S. Fotheringham, Taylor and Francis, London.

Wilson, J. P. and Gallant, J. C., 1996a. EROS: A grid-based program for estimating spatially distributed erosion indices. *Computers and Geosciences*, **22** (in press).

Wilson, J. P. and Gallant, J. C., 1996b. SRAD: A grid-based program for estimating radiation balances in complex terrain (submitted).

Wilson, J. P., Spangrud, D. J., Landon, M. A., Jacobsen, J. S. and Nielsen, G. A. 1994. Mapping soil attributes for site-specific management of a Montana field. In *Optics in Agriculture, Forestry, and Biological Processing*, eds G. E. Meyer and J. A. DeShazer, International Society for Optical Engineering Proceedings, Volume 2345, Bellingham, 324–335.

Wischmeier, W. H. and Smith, D. D., 1965. *Predicting Rainfall-Erosion Losses from Cropland East of the Rocky Mountains*, US Department of Agriculture, Agriculture Handbook No. 282, Washington, DC.

Wischmeier, W. H. and Smith, D. D., 1978. *Predicting Rainfall Erosion Losses: A Guide to Conservation Planning*, US Department of Agriculture, Agriculture Handbook No. 537, Washington, DC.

11 The Role of GIS in Watershed Analysis

DAVID R. MONTGOMERY[1], WILLIAM E. DIETRICH[2] and
KATHLEEN SULLIVAN[3]
[1] Department of Geological Sciences, University of Washington, USA
[2] Department of Geology and Geophysics, University of California, USA
[3] Weyerhaeuser Technology Center, Tacoma, USA

ABSTRACT

Watershed analysis methods characterize historical, current and potential geomorphological and ecological processes and conditions within a drainage basin in order to guide assessment of the ecological impacts of land management. Within the context of watershed analysis, Geographical Information Systems (GIS) offer advantages for: (i) predicting the spatial distribution of geomorphological processes; (ii) formulating spatially explicit hypotheses against which to compare historical reconstructions and field observations; (iii) denoting central tendencies in landscape properties; (iv) generalising from point field data to an entire watershed; and (v) exploring and illustrating potential effects of specific land management strategies. Evaluation of watershed conditions and potential response to land management decisions requires field data, however, and cannot consist entirely of GIS-driven models because landscape conditions reflect the effects of historic variability in geomorphological processes. Watershed-scale predictions of areas prone to shallow landsliding, of stream channel type, and of bed-surface grain size illustrate how GIS-driven analytical models can contribute to the analysis of geomorphological processes in mountain drainage basins. Although such spatially explicit predictions can provide hypotheses for landscape sensitivity, digital elevation models do not always adequately represent relevant topographic attributes (e.g. fine-scale hollows, channel slope and channel confinement). Also, mountain drainage basins are characterised by temporal variability in landscape characteristics that arise from disturbance and recovery processes. The framework outlined here provides a model for integrating GIS-based analyses into procedures for generating a landscape-scale understanding of geomorphological processes, and for linking such an understanding to land use decision making.

INTRODUCTION

More than 2000 years ago Plato (427–347 BC) recognised that human actions may degrade natural resources enough to limit economic productivity (Lee, 1977). In spite of such early insight, accelerated erosion and the associated downstream effects impoverished many great civilisations, with the legacy of barren upland soils

Landform Monitoring, Modelling and Analysis. Edited by S. N. Lane, K. S. Richards and J. H. Chandler.
© 1998 John Wiley & Sons Ltd.

around the Mediterranean attesting to the potential longevity of the resulting societal impacts (e.g. Loudermilk, 1950; Carter and Dale, 1955; Perline 1989; Hillel, 1991; Ponting, 1992). More recent examples of the social impacts of resource depletion include the profound influence of soil exhaustion in the eastern United States on socio-political changes in the 18th and 19th centuries (Craven, 1925); the role of farming practices on generating destructive dust storms in the Great Plains during the 1930s (Carter and Dale, 1955); and depletion of the Ogallala aquifer on contemporary depopulation of the arid midwest (Dallas, 1990). Although geomorphological processes shape the physical environment that produces and sustains natural resources upon which economies depend, societies still generally lack effective techniques for managing the long-term influence of human activity on environmental processes.

Approaches to land management and planning in Western cultures evolved in response to growing recognition of environmental constraints on human use of landscapes and advances in understanding landscape processes. Even in ancient agricultural societies, recognition that local actions can impact neighbouring or downslope areas led to restraints on land use, such as those recorded in the ancient Babylonian Code of Hammurabi (Biswas, 1967). The importance of water in early hydraulic civilisations also led to development of complex administrative systems for flow regulation and allotment, such as that for the Nile River in classical Egypt (Said, 1993). By the 14th century, English common law recognised that streamside land owners had obligations to downstream users, and European laws governing streamside land use and pollution control progressively developed during the Renaissance and industrial revolution (Parker, 1975; Newson, 1992). Based on exploration of arid regions in the western United States, Powell (1878) was among the first to argue for determining the style, intensity and distribution of land use by the capacity of the landscape to sustain such use. He also advocated a watershed basis for land management, foreshadowing renewed calls for such an approach (e.g. Likens and Bormann, 1974; Lotspeich, 1980; Montgomery et al., 1995a). Gilbert's (1917) classic study of the movement of sediment eroded by 19th century hydraulic mining in the Sierra Nevada demonstrated the potential for significant spatial and temporal separation between land use and downstream effects. Early in the 20th century, Leopold (1934) advocated preventive planning as the most cost-effective approach for society to address environmental degradation and downslope impacts from accelerated erosion. Subsequently, many river basin plans sought to coordinate regional environmental planning as increasing attention focused on the cumulative effects of distributed land use activities (Reid, 1993). Recent proposals for ecosystem management seek to implement ecologically oriented land management based on sustaining environmental values while deriving economic benefits from resource and land use (e.g. Norton, 1992; Slocombe, 1993).

Ecosystems are far too complex to manage directly, and ecosystem management is best thought of as the management of human actions so as to either minimally or predictably impact ecological functions and processes. Whether implemented from a biocentric or anthropocentric point of view, ecosystem management requires a solid understanding of landscape-level ecosystem processes, and in particular of the interaction of geomorphological, hydrological and biological processes (see Stanley

(1995) for a critique of anthropocentric views of ecosystem management). Although Ludwig *et al.* (1993) argue that politics and greed may ultimately compromise attempts to restrain resource depletion, more immediate problems remain for designing methods for integrating ecological considerations into land use planning and decision making.

At present, poor information about landscape-scale processes generally hinders assessment of the ecological consequences of human actions and helps institutionalise land use conflicts. Landform analysis can provide an understanding of geomorphological processes that influence ecological processes and systems. Environmental impact analysis protocols developed in response to environmental legislation generally focus on site-, ownership- or species-specific issues at scales inadequate for assessing ecosystem processes and conditions. In part reflecting these problems, the integrated effects of local land management decisions often fail to meet larger-scale societal objectives (e.g. species protection). Many environmental impacts of incremental or distributed actions become apparent at the landscape scale and current watershed analysis efforts in the western US (WFPB, 1993; USDA, 1994) arose from ongoing conflict between resource use and environmental values. The spatially explicit understanding developed through watershed analysis can facilitate minimising impacts, rather than simply striving for compliance with baseline environmental standards (e.g. Montgomery, 1995). Watersheds provide a logical basis for organising analyses of local environmental problems within a larger framework, even though some environmental issues (e.g. spotted owl abundance) are not well addressed on a watershed basis (see Smith (1969) for an overview of the role of drainage basins in shaping human activity, and Omernik and Griffith (1991) for a critique of using drainage basins as a basis for environmental management). Widespread acknowledgement that Geographical Information System (GIS) technology provides an ideal platform for organising, manipulating and presenting landscape-scale data often eclipses evaluation of the limitations and appropriate role of GIS for environmental analysis, modelling and assessment. Here we review the intent of watershed analysis and propose a general framework for the role of GIS in supporting watershed analyses and land management decision making.

STRUCTURE AND INTENT OF WATERSHED ANALYSIS

Watershed analysis aims to facilitate reconciling land management decisions with legal and ethical requirements for preservation of ecological systems by assessing the capacity of a landscape to sustain land use in an objective, scientific manner. The general approach rests on the belief that managing human impacts on environmental systems requires both significant understanding of the processes structuring the system, and feedback to assess the performance of past decisions. Lacking a landscape-scale understanding of biophysical processes, landscape management essentially reduces to an uncontrolled experiment designed to optimise one or at most several economic factors (e.g. timber production). The complex task of designing more holistic strategies for land management requires several elements: it should incorporate a scale large enough to consider routing processes and the cumulative impacts of distributed actions (Reid, 1993; Smit and Spaling, 1995); and it should determine the spatial distribution of landscape processes in order to evaluate potential consequences of planned or potential activity. The basic goal of watershed analysis is to generate spatially explicit understanding of a landscape and its ecosystems at a resolution sufficient to allow assessment of the integrated environmental consequences of inherently local land use practices (Montgomery *et al.*, 1995a).

Every watershed hosts a unique array of processes and spatial contexts that influence both the type and propagation of impacts resulting from particular styles of land management. A key tenet of watershed analysis is that each watershed supports a unique spatial context for interaction among key processes. Understanding this array requires interpreting the spatially distributed context of biological and geomorphological processes, and watershed analysis aims to provide information and understanding to tailor land use to particular landscapes. A general approach to watershed analysis therefore should incorporate theoretically based assessments of landscape attributes that define key system properties. Many current approaches to landscape analysis and assessment rely, however, on either *ad hoc* assumptions or simple empirical correlations. A solid theoretical foundation would provide a means of generalising field data, introduce an objective, repeatable approach, and encourage hypothesis testing in place of simple data-gathering exercises. Towards this end, models for predicting both the spatial extent and frequency of landscape processes (e.g. shallow landsliding) and landscape attributes (e.g. stream channel type) constitute important components of watershed analysis methods. Three key steps structure watershed analysis: (i) developing a conceptual framework to assess key processes; (ii) casting these processes as hypotheses in a spatial context; and (iii) making field observations to test these hypotheses.

Integrating such information into land management requires a symbiotic relationship among analysis, planning and monitoring. Procedures to integrate watershed analyses into decision making need to recognise that landscape processes differ dramatically from region to region, from basin to basin, and often within a watershed. The framework for watershed analysis therefore needs to define the general flow of logic necessary for analysing any landscape; more specific regional protocols can guide detailed analyses. A general structure of watershed analysis consists of (i) defining an initial model of landscape-level functions and processes, (ii) reconstruction of historical and observing contemporary conditions, and (iii) projection of potential future conditions under different management scenarios (Montgomery *et al.*, 1995a). The goal is to provide understanding of critical or important aspects of physical and biological processes that may or may not define issues identified at the onset of the analysis. The analysis both develops a synthetic understanding of landscape-scale processes for use in planning and provides a process-based framework within which to develop monitoring programmes (Figure 11.1). A programme to implement this system in Washington, USA, has so far focused on protecting aquatic ecosystems within the forest regulatory system (WFPB, 1993). While this system can better inform land managers and encourage the constraint of activities that exceed landscape capabilities, clearly defined priorities and objectives remain essential for guiding decisions based upon watershed analyses.

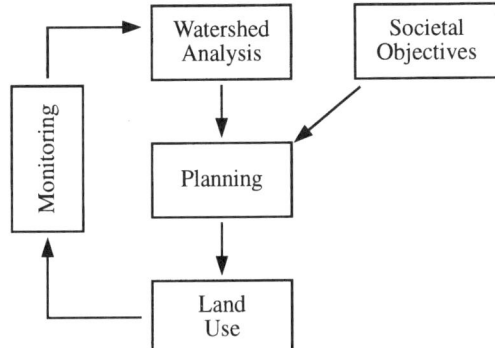

Figure 11.1 Relationship between watershed analysis, planning and monitoring

In mountain drainage basins, geomorphological processes are key drivers of biological processes (Swanson *et al.*, 1988). Although similar geomorphological and ecological processes may operate throughout a region, the nature of linkages among processes and their rates depends on local geology, topography, climate, and disturbance history. A process-based understanding of these processes and linkages is central to estimation of downstream effects of human actions distributed through space and time. Such insight is essential for interpreting past events in order to establish causality because the legacy of past events structures current conditions. Disturbance history can provide critical insight into ecosystem and landscape conditions and processes; many historical patterns of land use provide experiments that can yield insight into possible future responses. Definition of current conditions is essential for assessing system disturbance and for providing a reference against which to predict and interpret future changes; the nature of future response also may depend on current conditions. Because of these considerations, the evaluation of both the potential impacts of management and the benefits of restoration efforts requires an understanding of landscape dynamics and trends in ecosystem conditions.

Generation of a watershed-level model of environmental processes requires simplification of biophysical processes and relationships. Watershed analysis assumes that an understanding of key aspects, processes and linkages can guide representation of landscape processes and provide insight into the complex fabric of ecological systems. It is unreasonable to expect a single person to possess the expertise to tackle all aspects of the linked physical and biological system. An approach to watershed analysis that takes advantage of advanced training breaks up the analysis into modules that deal with specific attributes of the system. One example, the Washington watershed analysis method (WFPB, 1993), relies on a small set of analysis modules that assess mass wasting, surface erosion, hydrology, stream channels, riparian forests and fish. The goal of such modularisation is to develop an understanding of key processes. An alternative approach structures analyses around key processes either assumed or known to occur in the landscape of interest. In either approach, synthesis of the spatially explicit, process-based understanding developed in the analysis results in a characterisation of landscape-scale ecological processes.

The structure of each analysis follows the same general flow of logic. An initial model provides an expectation of landscape conditions and processes. This may involve, for example, predicting the probability of landslide initiation across the landscape. Such an initial hypothesis is tested through historical analyses and fieldwork to characterise both trends and current conditions. Observations are interpreted against the hypothesized condition to both assess trends in landscape conditions and revise the initial model to account for landscape-specific factors (e.g. influence of groundwater flow through bedrock fractures on debris flow initiation). The revised stratification and conceptual model provide a context for assessing potential responses to land management. A primary goal of the process described above is to systematise the assessment of the capacity of different portions of a landscape to sustain different land use practices. The inherent complexity of most ecosystems means that implementing ecosystem-oriented land management requires both monitoring programmes and feedback between monitoring data and future planning.

GIS AND WATERSHED ANALYSIS

A comparison of GIS capabilities with the structure of watershed analysis helps define and evaluate potential roles for GIS. Basic capabilities and strengths of GIS include: (i) compiling and overlaying landscape attributes; (ii) predicting the spatial distribution of environmental processes using process models; (iii) generalizing locally collected field data across the landscape; (iv) simulating potential future conditions; and (v) archiving data for monitoring efforts. The spatial data handling capabilities of GIS are ideal for organizing information on basic landscape attributes or current conditions (Figure 11.2A). Examination of the spatial correspondence between landslides and steep slopes, for example, involves straightforward overlays of landscape characteristics. While such correspondence can illuminate landscape patterns, the approach generally proves inadequate for predicting future response. Prediction of the spatial distribution of geomorphological processes using physically based models provides spatially explicit insight into the distribution of processes across a landscape. This involves combining landscape attributes with theoretically based models to predict landscape sensitivity or the spatial distribution and intensity of ecological or geomorphological processes (Figure 11.2B). Casting such predictions in terms of drainage area, slope and spatial data for other landscape attributes (e.g. soil strength or vegetation type) provides a ready format for GIS-based analyses. Calculating the topographic control on the relative potential for shallow landslide initiation illustrates the approach of forging process-based models into an index for landscape sensitivity (Montgomery and Dietrich, 1994). Such analytical tools, however, lack information on actual conditions. Using models to iteratively modify landscape attributes allows the exploration of dynamic scenarios for possible future conditions (Figure 11.2C). The capacity to readily create, update and access data linked to specific locations in a landscape allows the archiving, storage and retrieval of data from monitoring efforts (Figure 11.2D). Each of these distinct roles for GIS offers different potential contributions to each step of watershed analysis.

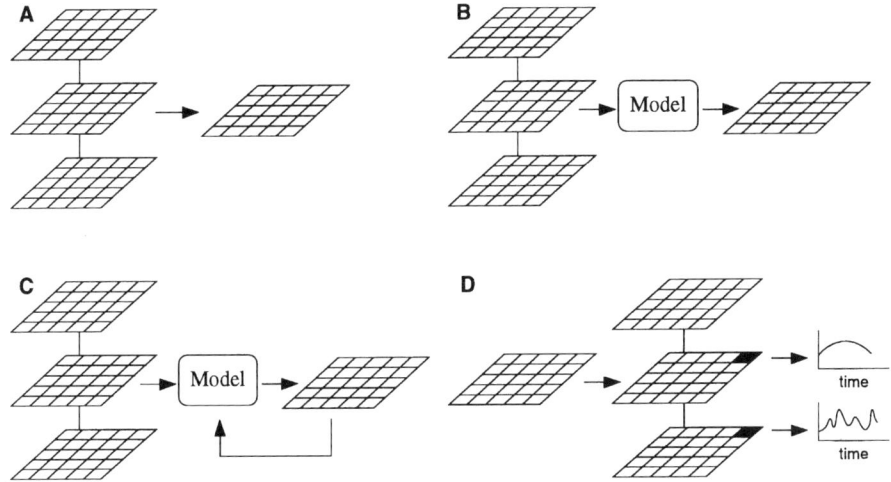

Figure 11.2 Distinct capabilities of GIS in watershed analysis: (A) landscape over-lays; (B) spatial pattern mapping using simple models; (C) scenario generation using dynamic process models; (D) a framework for monitoring

GIS-based analyses provide tools for formulating hypotheses about landscape states, but testing these hypotheses requires field investigations to assess history and current conditions. Lack of a process-based understanding plagues many watershed assessments. The prevailing approach of using sites assumed to represent desirable or "natural" conditions suffers from the classic problems of paired-basin studies: basin characteristics differ; it is hard to account for all potential factors; and unique events may dominate comparisons. An alternative approach is to use GIS-based analyses to develop spatially explicit hypotheses of landscape function that are tested against field data. These two approaches are not incompatible; sites assumed to represent natural conditions can be useful for developing and calibrating more general process models. Formulation of spatially explicit initial hypotheses can force comparison against field data and provide guidelines for identifying data worth collecting. Below, we consider potential contributions of GIS-based process models to the assessment of shallow landslide hazards and stream channel assessment in the context of watershed analysis.

SHALLOW LANDSLIDING

Landsliding is a dominant erosion processes in steep forested terrain throughout the Pacific Northwest, and addressing landslide-related impacts is a key concern of forest land managers. Failure to address these issues in the past has resulted in large influxes of sediment into stream channels and significant adverse impacts on fish habitat. Methods to predict sites susceptible to landslide initiation are essential for identifying measures to reduce landslide-related impacts of land use.

A common approach to determining landslide hazards consists of mapping historical and current landslides from sequential aerial photographs. A key shortcoming of this procedure is that the assessment of landslide failure potential relies primarily on historical patterns and topographic features, yet virtually no record is long enough to fully characterise the landslide threat in an area. This complicates extrapolation of future hazards based on patterns of landsliding observed in the historical record, which reflect the interaction of storm intensities and disturbances such as fire or land management (e.g. Dunne, 1991), as well as patterns of high-intensity rainfall cells that can localise shallow landsliding in mountainous terrain. Within this context, simple physically based models of the topographic control on the potential for shallow landslide initiation can provide spatially explicit hypotheses against which to interpret observed patterns of slope instability.

GIS-based approaches to predicting landslide hazards include: multiple regression correlations of landscape attributes associated with observed landslides (e.g. Carrara *et al.*, 1991); indices based on geology, slope and/or land form (e.g. Lanyon and Hall, 1983; Niemann and Howes, 1991; Ellen *et al.*, 1993); and physically based models (e.g. Okimura and Nakagawa, 1988; Dietrich *et al.*, 1992, 1993, 1995; van Asch *et al.*, 1993; Montgomery and Dietrich, 1994; Wu and Sidle, 1995). Empirically based models often are not transferable and require site-specific data typically unavailable at the start of watershed analysis. Physically based models rely on combining topographic attributes readily determined from digital elevation models (DEMs) (e.g. drainage area (a) and slope (S)) with soil and bedrock properties (e.g. soil depth (z), friction angle (ϕ) and hydraulic conductivity (K)) in a model to predict relative slope stability. The approach described below contrasts with stochastic simulations of slope stability (e.g. Hammond *et al.*, 1992) in which relative stability is determined based on sampling from distributions of parameters for each location in a catchment. While spatial variability of model parameters can be significant, we choose not to attempt to place some uncertainty on our stability analyses because we see no way to quantitatively assess this uncertainty (as it is not spatially constant) and we do not see that it would alter interpretations of relative slope stability across a landscape. We show below that physically based models for the topographic control on relative slope stability substantially refine identification of hazards beyond typical representations on slope maps.

Coupled hydrologic and slope stability models allow prediction of spatial patterns of relative landslide susceptibility from digital terrain data. A model for failure of cohesive soils overlying less conductive bedrock involves combining a relative soil moisture criterion for steady-state rainfall with the infinite slope stability model (Dietrich *et al.*, 1992). The soil wetness, W, is given by:

$$W = \frac{qa}{bT\sin\vartheta} \qquad (11.1)$$

where a/b is drainage area per unit contour length, ϑ is local ground slope (in degrees), T is the depth-integrated soil transmissivity, and q is the effective steady-state rainfall rate (O'Loughlin, 1986). This yields an expression for the critical steady-state rainfall rate (q_c) required to cause landsliding for each grid cell in a catchment:

$$q_c = \frac{Tb \sin \vartheta}{a} \left[\left(\frac{C'}{\rho_w gz \cos^2 \vartheta \tan \phi} \right) + \frac{\rho_s}{\rho_w} \left(1 - \frac{\tan \vartheta}{\tan \phi} \right) \right] \qquad (11.2)$$

where ρ_s and ρ_w are the bulk density of soil and water; C' is the effective cohesion of the soil and g is gravitational acceleration. For the case of cohesionless soils, equation (11.2) reduces to:

$$q_c = \frac{\rho_s Tb \sin \vartheta}{\rho_w a} \left(1 - \frac{\tan \vartheta}{\tan \phi} \right) \qquad (11.3)$$

Because of the steady-state hydrologic model, critical rainfall values predicted to induce slope instability may not readily apply to actual rainfall values necessary to trigger landsliding. Rather, critical rainfall values provide a relative index of the potential for shallow landsliding. Previous applications of equation (11.3) using high resolution DEMs of three small watersheds (i.e. < 2 km^2) revealed that more than 80% of shallow landslides originated in locations predicted to be least stable (Montgomery and Dietrich, 1994). Application of equation (11.3) to the Tolt River and Skokomish River watersheds using widely available 30 m DEMs (Plate I) reveals that even though each watershed exhibits a unique pattern of landslide susceptibility, the critical rainfall maps highlight steep, convergent headwater areas and inner gorges along larger incised channels as the most likely source of debris flow initiation.

Grid size influences the representation of drainage areas and local slopes derived from a DEM; contributing areas tend to increase and local slope decreases with increasing grid size (Jenson, 1991; Panuska et al., 1991; Moore et al., 1993; Zhang and Montgomery, 1994). Critical rainfall maps derived using equation (11.3) for 10 m, 30 m and 90 m grid size DEMs created from the contours of the Alleghany, Oregon, US Geological Survey 7.5' topographic quadrangle illustrate profound shifts in the spatial distribution of predicted landslide hazard (Plate II). Soil parameter values used in this example are those used by Montgomery and Dietrich (1994) in their analysis of Mettman Ridge, Oregon. At the finest grid size (10 m) the least stable areas (q_c of 0–50 mm/day) correspond to hollows at the head of valleys and steep side slopes, locations recognized as primary sources for debris flows (e.g. Dietrich et al., 1986; Reneau and Dietrich, 1987). This relation between fine-scale topographic features and the predicted pattern of relative slope stability becomes obscured with increasing grid size, especially for grid sizes larger than 30 m. The proportion of the catchment predicted to be unconditionally stable increases with increasing grid size, whereas the area with critical rainfall values less than 100 mm/day and the area of unconditionally unstable ground both generally decrease with increasing grid size (Figure 11.3). Grid sizes as fine as the topographic data allow are therefore preferable for physically based models of shallow landslide hazards, as coarse grid sizes will underestimate the extent of potential instability.

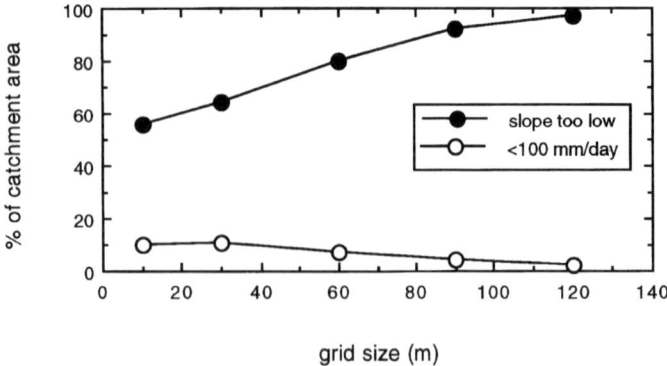

Figure 11.3 Percentage of landscape area too gentle to fail and with critical steady-state rainfall less than 100 mm/day for maps of the Alleghany quadrangle derived using 10 m, 30 m, 60 m, 90 m and 120 m grid sizes. Parameter values as for Plate II

GIS-driven models of slope stability can also portray landscape sensitivity to particular land use practices. The effective cohesion provided by tree roots, for example, decreases dramatically following timber harvesting (e.g. Burroughs and Thomas, 1977). Numerous studies document increased rates of shallow landsliding in steep terrain following forest clearance (Sidle *et al.* (1985) review many such studies). A comparison of the critical rainfall patterns derived using equation (11.2) for cohesion of 15 and 2 kN/m^2, values respectively approximating the root strength of mature Douglas fir (*Psuedotsuga menziesii*) and cut stumps (Burroughs and Thomas, 1977), illustrates the influence of clear-cut logging on the distribution of slopes prone to failure. Using a soil depth of 1 m and soil properties estimated from field measurements ($\rho_s/\rho_w = 2.0$; $T = 65$ m^2/day; $\phi = 33°$), the map pattern of the critical rainfall for the Alleghany quadrangle in the Oregon Coast Range (Plate III) shows that the greatest decrease in slope stability occurs in steep convergent topography, locations where shallow landsliding typically occurs after timber harvesting in steep forested landscapes (Reneau and Dietrich, 1987). Soil depths, however, vary systematically with topographic position: less stable deep soils occur in topographic hollows and more stable shallow soils occur on topographically divergent slopes. Inclusion of colluvial soil depths predicted by a process-based model further concentrates areas with a high potential for slope instability in steep topographically convergent areas at the tips of the drainage network (Dietrich *et al.*, 1995). The contrast in map patterns between the pre- and post-cutting scenarios therefore provides a conservative illustration of the sensitivity of steep landscapes to loss of root strength and provides guidance for interpreting spatial patterns of landscape sensitivity to timber harvesting.

Although integration with GIS allows incorporation of spatially distributed properties where such information is available, the data required to portray fine-scale spatial variation in soil properties are rarely available. Soil surveys, for example, seldom report either soil strength properties or depth-integrated transmissivity.

Moreover, the variance of these parameters over short length scales may approach that between different soil types. This suggests several strategies for process-based modelling: (i) treat soil properties as spatially uniform and model the topographic control on geomorphic processes (e.g. Montgomery and Dietrich, 1994); (ii) generalise soil properties within major lithologic or pedologic units (e.g. Tang and Montgomery, 1995); or (iii) develop models to predict the spatial variability of soil properties (e.g. Dietrich *et al.*, 1995). Incorporating spatial variability in soil hydraulic properties becomes increasingly important at larger spatial scales in watersheds with a variety of bedrock and soil types.

Consideration of the potential contributions to watershed analysis of GIS-based analyses of landslide hazards highlights the utility of using spatially explicit models to provide a framework for assessment of actual conditions. Incorporation of models such as those discussed above into initial hypotheses against which to interpret historical patterns of shallow landsliding can improve the predictive capacity and theoretical underpinning of watershed analyses. GIS-driven process models, however, cannot yield information about past history and observed rates of shallow landsliding. Hence GIS-based analyses cannot replace field analysis. Rather, the ideal role of GIS in mass wasting assessments is to define conceptual and spatial contexts that enhance the relevance of field observations. GIS also provides an ideal platform for archiving landslide locations. Using coupled field data and GIS analyses to update and revise the range of critical rainfall values considered hazardous can provide for ongoing evaluation of land management decisions.

STREAM CHANNELS

Stream channels receive and either transport or store material eroded from their watersheds. Extensive alteration of drainage timing, volume and patterns, runoff generation mechanisms, and sediment delivery to stream channels has adversely impacted aquatic ecosystems throughout the Pacific Northwest. The wide variety of channel morphologies and potential channel responses complicate the assessment of channel conditions in forested mountain drainage basins. Landscape managers, however, need methods to both decipher channel condition and predict relative sensitivity to changes in sediment supply, discharge and the supply of large roughness elements. GIS-based predictions of channel type and attributes can provide a context for interpreting field-based channel assessments.

Defining the extent of the channel network presents the first challenge to GIS-based analysis of stream channels. It is widely known that the blue lines on most US Geological Survey topographic maps poorly represent the channel network identifiable in the field (Morisawa, 1957; Coffman *et al.*, 1972; Mark, 1983) and that the portrayal of channel network extent varies with map scale (e.g., Eyles, 1966; Robert and Roy, 1990). Coupling independently derived hydrographic coverage with DEMs can exacerbate such problems, as registration and data quality problems may result in channels that are located on valley walls or that flow over drainage divides. However, blue line locations are essential for delineating channel paths in wide unentrenched valley bottoms where both contours and DEMs prove useless for locating channels.

Two distinct methods allow the estimation of channel network extent from DEMs. A number of workers offer techniques for estimating channel head locations using a constant critical support area (A) (e.g. Band, 1986, 1989; Jenson and Domingue, 1988; Morris and Heerdegen, 1988; Tarboton et al., 1991). Field data, however, indicate a slope dependence to the size of source areas, and thus channel network extent (Montgomery and Dietrich, 1988). Several studies discuss methods that incorporate a slope (S) dependence for determining channel network extent from DEMs (Dietrich et al., 1992, 1993; Montgomery and Foufoula-Georgiou, 1993). The approach used in watershed analysis needs to reflect the purpose of, and confidence required from, the analysis and GIS provides an efficient method for delineating channel network extent after establishing the appropriate method and resolution.

Estimation of channel network extent in the watershed of Finney Creek, Washington, illustrates the utility of using GIS to extrapolate minimal field data into basin-wide predictions. This example also illustrates the power of coupling general process models with local data for characterising landscape-wide phenomena. Source area sizes derived from field-mapped channel head locations imply a rough criterion of $AS^2 > 6000$ m^2 for representing the channel network identifiable in the field. The alternative approach of adjusting the channel initiation threshold to identify the most extensive network without a "feathering" of first-order channels (Montgomery and Foufoula-Georgiou, 1993) suggests that a criterion of $AS^2 = 45\,000$ m^2 portrays the network resolvable from the 30 m DEM. This example illustrates that the poor resolution of high-frequency topography typical of 30 m DEMs (Zhang and Montgomery, 1994) results in underestimation of the channel network identifiable in the field using DEM-based approaches. Consequently, the networks so derived represent hypotheses to be tested, and probably revised, during watershed analysis.

Many stream channel classifications impose order on a wide range of natural channel types (see reviews by Hawkes (1975), Mosely (1987) and Church (1992)). Channel morphology, slope and confinement provide primary indicators of potential channel response to land management in forested watersheds, and stratification of expected channel morphology by gradient and confinement provides a simple classification related to channel response potential for distinct channel types identified through fieldwork in Washington, Oregon and Alaska (Montgomery and Buffington, 1993). In forested environments, however, large woody debris can dramatically alter channel morphology and response potential by forcing pool formation (e.g. Swanson et al., 1976; Montgomery et al., 1995b) and by modifying bed-surface grain size through roughness effects (Buffington, 1995). Such influences can govern the channel type present in the field; channel types predicted from map- or DEM-derived slopes can only provide hypotheses against which to interpret actual conditions.

In some cases it is possible to discern channel confinement from topographic map inspection, but our fieldwork in Pacific Northwest channels indicates that such interpretations are often erroneous. Although experience can guide interpretation of

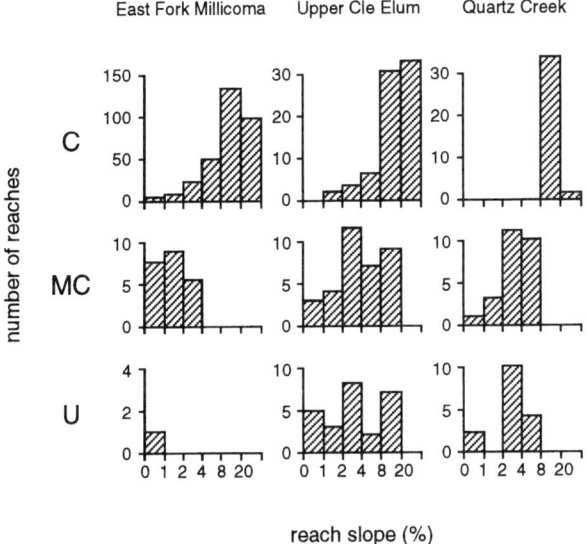

Figure 11.4 Frequency distributions of reach slope for confined (C), moderately confined (MC) and unconfined (U) channels, as identified in watershed analyses of East Fork Millicoma River, Oregon, Upper Cle Elum River, Washington, and Quartz Creek, Washington

channel type from maps or DEMs, observed correlations with slope provide an objective approach to the prediction of expected channel types throughout a watershed. Channel slope, however, provides only a general indication of both confinement and channel type. Different watersheds have widely ranging distributions of gradient and confinement, but data compiled from watershed analyses in Washington illustrate that steeper channels tend to be more confined (Figure 11.4). Channel type predictions based on reach slope provide reasonable hypotheses that should be field tested prior to evaluating potential impacts of management decisions.

Although the distance between successive contours provides a simple, albeit tedious, method for mapping channel slope from topographic maps, a number of issues complicate the definition of channel slope from digital elevation data. Fine-scale variations in channel slopes derived from DEMs (Plate IV) arise due to pit filling, poor fine-scale representation of topography, and because the slope defined by the direction of steepest descent in a grid cell may reflect the hillslope gradient rather than that of the channel. Channel slopes in the Finney Creek watershed, for example, vary by up to a factor of 4 due to this effect. The practice of rounding grid elevations to the nearest metre presents an additional problem for low-gradient channels. With 1 m vertical resolution, the minimum resolvable grid-cell slope in a 30 m DEM is 0.033, a relatively steep slope for alluvial stream channels. In addition, the resulting 1 m high digital waterfalls dominate the elevation drop in low-gradient reaches. Variations in the slope of the low-gradient reaches of Finney Creek illustrate such artefacts. Averaging slopes over longer reaches, such as

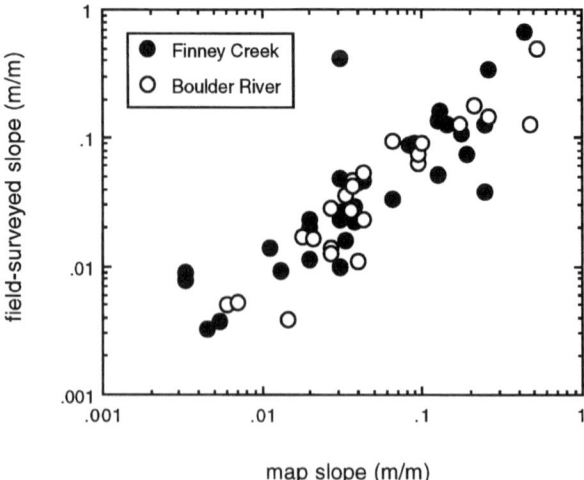

Figure 11.5 Field-surveyed versus map-derived slope for channel reaches in Finney Creek and Boulder River watersheds. Reach slopes were surveyed over reaches defined by similar channel morphology

between tributary confluences, can mask such problems (Plate IVB), but may also obscure short stretches of channel with anomalous gradients. These considerations indicate that reach-average slopes are more appropriate than pixel slopes for watershed-scale representations, although short reaches of significantly higher or lower slopes are to be expected in the field.

Another issue involves the relation between field- and DEM-derived slopes. Considerable scatter characterizes the relation between slopes measured from the distance between contours on 1:24 000 scale topographic maps and channel reach slopes surveyed over distances of 10 to 20 channel widths in the Finney Creek and adjacent Boulder River watersheds (Figure 11.5). Although the general trend of map-derived slopes is correlated with surveyed reach slopes, the variance about this trend can result in serious misestimation of channel slope, and thereby channel type. This discrepancy reflects both significant reach-scale variation of field slopes and the fact that high-frequency noise can dominate DEM-derived slopes over short length scales. Significant variability in channel morphology should be expected in forested mountain watersheds, and reach-level predictions of channel type provide only hypotheses for finer-scale channel attributes and response potential.

Temporal variations in channel type impart an inherent unpredictability to the actual channel type at any point in a channel network at a single point in time. Scour by debris flow, sequential aggradation and degradation by passage of sediment waves, and changes in the supply of in-channel large woody debris can all alter channel type (e.g. Benda, 1990). Reductions in large woody debris loading, for example, can change a forced pool-riffle channel into a plane-bed channel (Montgomery *et al.*, 1995b) or a forced alluvial channel into a bedrock channel. None of these differences, which have important implications for interpreting channel condition, response potential and ecological conditions, can be derived from automated landform analyses. Hence, GIS-derived predictions of channel type can provide only a guide to actual conditions.

Bed-surface grain size is a channel characteristic that strongly influences ecological systems and is also an indicator of channel condition and response potential. A model for bed-surface grain size illustrates the approach of generating distributed hypotheses for landscape states against which to compare and interpret field observations. We assume that channel slope, bankfull width and depth, and grain size on the bed surface mutually adjust such that the median grain size of the bed just starts to move at bankfull discharge. This is a good approximation for gravel-bedded rivers receiving a low sediment supply and composed of gravel banks (see discussions in Parker (1978) and Buffington (1995)). Calculations based on this assumption may provide a means for detecting the effects of sediment supply, roughness due to bedforms and wood debris, and rare flow events (see Buffington (1995) for detailed analysis of this hypothesis).

Combining an expression for basal shear stress:

$$\tau_{ob} = \rho_w gDS \tag{11.4}$$

(where τ_{ob} is the bankfull shear stress, D is the bankfull flow depth and S is the reach slope) with Shields' critical shear stress relation:

$$\tau_{oc50} = \tau^*(\rho_s - \rho_w)gd_{50} \tag{11.5}$$

(where τ_{oc50} is the critical shear stress for the median grain size, τ^* is the dimensionless critical shear stress, and d_{50} is the median grain size on the bed surface) yields an expression for the predicted median grain size of a threshold channel (Buffington, 1995):

$$d_{50} = \frac{\rho_w}{\rho_s - \rho_w} \frac{DS}{\tau^*} \tag{11.6}$$

Assuming that $\rho_s = 2650 \text{ kg/m}^3$, $\rho_w = 1000 \text{ kg/m}^3$, and $\tau^* = 0.045$ (Miller et al., 1977), and combining with hydraulic geometry relations between bankfull flow depth and drainage area for Finney Creek (Figure 11.6):

$$D = 0.0027 \, A^{0.33} \qquad (R^2 = 0.81) \tag{11.7}$$

(where D is in metres and A is in square metres) yields an expression for the predicted grain size for threshold channels in the Finney Creek watershed:

$$d_{50} = 0.0365 \, A^{0.33} S \tag{11.8}$$

where d_{50} is in metres and A is in square metres. The map pattern of predicted grain size (Plate IVC) provides watershed-wide predictions against which to interpret field observations.

Generalization of predicted grain size into dominant bed-surface textures (i.e. sand, fine and coarse gravel, cobble and boulder) corresponds to an ecologically important physical characteristic of stream channels. These general bed-surface textures relate to an ecological stratification for predicting the type of organisms

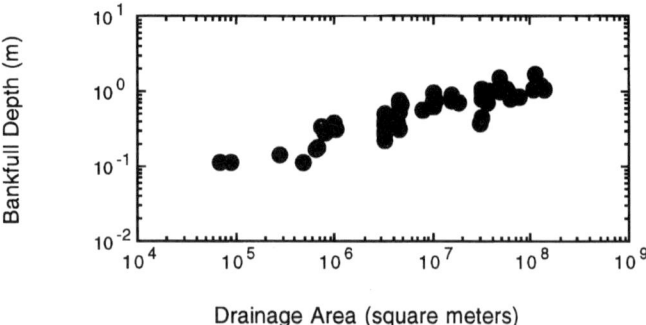

Figure 11.6 Bankfull depth versus drainage area for channel reaches in the Finney Creek and Boulder River watersheds in the Washington Cascades

and communities using different parts of a channel network (Hawkes, 1975). Moreover, the predicted grain size facilitates the interpretation of potential eco-logical changes in response to the addition or removal of roughness elements, which can significantly alter d_{50} (Buffington, 1995).

Many factors influence channel condition and response potential; channel type suggests the likely range of probable channel response and expected channel con-ditions. Ecologically important changes in channel condition occur without changing channel type, which may be relatively resilient to all but massive disturbance. Evaluation of potential responses to land management requires the assessment of these often more subtle channel attributes and conditions, and the wide variety of potential influences and responses suggests a diagnostic approach. Adequately diag-nosing channel condition involves considering a variety of stream bed, active channel and floodplain characteristics. GIS-based analyses can predict channel type, the basic architecture of the network, and grain size, but temporal variability in sediment delivery, the influence of large woody debris, and DEM resolution impart substantial uncertainty to these predictions. Confident assessment of channel response potential therefore requires field assessment. GIS-based analyses can formalize an initial hypothesis for properties such as channel type or grain size against which to interpret or explain field observations, but such analyses cannot provide information on current conditions. Consequently, appropriate roles for GIS in the channel assess-ment module include stratifying the channel network, generating hypotheses against which to interpret field observations, and organising presentation, storage and retrieval of data.

DISCUSSION

An often neglected issue concerning the role of GIS in watershed analysis involves how well we can hope to characterise landscape attributes and processes from DEM-based analyses. An important aspect of this issue is that DEM grid size fundamentally constrains the role of physically based models in real landscapes. A more subtle issue involves landscape complexity and the resolution and/or confidence with which we can predict temporal and spatial variability in landscape attributes or sediment transport processes. The heterogeneity of real landscapes probably limits the degree to which we can use process understanding to predict spatial associations and trends. Grayson *et al.*'s (1992) discussion of the limitations of physically based hydrologic modelling provides a pertinent perspective on this issue.

GIS-based predictions provide only a hazy image of reality, as spatial and temporal variability of landscape processes, materials and attributes can confound model predictions. Both the sequencing of impacts and existing conditions can influence landscape response to land management, implying that model predictions are themselves hypotheses. In the context of watershed analysis, models may be inappropriate for the landscape under consideration because of either inadequate input data or neglect of important processes. Application of the model for topographic control on shallow landsliding in a landscape where geologic structure favours extensive deep-seated landsliding illustrates the latter case. As an example of the former case, consider that even minor misrepresentations of topography, common in many digital terrain models, can seriously impact the predictions of process models. The headwaters of Finney Creek in the US Geological Survey 30 m DEM, for example, flow into a different watershed; reducing the elevation of a single pixel in the original DEM corrected the misrepresentation. Even relatively modest field efforts can address many such problems.

Although confident predictions of specific landscape states under particular land management scenarios may prove difficult, a simple framework may provide for a practical engagement between GIS, watershed analyses and fieldwork. We foresee eventual integration of field and GIS-based analyses, monitoring data, and GIS-based archiving of data into a process of iterative data collection and interpretation. In the case of bed-surface grain size, for example, comparison of observed and predicted grain sizes may allow detection of changes in sediment supply or influence of local sources of high supply. Acquisition of digital elevation data that adequately portrays landscape features is a crucial initial step for modelling landscape processes. The second step involves using GIS-driven models to make predictions about important landscape processes and attributes. These predictions become hypotheses. Field and laboratory analyses test these predictions to provide a basis for diagnosing landscape conditions and response potential. Georeferencing field observations and data using Global Positioning System units would enhance re-evaluation of models and hypotheses of landscape function and condition as new information arises during periodic watershed analyses. Each of the components of this framework exists in practice; integrating current capabilities into cohesive science-based programmes to inform more environmentally oriented land management awaits institutional commitment to programmes along the lines of this vision.

CONCLUSIONS

The structure and intent of watershed analysis methods imply complementary and symbiotic roles for GIS- and field-based analyses. An ideal role for GIS-based process models in watershed analysis involves generating spatially explicit hypotheses against which to test, organise and interpret field observations. Initial hypotheses allow stratification of field observations, provide a background interpretation of response potential, and define a tool for simulating potential impacts of various land management options. GIS-based analyses, however, reveal neither historic nor current landscape-level conditions and processes, information crucial to the process of watershed analysis. Development of stochastically driven simulations of geomorphological processes may help to understand rates and ranges of possible conditions under different environmental conditions or management regimes. Typical GIS-based analyses also cannot directly evaluate relations to previous land management, as interpreting or establishing causality often rests on a historical perspective. Conversely, field assessments reveal little about potential future response to management activity unless coupled to either conceptual understanding or a more explicit process model. Watershed analysis requires both approaches: neither theoretical analyses nor fieldwork alone suffice. Although GIS provides a powerful tool for use in natural resource management and watershed analysis, automated predictions do not by themselves constitute viable analyses for informing decision makers of landscape-scale environmental conditions and potential impacts of land use.

ACKNOWLEDGEMENTS

This work was supported by the Co-operative State Research Service, US Department of Agriculture, under Agreement No. 94-37101-0321, and by a gift from the Weyerhaeuser Company. Harvey Greenberg provided cartographic, programming, graphics and analysis support. Peerless Management Systems of Eugene, Oregon, provided the 10 m DEM of the Alleghany quadrangle.

REFERENCES

Band, L. E., 1986. Topographic partition of watersheds with digital elevation models. *Water Resources Research*, **22**, 15–24.
Band, L. E., 1989. A terrain-based watershed information system. *Hydrological Processes*, **3**, 151–162.
Benda, L., 1990. The influence of debris flows on channels and valley floors in the Oregon Coast Range, U.S.A.. *Earth Surface Processes and Landforms*, **15**, 457–466.
Biswas, A. K., 1967. Hydraulic engineering prior to 600 B.C.. *Proceedings of the American Society of Civil Engineers, Journal of the Hydraulics Division*, **HY5**, 115–135.
Buffington, J. M., 1995. *Effects of Hydraulic Roughness and Sediment Supply on Surface Textures of Gravel-Bedded Rivers*. MS Thesis, University of Washington, 184pp.
Burroughs, E. R., Jr and Thomas, B. R., 1977. *Declining Root Strength in Douglas Fir After Felling as a Factor in Slope Stability*. US Department of Agriculture, Forest Service, Intermountain Forest and Range Experiment Station, Research Paper INT-190, 27pp.
Carrara, A., Cardinali, M., Detti, R., Guzzetti, F., Pasqui, V. and Reichenback, P., 1991. GIS techniques and statistical models in evaluating landslide hazard. *Earth Surface Processes and Landforms*, **16**, 427–445.
Carter, V. G. and Dale, T., 1955. *Topsoil and Civilisation*. University of Oklahoma Press, Norman, Oklahoma, 292pp.

Church, M., 1992. Channel morphology and typology. In *The Rivers Handbook*, eds by P. Callow and G. E. Petts, Blackwell Science, Oxford, 126–143.

Coffman, D. M., Keller, E. A. and Melhorn, W. N., 1972. New topologic relationship as an indicator of drainage network evolution. *Water Resources Research*, **8**, 1497–1505.

Craven, A. O., 1925. Soil exhaustion as a factor in the agricultural history of Virginia and Maryland. 1606–1860. *University of Illinois Studies in the Social Sciences*, **13**, 9–177.

Dallas, R., 1990. The agricultural collapse of the arid midwest. *Geographical Magazine*, **62**, 16–20.

Dietrich, W. E., Wilson, C. J., and Reneau, S. L., 1986. Hollows, colluvium, and landslides in soil-mantled landscapes. In *Hillslope Processes*, ed. A. D. Abrahams, Allen and Unwin, Winchester, 361–388.

Dietrich, W. E., Wilson, C. J., Montgomery, D. R., McKean, J. and Bauer, R., 1992. Channelization thresholds and land surface morphology. *Geology*, **20**, 675–679.

Dietrich, W. E., Wilson, C. J., Montgomery, D. R. and McKean, J., 1993. Analysis of erosion thresholds, channel networks and landscape morphology using a digital terrain model. *Journal of Geology*, **101**, 259–278.

Dietrich, W. E., Reiss, R., Hsu, M.-L. and Montgomery, D. R., 1995. A process-based model for colluvial soil depth and shallow landsliding using digital elevation data. *Hydrological Processes*, **9**, 383–400.

Dunne, T., 1991. Stochastic aspects of the relations between climate, hydrology and landform evolution. *Transactions of the Japanese Geomorphological Union*, **12**, 1–24.

Ellen, S. D., Mark, R. K., Cannon, S. H. and Knifong, D. L., 1993. *Map of debris-flow hazard in the Honolulu District of Oahu, Hawaii*. US Geological Survey Open-File Report 93-213, 25pp.

Eyles, R. J., 1966. Stream representation on Malaysian maps. *Journal of Tropical Geology*, **22**, 1–9.

Gilbert, G. K., 1917. *Hydraulic-mining debris in the Sierra Nevada*. US Geological Survey Professional Paper 105, 154pp.

Grayson, R. B., Moore, I. D. and McMahon, T. A., 1992. Physically based hydrologic modelling 2. Is the concept realistic? *Water Resources Research*, **28**, 2659–2666.

Hammond, C., Hall, D., Miller, S. and Swetik, P., 1992. *Level 1 stability analysis (LISA) documentation for version 2.0*. General Technical Report INT-285, US Department of Agriculture Forest Service, Intermountain Research Station, Ogden, Utah, 190pp.

Hawkes, H. A., 1975. River zonation and classification. In *River Ecology*, ed. B. A. Whitton, University of California Press, Berkeley, 312–374.

Hillel, D., 1991. *Out of the Earth: Civilization and the Life of the Soil*. University of California Press, Berkeley, 321pp.

Jenson, S. K., 1991. Application of hydrologic information automatically extracted from digital elevation models. *Hydrological Processes*, **5**, 31–44.

Jenson, S. K. and Domingue, J. O., 1988. Extracting topographic structure from digital elevation data for geographic information systems analysis. *Photogrammetric Engineering and Remote Sensing*, **54**, 1593–1600.

Lanyon, L. E. and Hall, G. F., 1983. Land surface morphology 2: Predicting potential landscape instability in eastern Ohio. *Soil Science*, **136**, 382–386.

Lee, D. (translator), 1977. Plato (427–347 BC), Critias 3, 111, in *Timaeus and Critias*. Penguin Classics, London, 133–134.

Leopold, A., 1934. Conservation economics. *Journal of Forestry*, **32**, 537–544.

Likens, G. E. and Bormann, F. H., 1974. Linkages between terrestrial and aquatic ecosystems. *BioScience*, **24**, 447–456.

Lotspeich, F. B., 1980. Watersheds as the basic ecosystem: this conceptual framework provides a basis for a natural classification system. *Water Resources Bulletin*, **16**, 581–586.

Loudermilk, W. C., 1950. *Conquest of the Land through Seven Thousand Years*. US Department of Agriculture, Soil Conservation Service, MP-32.

Ludwig, D., Hilborn, R. and Walters, C., 1993. Uncertainty, resource exploitation, and conservation: lessons from history. *Science*, **260**, 17, 36.

Mark, D. M., 1983. Automated detection of drainage networks from digital elevation models. *Auto-Carto*, **6**, 169–178.

Miller, M. C., McCave, I. N. and Komar, P. D., 1977. Threshold of sediment motion under unidirectional currents. *Sedimentology*, **24**, 507–527.

Montgomery, D. R., 1995. Input- and output-oriented approaches to implementing ecosystem management. *Environmental Management*, **19**, 183–188.

Montgomery, D. R. and Buffington, J. M., 1993. *Channel Classification. Prediction of Channel Response and Assessment of Channel Condition*. Washington State Department of Natural Resources Report TFW-SH10-93-002, 86pp.

Montgomery, D. R. and Dietrich, W. E., 1988. Where do channels begin? *Nature*, **336**, 232–234.

Montgomery, D. R. and Dietrich, W. E., 1994. A physically based model for the topographic control on shallow landsliding. *Water Resources Research*, **30**, 1153–1171.

Montgomery, D. R. and Foufoula-Georgiou, E., 1993. Channel network source representation using digital elevation models. *Water Resources Research*, **29**, 3925–3934.

Montgomery, D. R., Grant, G. E. and Sullivan, K., 1995a. Watershed analysis as a framework for implementing ecosystem management. *Water Resources Bulletin*, **31**, 369–386.

Montgomery, D. R., Buffington, J. M., Smith, R., Schmidt, K. and Pess, G., 1995b. Pool spacing in forest channels. *Water Resources Research*, **31**, 1097–1105.

Moore, I. D., Turner, A. K., Wilson, J. P. Jenson, S. K. and Band, L. E., 1993. GIS and land-surface-subsurface process modelling. In *Environmental Modelling with GIS*, eds M. F. Goodchild, B. O. Parks and L. T. Steyaert, Oxford University Press, New York, 196–230.

Morisawa, M. E., 1957. Accuracy of determination of stream lengths from topographic maps. *EOS, Transactions of the American Geophysical Union*, **38**, 86–88.

Morris, D. G. and Heerdegen, R. G., 1988. Automatically derived catchment boundaries and channel networks and their hydrological applications. *Geomorphology*, **1**, 131–148.

Mosley, M. P., 1987. The classification and characterisation of rivers. In *River Channels: Environment and Process*, ed. K. S. Richards, Blackwell, Oxford, 294–320.

Newson, M., 1992. *Land, Water and Development*. Routledge, London and New York, 351pp.

Niemann, K. O. and Howes, D. E., 1991. Applicability of digital terrain models for slope stability assessment. *ITC Journal*, **1991**, 127–137.

Norton, B. G., 1992. A new paradigm for environmental management. In *Ecosystem Health*, eds R. Costanza, B. G. Norton and B. D. Haskell, Island Press, Washington, DC, 23–41.

Okimura, T. and Nakagawa, M., 1988. A method for predicting surface mountain slope failure with a digital landform model. *Shin Sabo*, **41**, 48–56.

O'Loughlin, E. M., 1986. Prediction of surface saturation zones in natural catchments by topographic analysis. *Water Resources Research*, **22**, 794–804.

Omernik, J. M. and Griffith G. E., 1991. Ecological regions versus hydrologic units: frameworks for managing water quality. *Journal of Soil and Water Conservation*, **46**, 334–340.

Panuska, J. C., Moore, I. D. and Kramer, L. A., 1991. Terrain analysis: integration into the agriculture nonpoint source (AGNPS) pollution model. *Journal of Soil and Water Conservation*, **46**, 59–64.

Parker, G., 1978. Self-formed straight rivers with equilibrium banks and mobile bed. Part 2. The gravel river. *Journal of Fluid Mechanics*, **89**, 127–146.

Parker, R., 1975. *The Common Stream*. Collins, London, 283pp.

Perline, J., 1989. *A Forest Journey: The Role of Wood in the Development of Civilization*. Harvard University Press, Cambridge, 445pp.

Ponting, C., 1992. *A Green History of the World: The Environment and the Collapse of Great Civilizations*. St. Martins Press, New York, 432pp.

Powell, J. W., 1878. *Report on the Lands of the Arid Region of the United States*. 45th Congress 2nd Session H.R. Exec. Doc. 73.

Reid, L. M., 1993. *Research and Cumulative Watershed Effects.* US Department of Agriculture, Forest Service, General Technical Report PSW-GTR-141, 118pp.

Reneau, S. L. and Dietrich, W. E., 1987. *Size and Location of Colluvial Landslides in a Steep Forested Landscape.* International Association of Hydrological Sciences Publication 165, 39–49.

Robert, A. and Roy, A., 1990. On the fractal interpretation of the mainstream length-drainage area relationship. *Water Resources Research*, **26**, 839–842.

Said, R., 1993. *River Nile: Geology, Hydrology, and Utilization.* Pergamon, Oxford, 320pp.

Sidle, R. C., Pearce, A. J. and O'Loughlin, C. L., 1985. *Hillslope Stability and Land Use.* American Geophysical Union Water Resources Monograph 11, American Geophysical Union, Washington, DC.

Slocombe, D. S., 1993. Implementing ecosystem-based management. *BioScience*, **43**, 612–622.

Smit, B. and Spaling, H., 1995. Methods for cumulative effects assessment. *Environmental Impact Assessment Review*, **15**, 81–106.

Smith, C. T., 1969. The drainage basin as an historical basis for human activity. In *Water, Earth, and Man*, ed. R. J. Chorley, Methuen & Co., London, 101–110.

Stanley, T. R., Jr, 1995. Ecosystem management and the arrogance of humanism. *Conservation Biology*, **9**, 255–262.

Swanson, F. J., Lienkaemper, G. W. and Sedell, J. R., 1976. *History, physical effects, and management implications of large organic debris in western Oregon streams.* USDA Forest Service General Technical Report, PNW-GTR-56, 15pp.

Swanson, F. J., Kratz, T. K., Caine, N. and Woodmansee, R. G., 1988. Landform effects on ecosystem patterns and processes. *BioScience*, **38**, 92–98.

Tang, S. M. and Montgomery, D. R., 1995. Riparian buffers and potentially unstable ground. *Environmental Management*, **19**, 741–749.

Tarboton, D. G., Bras, R. L. and Rodriguez-Iturbe, I., 1991. On the extraction of channel networks from digital elevation data. *Hydrological Processes*, **5**, 81–100.

USDA (United States Department of Agriculture), 1994. *A Federal Agency Guide for Pilot Watershed Analysis*, 49pp.

van Asch, T., Cupers, B. and van der Zanden, D. J., 1993. An information system for large scale quantitative hazard analyses of landslides. *Zeitschrift für Geomorphologie, N. F., Supplementband*, **87**, 133–140.

WFPB (Washington Forest Practice Board), 1993. *Standard Methodology for Conducting Watershed Analysis.* Version 2.0, 85pp.

Wu, W., and Sidle, R. C., 1995. A distributed slope stability model for steep forested basins. *Water Resources Research*, **31**, 2097–2110.

Zhang. W. and Montgomery. D. R., 1994, Digital elevation model grid size, landscape representation and hydrologic simulations. *Water Resources Research*, **30**, 1019–1028.

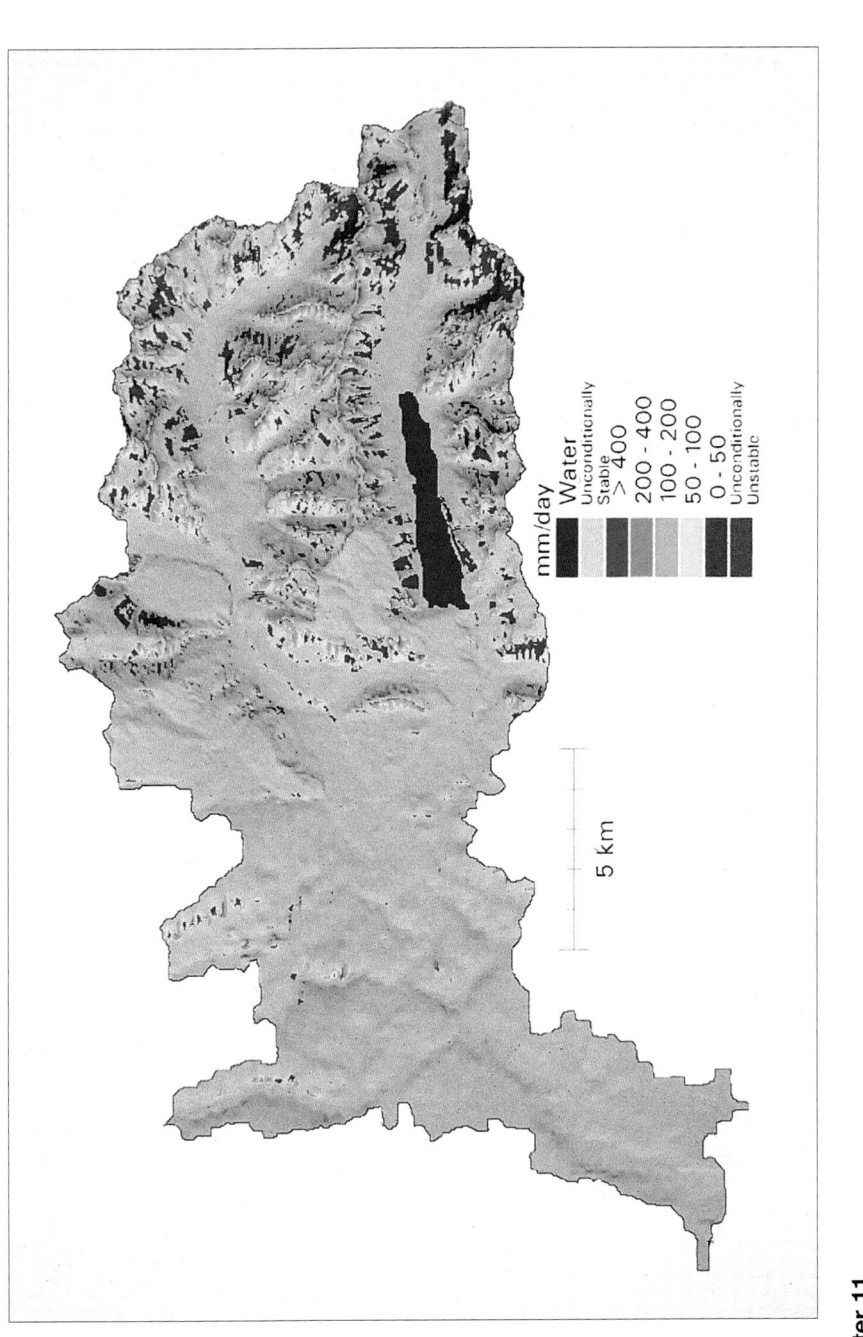

Chapter 11
Plate I Critical steady-state rainfall maps developed from 30m DEMs of the Tolt (a) and Skokomish (b) watersheds in Washington using $\rho_s/\rho_w = 2$, $T = 65m^2/day$, $f = 45°$, and $C = 0$

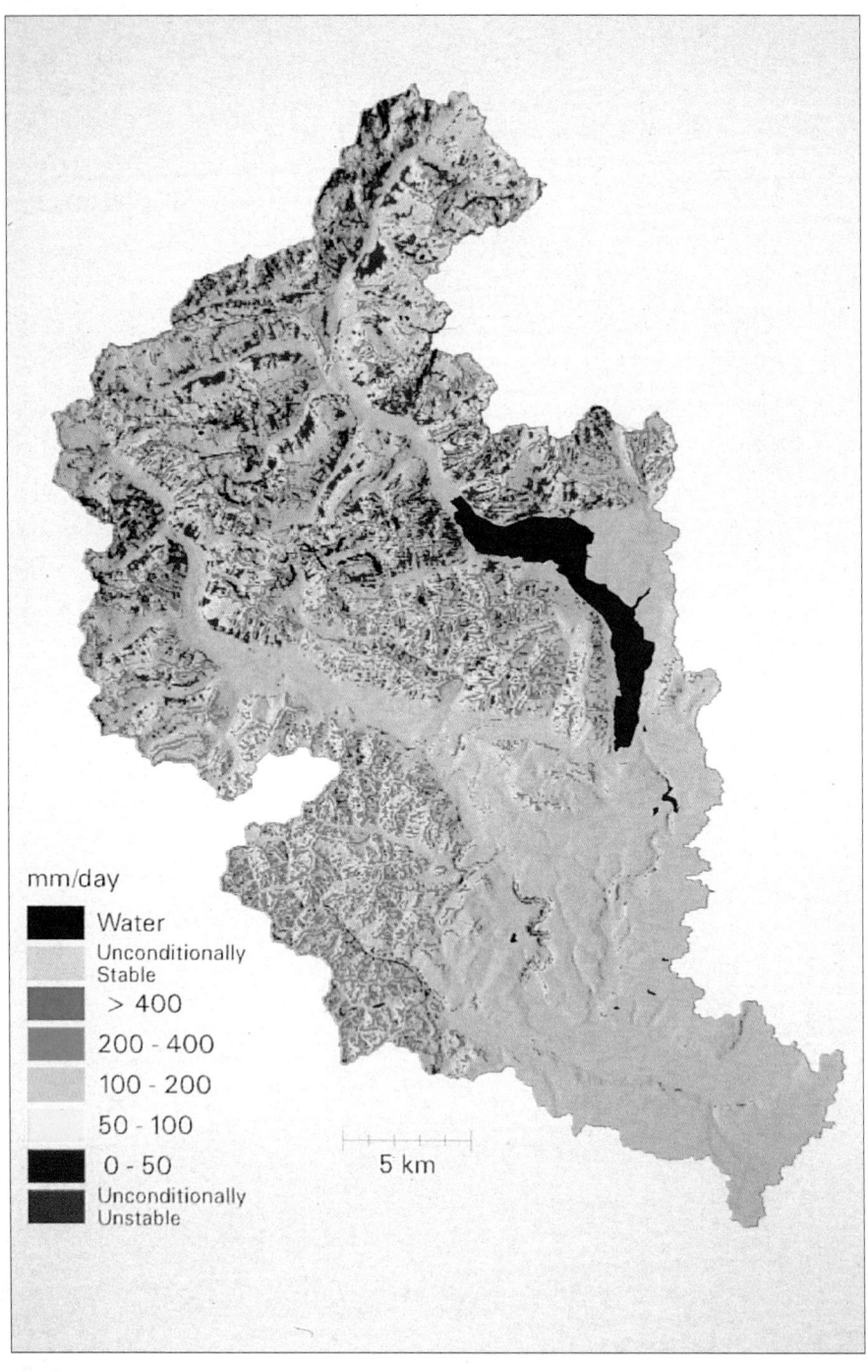

mm/day

- ■ Water
- Unconditionally Stable
- > 400
- 200 - 400
- 100 - 200
- 50 - 100
- 0 - 50
- Unconditionally Unstable

5 km

Chapter 11
Plate I(b)

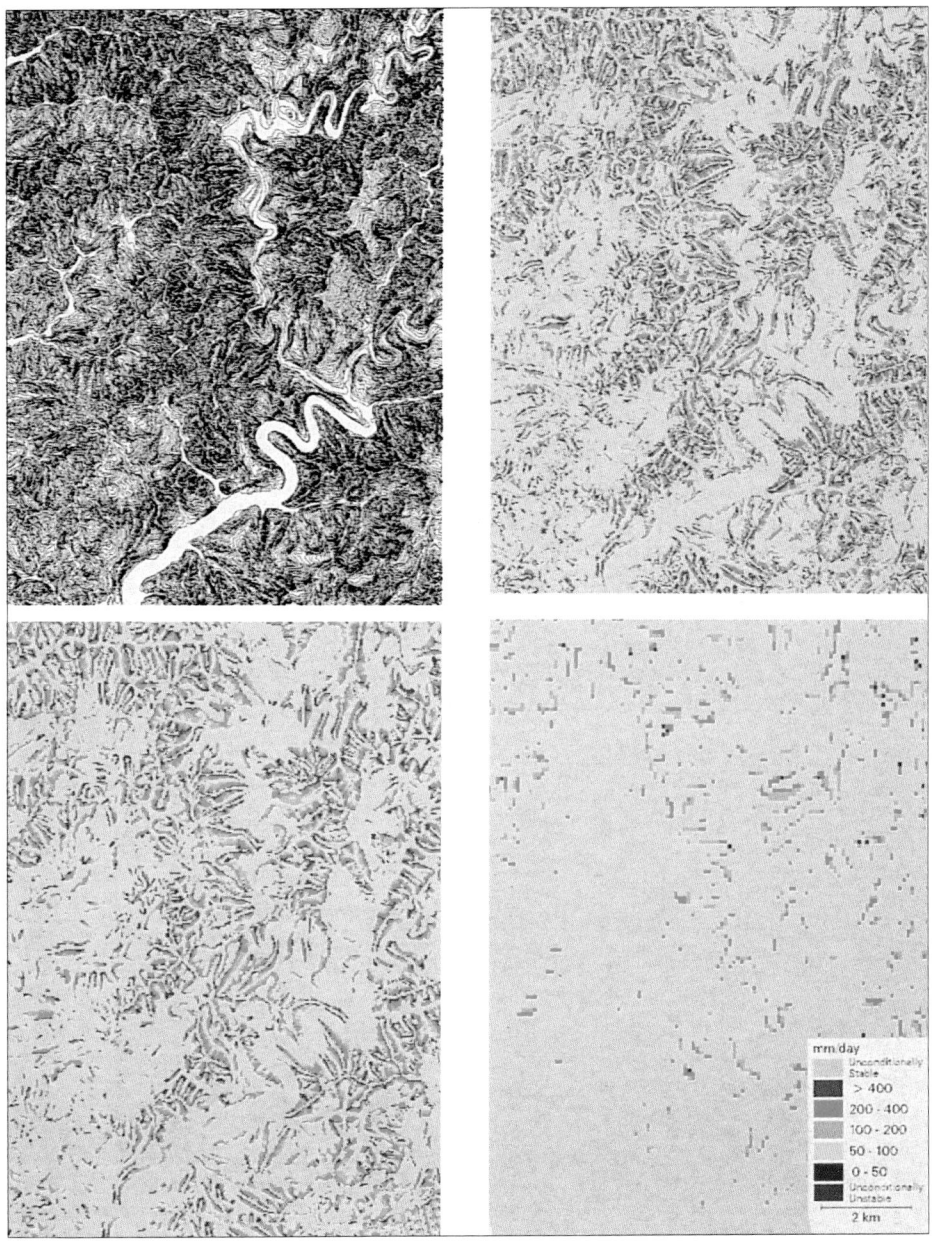

Chapter 11
Plate II Topographic and critical steady-state rainfall maps derived from an original 10m DEM of the Alleghany quadrangle, Oregon, USA, using $\rho_s/\rho_w = 2$, $T = 65m^2$/day, $f = 45°$, and $C = 0$, for a 10m contour interval (upper left) and grid sizes of 10m, (upper right) and 30m, (lower left) and 90m, (lower right)

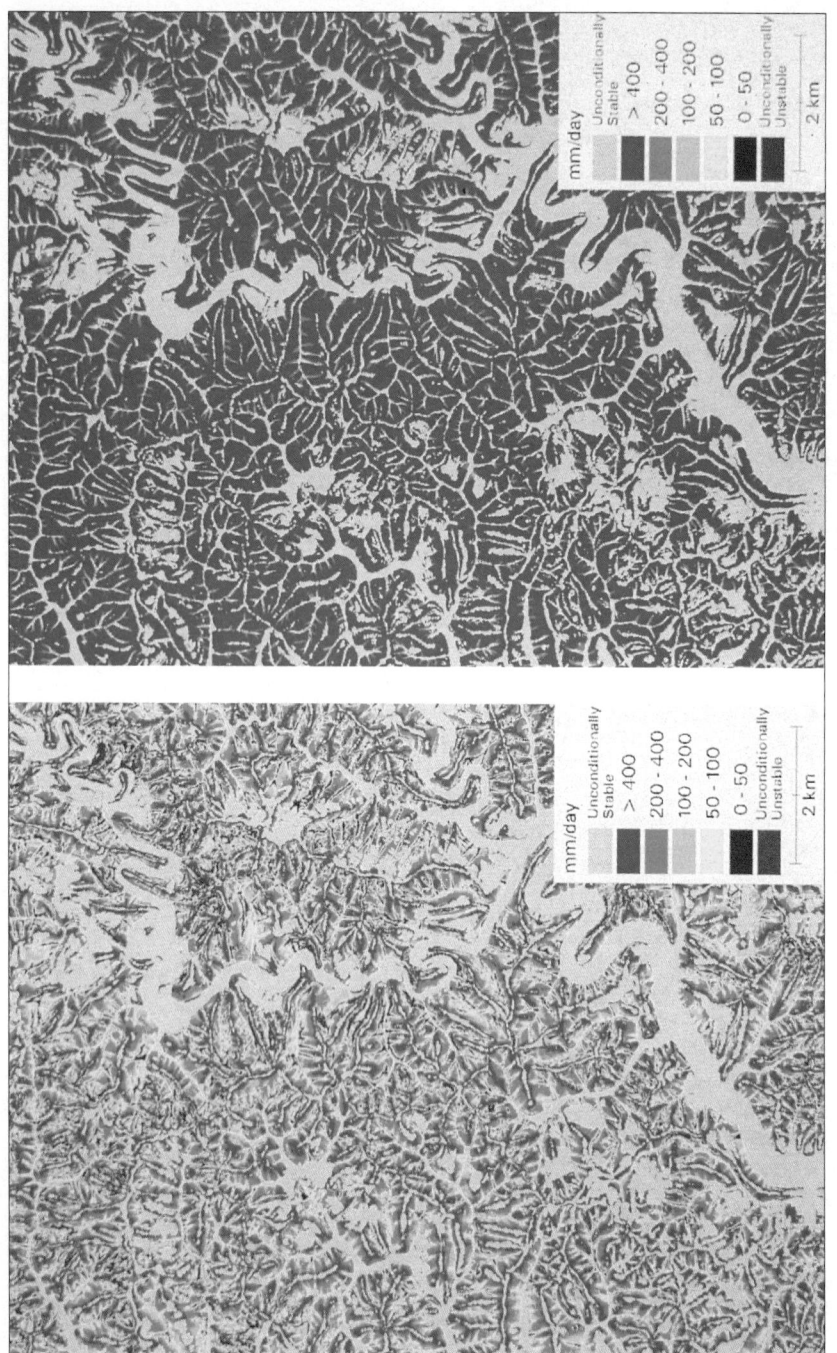

Chapter 11
Plate III Critical steady-state rainfall maps of the Alleghany quadrangle derived using $\rho_s/\rho_w = 2$, $T = 65m^2/day$, $f = 33°$, $z = 1m$, and $C = 2kN/m^2$ (left) and $15kN/m^2$ (right)

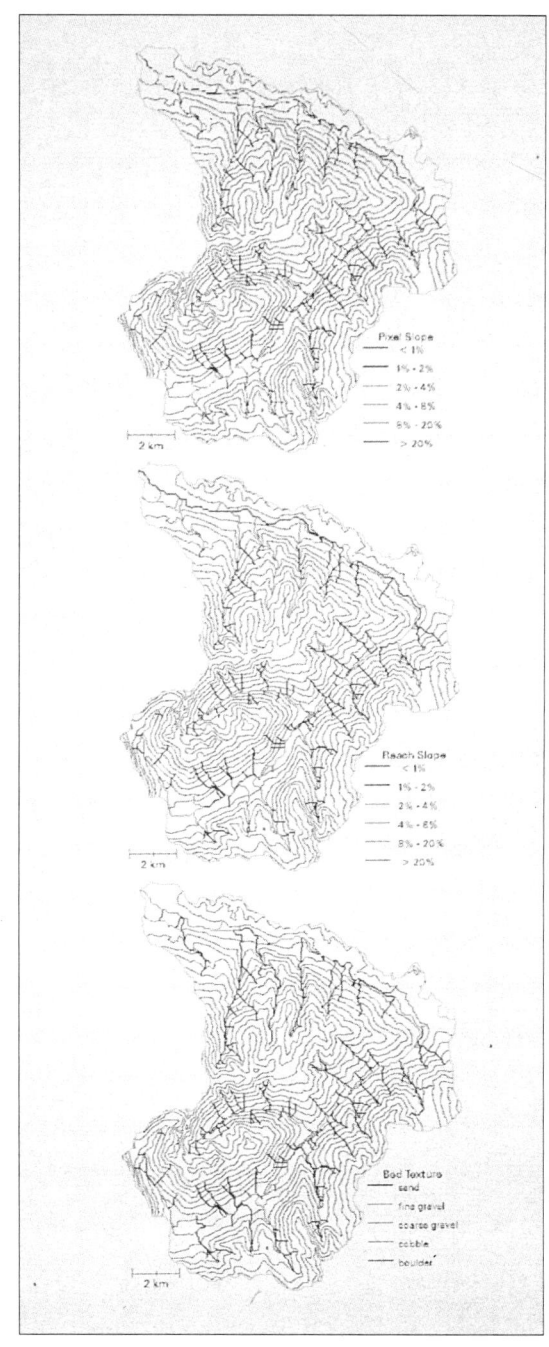

Chapter 11
Plate IV Channel network maps for Finney Creek: (a) pixel slope; (b) reach slope between tributary junctions; and (c) dominant bed-surface texture predicted from equation (8)

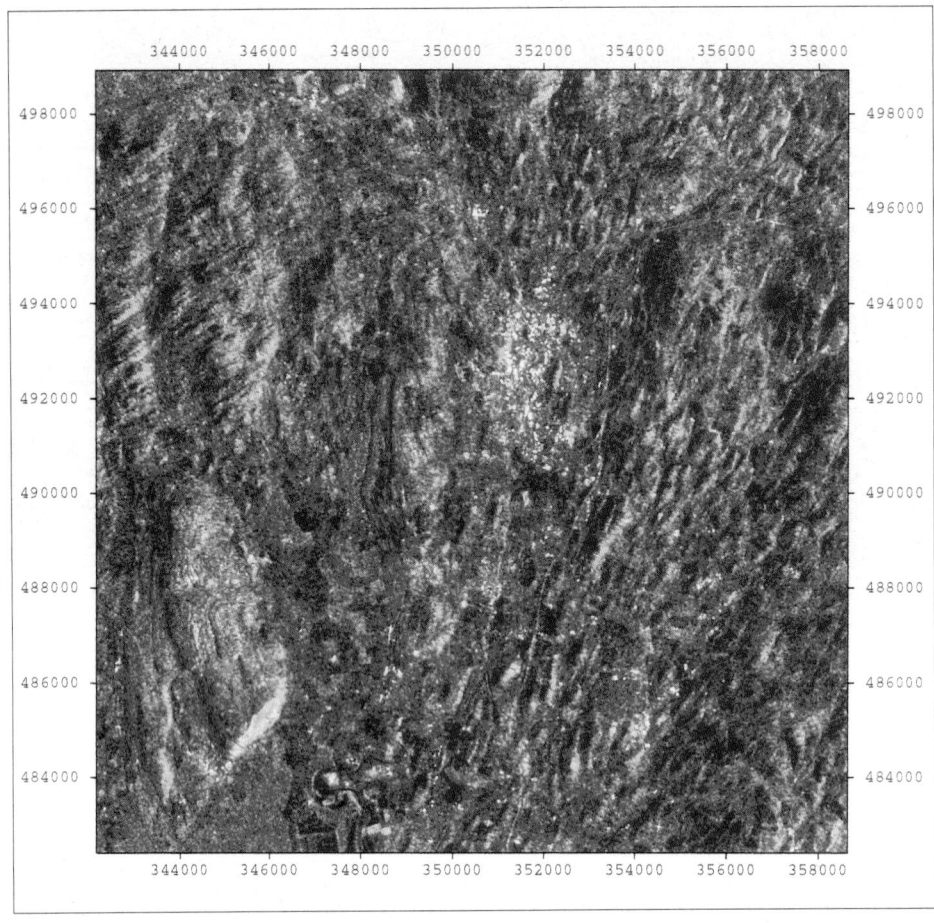

Chapter 8
Plate I Colour composite of two SAR images (same area as in Figure 8.1) with different look directions. The image with look direction towards 290° is yellow; the other has a look direction of 70° and is blue. Interpretation of the colours reveals differences in lineament information content between the two images

12 A Generalised Topographic–Soils Hydrological Index

ROBERT LAMB[1], KEITH BEVEN[1] and STEINAR MYRABØ[2]
[1] *Centre for Research on Environmental Systems, Lancaster University, UK*
[2] *Institute of Geophysics, University of Oslo, Norway*

ABSTRACT

A minimally parameterised distributed hydrological model may be realised using the $a/\tan\beta$ similarity index (a = accumulated upslope contributing area, $\tan\beta$ = slope angle) to determine local sub-surface storage deficits. These may be used to generate distributed water table predictions, but it is often impossible to test such predictions owing to a lack of detailed water table observations. The Seternbekken MINIFELT is a small catchment in Norway which has been densely instrumented, allowing water table patterns to be recorded at a relatively high spatial resolution. The $a/\tan\beta$ index, calculated from a digital elevation model (DEM) of the catchment has been used within the hydrological model TOPMODEL in a spatial simulation mode to predict local water table depths. Given model assumptions, errors in the predictions may be interpreted in terms of a local apparent transmissivity distribution. An approximate, non-linear relationship has been found between apparent transmissivity and $a/\tan\beta$, suggesting a simple functional modification of the $a/\tan\beta$ index which is shown to give improved water table predictions. The resulting distribution of apparent transmissivity is difficult to explain in physical terms. However, the modification of the topographic $a/\tan\beta$ index distribution may in fact represent errors in the estimation of drained area and local hydraulic gradients from the catchment digital elevation model. These may arise if the effective area of the catchment contributing to flow is not coincident with the topographically-defined area and if the water table is not parallel to the surface.

INTRODUCTION

A popular approach to distributed hydrological modelling is based upon the TOPMODEL theory of Beven and Kirkby (1979) which uses the topographic index $a/\tan\beta$ of Kirkby (1975), where a (dimension L) is the upslope area contributing flow per unit contour length and $\tan\beta$ is the local slope. The distributed nature of the $a/\tan\beta$ index means that it can be used in TOPMODEL to calculate the pattern of local saturated zone storage deficits or water table depths. The distribution of $a/\tan\beta$ can be calculated from a catchment digital elevation model (DEM) using

Landform Monitoring, Modelling and Analysis. Edited by S. N. Lane, K. S. Richards and J. H. Chandler.
© 1998 John Wiley & Sons Ltd.

automated digital terrain analysis (DTA) procedures such as those described by Quinn (1991) and Quinn *et al.* (1991, 1995).

Terrain analysis therefore plays an important role in the application of TOPMODEL. In particular, the calculation of the drained area, a, depends upon the sub-surface flow pathways inferred from surface slopes during the terrain analysis procedure. The nature of this inference depends upon the weighting given to different potential flow directions by the DTA algorithm and upon the grid scale of the DEM. Recent studies have investigated the theoretical effects of different flow apportioning assumptions and DEM resolutions on calculated distributions of $a/\tan\beta$ and calibrated TOPMODEL parameters (Bruneau *et al.*, 1995; Quinn *et al.*, 1995; Wolock and McCabe, 1995; Wolock and Price, 1994; Holgrem, 1994). However, there have been few opportunities to test spatially distributed predictions made using $a/\tan\beta$ within TOPMODEL theory owing to a lack of suitable data.

This paper presents the application of TOPMODEL theory in an instantaneous spatial prediction mode to an unusually detailed set of spatially distributed water table measurements. We will use these data to infer an apparent distribution of the TOPMODEL parameter T_0 which represents the local saturated transmissivity of the soil and may be brought within the topographic–soils index, $a/T_0 \tan\beta$, of Beven (1986).

DISTRIBUTED MEASUREMENTS FROM THE SETERNBEKKEN CATCHMENT

The Seternbekken MINIFELT is a small (*c.* 7500 m^2) experimental catchment located approximately 10 km west of Oslo, Norway. Time series measurements of the main hydrological variables have been made in the MINIFELT since 1986 as part of an automated telemetry and logging project described by Myrabø (1988, 1994), where the main objective was to gain a deeper insight into runoff processes in natural till and forest catchments. These time series data include water levels recorded in four boreholes at locations shown in Figure 12.1.

In addition, there are 105 piezometers distributed across the catchment (Figure 12.1), allowing an exceptionally detailed picture of the distributed water table to be obtained. These instruments are not continuously logged, but simultaneous recordings have been taken for a range of known discharges at the outlet. We will concentrate on five spatial data sets recorded when the outflow discharge Q (expressed as an areally averaged depth of flow) was equal to 0.1, 0.5, 0.61, 4.9 and 6.8 mm/h, referred to hereafter as cases $Q_{0.1}$, $Q_{0.5}$, $Q_{0.6}$, $Q_{4.9}$ and $Q_{6.8}$ respectively.

In an earlier study, Erichsen and Myrabø (1990) investigated the relationship between water table data measured in the MINIFELT and catchment topography. In their study, topography was interpolated from point survey data onto a regular 2×2 m grid, and slope, aspect, plan curvature (at 2 m and 10 m resolution), profile curvature (at 2 m and 10 m resolution) and drainage area were calculated. Erichsen and Myrabø (1990) then applied cluster analysis to these variables to spatially delimit distinct topographic categories, which were compared with a water

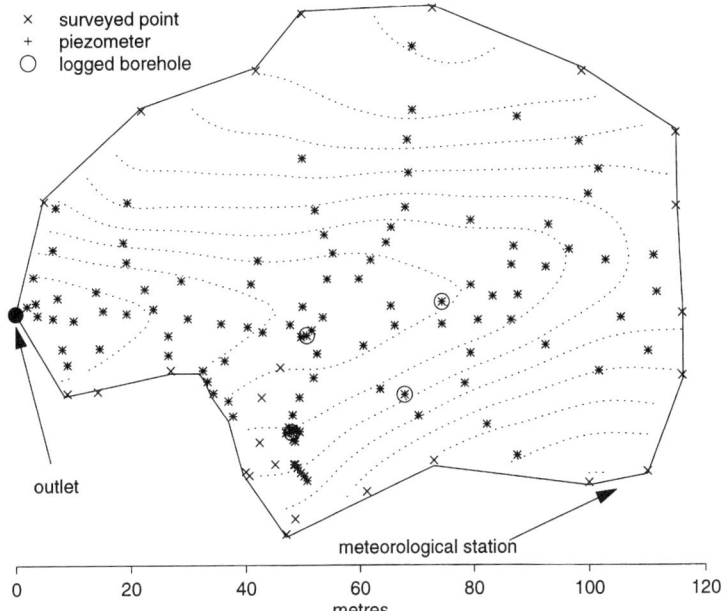

Figure 12.1 Contour map of the Seternbekken MINIFELT catchment showing the location of instruments. Contours are at 1 m intervals. Note that in the topographic survey, elevation was measured at each piezometer location

table map (constructed by interpolating point piezometer measurements onto a regular grid). The mapped topographic clusters were interpreted as wetness indicators and found to lie in a pattern that reproduced some, although not all, of the spatial variation in observed relative depths to the water table. The major differences were concluded to be due to the fact that soil characteristics were not taken into consideration in the cluster analysis.

TOPMODEL IN A SPATIAL PREDICTION MODE

A limitation of the cluster analysis approach is the lack of a theory to allow quantitative predictions of water table position to be made. TOPMODEL theory allows such predictions to be obtained and also allows the observed data to be used in a calibration sense to help determine the distribution of the local transmissivity T_0. An assumption of TOPMODEL is that the dynamics of the catchment saturated zone can be approximated by a succession of steady states. This assumption is convenient because it allows an instantaneous spatial prediction of water table depths to be treated independently of the full hydrological model, given a knowledge of the averaged state of the saturated zone.

Distributed Water Table Predictions from TOPMODEL

TOPMODEL will not be described in detail here. A full review is given by Beven *et al.* (1995). The steady-state assumption made in TOPMODEL allows the topographic–soils index $a/T_0 \tan\beta$ to be combined with an exponential transmissivity profile to give an expression for the local saturated zone storage deficit at a point i:

$$s_i = \bar{S} + m \left(\lambda - \ln \frac{a}{T_0 \tan\beta} - \ln T_e \right) \tag{12.1}$$

where $\bar{S}$ (dimension L) is the areally averaged storage deficit in the saturated zone, m (L) controls the steepness of the exponential profile, T_e is the areal average of T_0 (L^2/T) and λ is the areal average of $\ln(a/\tan\beta)$.

The average storage deficit $\bar{S}$ can be shown (Beven *et al.*, 1995) to be a function of the saturated zone discharge Q_b:

$$\bar{S} = -m \ln \frac{Q_b}{Q_0} \tag{12.2}$$

where:

$$Q_0 = A \, e^{(\ln T_e - \lambda)} \tag{12.3}$$

and A (L^2) is the catchment area. In most applications of TOPMODEL in the past it has been necessary to make the assumption that the transmissivity T_0 is spatially uniform, hence $T_e = T_0$.

The simplest way to calculate water table depth from the local storage deficit, s_i, is to assume that the local saturated zone deficit represents gravity drainage through the macroporosity between saturation and some "field capacity" and that this effective porosity, $\delta\theta$, is constant with depth. Under these assumptions, the local water table depth z_i is given by:

$$z_i = \frac{s_i}{\delta\theta} \qquad (s_i > 0) \tag{12.4}$$

Using equations (12.1) and (12.4), local water table depths may be obtained given a knowledge of $\bar{S}$, the parameters m, T_0 and $\delta\theta$, and the local value of the topographic index $\ln(a/\tan\beta)_i$. $\bar{S}$ may be calculated using equation (12.2) from the flow at the catchment outlet as long as this can be held to represent the saturated zone discharge, Q_b, alone.

Calculating ln(a/tan β)

Raw topographic data for the MINIFELT consisted of 133 spot heights surveyed at the locations shown in Figure 12.1. These spot heights were interpolated onto a 2 m^2 grid using the GINTP1 bilinear interpolation routine from the AGL UNIRAS library. Given the small size of the MINIFELT, the raster DEM could be qualitatively

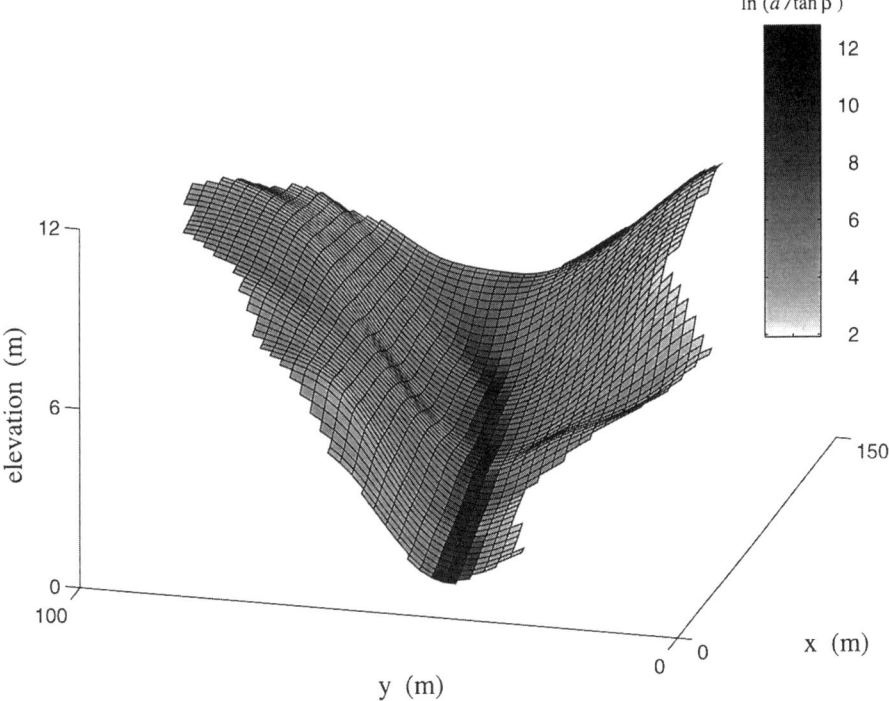

Figure 12.2 Catchment DEM showing superimposed distribution of ln(a/tanβ). Vertical exaggeration approximately ×5. Elevation relative to the outflow

validated and appeared to be an acceptable representation of the catchment topography. It may be noted that the spatial distribution of spot heights is not regular, but rather concentrates on the main topographic features in the catchment.

A distribution of $a/\tan\beta$ was calculated for the MINIFELT catchment using the multiple flow direction algorithm of Quinn *et al.* (1991). The resulting raster image of $a/\tan\beta$ is shown superimposed upon a surface representation of the DEM in Figure 12.2 where it can be seen that $\ln(a/\tan\beta)$ increases towards the valley bottom. Local values of $\ln(a/\tan\beta)$ were then obtained for the piezometer and borehole sites by linear interpolation within the $\ln(a/\tan\beta)$ raster map.

Parameter Identification

In order to make spatial predictions, the parameters m, T_e and $\delta\theta$ have to be identified. Monte Carlo simulations of discharge and borehole time series were used to calibrate these parameters by randomly sampling parameter values over an arbitrarily wide range and identifying the values corresponding to the best simulations. The full results of these calibrations are not presented here because they were carried out only to obtain useful estimates for the parameters required to make spatial predictions.

The global parameters m and T_e can be estimated by calibration on flow data, since discharges at the outlet are most representative of the areally integrated behaviour of the catchment. In this case, m effectively controls the shape of the simulated recession curve; T_e, together with m, determines the surface runoff generated by TOPMODEL from saturated areas (for which $s_i < 0$). These may be determined by applying equation (12.1) at each time step to an areal $\ln(a/\tan\beta)$ distribution function.

Ten thousand Monte Carlo simulations of a 900 h period in autumn 1987 were carried out. Discharge simulation efficiencies were appraised using the coefficient of determination or efficiency measure, E, of Nash and Sutcliffe (1970). This returns a value of unity for an error-free simulation of the observed series. The best simulation achieved an efficiency of 0.89 and suggested parameter values of $T_e = 1.3$ m^2/h and $m = 5.3$ mm.

The effective macroporosity, $\delta\theta$, was estimated from Monte Carlo simulations of the borehole time series, obtained by combining equations (12.1) and (12.4) for the $\ln(a/\tan\beta)$ values at each borehole site. For every simulation, four values of E were therefore obtained corresponding to the four borehole series. Simulations for which $E < 0$ were rejected as "non-behavioural". Efficiencies from the remaining simulations were combined by summation. No clear optimum simulation could be identified from the combined efficiencies. However, good simulations could be obtained everywhere in the range $0.02 < \delta\theta < 0.1$. The cumulative distribution of combined efficiencies was calculated over this range and the median value of this distribution used to identify an estimate of 0.06 for $\delta\theta$. This value appeared to be robust in that it gave rise to reasonably good simulations of each individual borehole series.

Spatial Simulations

Figure 12.3 shows observed water table depths as a function of the local $\ln(a/\tan\beta)$ at each piezometer. Simulated water table depths, obtained using equations (12.1) and (12.4) with the parameters identified above, are also shown. The catchment average storage deficit ($\bar{S}$) was calculated using equation (12.2) where it was assumed that the discharges $Q_{0.1}, \ldots, Q_{6.8}$ represent saturated zone discharge.

It can be seen that there is a high degree of scatter in the observed data, although the expected decrease in water table depths with increasing catchment discharge is observed. As this occurs, the number of observed water table depths reduces due to some piezometer sites becoming saturated at the surface. For the global transmissivity case, $T_0 = T_e$ everywhere and equation (12.1) reduces to:

$$s_i = \bar{S} + m \left(\lambda - \ln\frac{a}{\tan\beta} \right) \qquad (12.5)$$

It follows from equations (12.4) and (12.5) that predicted water table depth is linearly related to $\ln(a/\tan\beta)$ for the global transmissivity case with constant $\delta\theta$. The simulations shown in Figure 12.3 hence plot as straight lines. However, where the water table was predicted to be above the surface ($s_i < 0$), the porosity $\delta\theta$ was

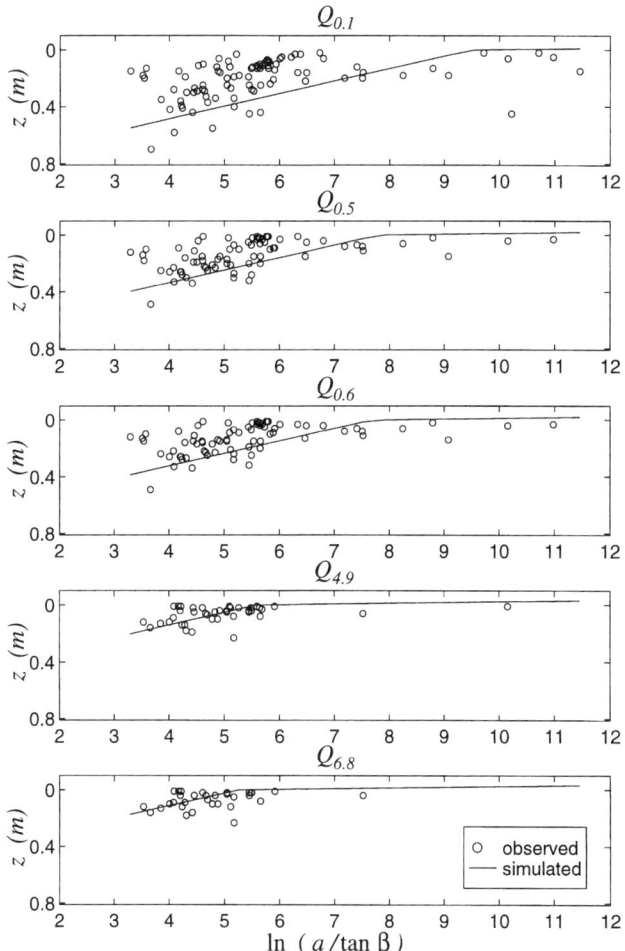

Figure 12.3 Spatial simulation results using parameters derived from time series calibrations. z is the depth to the water table in metres

set to unity giving rise to the change in the slope of the simulated water tables for $z < 0$.

Comparing observed and simulated data, it is clear that the model theory used here fails to capture the full heterogeneity of the hydrological processes in the field. However, the predictions are scaled to about the right magnitude by the calibrated values of T_e, m and $\delta\theta$. The summed absolute simulation errors are given in Table 12.1.

These results have been obtained under the assumption of a uniform soil where the saturated lateral transmissivity T_0 is everywhere equal to the global effective value T_e. The distribution of the predicted water table depths is therefore controlled solely by the distribution of the topographic index $a/\tan\beta$.

Table 12.1 Summed absolute errors (metres) for spatial simulations assuming a global effective saturated transmissivity (T_e)

Discharge	$Q_{0.1}$	$Q_{0.5}$	$Q_{0.6}$	$Q_{4.9}$	$Q_{6.8}$
Summed absolute error	15.8	8.3	8.1	1.8	1.5

A more realistic hydrological index would take into account soil properties as well as topography. The topographic–soils index $a/T_0 \tan\beta$ (Beven, 1986) allows local transmissivities to be used in a distributed sense within TOPMODEL via equation (12.1). The data required to specify the distribution of T_0 are rarely available. However, water table data can be used to infer the distribution of T_0 by model inversion.

CALCULATING LOCAL APPARENT TRANSMISSIVITIES

Rearranging equation (12.1) we may obtain an expression for the natural logarithm of the local saturated transmissivity:

$$\ln T_0 = \frac{s_i - \bar{S}}{m} - \lambda + T_e + \ln \frac{a}{\tan\beta} \tag{12.6}$$

Note that this relationship implies that the *effective* global transmissivity T_e has been identified, in this case by calibration on flow data, independently of the local values (T_0).

Equation (12.6) was applied to the five Seternbekken spatial data sets. In each case, s_i was determined by rearranging equation (12.4) to give:

$$s_i = z_i \cdot \delta\theta \tag{12.7}$$

where z_i was taken to be the observed water table depth at each measurement site.

The resulting apparent log transmissivities are plotted as a function of the local $\ln(a/\tan\beta)$ at each site in Figure 12.4. Although these show considerable scatter for small $\ln(a/\tan\beta)$, there are clear linear trends between $\ln T_0$ and $\ln(a/\tan\beta)$, especially for larger $\ln(a/\tan\beta)$ values and wetter conditions. Linear relationships have been fitted through the data by least squares and are shown in Figure 12.4. The spatial distribution of differences between local apparent log transmissivities ($\ln T_0$) and the effective global value ($\ln T_e$) is shown for the median discharge $Q_{0.6}$ in Figure 12.5, where the map of $\ln T_0$ was obtained by the same interpolation routine used to generate the catchment DEM.

A POWER LAW TOPOGRAPHIC–SOILS INDEX?

The log–log linear relationships found between apparent transmissivity and $a/\tan\beta$ may be written:

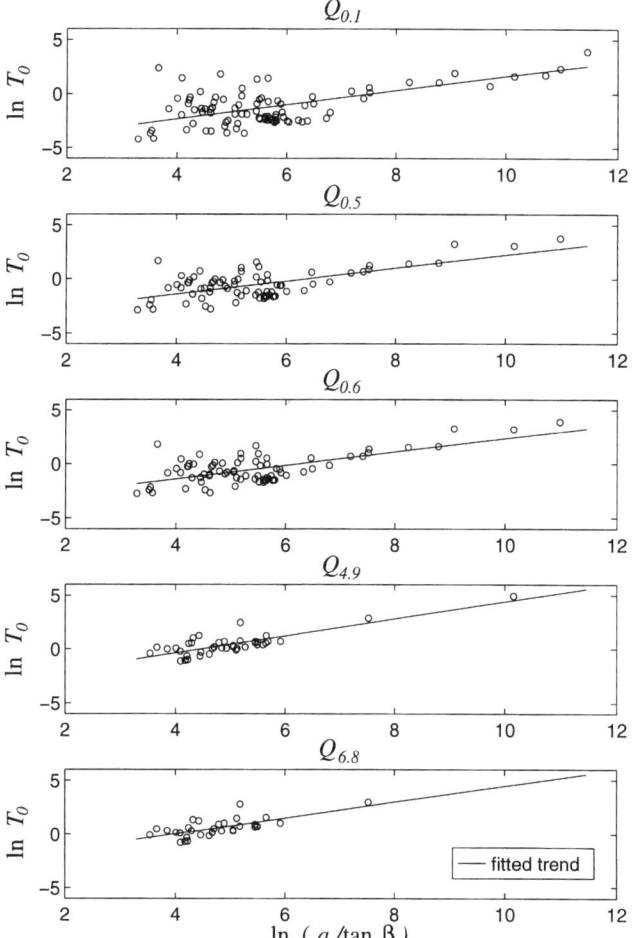

Figure 12.4 Apparent local saturated transmissivities (in log) calculated using observed water table data. Trends were fitted by least squares linear regression

$$T_0 = c \left(\frac{a}{\tan \beta} \right)^p \tag{12.8}$$

Substituting this expression for T_0 into the combined topographic–soils index $a/T_0 \tan \beta$ suggests that $a/T_0 \tan \beta$ can be related to the topographic $a/\tan \beta$ index by a power law:

$$\frac{a}{T_0 \tan \beta} = \frac{a}{c \left(\dfrac{a}{\tan \beta} \right)^p \tan \beta} = C \left(\frac{a}{\tan \beta} \right)^P \tag{12.9}$$

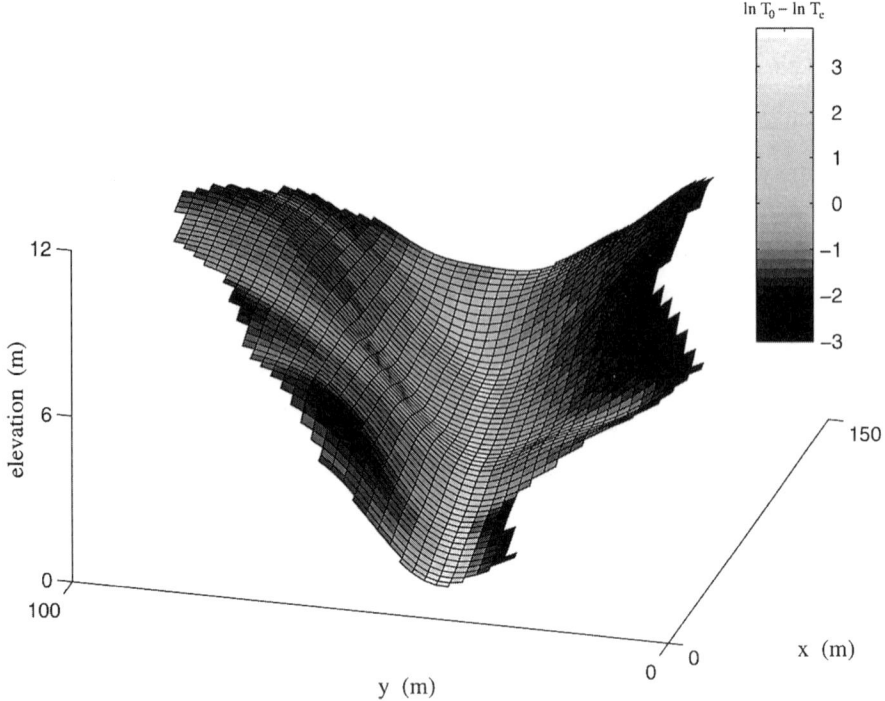

Figure 12.5 Differences between the local apparent log transmissivity to T_0 for $Q_{0.6}$ and the effective global log transmissivity in T_e

where

$$C = 1/c \tag{12.10}$$

and

$$P = (1 - p) \tag{12.11}$$

The parameters c and p were estimated by regression from the relationships shown in Figure 12.4. Table 12.2 gives the values of C and P thus obtained.

Water table depth predictions were obtained using the distributed power law topographic–soils index in equation (12.1). These are shown in Figure 12.6. Also shown for comparison are the simulations obtained earlier using a global transmissivity. Predictions made using the power law $a/T_0 \tan \beta$ pass more centrally through the scattered observed water table depths in each set of observations. This reflects the fact that the parameters C and P have been effectively calibrated for each spatial data set by the procedures described above. Summed absolute errors for the power law index, shown in Table 12.3, are smaller than the errors produced using a global transmissivity (Table 12.1).

We may interpret the P parameter as a control on the form of the distribution of the power law $a/T_0 \tan \beta$ index within the catchment. This distribution is then scaled

Table 12.2 Parameters of the empirical power law index

Discharge	$Q_{0.1}$	$Q_{0.5}$	$Q_{0.6}$	$Q_{4.9}$	$Q_{6.8}$
C	141.4	46.8	49.2	34.8	18.9
P	0.34	0.39	0.38	0.20	0.26

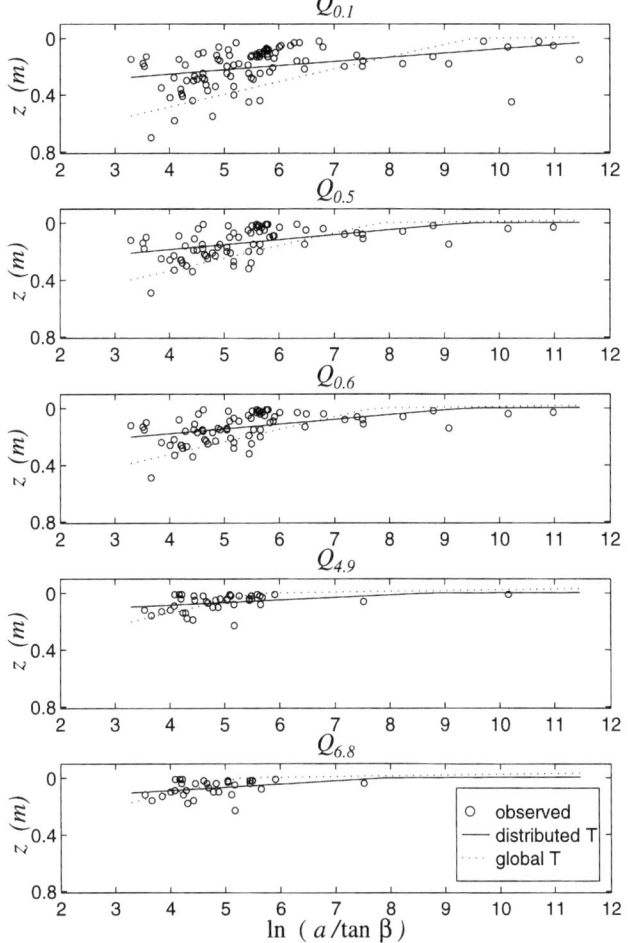

Figure 12.6 Spatial simulations using the power law $a/T_0 \tan\beta$ index. Simulations obtained assuming a uniform effective transmissivity (Figure 12.3) are shown again for comparison

Table 12.3 Summed absolute errors (metres) for spatial simulations using the power law $a/T_0 \tan \beta$ index

Discharge	$Q_{0.1}$	$Q_{0.5}$	$Q_{0.6}$	$Q_{4.9}$	$Q_{6.8}$
Summed absolute error	9.2	5.8	5.5	1.6	1.3

by C. Table 12.2 shows that the inferred value of C decreases when wetter catchment conditions are considered. The reason for this may be connected to the assumption that the recorded flow corresponding to each spatial data set represents solely the sub-surface, saturated zone discharge, Q_b. In fact, the recorded flow may be an underestimate of Q_b for the driest conditions, due to the effects of evaporation behind the weir or underflow. During wetter conditions the recorded flow may be an overestimate of the sub-surface drainage component due to the presence of surface storm flows in the total outflow at the weir. The variation in calculated values of C may therefore by compensating, at least in part, for errors in the estimation of Q_b (and hence $\bar{S}$ in equation (12.2)).

POSSIBLE IMPLICATIONS FOR THE ESTIMATION OF $a/\tan\beta$ BY TERRAIN ANALYSIS

Although the power law $a/T_0 \tan \beta$ index cannot reproduce all of the local variability in water table depths observed in the field, it does improve upon the predictions obtained under an assumption of uniform transmissivity. This improvement is brought about by modification of both the scaling and the form of the $a/T_0 \tan \beta$ distribution within the catchment. With a uniform transmissivity, only the scaling can be modified by the adjustment of T_e.

Distributions of the topographic–soils index can be compared by plotting the cumulative fractional area of the catchment associated with increasing values of the index. Figure 12.7 shows the distribution functions for $a/T_0 \tan \beta$ under the assumption of a global transmissivity ($T_e = 1.3$) and for the power law index calibrated using water table data $Q_{0.1}$, $Q_{0.6}$ and $Q_{6.8}$. The power law distributions are scaled by the C parameter and therefore have different median values. However, the consistency in the estimated values of P (Table 12.2) means that the distributions are each of similar form, having a narrower range than the topographic $a/\tan\beta$ distribution (and hence a narrower range than the global transmissivity case).

The trend in apparent transmissivities has the effect of modifying the form of the $a/T_0 \tan \beta$ index such that higher index values are adjusted downwards whilst lower values are adjusted upwards. There are a number of possible explanations for this behaviour.

The reduction of higher $a/\tan\beta$ values occurs due to the increase in apparent T_0 with $a/\tan\beta$ found as a result of the model inversion. This increase in T_0 might be explained physically by an increase in soil depth or conductivity downslope, where higher values of $a/\tan\beta$ occur. Soil depth has been measured at piezometer locations

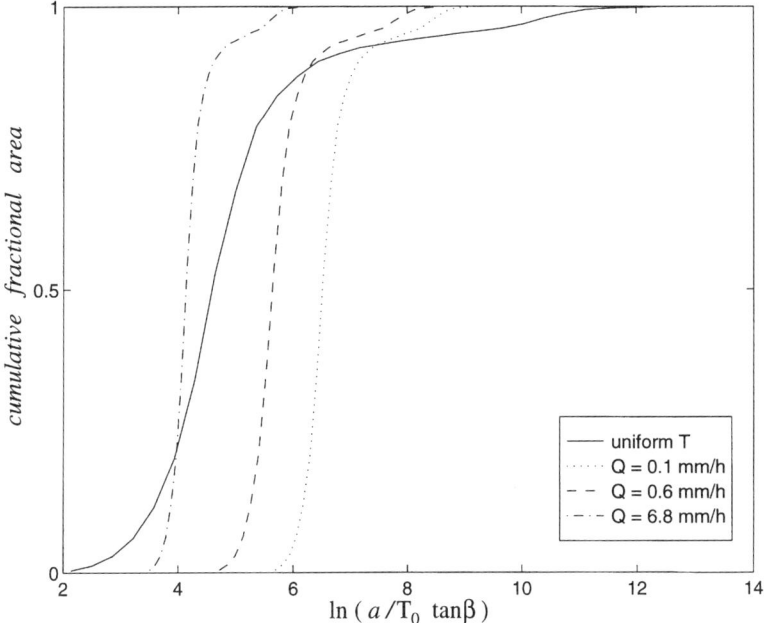

Figure 12.7 Distribution functions for $\ln(a/T_0 \tan\beta)$ calculated for a uniform global transmissivity and for $Q_{0.1}$, $Q_{0.6}$ and $Q_{6.8}$ using the power law form

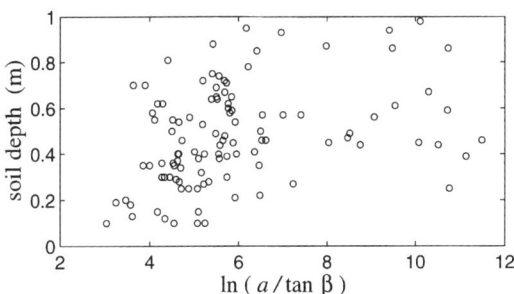

Figure 12.8 Soil profile depth as a function of $\ln(a/\tan\beta)$ at each piezometer site

throughout the catchment, allowing its relationship to $a/\tan\beta$ to be investigated. Figure 12.8 shows the soil depth at each piezometer site compared to the local value of $\ln(a/\tan\beta)$. There is no clear relationship between the two variables that could account for the trend in apparent T_0, although the variation in soil depth may nevertheless be a factor in the scatter of the apparent T_0 values. The hypothesis that conductivity may increase downslope cannot be tested, given currently available data.

An alternative interpretation of the power law $a/T_0 \tan \beta$ is that the modification of the topographic index distribution reflects errors in the local values of a and $\tan \beta$ estimated from the catchment DEM. The reduction of high index values is consistent with an overestimation of the effective drained area of the catchment for downslope areas where a, and hence $a/\tan \beta$, tends to be large. The reduction of high index values could also be consistent with a water table drawdown close to stream channels such that the effective $\tan \beta$, representing the hydraulic gradient, would be larger than the surface slope. However, the absence of permanent channels in most of the catchment makes this latter explanation unlikely.

For lower $a/\tan \beta$ values, the power law modification of the index is consistent with an underestimation of the local drained area or an overestimation of the effective $\tan \beta$ by the surface slope. Both effects are plausible. Low values of $a/\tan \beta$ tend to occur close to the catchment boundary where supply limitations due to the lack of upslope flow contributions may cause the water table to be sub-parallel to the surface slope. A sub-parallel water table can be incorporated into TOPMODEL theory using the reference level concept of Quinn et al. (1991). However, the underestimation of a from the catchment DEM is quite likely for the MINIFELT catchment since almost half of its perimeter consists of flat interfluves which makes the definition of the catchment boundary difficult. It is therefore possible that some areas not included in the DEM may in fact be draining into the catchment.

CONCLUSIONS

The Seternbekken MINIFELT water table data have allowed us to test the effectiveness of the $a/\tan \beta$ index, calculated by DTA procedures, in simulating the water table in a distributed sense. Under the assumption of a uniform global transmissivity, the $a/\tan \beta$ index fails to reproduce all the heterogeneity observed in the field. By inversion of TOPMODEL theory, the distribution of apparent local transmissivities has been calculated from the measured water table depths. There is considerable variability in the estimated local transmissivities which may reflect the heterogeneity of soil properties in the field. Despite this variability, trends appear to exist such that apparent transmissivity increases with increasing $a/\tan \beta$.

It is difficult to see why this should be the case. A relationship between transmissivity and $a/\tan \beta$ might occur if soil depth, which is a primary control on transmissivity, were to increase with $a/\tan \beta$. However, soil depth data for the catchment indicate no strong relationship. The non-linearity in the relationships between transmissivity and $a/\tan \beta$ suggests a modification of the form of the $a/\tan \beta$ distribution. The form of the power law $a/T_0 \tan \beta$ distributions obtained under different conditions is a consistent modification of the form of the $a/\tan \beta$ distribution.

If this modification of $a/\tan \beta$ implied by the water table data does not reflect a trend in transmissivities, it may instead be interpreted as a correction for errors in the estimation of $a/\tan \beta$ from a DEM. In particular, the drained area term a may be underestimated towards the catchment boundary but over-estimated further downslope, and the effective $\tan \beta$ may be overestimated by the topographic slope towards the catchment boundary.

Such errors are not due to the terrain analysis procedure itself, but rather the hydrological assumptions that go into it, especially that the area contributing to flow at any point within the catchment is equal to the upslope surface area and that the DEM represents the effective area of the catchment. In some cases, the steady-state assumption that a is constant in time may also be unrealistic, requiring dynamic modification of the topographic index (e.g. Barling *et al.*, 1994). The results obtained in this study suggest that an empirical modification of the $aT_0 \tan\beta$ index, in this case a power law, may be appropriate in the steady-state case to correct for errors introduced when the assumptions underlying the $a/\tan\beta$ index do not hold.

ACKNOWLEDGEMENTS

This work has been supported by UK NERC research studentship GT4/92/174/P awarded to Robert Lamb. The original field study was supported by The Norwegian National Committee for Hydrology (NHK) and the Norwegian Research Council.

REFERENCES

Barling, R. D., Moore, I. D. and Grayson, R. B., 1994. A quasi-dynamic wetness index for characterizing the spatial distribution of zones of surface saturation and soil water content. *Water Resources Research*, **30**, 1029–1044.

Beven, K. J., 1986. Hillslope runoff processes and flood frequency characteristics. In *Hillslope Processes*, ed. A. D. Abrahams, Allen and Unwin, Boston, 187–202.

Beven, K. J. and Kirkby, M. J., 1979. A physically based, variable contributing area model of basin hydrology. *Hydrological Science Bulletin*, **24**, 43–69.

Beven, K. J., Lamb, R., Quinn, P., Romanowicz, R and Freer, J., 1995. TOPMODEL. In *Computer Models of Watershed Hydrology*, ed. V. P. Singh, Water Resources Publications, Highlands Ranch, Colorado, Ch. 18.

Bruneau, P., Gascuel-Odoux, C., Robin, P., Merot, P. and Beven, K. J., 1995. Sensitivity to space and time resolution of a hydrological model using digital elevation data. *Hydrological* Processes, **9**, 69–81.

Erichsen, B. and Myrabø, S., 1990. Studies of the relationship between soil moisture and topography in a small catchment. In *Proceedings of 8th International Conference on Computational Methods in Water Resources*, eds G. Gambolati *et al.*, Springer-Verlag, Computational Mechanics Publications, Southampton, 551–60.

Holgrem, P., 1994. Multiple flow direction algorithms for runoff modelling in grid-based elevation models: an empirical evaluation. *Hydrological Processes*, **8**, 327–334.

Kirkby, M., 1975. Hydrograph modelling strategies. In *Processes in Physical and Human Geography*, eds R. Peel, M. Chisholm, and P. Haggett, Heinemann, London, 69–90.

Myrabø, S., 1988. Automation in hillslope hydrology. *NHP-rapport*, **22**, 36–45.

Myrabø, S., 1994. *Sæternbekken Forsøksfelt*. Technical Report 2–94, Norwegian Water Resources and Energy Administration, NVE, Middelthuns gate 29, PO Box 5091, Maj. N-0301, Oslo 3, Norway.

Nash, J. E. and Sutcliffe, J. V., 1970. River flow forecasting through conceptual models 1. A discussion of principles. *Journal of Hydrology*, **10**, 282–290.

Quinn, P. F., 1991. *The Role of Digital terrain analysis in hydrological modelling*. PhD thesis, Lancaster University, Lancaster, UK.

Quinn, P. F., Beven, K. J., Chevallier, P. and Planchon, O., 1991. The prediction of hillslope flow paths for distributed hydrological modeling using digital terrain models. *Hydrological Processes*, **5**, 59–79.

Quinn, P. F., Beven, K. J. and Lamb, R., 1995. The ln(a/tanβ) index: How to calculate it and how to use it within the TOPMODEL framework. *Hydrological Processes*, **9**, 161–182.

Wolock, D. and McCabe, G., 1995. Comparison of single and multiple flow direction algorithms for computing topographic parameters in TOPMODEL. *Water Resources Research*, **31**, 1315–1324.

Wolock, D. and Price, C., 1994. Effects of digital elevation model map scale and data resolution on a topography-based watershed model. *Water Resources Research*, **30**, 3041–3052.

13 Terrain Information in Geomorphological Models: Stability, Resolution and Sensitivity

P. D. BATES, M. G. ANDERSON and M. HORRITT
Department of Geography, University of Bristol, UK

ABSTRACT

Topographic data are among the few input parameters for distributed flow and sediment transport models that can be accurately collected at a spatial resolution commensurate with the model grid size. More frequently, very few data are available for model parameterisation and the problems encountered typically involve decisions regarding the spatial averaging to be employed rather than questions of stability, resolution and sensitivity. In this respect, study of the assimilation of terrain information into distributed models allows consideration of a unique set of general issues which may become more prevalent in modelling research as distributed field data collection techniques are developed for further model parameters. This chapter seeks to begin this process of exploring the consequences of terrain data assimilation into distributed models and to highlight the complexity of model response. In particular, three examples of such complexity are chosen: (i) the presence of numerical stability limits that can be expressed in terms of topographic criteria; (ii) the interaction between model grid size and digital terrain model (DTM) resolution; and (iii) the impact of errors in the topographic specification on model response. These issues are illustrated by examples drawn from hydraulic and hydrology modelling.

INTRODUCTION

There is increasing awareness of the critical role of topography as a boundary condition in distributed flow and sediment transport models used for geomorphological purposes (Lane *et al.*, 1994a). The stimulus for this research derives from two separate sources: process and modelling studies of topographic heterogeneity, and improvements in the available range of sources of topographic information. As a result of these studies, geomorphologists are beginning to acknowledge the complexity of distributed model response to terrain information, although a full appreciation of the implications of these findings is only now beginning to emerge. The focus of this chapter will therefore be to identify the major issues for future research into the relationship between terrain information and distributed model behaviour, and to develop an integrated research design for their investigation.

Landform Monitoring, Modelling and Analysis. Edited by S. N. Lane, K. S. Richards and J. H. Chandler.
© 1998 John Wiley & Sons Ltd.

Process and Modelling Studies of Topographic Heterogeneity

Both process and modelling studies have demonstrated significant variations in measured or predicted variables in the presence of topographic heterogeneity. For example, studies of floodplain sedimentation processes (Walling *et al.*, 1986; Simm, 1993) using sediment traps have recorded variations in suspended sediment deposition rates for individual flood events over length scales as small as 10–20 cm. These have argued that spatial variations in deposition at this scale could only be attributed to variations in floodplain micro-topography and vegetation creating local differences in hydraulic conditions. This work has subsequently been extended to demonstrate sediment responses to topographic variation over medium (10–30 year) time scales (Walling *et al.*, 1992) using caesium-137 inventories of floodplain deposition rates. Studies of convergent and divergent hillslope flow processes provide a further example of the influence of topography both from process studies (e.g. Anderson and Burt, 1977; Anderson and Kneale, 1982) and modelling (e.g. Anderson, 1982). Anderson and Burt (1977) used an automatic tensiometer system to monitor spatial variations in soil moisture in a hillslope spur and hollow. Their results demonstrated that topography exerted a significant control on soil moisture conditions and resulting stream flow response. Modelling water movement on a range of hillslope topographies, Anderson (1982) demonstrated that only in specific cases (hydraulic conductivity $> 10^{-4}$ cm s^{-1} and slopes $> 25°$) did soil water always converge into hillslope hollows. For all other topography conditions, model behaviour included complex patterns of cross hollow–spur movements. This has led, in turn, to the recognition of topography as a major parameter controlling geotechnical slope stability (e.g. Anderson and Lloyd, 1991). In this study, a two-dimensional finite difference hillslope hydrology model was used to develop design charts for field slope stability assessment based on slope form.

Distributed Model Response to Terrain Information

It can therefore be seen that topography is a key variable in a wide range of environmental processes that we may wish to model, and that process response to topographic variation can be complex. Moreover, terrain data are increasingly available to a high degree of accuracy and at a scale commensurate with recent high resolution distributed modelling (e.g. Krabill *et al.*, 1984; Chandler *et al.*, 1989; Lane *et al.*, 1994a; Kwoh *et al.*, 1994; Dixon *et al.*, Chapter 4; Vencastasawmy *et al.*, Chapter 8). The impact on model results of these developments is, however, less than clear and there is a fundamental need in geomorphological modelling to investigate the issues raised by the increased availability of topographic information. This is the focus of this chapter, where we seek to illustrate the complexity of distributed model response to topographic parameters with reference to examples drawn from our own and others' research in fluvial hydraulics and hillslope hydrology. Although new topographic data sources may have an impact on the whole of the modelling process, this review will limit itself to an examination of three specific issues: model stability, model discretisation resolution and model sensitivity. Consideration of these areas will be seen to raise a number of issues for geomorphological modelling that future research should seek to address.

MODEL STABILITY

Numerical models may have stability limits that can be expressed in terms of topographic criteria. Such limits may derive from either the model equation base itself or the numerical approximation procedure (finite difference, finite element, etc.) used to solve the controlling equations. In the former case, particular concepts or assumptions in the model structure may be contradicted for given topographies. In the latter, particular topography/discretisation configurations may force rates of change in model variables such that the numerical approximation procedure may find it impossible to project forward to a unique solution of the controlling equations in either space or time.

Model Equations and Model Stability

A number of examples can be given of particular model structures which have limits on their range of application imposed by topography. One such, the Dupuit–Forcheimer assumption, is a simplification often made in hillslope sub-surface hydrology models. Here, it is assumed that flow lines are horizontal and of uniform velocity in any given vertical section. This method is not strictly possible in nature but gives acceptable results where the flow region is of a large horizontal extent relative to depth. In a similar manner, studies have shown that the assumption that the local surface gradient is also the hydraulic gradient (such as in TOPMODEL and the resulting index $\ln(a/\tan \beta)$, where a is the area drained per unit contour and β is the local surface gradient) can be incompatible with known soil water process behaviour. For example, Anderson and Kneale (1982) showed that in cases of densely instrumented hillslope hollows, the correspondence between $\ln(a/\tan \beta)$ and the maximum extent of the zero pore water pressure line was significantly less good for a shallow 6° slope than for a 26° slope. They concluded that such results may restrict the applicability of static topographic indices for certain topographic configurations (see Figure 13.1). With the use of increasingly complex terrain information, there is a greater chance that model assumptions that have a topographic component may be breached. The solution to this merely involves encouraging a greater awareness of the need to check for violations of model assumptions.

Numerical Approximation Procedures and Model Stability

More problematic are stability limits inherent to particular numerical approximation procedures, as these may not stem from obvious theoretical considerations and may in part be dependent on the particular grid discretisation chosen. An example of such an effect is provided by RMA-2, a two-dimensional finite element model for depth-averaged free surface flows (King and Norton, 1978). This model solves the

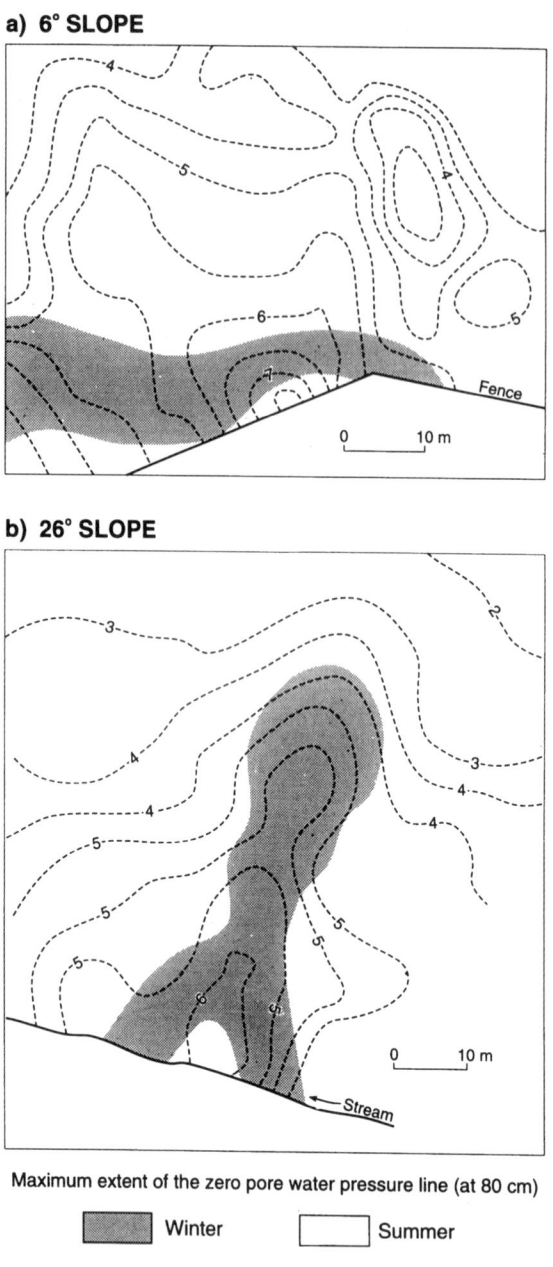

Figure 13.1 Relationship between ln(a/tanβ) and zero pore pressure lines for (a) a shallow 6° slope and (b) a 26° slope (after Anderson and Kneale, 1982)

depth-averaged St. Venant equations, derived from the full three-dimensional Navier–Stokes equations, for a continuum of triangular and quadrilateral finite elements using a fully implicit implementation of the Galerkin weighted residual technique. In the generalised weighted residual method, the exact solution to a problem is defined as being formed by a trial solution, composed of a combination of the model controlling equations and the finite element basis functions, which describe the variation of the predicted quantities across an element, and some residual. A series of weights are then chosen which attempt to force this residual to zero in the average sense at the nodal points (Pinder and Gray, 1977). Galerkin weighted residuals assume that the best approximation to the true solution is found when the finite element basis functions themselves are chosen as the weighting functions. Owing to the extreme non-linearity of the controlling equations, the numerical integration for the Galerkin procedure is performed iteratively using a Newton–Raphson type solver. When applied to structural or heat conduction problems, this finite element method leads to symmetric stiffness matrices (Brookes and Hughes, 1982). In this case, the difference between the finite element and the exact solution is minimised with respect to some norm and the numerical result is termed a "best approximation" for the problem. The success of the method was found to be largely due to its "best approximation" property. In fluid flows this property is lost, as the convective term matrix is non-symmetric. Brookes and Hughes (1982) have noted that for fluid flows the solution may be corrupted by spurious node-to-node oscillations when one tries to force rapid changes in the solution. They found that the only way to eliminate such problems while using a Galerkin solver was to employ a highly refined mesh, even in cases where such refinement was not necessary for an accurate representation of the flow field.

In the case of Galerkin finite element simulations of river channel/floodplain flows, planform and topographic complexity combine to give a problem that approaches the limit of current computational resources. Thus, mesh refinement of the type outlined above may not always be possible. However, topographic complexity in the channel cross-section can, along with such processes as the momentum exchange mechanism between main channel and floodplain flows (Knight and Shiono, 1990; Shiono and Knight, 1991), generate large differences in flow velocity over very short distances. This forcing of rapid change can lead to the development of the node-to-node oscillations noted above. In extreme situations this may lead to instability in the iterative numerical procedure whereby the solution "explodes". Here, values of the predicted variables grow without limit instead of gradually converging to a single figure (Abbott, 1991, p. 112).

Illustrating Topograph-Generated Model Stability Problems

An application of RMA-2 to an 11 km reach of the River Culm, Devon, UK (see Figure 13.2) was constructed in order to explore the potential utility of the Galerkin approach and define its stability limits for typical floodplain finite element discretisations. A finite element mesh was constructed for this reach (see Figure 11.3) consisting of 3655 nodes and 1090 triangular and quadrilateral elements with topography obtained from UK Ordnance Survey 1:2500 maps and a limited field

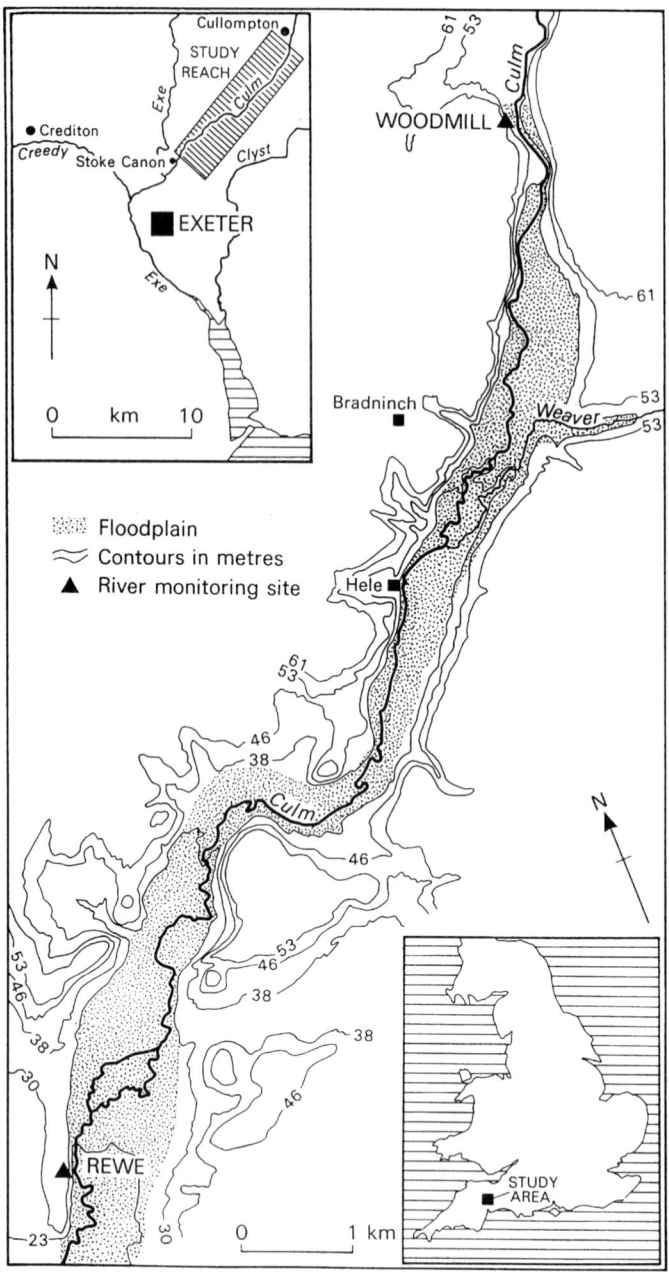

Figure 13.2 The River Culm study reach, Devon, UK. The finite element model covers the 11 km of channel and floodplain between the gauging stations at Woodmill (upstream) and Rewe (downstream)

survey. The discretisation was linear for water depth and quadratic for flow velocity, giving six or eight nodes per triangular or quadrilateral element respectively. This reach included such features as channel bifurcations, embankments on the floodplain and mill races which, it was felt, would provide a rigorous test for the model. A basic requirement for this mesh was that it should include sufficient density of elements in channel and near-channel areas to correctly represent the channel cross-sectional topography and the bulk flow characteristics of the shear layer between main channel and floodplain flows. Computational resources dictated that a maximum of five elements across the channel could be used, giving a trapezoidal cross-section with an additional element on either side of the channel to represent near-bank effects. In attempting dynamic and steady-state simulations with this mesh configuration, instabilities developed in specific locations (Bates *et al.*, 1992). Moreover, the existence and precise values of these thresholds could not have been predicted directly from Galerkin finite element theory. These locations were found to be characterised by sharp changes in topography which generated gradients of the predicted variables beyond some unknown stability threshold imposed by the discretisation. Examples of such situations included areas of steep lateral or longitudinal slope and sharp changes in the direction of flow. By examination and classification of these areas, the stability limits for this application were found to be (Bates and Anderson, 1993): (i) a maximum permitted longitudinal slope of 0.002, (ii) a maximum permitted lateral slope of 0.03, and (iii) a maximum channel angular deviation from downstream direction in a single element of 70°.

This hypothesised stability domain for the model was independently tested on a 3 km extension of the River Culm model (Anderson and Bates, 1994) whose topographic criteria were close to, but within, the upper limits noted above (a longitudinal slope of 0.0016, a lateral slope of 0.0015 and a maximum angular deviation of 60°). A similar discretisation resolution was used for this model extension in conjunction with identical topographic data sources. The extended model achieved stable numerical convergence to ±0.0001–0.2% of the unique solution within six iterations per time step, thus confirming the proposed model stability limits.

To solve instability problems in the original model, two choices were available: mesh refinement or topographic smoothing. Owing to computational constraints, the latter option was chosen. Unstable areas were identified and the particular set of features present isolated. The mesh topographic representation was then gradually relaxed until a stable solution was found (Bates *et al.*, 1992). Although research is currently underway to define a series of explicit rules for this procedure, a number of issues are raised. Firstly, it is unclear whether the ability of the model to correctly represent flow processes is compromised by the removal of topographic information. This is in part due to a lack of information on model sensitivity to various levels of topographic information provision. Secondly, although the model stability domain defined for this study is sufficiently wide to enable simulations to be developed for a reasonable range of floodplain environments, the limitations mean that a Galerkin finite element solution can only represent the overall topographic structure (features in the size range 10–100 m) rather than micro-scale variations (features in the range 1–10 m) known to be important in determining floodplain sediment deposition patterns (Walling *et al.*, 1986).

Clearly, a more acceptable solution to this problem would be achieved by using an oscillation-free numerical solution technique. This would enable stable solutions to be developed for complex topographies and planforms without recourse to either excessive mesh refinement or smoothing. One method with potential in this area is the Streamline Upwind/Petrov Galerkin (SUPG) scheme developed by Brookes and Hughes (1982). Here, upwind differencing techniques are employed to obtain their well known oscillation-free properties. However, upwinding has been extensively criticised (e.g. Davis and Mallinson, 1976) as it results in a loss of accuracy that can only be overcome by adding artificial diffusion (viscosity). In the SUPG technique this criticism is overcome by modifying standard Galerkin weighting functions with the addition of viscosity in the flow (streamline) direction only. When applied to all terms in the flow equation, this gives a consistent weighted-residual formulation. This has the effect of giving the oscillation-free advantages of upwind differencing methods whilst avoiding an artificially diffuse solution due to the extra viscosity.

Brookes and Hughes (1982) have tested the SUPG method for the linear advection–diffusion equation and demonstrated it to be accurate on a variety of test problems. For floodplain flows, however, such testing has only recently been completed (Bates *et al.*, 1994, 1995). An SUPG solver was implemented in the two-dimensional finite element code TELEMAC-2D (Hervouet and Janin, 1994), originally developed by the Laboratoire National d'Hydraulique in Paris (Hervouet, 1989, 1993). This scheme also solves the depth-averaged St. Venant equations but differs from RMA-2 by using a discretisation consisting of linear triangular elements with three nodes per element. To solve these equations, TELEMAC-2D uses a fractional step method (Marchuk, 1975) where advection terms are solved initially, separately from propagation, diffusion and source terms which are solved together in a second step. For the advection step several schemes could be selected, with the Method of Characteristics chosen for the momentum equation for all the simulations reported here. To ensure mass conservation and an oscillation-free solution, the SUPG technique was applied for the advection of water depth, h, in the continuity equation. The propagation step makes use of an implicit time discretisation and solves the resulting system with a conjugate gradient type method.

This scheme has also been applied to the River Culm test case and compared to previous results obtained with RMA-2 (Bates *et al.*, 1995). To obtain an accurate comparison, the TELEMAC-2D finite element mesh was developed by cutting each quadrilateral element in the mesh shown in Figure 13.3 into two triangular elements and using the same topographic data base. As the approach to wetting and drying of the finite element mesh taken by both models is based on calculations made at the nodal points, no additional complications should be generated by the element division process as no new nodes are created. No further stability problems were noted with this model, and a comparison of model-predicted discharge with gauging station observations taken at the downstream extremity of the reach showed a significant improvement over the RMA-2 scheme (Figure 13.4). In addition, numerical oscillations on the rising limb of the RMA-2 predicted hydrograph disappear when using TELEMAC-2D.

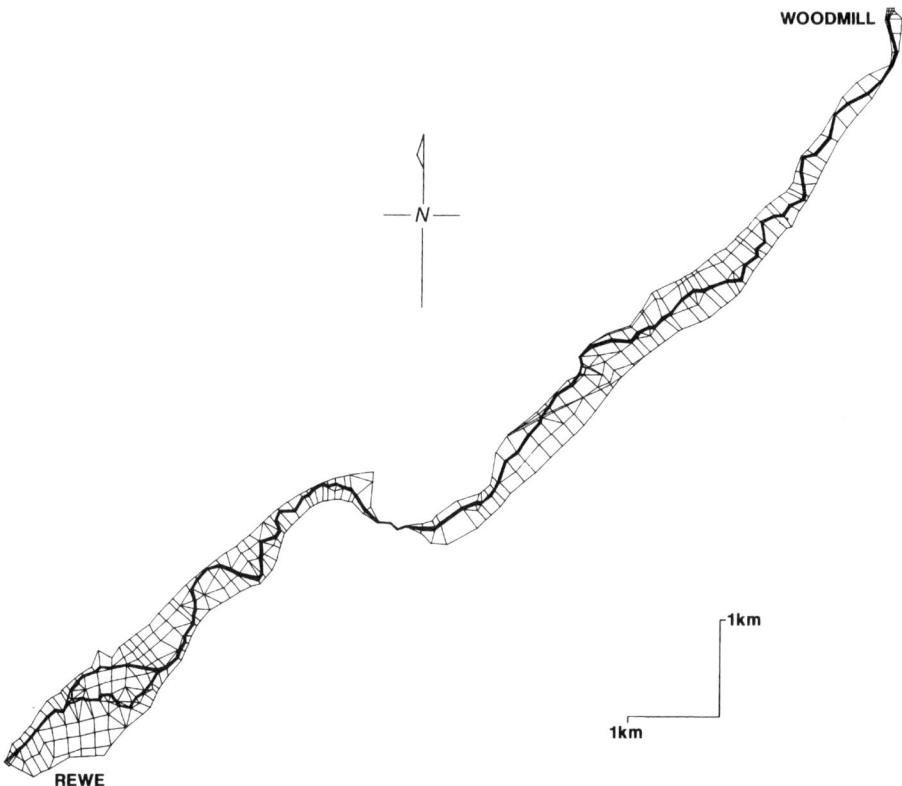

Figure 13.3 Finite element mesh constructed for the River Culm study reach

As other differences between the codes, in terms of the boundary condition set-up and the approach taken to element wetting and drying, could account for some of the prediction improvement, a further TELEMAC-2D application was made to a 12 km reach of the River Stour, Dorset, UK, using a more complex topographic data set. This was derived by combining approximately 25 surveyed channel cross-sections made available by NRA South West, and an experimental 10×10 m resolution digital terrain model (DTM) developed by the UK Ordnance Survey. For this reach, a finite element mesh was constructed consisting of 4396 elements and 2481 nodes (Figure 13.5) which attempted to give a smoother transition between areas of high and low element density. The topography data was then interpolated onto this grid to obtain a full specification of the finite element domain. Figure 13.6 shows the complexity of this topographic specification compared to the River Culm study, where the floodplain is broadly flat. Even with such topographic complexity, the model behaves in a stable fashion for steady-state and dynamic flow, and provides a reasonable match to available discharge data for a 1-in-4-year recurrence interval event (Figure 13.7). In particular, the SUPG method gives stable results at

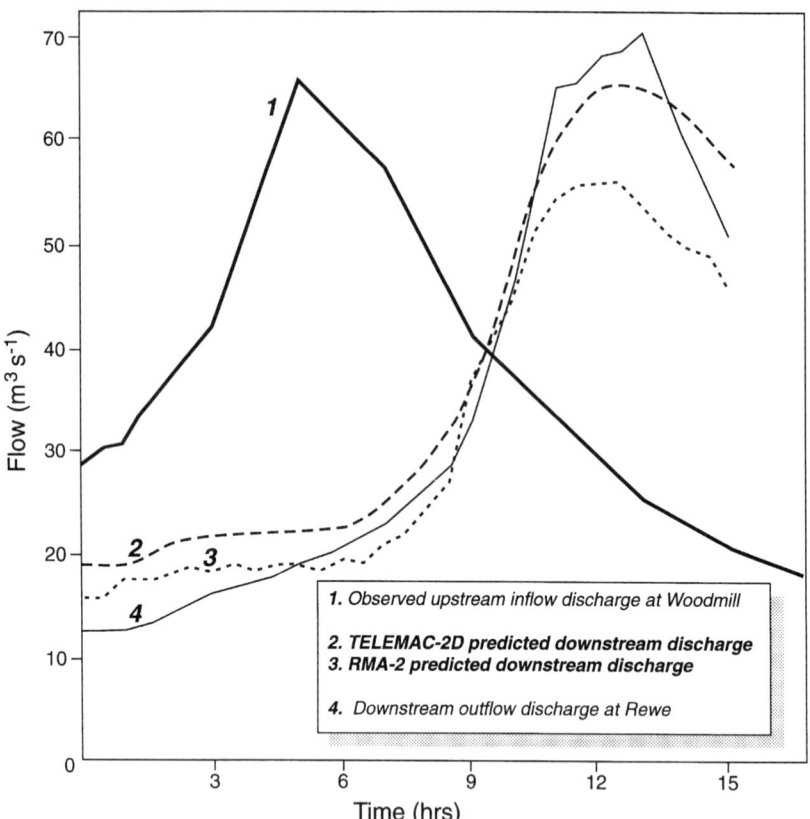

Figure 13.4 A comparison of downstream discharge predictions derived from the RMA-2 and TELEMAC-2D two-dimensional finite element models against observed field data for a 1-in-1-year River Culm flood event. Note the improvement in predictive accuracy given by TELEMAC-2D

flow rates well below bankfull discharge. This allows a simulation of flood inundation extent to be developed, commencing with a completely dry floodplain and continuing until the floodwave has completely receded (Figure 13.8). In the case of RMA-2, instability at low flows (Bates, 1992) necessitates beginning the simulation with flow at, or slightly above, bankfull discharge. This results in a partially wet floodplain (c. 35% of total floodplain area) at the beginning of the simulation (Bates *et al.*, 1995). The SUPG method can therefore be seen to allow stable solutions to be developed at a particular resolution without recourse to topographic smoothing, and holds out the possibility of moving towards the finite element simulation of micro-scale topographic features where rapid changes in flow variables need to be represented.

The example of two-dimensional finite element modelling of river channel/floodplain flows demonstrates the complexity of topography-generated model

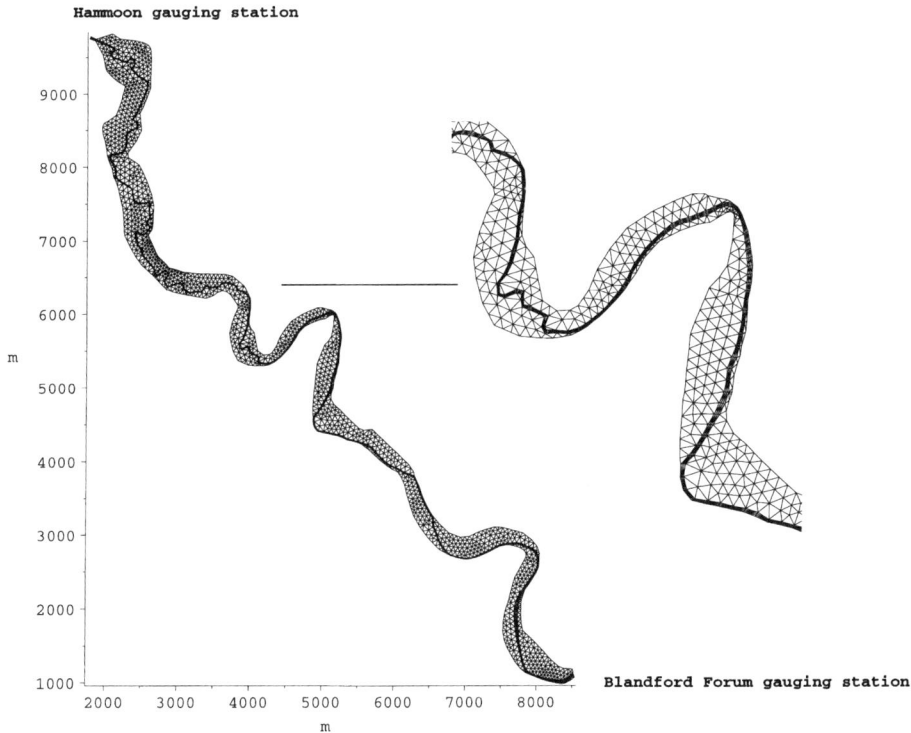

Figure 13.5   A finite element mesh constructed for a 12 km reach of the River Stour, Dorset, UK, between the gauging stations at Hammoon (upstream) and Blandford Forum (downstream)

stability problems. Frequently, the explanation and solution of such effects require a rigorous examination of both the numerical approximation procedure used by the model and a knowledge of the model sensitivity to changes in the domain discretisation. However, clear scope now exists for the development of models which can accommodate high resolution topography data in a stable manner whilst avoiding overspecification of mesh resolution.

MODEL RESOLUTION

Achieving Compatible Model and Topographic Resolution

Moving towards the inclusion of greater levels of topographic information in distributed geomorphological models requires model resolution to be increased such that it is commensurate with the new data sets. Moreover, a retrospective assessment of the density of topographic information is required to achieve adequate process representation. This can be achieved in a number of ways: (i) computational

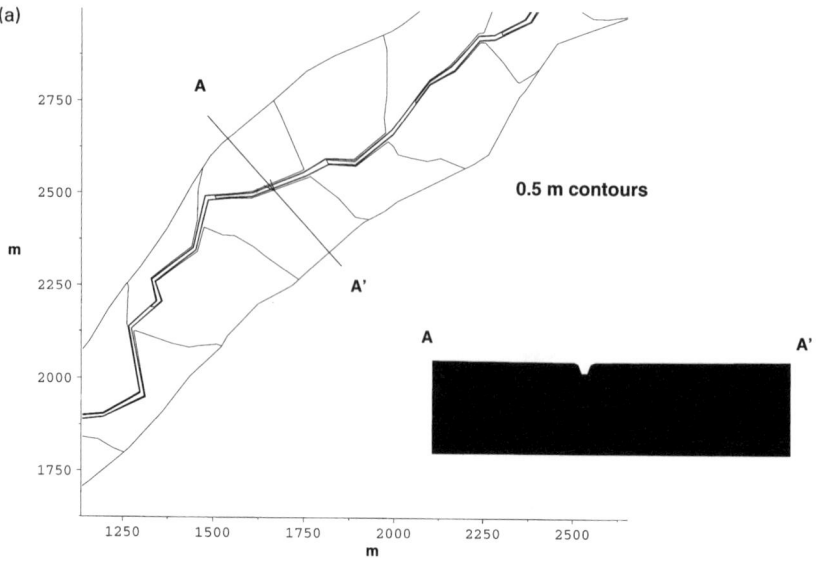

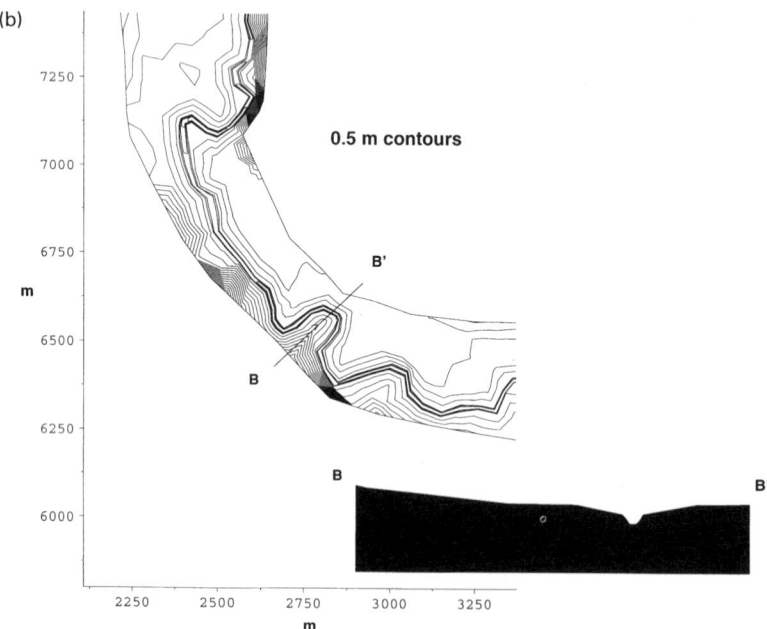

Figure 13.6 A comparison of the typical topographic surface complexity used to parameterise (a) the River Culm finite element model and (b) the River Stour finite element model

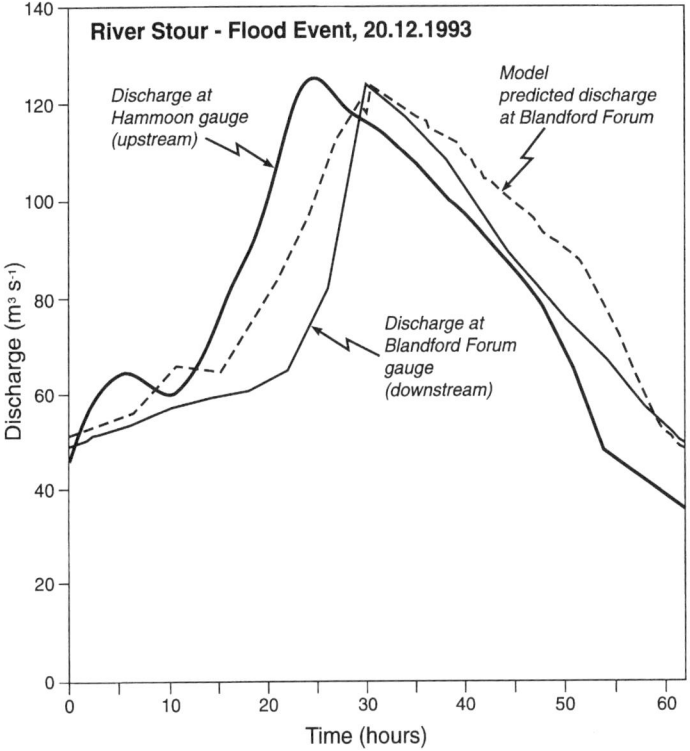

Figure 13.7 A comparison of downstream discharge predictions derived from the TELEMAC-2D two-dimensional finite element model against observed field data for a 1-in-4-year recurrence interval River Stour flood event. Note the ability of the model to simulate flow over a complex topography in a stable and accurate fashion

resources used for a particular problem can be increased; (ii) advantage can be taken of computing developments; or (iii) model developments can be adopted that increase computational efficiency. In terms of the latter option, small improvements in model efficiency can be obtained through the use of particular numerical algorithms. For example, two TELEMAC-2D simulations of a 1-in-1-year flood event were conducted, which varied only in the numerical method used to treat the advection of water depth, h, in the continuity equation (Bates *et al.*, 1994). The methods used were a hybrid scheme consisting of a combination of the Method of Characteristics and centred differences, and the SUPG technique described above. The simulations demonstrated that the SUPG technique gave a 20% reduction in computation time over the alternative scheme, with no loss in predictive accuracy. More significant efficiency gains usually result from improvements to matrix handling procedures within the numerical model. This is particularly so for finite element techniques where the matrices of the linear system assembled and solved by the model are large, sparse and contain relatively few non-zero terms. Early solutions to this problem focused on algorithms which could optimally reorder the

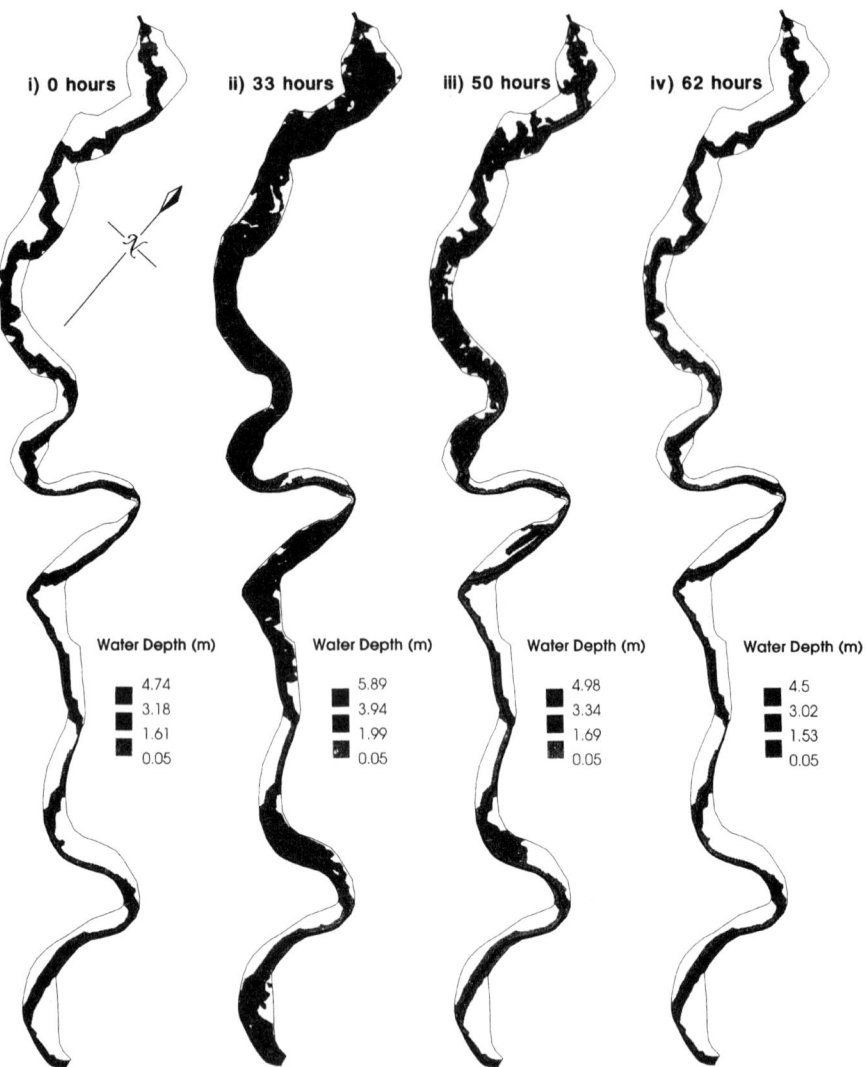

Figure 13.8 Flood inundation extent predicted by the TELEMAC-2D model at four time intervals during the 1-in-4-year River Stour flood event shown in Figure 13.7. Shown here are: (i) simulation initial conditions; (ii) peak inundation extent; (iii) hydrograph recession accompanied by draining of the floodplain; and (iv) inundation extent at the end of the simulation. Note the ability of the model to simulate a flood event commencing with initially dry floodplains followed by widespread inundation and complete drainage

Table 13.1 Examples of space/time resolution of two-dimensional finite element hydraulic simulations of a 1-in-1-year recurrence interval flood event performed for the River Culm test case using the RMA-2 (direct solver) and TELEMAC-2D (element-by-element solver) models. Note the large increase in resolution achieved with element-by-element techniques. The Courant number is given as $Cr = (u\Delta t/\Delta x)$, where Cr is the Courant number, u is the flow velocity, t is the time step and x is the mesh size

Model	Number of elements	Simulation length (h)	Time step (s)	Number of time steps	Total computation time (min)	Courant number
RMA-2 (direct solver)	1090	15	1800	30	265	540
TELEMAC-2D: low resolution (EBE solver)	2040	15	2	27 000	320	0.6
TELEMAC-2D: high resolution (EBE solver)	9600	15	2	27 000	2188	0.2

sequence of elements used to build the matrix in order to minimise its bandwidth (King, 1970; Sloan and Randolph, 1983; Sloan, 1986). Such strategies attempt to cluster non-zero terms along the diagonal of the matrix and then exploit the banded nature of the resulting structure to restrict the number of calculations that have to be performed. This is the approach, sometimes termed a direct solution algorithm, adopted by the RMA-2 code described above. More recently, element-by-element techniques have been developed. Here, the matrices of the linear system are stored in their elementary form and are never fully assembled (e.g. Carey and Jiang, 1986). This has a number of advantages for computer memory usage and processing time, as well as allowing easier vectorisation of the resultant code. Binley and Beven (1993) have implemented this approach to enable the development of a sophisticated three-dimensional simulation of hillslope hydrology. This code was applied to Darcian flow in a small headwater catchment with reasonable results. Element-by-element techniques have also been applied to the two-dimensional finite element hydraulic code TELEMAC-2D described above (Hervouet, 1992). This has enabled a considerable increase in the space and time resolution at which problems can be treated (see Table 13.1).

Table 13.1 demonstrates that for simulations of a 1-in-1-year recurrence interval flood event on the River Culm test reach a large increase in the space/time resolution of the simulation can be achieved through the use of element-by-element solvers. Very large finite element meshes of the order of 10 000 elements (Figure 13.9) are now feasible, even on medium-sized workstations, whilst simultaneously allowing a reduction of the Courant number to less than 1. This is significant as although both RMA-2 and TELEMAC-2D utilise fully implicit solvers which do not exhibit time step-related stability problems, the presence of high Courant numbers can imply problems with the model representation (Hervouet and Janin, 1994).

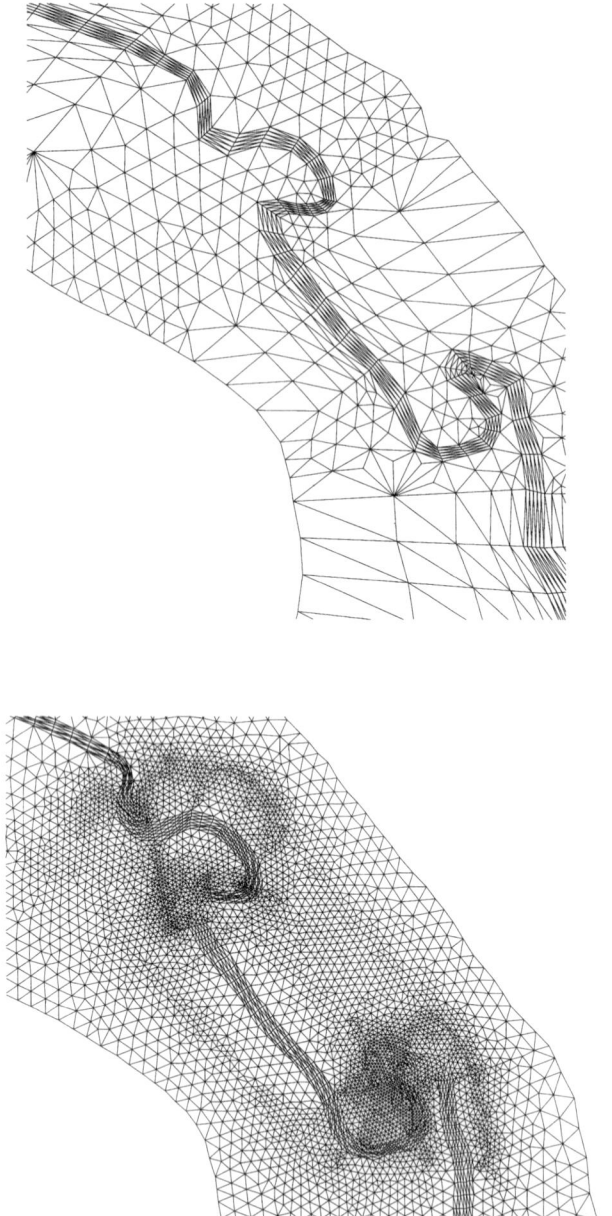

Figure 13.9 Examples of high resolution finite element discretisations of the specific area of the River Culm model shown in Figure 13.3. This demonstrates the ability of new numerical algorithms to enable a model resolution commensurate with that of new topographic data sources

LIVERPOOL HOPE U. TY COLLEGE

Relationships Between DTM and Model Grid "Filters"

Scope therefore exists to utilise a number of developments in computing and numerical analysis to facilitate an increase in the space/time resolution of distributed models commensurate with new topographic data sets. This does, however, raise a number of issues relating to the scaling of terrain information in geomorphological models. Band and Moore (1995) have described how the grid size of the DTM used to describe a land surface prior to assimilation into a model acts as a filter on topographic information content. As terrain is sampled at progressively lower resolutions, the higher frequency information is lost. The impact of this filtering process on model results has been examined to determine the sensitivity of the model to terrain information resolution. To date, this has been attempted only with semi-distributed schemes such as TOPMODEL (e.g. Quinn et al., 1993). These studies have shown that catchment behaviour predicted by the model can be sensitive to the resolution of the DTM used to generate the hydrological similarity index, $\ln(a/\tan\beta)$. For fully distributed models, however, the grid size resolution may, if set at a resolution lower than that of the DTM, constitutes a second filter on topographic information (Figure 13.10) thereby adding additional complexity to the model response. The solution here would be to make grid resolution progressively finer, both to ensure minimum data redundancy and to generate a numerical approximation closer to the true solution (as in the limit where the grid size tends to an infinitely small value, the model prediction tends to the true solution). In most hydrological and geomorphological applications, however, this is not an option owing to the fractal nature of topography (e.g. Culling and Datko, 1987; Huang and Turcotte, 1989; Moore et al., 1993). This results in a significant practical problem in any attempt to identify general rules for choice of model grid resolution with respect to terrain information.

A further problem created by altering model resolution has been noted by Grayson et al. (1993), who stated that as the information content of the DTM drops below some threshold certain assumptions in distributed hydrology models, particularly those concerned with flow routing, may not be met. This argument can be extended by considering model response to increases in DTM resolution. This will involve both an increase in the simulation complexity and an increase in model grid resolution to allow greater levels of terrain information to be included. In such situations, the model process representation may become insufficient to adequately characterise the physics of the problem being studied, and the model may be rendered invalid. For example, if we consider two-dimensional fluid and sediment transport modelling, over complex topography and at high spatial resolutions the flow may exhibit significant vertical accelerations that are not represented in depth-averaged models. For three-dimensional models of strongly turbulent flows, high spatial and topographic resolution may require improvements to the turbulence closure scheme employed to account for such effects. There is thus a sense in which any particular model structure can be said to have an optimal range of scales at which it is applicable.

At present, no studies have attempted to analyse the interaction of DTM resolution and model grid size resolution filters on model response, or attempted to

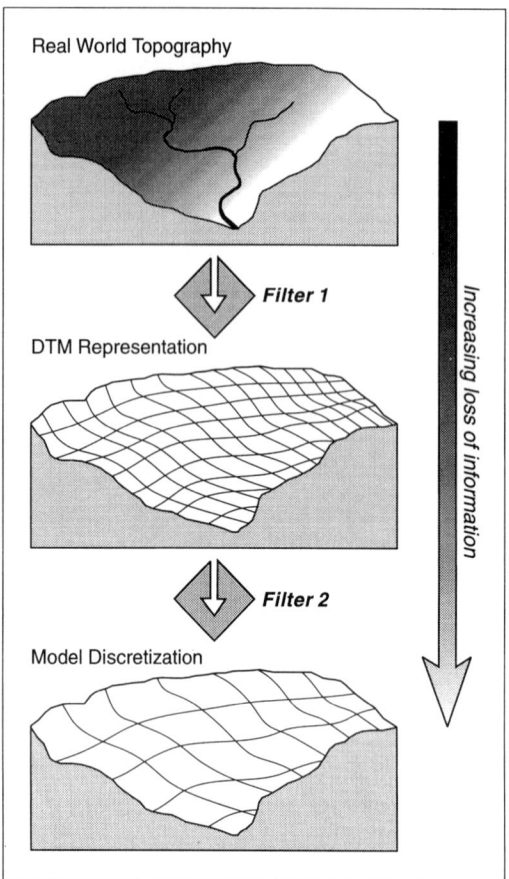

Figure 13.10 Schematic diagram showing the loss of topographic information resulting from the filtering effect of DTM and model discretisation construction

determine scale limits to model application. Furthermore, no studies have included consideration of those aspects of model parameterisation that have a significant sub-grid-scale component practically dependent on topography, such as surface friction in hydraulic or overland flow models. This need is analogous to the call by Beven (1989, 1995) to develop a methodology to estimate the "value" of different types of data. Uncertainty analysis has been suggested as an appropriate tool with which to investigate this problem, and this can be extended to consider the two additional problems outlined above. In terms of terrain information, this requires a high resolution data set which can be progressively degraded to examine model response to loss of information. As an initial step there would appear to be considerable merit in the use of synthetic topographic data, as at present only a few field data sets are available at an appropriate resolution. Synthetic data sets would allow a wider range of simulation experiments to be conducted and allow the requirements for actual field data to be more rigorously defined. At present, very little research has been conducted explicitly on developing synthetic topographies, although a number of techniques do present themselves. Three potential methods for developing synthetic topographies exist, which rely on various amounts of "seed" data for success. At the most basic level, a methodology can be developed that has no reliance on field data but relies merely on an appreciation of the typical length scales of topographic variation for particular environments. Each typical length scale may then be transformed into a single sine function with given amplitude and wavelength. The individual sine functions may then be combined to create pseudo-realistic variations in a topographic surface. Secondly, kriging techniques (Gambolati and Volpi, 1979; Hughes and Lettenmaier, 1981; Clarke, 1994) could be used to stochastically interpolate realistic topographic surfaces from sparse field surveys. Ultimately, however, techniques involving the statistical analysis of scaling in topographic data could be used to develop laws of self-similarity in terrain information which would categorise terrain heterogeneity over a broad range of scales (e.g. Culling and Datko, 1987). This is the approach taken here, where we have used an analysis of the fractal dimension of real floodplain terrain to generate a synthetic DTM for hydraulic model parameterisation. The method adopted was that given by Turcotte (1992). Features with a length scale of less than 200 m in the River Stour 10×10 m DTM, discussed above, were analysed by determining the variogram for lags from 0 to 200. These features were found to have a fractal dimension of 2.4 and a variance of 0.4. These are higher than the values found by Culling and Datko (1987); however, the variogram analysis used did not address such issues as anisotropy and spatial variations in the correlation surface which would be required for a more complete analysis. A regular grid of dimensions 512×128 was constructed, and a height at each node set to a random value using a normal distribution and unit variance. A fast Fourier transform was performed on the grid of random numbers and each component in the Fourier space multiplied by a factor representing a function of the radial wave number, k. For large k values, corresponding to wavelengths less than 200 m, the function is $k^{-(4-D)}$, where D is the fractal dimension required. For small k values the function tends to 1. These data were transformed back into real space using a further fast Fourier transform and the resulting data set was normalised to give the required variance. This gave a DTM with a ground resolution of approximately 10×10 m. This generated a random perturbation at each DTM grid point with a root mean square height of 0.223 m which could be used to generate "real" topographic variation at a range of length scales onto simple model topographies. To begin to investigate the interaction of mesh and DTM resolution of model output, the DTM was sampled onto two finite element discretisations. Each domain consisted of a sinuous channel, 20 m wide and 2 m deep, flowing down the centre of a hypothetical domain 2000 m long and 800 m wide. The two finite element meshes used consisted of 2284 and 7310 elements (Figure 13.11a and 13.11b) and consisted of a planar floodplain onto which the above DTM was sampled by a nearest neighbour method. The 512×128 DTM was then degraded to a resolution of 256×128 and resampled onto each finite element mesh. A single flood hydrograph scaled from real data, lasting 20 h and with a peak discharge of 40 m^3s^{-1} was then simulated over each mesh/topography combination giving a total of four simulations in all.

(a)

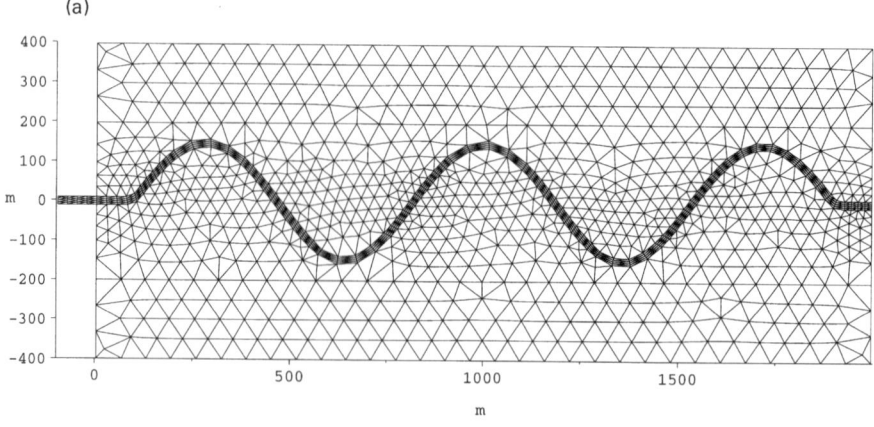

(b)

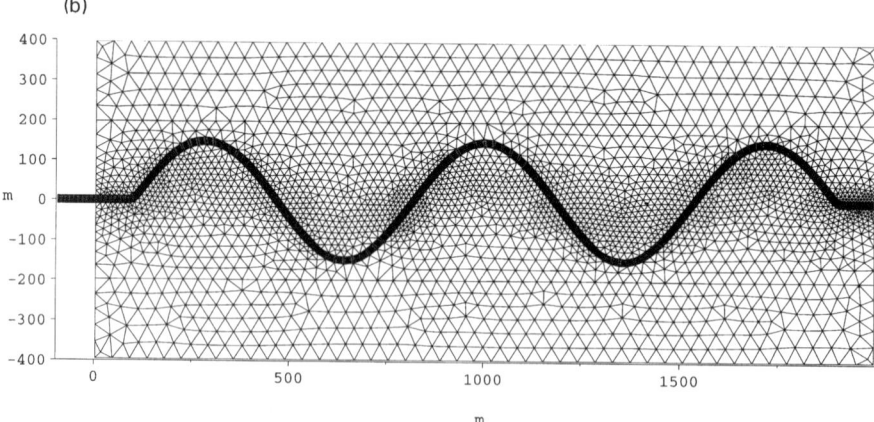

Figure 13.11 High and low resolution finite element discretisations developed for a sinuous channel embedded within a floodplain domain 2000 m long by 800 m wide. Two discretisations are shown comprising (a) 2284 and (b) 7310 elements

Results from the simulations show that for the discretisations studied, decreasing topographic resolution has virtually no impact on two-dimensional bulk flow predictions of outflow discharge (Figure 13.12) or percentage inundation extent (Figure 13.13). For such variables, the results indicate that mesh resolution is more important in conditioning the solution obtained. An impact is, however, noted in terms of local hydraulic conditions. Figure 13.14 shows the difference surface for predicted water depth for the high resolution mesh at the two different DTM resolutions; this shows that variations in floodplain water depth of up to ±10 cm can be generated. This is in line with the RMS height of the fractal perturbation superimposed on the initial planar floodplain. These results raise the possibility that different optimum levels of topographic provision exist in order to achieve different simulation objectives. For example, the results indicate that predicting bulk flow over the reach would require different topographic information than a

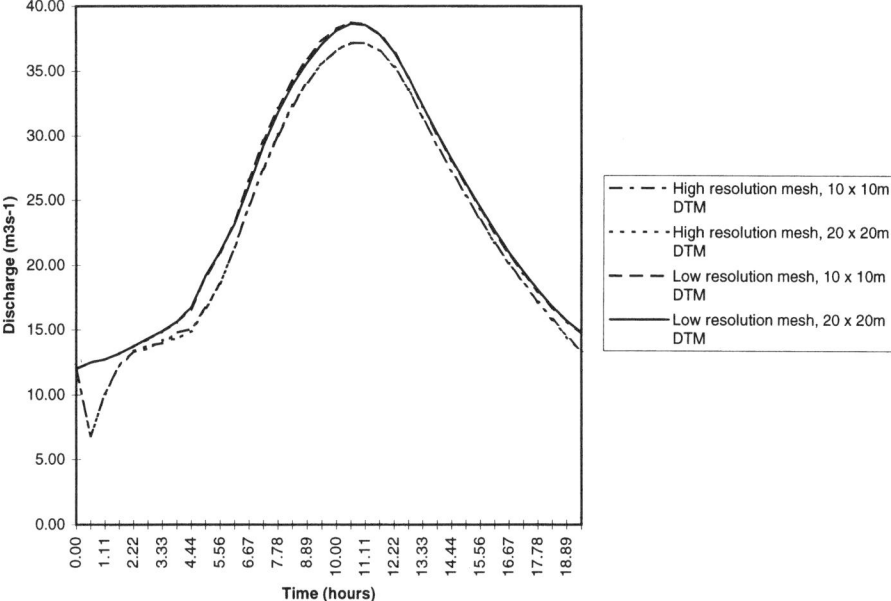

Figure 13.12 Differences in predicted reach outflow discharge for a small flood event simulated using the finite element meshes shown in Figure 13.11 for DTM resolutions of 10 × 10 m and 20 × 20 m.

sediment transport simulation, where local hydraulic variations may be important (cf. Walling *et al.*, 1992).

The above discussion has highlighted the need to develop high resolution distributed modelling at a scale commensurate with data from new topographic information sources. We have attempted to show that such developments raise a number of issues concerning the physical representation employed by the model, and that there is a need for an explicit consideration of the data assimilation process to avoid a loss of potentially critical information (Figure 13.10). From this it should be clear that the use of ever higher resolution models may not generate a similar improvement in system definition, and it will thus be necessary to determine precisely how meaningful particular resolutions are. The move to high resolution modelling will therefore create a number of new research questions that need to be resolved before the utility of such improvements can be fully assessed.

MODEL SENSITIVITY

The Need to Determine Model Sensitivity to Levels of Topographic Information Provision

To date, relatively few studies have attempted to examine the effects of topographic estimation errors on the output from physically based distributed models. This

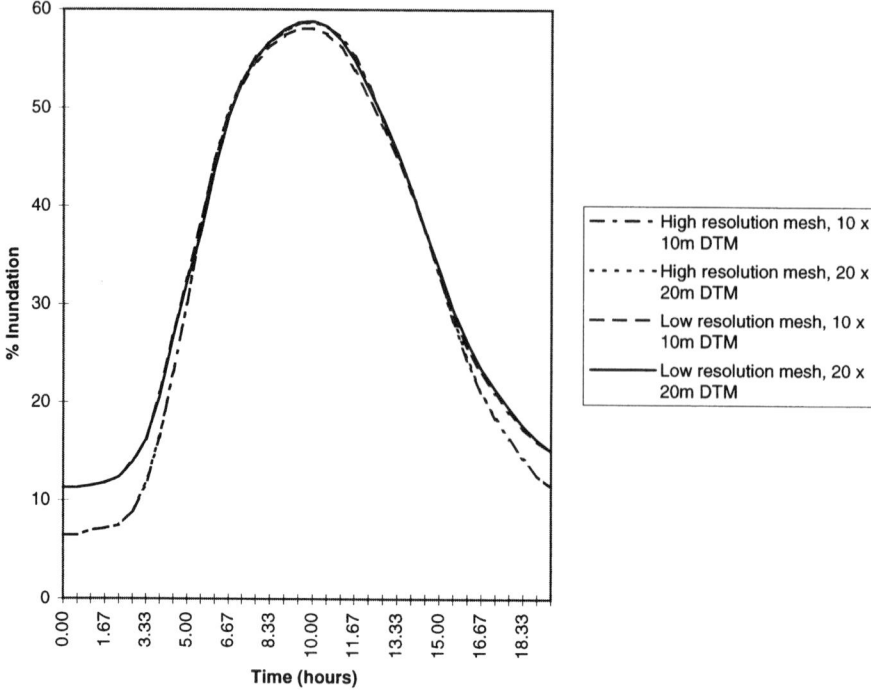

Figure 13.13 Differences in predicted percentage inundation extent for a small flood event simulated using the finite element meshes shown in Figure 13.11 for DTM resolutions of 10 × 10 m and 20 × 20 m

is due partially to current sensitivity analysis methodology which, as typically employed in hydrology and hydraulics, is unsophisticated compared to the space/time resolution of present generation distributed modelling schemes. Sensitivity analysis was developed for the previous generation of lumped hydrology and hydraulic models and has proved adequate for the task of analysing such simple modelling systems. Despite a move towards high space/time resolution modelling, no commensurate development of sensitivity analysis techniques has been made (Lane *et al.*, 1994b). This is a particular problem for analysing model response to errors in a topographic surface due to the distributed nature of the variable itself. In this section we therefore present results from a distributed sensitivity analysis of topographic parameterisation accuracy on flood inundation extent predicted by a two-dimensional finite element hydraulic model. Initially this analysis has been performed using the RMA-2 model described above because its low computational cost (Table 13.1) and the large number of simulations involved. As this is a standard tool for this class of problem, a knowledge of its behaviour is critical. Moreover, the results on model stability, presented above, show it to be capable of adequately simulating flow in floodplain environments and thus able to model the response to small variations in topographic parameterisation accuracy. Ultimately, however, the analysis will need to be repeated for more sophisticated and

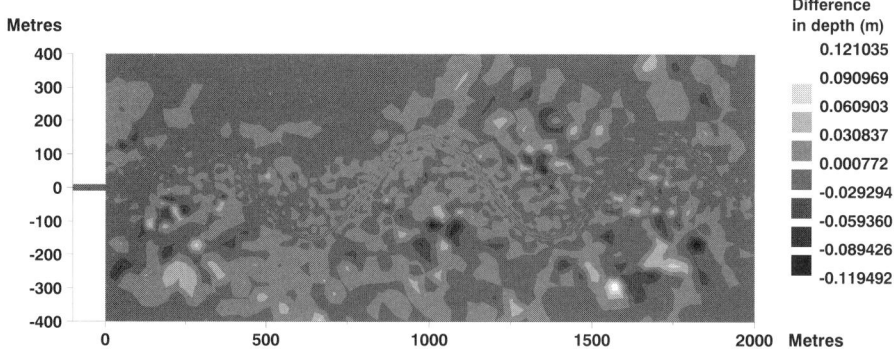

Figure 13.14 Difference surface of predicted water depth at peak inundation for a finite element mesh consisting of 7310 elements (Figure 13.11b) between simulations parameterized with terrain taken from a 10 × 10 m DTM and from the same DTM with information content degraded by 50% to 20 × 20 m

computationally demanding schemes such as the TELEMAC-2D model also described in this chapter.

Illustrating the Effect of Topographic "Error" on Model Response

The effect of topographic parameterisation accuracy has been examined for the River Culm finite element mesh, shown in Figure 13.3, for two storm events: the 1-in-1-year recurrence interval event shown in Figure 13.4, and a further double-peaked event consisting of 1-in-1 and 1-in-5-year recurrence interval floods. These events were chosen as field data and other documentary sources such as air and ground photos indicated flow conditions ranging from partial (63%) to near-complete (92%) maximum flood inundation extent respectively. This therefore covers the observed range of possible floodplain inundation states for this reach. For each of these floods, a control simulation was conducted using calibrated roughness parameters to achieve minimum phase error between model-predicted downstream discharge and gauging station observations taken at the same point. Keeping all other conditions, including channel topography, constant, four simulations were conducted for each flood event at various levels of floodplain topographic parameterisation error. As a first step, only uniform error surfaces were applied to the whole floodplain domain as we wished to determine how a simple perturbation of topography propagated errors through a complex spatially distributed modelling system under dynamic conditions. The topography specified for this model was obtained by combining terrain information from UK Ordnance Survey 1:2500 sheets and a limited field survey. In terms of accuracy, Ordnance Survey bench marks and topographic survey are typically reliable to better than

±1 cm. However, contour information, especially in areas of low relief, is considerably less reliable (Fryer *et al.*, 1994; McCullagh, Chapter 5; Wise, Chapter 7). We therefore estimated that a combination of these data sources allowed floodplain topography to be specified to within an average error of ±10 cm. More complex specifications of the error probability distribution could be employed, such as a Gaussian distribution with standard deviation based on Ordnance Survey accuracy specifications. It was felt, however, that the simple linear distribution used here is sufficient to capture the upper and lower boundaries of an uncertainty field prior to generating intermediate behaviour with a more sophisticated technique. Two positive and two negative treatments were then selected up to this maximum variability. Specifically, values of –10 cm, –5 cm, +5 cm and +10 cm were chosen to represent potential errors in the parameterisation of the floodplain surface.

For each simulation, a time sequence of maps was constructed showing flood inundation extent at 3 h intervals. For each map, a series of 20 evenly spaced cross-sections was superimposed and the percentage inundation extent at each cross-section calculated. The change in inundation extent from the control simulation at each cross-section was then calculated for each time increment. The variation of this quantity (the model sensitivity) in both time and space could thus be obtained. Figures 13.15 and 13.16 show the sensitivity surface obtained for the maximum positive and maximum negative changes in floodplain topography for the 1-in-1-year event and the double peaked 1-in-5-year event respectively. These show model sensitivity to topographic parameterisation error to be variable in both time and space. This variation does not appear to be random, with recognisable features, principally linear ridges and troughs, present within the data. A correlation analysis was conducted to determine whether the observed features were representative of realistic model behaviour or merely artefacts resulting from site-specific attributes of the River Culm discretisation. Specifically, the correlation of high sensitivity with low floodplain occupancy, high floodplain width and low floodplain lateral slope was tested, as each of these factors could give a propensity towards large changes in flood inundation extent with relatively small hydraulic changes. No correlation was shown in any of these tests and the presence of any simple mechanism generating the response was therefore discounted, leaving the likelihood that the observations show actual model behaviour.

This study has shown that a uniform change in floodplain topography can produce a distributed model response that is complex and non-uniform in both space and time. This raises the possibility that an uneven density of initial data may be necessary to achieve an even model robustness in respect of inundation predictions. Although unexpected, this result is possibly not surprising given that we consider a dynamic simulation of river flood flow developed using a highly non-linear equation set. The pattern and response level of the model for each applied error surface have been shown to be consistent within treatments, as evidenced by the strong ridges in the data. These patterns are complicated by the dynamic nature of the simulation, with sensitivities being observed at particular space/time locations in the opposite sense to that expected from a more simplistic analysis. Furthermore, significant variations in time/space response are shown between treatments and between flood events. The use of more realistic topographic error surfaces in the

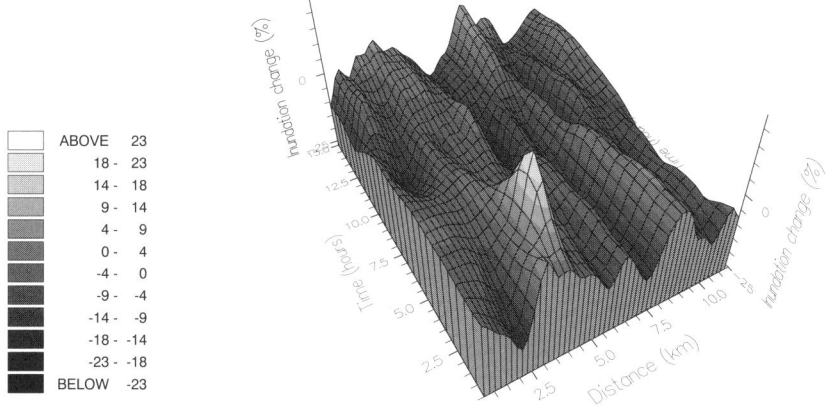

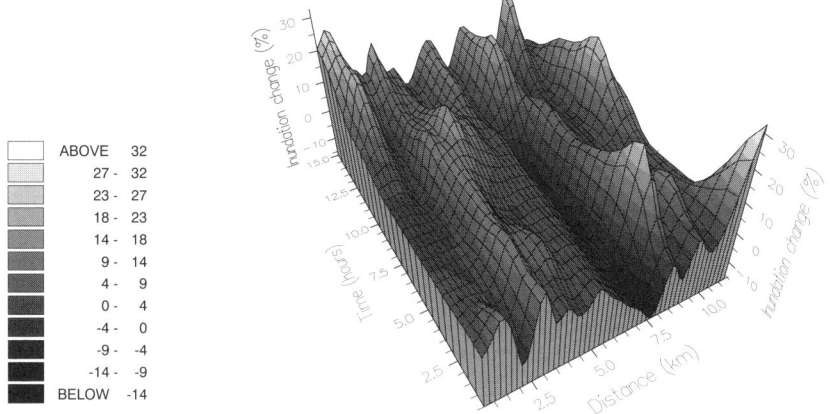

Figure 13.15 Time/space distributed surfaces showing the sensitivity of inundation extent predicted by the RMA-2 model to (a) a 10 cm increase in floodplain topography and (b) a 10 cm reduction in floodplain topography for the 1-in-1-year recurrence interval flood event shown in Figure 13.4

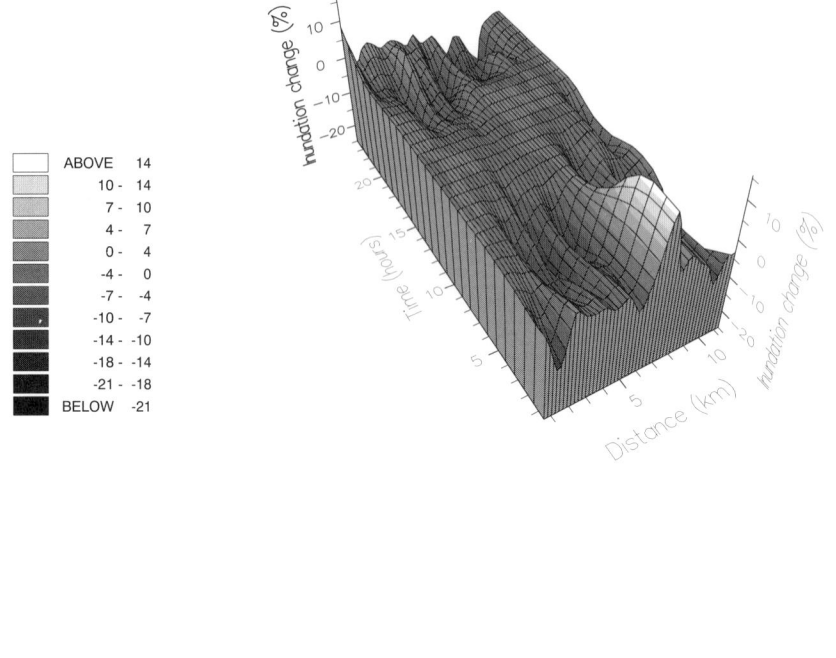

ABOVE   14
10 -   14
7 -   10
4 -   7
0 -   4
-4 -   0
-7 -   -4
-10 -   -7
-14 - -10
-18 - -14
-21 - -18
BELOW   -21

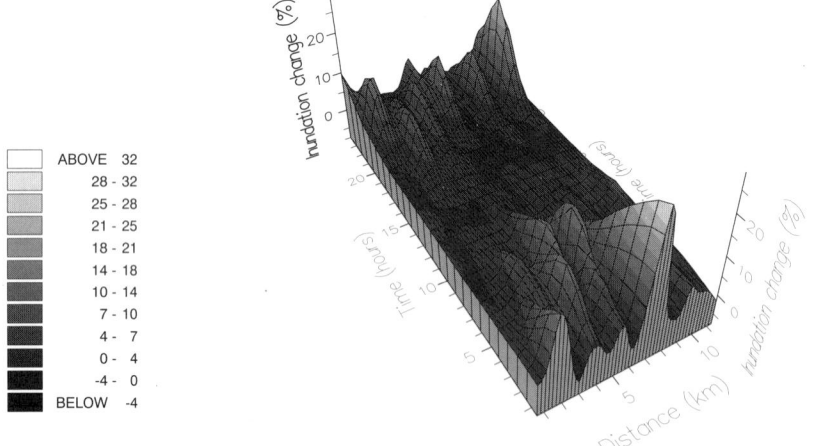

ABOVE   32
28 - 32
25 - 28
21 - 25
18 - 21
14 - 18
10 - 14
7 - 10
4 -   7
0 -   4
-4 -   0
BELOW   -4

Figure 13.16   Time/space distributed surfaces showing the sensitivity of inundation extent predicted by the RMA-2 model to (a) a 10 cm increase in floodplain topography and (b) a 10 cm reduction in floodplain topography for a double-peaked 1-in-1 and 1-in-5-year recurrence interval flood event

analysis will further complicate this pattern, thus highlighting the need to conduct a structured series of tests commencing with very simple variations. The results imply that any research to determine the impact of terrain information on distributed models cannot restrict itself to changes at a single point, such as the catchment outflow, as significant model prediction sensitivity may occur at all points on the network.

A RESEARCH DESIGN FOR INVESTIGATING MODEL RESPONSE TO TERRAIN INFORMATION

This review has attempted to highlight a number of issues arising from recent growing awareness of the critical role of topography as a boundary condition in distributed flow and sediment transport models used for geomorphological purposes. We have sought to demonstrate that the numerical procedures used in such models may have complex stability limits that can be defined in terms of topographic criteria, that both increasing or decreasing levels of topographic data provision may render particular model structures invalid as they no longer adequately represent processes at the required scale, and finally that model sensitivity to errors in the specification of topographic surfaces may be highly complex. In terms of topography-generated stability limits, a study has been presented to demonstrate the identification and resolution of such problems. This has been shown to necessitate a detailed examination of numerical analysis theory and a recognition of the dependency of such problems on the particular domain discretisation employed. Achieving commensurate model and topographic data resolution has also been shown to necessitate numerical developments, but two potential problems have been highlighted which require investigation. Firstly, most studies have implicitly assumed that terrain data are correctly assimilated into distributed models. This is not necessarily the case, owing to the interaction of the DTM resolution and the domain discretisation as "filters" on the data flow. At each stage in the assimilation process data may therefore be lost, although, to date, explicit research on such effects has only been conducted on semi-distributed models such as TOPMODEL (Quinn et al., 1993) where only the former of the above two "filters" applies. There is thus a need to pursue this research in the context of distributed models. Secondly, the impact of such errors is likely to be complicated by the sensitivity relationships shown to occur with dynamic, non-linear distributed models. An example has been presented which shows the sensitivity of a two-dimensional finite element model for river flood inundation to topographic parameterisation errors to be highly variable in both time and space. Aside from the implications for model calibration or stability, this last consideration necessitates the development of an integrated research design to fully examine all the above interactions.

We have therefore identified a need to examine the relationships between model dimensional representation, discretisation resolution, sensitivity and stability in response to improvements in topographic data collection techniques. Contrary to previous approaches, these areas are now acknowledged to interact in a complex and significant fashion. A fundamental question is therefore how to conduct an appropriate analysis. As a starting point, we can identify those elements that any research design would need to include. Taking the example of modelling flood inundation, these would be access to one-, two- and three-dimensional models for this particular physical problem, construction of a number of domain discretisations for each study reach, and availability of a high resolution topographic data set that could be degraded to give varying levels of data provision. This latter prerequisite could be obtained from field data sources or, as an initial step, use could be made of a synthetic topographic data set prior to defining requirements for a field data capture programme. The effect of calibration on each of the model results would also need to be assessed to determine whether model response to topographic information is masked by uncertainties in parameters such as friction coefficients. Although complex, such a design represents the minimum commitment necessary to fulfil current research needs in this area of geomorphological modelling.

Understanding the role of terrain information in geomorphological models will be a key requirement for future research if the advances made in the acquisition of terrain data are to be capitalised upon. The potential to resolve significant current barriers to progress, such as the need to acquire wide scale data on rapid landform change for model validation purposes or to better parameterise distributed models, would appear to be good. Technological advances can, however, create additional problems that require identification and investigation. In the case of terrain information, this now appears to be a less straightforward task than originally assumed and would seem to require a wide range of modelling resources to be employed to even begin the investigation. Nevertheless, such research is now becoming possible and should be the focus of a significant effort in the future.

ACKNOWLEDGEMENTS

This research has been supported by the UK Natural Environment Research Council Grant No. GR3/8633, Electricité de France Direction des Etudes et Recherches, and the Hewlett Packard European Research Laboratory, UK. The authors are grateful for data provided by the UK Ordnance Survey and the UK National Rivers Authority. Matthew Horritt was supported by a studentship funded by NRSC Ltd.

REFERENCES

Abbott, M. B., 1991. *Hydroinformatics: Information Technology and the Aquatic Environment*, Avebury Technical, Aldershot, 145pp.

Anderson, M. G., 1982. Modelling hillslope soil water status during drainage. *Transactions of the Institute of British Geographers*, 7, 337–353.

Anderson, M. G. and Bates, P. D., 1994. Initial testing of a two dimensional finite element model for floodplain inundation. *Proceedings of the Royal Society of London, Series A*, 444, 149–159.

Anderson, M. G. and Burt, T. P., 1977. Automatic monitoring of soil moisture conditions in a hillslope spur and hollow. *Journal of Hydrology*, 33, 27–36.

Anderson, M. G. and Kneale, P. E., 1982. The influence of low angled topography on hillslope soil-water convergence and stream discharge. *Journal of Hydrology*, 57, 65–80.

Anderson, M. G. and Lloyd, D. M., 1991. Using a combined slope hydrology–stability model to develop cut slope design charts. *Proceedings of the Institute of Civil Engineers*, **91**, 705–718.

Band, L. E. and Moore, I. D., 1995. Scale: landscape attributes and Geographical Information Systems. In *Scale Issues in Hydrological Modelling*, eds J. D. Kalma and M. Sivapalan, John Wiley and Sons, Chichester, 159–180.

Bates, P. D., 1992. *Finite element modelling of floodplain inundation*. Unpublished PhD Thesis, University of Bristol, 220pp.

Bates, P. D. and Anderson, M. G., 1993. A two dimensional finite element model for river flow inundation. *Proceedings of the Royal Society of London, Series A*, **440**, 481–491.

Bates, P. D., Anderson, M. G., Baird, L., Walling, D. E. and Simm, D. E., 1992. Modelling floodplain flows using a two dimensional finite element model. *Earth Surface Processes and Landforms*, **17**, 575–588.

Bates, P. D., Anderson, M. G. and Hervouet, J.-M., 1994. Computation of a flood event using a two dimensional finite element model and its comparison to field data. In *Modelling Flood Propagation over Initially Dry Areas*, eds P. Molinaro and L. Natale, American Society of Civil Engineers, New York, 243–256.

Bates, P. D., Anderson, M. G. and Hervouet, J.-M., 1995. An initial comparison of two 2-dimensional finite element codes for river flood simulation. *Proceedings of the Institute of Civil Engineers: Water, Maritime and Energy*, **112**, 238–248.

Beven, K. J., 1989. Changing ideas in hydrology: the case of physically based distributed models. *Journal of Hydrology*, **105**, 157–172.

Beven, K. J., 1995. Linking parameters across scales: subgrid parameterizations and scale dependent hydrological models. In *Scale Issues in Hydrological Modelling*, eds J. D. Kalma and M. Sivapalan, John Wiley and Sons, Chichester, 263–281.

Binley, A. and Beven, K., 1993. Three dimensional modelling of hillslope hydrology. In *Terrain Analysis and Distributed Modelling in Hydrology*, eds K. J. Beven and I. D. Moore, John Wiley and Sons, Chichester, 107–119.

Brookes, A. N. and Hughes, T. J. R., 1982. Streamline Upwind/Petrov Galerkin formulations for convection dominated flows with particular emphasis on the incompressible Navier–Stokes equations. *Computer Methods in Applied Mechanics and Engineering*, **32**, 199–259.

Carey, G. F. and Jiang, B. N., 1986. Element-by-element linear and non-linear solution schemes. *Communications in Applied Numerical Methods*, **2**, 145–153.

Chandler, J. H., Cooper, M. A. R. and Robson, S., 1989. Analytical aspects of small format surveys using oblique aerial photographs. *Journal of Photographic Science*, **37**, 235–240.

Clarke, R. T., 1994. *Statistical Modelling in Hydrology*, John Wiley and Sons, Chichester, 412pp.

Culling, W. E. H. and Datko, M., 1987. The fractal geometry of the soil covered landscape. *Earth Surface Processes and Landforms*, **12**, 369–385.

Davis, G. D. and Mallinson, G., 1976. An evaluation of upwind and central difference approximations by a study of recirculating flow. *Computers and Fluids*, **4**, 29–43.

Fryer, J. G., Chandler, J. H. and Cooper, M. A. R., 1994. On the accuracy of heighting from maps and aerial photographs: implications for process modellers. *Earth Surface Processes and Landforms*, **19**, 577–583.

Gambolati, G. and Volpi, G., 1979. A conceptual deterministic analysis of the kriging technique in hydrology. *Water Resources Research*, **15**, 625–629.

Grayson, R. B., Blöschl, G., Barling, R. D. and Moore, I. D., 1993. *Process, Scale and Constraints to Hydrological Modelling*. IAHS Publication 211, 83–92.

Hervouet, J.-M., 1989. Comparison of experimental data and laser measurements with computational results of the TELEMAC-2D code (shallow water equations). In *Computational Modelling and Experimental Methods in Hydraulics (HYDROCOMP 89)*, eds C. Maksimovic and M. Radojkovic, Elsevier, Amsterdam, 237–242.

Hervouet, J.-M., 1992. Element-by-element methods for solving shallow water equations with

FEM. *Proceedings of the 9th International Conference on Computational Methods in Water Resources*, Denver, USA.

Hervouet, J.-M., 1993. Validating the simulation of dam breaks and floods. *Advances in Hydro-Science and Engineering,* Volume 1 Part A, Washington, USA, 754–761.

Hervouet, J.-M. and Janin, J.-M., 1994. Finite element algorithms for modelling flood propagation. In *Modelling Flood Propagation over Initially Dry Areas*, eds P. Molinaro and L. Natale, American Society of Civil Engineers, New York, 102–113.

Huang, J. and Turcotte, D. L., 1989. Fractal mapping of digitised images: application to the topography of Arizona and comparisons with synthetic images. *Journal of Geophysical Research*, **94**, 7491–7495.

Hughes, J. P. and Lettenmaier, D. P., 1981. Data requirements for kriging estimation and network design. *Water Resources Research*, **17**, 1641–1650.

King, I. P., 1970. An automatic reordering scheme for simultaneous equations derived from network systems. *International Journal of Numerical Methods in Engineering*, **2**, 523–533.

King, I. P. and Norton, W. R., 1978. Recent applications of RMAs finite element models for two dimensional hydrodynamics and water quality. In *Proceedings of the Second International Conference on Finite Elements in Water Resources*, eds C. A. Brebbia, W. G. Gray and G. F. Pinder, Pentech Press, London, 81–99.

Knight, D. W. and Shiono, K., 1990. Turbulence measurements in a shear layer region of a compound channel. *Journal of Hydraulic Research*, **28**, 175–196.

Krabill, W. B., Collins, J. G., Link, L. E., Swift, R.N and Butler, M. L., 1984. Airborne laser topographic mapping results. *Photogrammetric Engineering and Remote Sensing*, **50**, 685–694.

Kwoh, L. K., Chang, E. C., Heng, W. C. A. and Lim, H., 1994. DTM generation from 35–day repeat pass ERS-1 interferometry. *Proceedings of the International Geoscience and Remote Sensing Symposium*, 2288–2290.

Lane, S. N., Chandler, J. H. and Richards, K. S., 1994a. Developments in monitoring and modelling small-scale river bed topography. *Earth Surface Processes and Landforms*, **19**, 349–368.

Lane, S. N., Richards, K. S. and Chandler, J. H., 1994b. Application of distributed sensitivity analysis to a model of turbulent open channel flow in a natural river channel. *Proceedings of the Royal Society of London, Series A*, **447**, 49–63.

Marchuk, G. I., 1975. *Methods of Numerical Mathematics*. Springer-Verlag, New York, 316pp.

Moore, I. D., Grayson, R. B. and Ladson, A. R., 1993. Digital terrain modelling: a review of hydrological, geomorphological and biological implications. In *Terrain Analysis and Distributed Modelling in Hydrology*, eds K. J. Beven and I. D. Moore, John Wiley and Sons, Chichester, 7–34.

Pinder, G. F. and Gray, W. G., 1977. *Finite Element Simulation in Surface and Subsurface Hydrology*, Academic Press, New York, 295pp.

Quinn, P., Beven, K., Chevallier, P. and Planchon, O., 1993. The prediction of hillslope flow paths for distributed hydrological modelling using digital terrain models. In *Terrain Analysis and Distributed Modelling in Hydrology*, eds K. J. Beven and I. D. Moore, John Wiley and Sons, Chichester, 63–83.

Shiono, K. and Knight, D. W., 1991. Turbulent open channel flow with variable depth across the channel. *Journal of Fluid Mechanics*, **222**, 617–646.

Simm, D. J., 1993. *The deposition and storage of suspended sediment in contemporary floodplain systems: a case study of the River Culm, Devon.* Unpublished PhD Thesis, University of Exeter.

Sloan, S. W., 1986. An algorithm for profile and wavefront reduction of sparse matrices. *International Journal for Numerical Methods in Engineering*, **23**, 239–251.

Sloan, S. W. and Randolph, M. F., 1983. Automatic element reordering for finite element analysis with frontal solution schemes. *International Journal for Numerical Methods in Engineering*, **19**, 1153–1181.

Turcotte, D. L., 1992. *Fractal Models in Geology and Geophysics*, Cambridge University Press, Cambridge, 221pp.

Walling, D. E., Bradley, S. B. and Lambert, C. P., 1986. *Conveyance Loss of Suspended Sediment Within a Floodplain System.* IAHS Publication 159, 119–132.

Walling, D. E., Quine, T. A. and He, Q., 1992. Investigating contemporary rates of floodplain sedimentation. In *Lowland Floodplain rivers: a Geomorphological Perspective*, eds G. E. Petts and P. A. Carling, John Wiley and Sons, Chichester, 165–184.

14 The Use of Digital Terrain Modelling in the Understanding of Dynamic River Channel Systems

STUART N. LANE

Department of Geography, University of Cambridge, UK

ABSTRACT

This paper reviews the use of digital terrain modelling in the understanding of dynamic river channel behaviour using data from a case study reach of actively braiding proglacial stream in the Swiss Alps. The paper begins by identifying the importance of studies of channel form to the fluvial geomorphologist and by reviewing traditional approaches to its measurement and representation. There has been a historical emphasis upon river channel cross-sections to the detriment of full three-dimensional representation of river channel form and the wider topographical setting of which it forms a part. Use of digital elevation models (DEMs) based upon the Delaunay triangulation recognises this explicitly, making data collection more flexible, and resulting in better landform representation (e.g. the inclusion of breaks of slope) and easier means of interpretation and analysis than would otherwise be the case. Assessment of surface quality provides important information concerning both traditional and DEM-based methods. It suggests that close attention has to be given to: (i) the incorporation of breaks of slope; (ii) the problem of edge effects; (iii) the impact of point density and distribution upon topographic representation; and (iv) temporal resolution. However, each of these needs evaluation with respect to both the method used to collect the data and the use to which the data will be put. Application of digital terrain modelling methods to river channel research can be classified into: (i) the direct use of a DEM, and (ii) its indirect use to determine derivative attributes from the model which may have a direct link with process. Both (i) and (ii) may be applied to just a single DEM, or to the DEM of difference obtained from comparing two consecutive DEMs, and typical applications are illustrated. The paper concludes by identifying those issues that need further research.

INTRODUCTION: THE IMPORTANCE OF FORM TO THE FLUVIAL GEOMORPHOLOGIST

Fluvial geomorphology has long recognised the importance of river channel form as a control on river channel processes, and hence changes in form. An understanding of how form controls process, and results in changes in channel form through time, is central to attempts to understand and manage both small- and large-scale river

Landform Monitoring, Modelling and Analysis. Edited by S. N. Lane, K. S. Richards and J. H. Chandler.
© 1998 John Wiley & Sons Ltd.

systems. At the drainage basin scale, recent research has made use of catchment-scale digital terrain models to illustrate how the nature and precise location of geomorphological processes, including particular river channel types, is conditioned by drainage basin morphology (e.g. Montgomery *et al.*, Chapter 11). Applications of similar techniques at the within-reach scale are much less common, despite increased investigation at this scale (e.g. Dietrich and Smith, 1984; Ashworth and Ferguson, 1986; Bridge, 1993) and the clear recognition of the importance of distributed morphological parameters as a control on geomorphological processes. For instance, Ashworth and Ferguson (1986) and Ferguson *et al.* (1989) have described topographically induced flow divergence and convergence, Ashmore *et al.* (1992) have discussed the importance of confluence morphology as a control on the form and strength of secondary circulation, the latter decreasing in strength in the downstream direction from the maximum location of confluence scour, and Dietrich and Whiting (1989) have described topographical influences on the flow field associated with alternate bar forms, where changes in downstream velocity magnitudes could be associated with both rapid shoaling and increases in channel depth. Morphological controls on sediment transport patterns have also been described (e.g. Ferguson *et al.*, 1992).

These processes will be responsible for changes in river channel form in a manner that reflects both the magnitude–frequency regime and the sensitivity to change of the environment under consideration. In particularly dynamic systems, the changes may be continual, with a spatially distributed character: spatial variation in bed elevation will cause spatial variation in process and hence spatially variable changes. The latter are further complicated by mass continuity which means that changes in one location will influence future changes in other locations. Recent research has emphasised the importance of history in this process, with morphology exerting a conditioning influence upon patterns of river channel change (Lane and Richards, 1997). For instance, river channel avulsion into a new section of floodplain will leave a morphological legacy in the old channel that may condition future interactions between process and morphology (Figure 14.1). Implicit to the idea of morphological conditioning is the importance of viewing the river channel as one part of a continually varying morphology that includes surrounding topography.

In addition to a current emphasis on the substantive importance of morphology as both a control and consequence of river channel processes, there is a growing recognition of the utility of morphological information as a means of understanding river channel processes. There are three reasons for this.

(i)   Observation of river channel morphology and its changes through time has provided important information regarding the nature of river channel change. For instance, flume and field observations have allowed Ashmore (1991) and Ferguson (1993) to classify styles of channel change, and make tentative statements as to the likelihood of their occurrence.

(ii) Process-based studies have only just begun to attempt to simulate *numerically* the nature of the interaction between river channel form and process (e.g. Nelson and Smith, 1989; Andrews and Nelson, 1989; Lane *et al.*, 1995a,b; Murray and Paola, 1994), a method that involves coupling river channel topography to two- and three-dimensional models of river channel flow and sediment transport processes.

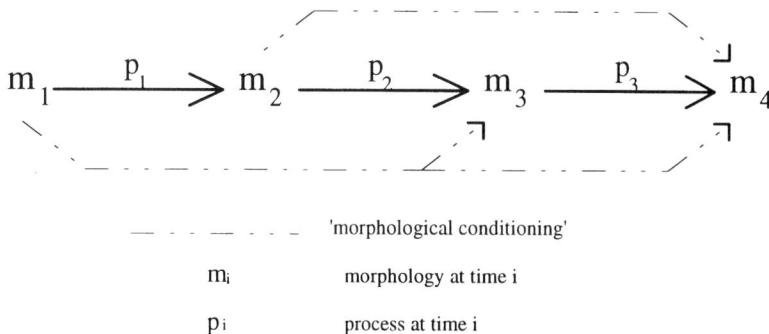

'morphological conditioning'

m_i morphology at time i

p_i process at time i

Figure 14.1 Form–process feedback and "morphological conditioning"

Such a methodological development is important because it may provide: (a) an alternative methodology for understanding the dynamics of river channel change, where the effects of combinations of different boundary conditions on river channel development may be explored through numerical simulation; and (b) the means of predicting short and medium time scale river channel dynamics in response to externally imposed boundary conditions. However, in the first instance, such models require measurement of the distributed form of the river channel, as well as other parameters such as perimeter sedimentology and river channel discharge.

(iii)   It may be the case that better estimates of certain river channel processes may come from the measurement of river channel form and its change through time, than from direct measurement of the processes themselves (Carson and Griffiths, 1989). For instance, spatially distributed measurement of bedload transport rates remains confounded by the problems of adequately characterising transport rates and fluctuations at single points (Gomez et al., 1990), before they are even extended to the mapping of spatial variations. One alternative to this problem is to estimate time-integrated bedload transport rates from observed changes in river channel form (Hubbell, 1964; Davies, 1987; Ferguson and Ashworth, 1992; Goff and Ashmore, 1994; Lane et al., 1995b), a method which Ashmore and Church (1995) describe as a "new paradigm for study".

It follows from the above that distributed information on river channel form is central to current thinking in fluvial geomorphology, at least in terms of attempts to understand the behaviour of dynamic river systems. The aim of this paper is to illustrate the contribution that digital terrain modelling methods can make to this research. The paper will: (i) review traditional means of monitoring such rivers, emphasising the advantage that comes from a fully distributed approach; (ii) consider a range of methodological issues that arise from the use of morphological information in this context; (iii) illustrate the application of such methods to a particular case study; and (iv) conclude by identifying those areas that merit further research.

The substantive and methodological examples herein are based upon field data collected from a reach of gravel-bedded proglacial stream in the Swiss Alps in the summers of 1992 and 1995. Morphological data were acquired photogrammetrically for exposed areas and using total-station survey for sub-aqueous zones. The methodology is detailed in full in Lane (1994).

TRADITIONAL APPROACHES: FROM CROSS-SECTIONS TO DIGITAL ELEVATION MODELS

Perhaps one of the legacies of hydraulic geometry (Leopold and Maddock, 1953) is that descriptions of river channel morphology are provided by monitoring one or a number of river channel cross-sections through time (e.g. Ferguson and Ashworth, 1992). This may be combined with some form of field mapping of river boundaries to provide information on planform change. If longer time series of channel change are required, it has proved possible to use aerial photographs for the same purpose (e.g. Laronne and Duncan, 1992). These methods need consideration in terms of the extent to which they provide information of both appropriate quality and appropriate nature.

In terms of data quality, it is possible to distinguish between the quality of individual points (the x,y positioning of elevation), and the quality of the full set of information acquired to represent the landform surface. The latter is a product of the former, but includes the spatio-temporal resolution associated with the field mapping methods used. The spatio-temporal resolution will often involve a trade-off between three factors: (i) the extent to which the techniques used limit the area from which information can be obtained; (ii) the density of information acquisition possible within this area; and (iii) the rate at which it is possible to resurvey an area to detect topographic change through time. The relative importance of these factors will be a function of field environment. For instance, time spent collecting data over a wider area or at a greater density will be at the expense of reduced frequency of return to the same points to measure landform change. The more rapidly the landform is changing the more serious this problem becomes.

Table 14.1 provides a summary of recent attempts to monitor river channel topography in dynamic glacial and glacier-fed streams with reference to these three variables. A number of points arise. First, research of this type has been dominated by repeat levelling (for height information), stadia tacheometry (to position those heights along a transect) and plane table mapping. Levelling generally requires hand booking in the field, which decreases the rate at which it is possible to obtain information. Subsequent data analysis is relatively slow, although electronic levels that automatically record information in the field can increase both the booking and the analysis rates. Positioning of heights has made use of stadia tacheometry, where the simple geometry of stadia hairs is used to calculate distance. Assuming a quality of elevation determination of ±1 cm, this means that the quality of distance determination could be as poor as ±1 m (the distance of an elevation sighting is $100\Delta x$, where x is the difference between the upper and lower stadia hairs). Reference to a standard surveying text book (Bannister et al., 1992) finds no mention of the use of such distance measurements to establish the position of heights, presumably because of their poor quality. Rather, distances from an instrument are determined using tape measurements to pegged out points whose

Table 14.1 A summary of recent attempts to monitor river channel form in dynamics alluvial channels

Researchers	River	Type	Method	Area extent (m²)	Spatial density of points (points/m²)	Frequency of resurvey (h)
Ashworth and Ferguson (1986)	Lyngsdalselva, North Norway	Proglacial, braided	Levelling Plane-table mapping	1150	0.2	24
Lane (1991)	Haut Glacier d'Arolla, Switzerland	Proglacial, braided	Levelling Plane-table mapping	960	0.1	48
Ferguson *et al.* (1992)	Sunwapta, Canada	Glacier-fed, braided	Levelling Plane-table mapping	499	0.2	24
Ferguson and Ashworth (1992)	White River, northwest USA	Glacier-fed locally braided	Levelling Plane-table mapping	1980	0.2	72
Goff and Ashmore (1994)	Sunwapta, Canada	Glacier-fed, braided	Levelling Oblique photography	700	0.1	24

elevations alone are determined by levelling. Levelling also requires instrument checks to detect any misalignment of cross-hairs with the central axis of the telescope (the collimation error).

Second, there has been little consideration of the effects of propagation of errors on the derived data, such as estimation of volumes of erosion and deposition (e.g. Ferguson and Ashworth, 1992; Goff and Ashmore, 1994). Errors in these estimates will be inevitable owing to the data quality problems described above, but will also result from sampling design. Their magnitude will reflect the level of spatial variability in topographic form and the rate at which it is changing. Considerable error can arise from staff positioning on the river bed, essentially a sampling problem. Gravel-, cobble- and boulder-bedded streams may be have local elevation variations associated with either the grain or bedform scale, or both. Physical constraints (both time and the dimensions of the staff base) prevent sampling at a density that can describe these scales. This results in a form of aliasing (Shannon, 1949), characteristic of time series analysis, but in this case in the spatial domain, where sampling at a lower frequency than is contained in the true surface has consequences for sample-based estimates of properties of that surface. This problem is reinforced when considering the typical emphasis of sampling upon the cross-stream variability in river channel form (1 m spacing between points) rather than downstream variability in form (the smallest cross-sectional spacing is 5 m, reported by Ashworth and Ferguson (1986), Ferguson et al., (1992) and Ferguson and Ashworth (1992)). Although the structure of a river channel bed naturally implies greater morphological variation in the cross-stream sense, thus justifying the higher spatial resolution in the cross-stream direction, there is growing recognition of the importance of sedimentological structures on river channel processes which have a distinctive downstream variation (e.g. Naden and Brayshaw, 1987; Hassan and Reid, 1991). These imply downstream as well as cross-stream changes in bed elevation and this necessarily requires increased emphasis on monitoring downstream morphological variation.

Third, morphological changes can only be detected if the temporal frequency of resurvey is commensurate with the rate at which those morphological changes occur, which in turn must be related to the rate of change of governing processes. In the rivers described in Table 14.1, the best reported frequency of resurvey is 24 h, with levelling undertaken during periods of constant stage, generally close to the discharge peak (Ferguson and Ashworth, 1992). On such a time scale, estimates of erosion and deposition may be obscured by the effects of temporal variation in morphological change. Most of the catchments described will be dominated by diurnal discharge hydrographs in which the discharge may rise two- or three-fold during a 24 h period (Richards et al., 1992). River reaches with such regimes may exhibit a complicated pattern of erosion and deposition *within* the diurnal cycle, in addition to diurnal variation in sediment supply effects, partly related to discharge variation, but also connected with other upstream processes. Thus 24 h surveys may allow day-to-day comparison of the net change in the state of the reach, such as whether over a few days it is aggrading or degrading (e.g. Goff and Ashmore, 1994). However, these time-integrated estimates of erosion and deposition are often related to instantaneous measures of process such as within-channel flow and sediment transport patterns (e.g. Ferguson and Ashworth, 1992; Goff and Ashmore, 1994). It is only possible to make such links when the process measurements are directly linked to the morphological changes with which they are associated, and not the morphological changes defined by the sampling strategy adopted. This requires: (i) improvement in the rate of acquisition of topographic information; and (ii) assessment of the need for such improvements so as to allow effective sampling strategies to be designed.

Finally, the emphasis upon repeat levelling necessarily implies a commitment to return to the same cross-sections. Research design is therefore constrained by the initial evaluation of the field site rather than being able to evolve as the landform does. Better landform representation will arise from inclusion of topographically defined sampling (e.g. breaks of slope) rather than from monumented cross-sections, which at best can only be guaranteed to effectively represent initial topographic variability.

Progress in the areas outlined above would seem to come from the recognition of the three-dimensional nature of river channel form, its dynamic nature, and the need for distributed information on land surface elevation. The use of a digital elevation model (DEM)-based approach has a number of important advantages: (i) it places fewer constraints upon the monitoring process as data collection need not be restricted to cross-sections, allowing topographically defined data collection and hence resulting in better surface representation; (ii) with growing calls for more system-scale studies of river channel behaviour, in which river channels are situated within their wider geomorphological setting (e.g. Montgomery et al., Chapter 11), DEM-based approaches allow the extension of data collection to hillslope zones; (iii) three-dimensional visualisation of topography and topographic change becomes possible, allowing qualitative statements regarding the relationship between river channel form and process; (iv) specifying river channel topography as a DEM allows the application of the large body of DEM-based terrain analysis methods (e.g. Evans, Chapter 6), which may become the basis of alternative means of estimating river channel behaviour; and (v) DEMs, after interpolation, can provide fully distributed topographic data which is a critical input to distributed flow and sediment transport models. The remainder of this paper will explore methodological issues associated with the use of a DEM-based approach in fluvial geomorphology and will illustrate how DEMs may be used to address specific fluvial problems.

METHODOLOGICAL ISSUES: DATA QUALITY AND ITS IMPLICATIONS

The term "quality" is used in this paper to refer to error that arises in the representation of a terrain surface. In the past, there has been extensive reference in the digital terrain model (DTM) literature to DTM accuracy (e.g. Ackermann, 1978; Li, 1992; Shearer, 1990), implying the extent to which a DTM surface is representative of the real surface. However, the use of the term "accuracy" requires caution. In practice it is more appropriate to refer to terrain surface quality, which is controlled by the precision, reliability and accuracy of individual points (see

Cooper, Chapter 2), and the spatial coverage of those points used to reconstruct the surface.

The quality of a DTM will be the product of two controls (Ackermann, 1978): (i) the means of data acquisition; and (ii) the nature of data processing. The data acquisition method should provide information that is of an appropriate density, distribution and quality for the spatial and temporal resolution that the model application requires. In practice, these parameters are also controlled by the nature of the terrain (Li, 1992), as more irregular terrain will require a greater density of points, and/or a carefully designed point distribution (e.g. break of slope inclusion). The nature of data processing must consider the way the three-dimensional form of the landform surface is reconstructed on the basis of the distributed height information. This section will not focus on issues surrounding the quality of individual points, as Cooper (Chapter 2) contains a useful discussion of controls on point quality in general, whilst Lane (1994) illustrates the different means by which it may be possible to determine point quality in the case of data acquired from river channel survey (both photogrammetrically and tacheometrically). This discussion will focus upon the surface quality as a whole.

Assessing terrain surface quality is difficult because the "real" surface is rarely known, and has to be based upon exploring the effects of sampling from an existing surface on either surface parameters or surface derivatives. Three issues surrounding surface quality must be explored: (i) the quality of the surface fitted to the data; (ii) the effects of point density on DTM quality; and (iii) the effects of point distribution on DTM quality.

The Quality of the Surface Fitted to the Data

The quality of the surface fitted to the data relates to the effectiveness with which the data points are used to construct a surface. With randomly distributed data, the Delaunay triangulation offers the most effective solution (see McCullagh, Chapter 5). Two aspects of surface fitting will be central to the evaluation of surface quality in river channel research: (i) the performance of surface fitting algorithms in the presence of breaks of slope (e.g. bank tops and toes); and (ii) edge effects in the boundaries of the study area.

Research has shown that effective representation of river channel morphology requires explicit inclusion of breaks of slope (Lane et al., 1994). In this study, in addition to randomly distributed point data from both sub-aqueous and exposed areas, photogrammetric data acquisition included water edge, bank top and bank toe information. These were incorporated into the surface as breaklines across which the triangulation algorithm was not permitted to place triangles. The importance of breakline information can be illustrated by comparing volumes of erosion and deposition calculated by comparing two consecutive DEMs, one comparison using DEMs with breaklines and the other using DEMs without breaklines, revealing that a failure to include such breaklines could have a substantial effect on calculated volumes (Table 14.2).

The second problem concerns edge effects occurring through triangulation along the boundaries of the area of data collection. This problem can be reduced in two

Table 14.2 An assessment of the effects of incorporating breaklines on surface representation; based on Lane *et al.* (1994). A small sub-area of the full surfaces was used for comparison

Comparison	Erosion (m^3)	Deposition (m^3)	Net change (m^3)
13 July compared to 16 July – without breaklines	5.03	10.88	−5.85
13 July compared to 16 July – with breaklines	2.09	13.27	−11.17
Percent difference	−58.4	+18.0	−91.1

Table 14.3 An assessment of the effects of edge effects on surface representation; based on Lane *et al.* (1994)

Comparison	Erosion (m^3)	Deposition (m^3)	Net change (m^3)
28 June compared to 5 July – with edge effects	35.78	27.85	7.93
28 June compared to 5 July – with edge effects removed	30.17	26.09	4.08
Percent difference	−15.7	−6.3	−48.5

ways. Ideally, data can be collected from areas that lie outside the area of immediate interest so moving edge effects to an area where representation is less critical. Secondly, once a surface has been triangulated, visualisation of the triangles and the data points used in generating the surface concurrently can be used to identify and delete triangles that will be associated with an edge effect. Again, the importance of this for individual surface comparison is illustrated by a volume comparison, this time for the epochs of 28 June and 5 July (Table 14.3). Removal of edge effects tends to reduce the amount of apparent volume change, although the differences are not as great as those relating to the use of breaklines.

Point Density and Surface Quality

The representation of channel form is a product of both the spatial density and the distribution of surface points. A continuum of within-channel roughness elements exists, related to the scale of feature under consideration. These forms may be loosely described as grain, micro-form and macro-form roughness (grain-scale roughness corresponds to that associated with individual particles; micro-form roughness corresponds to that associated with the grouping of particles into sedimentological structures; macro-form roughness corresponds to larger-scale channel geometry, such as meander bends and barforms; (Prestegaard, 1983; Clifford *et al.*, 1992). For perfect representation of river channel topography at the grain scale, it is necessary to measure the position and dimensions of every clast on the river bed surface. Earlier research (Lane *et al.*, 1994), based upon intensive

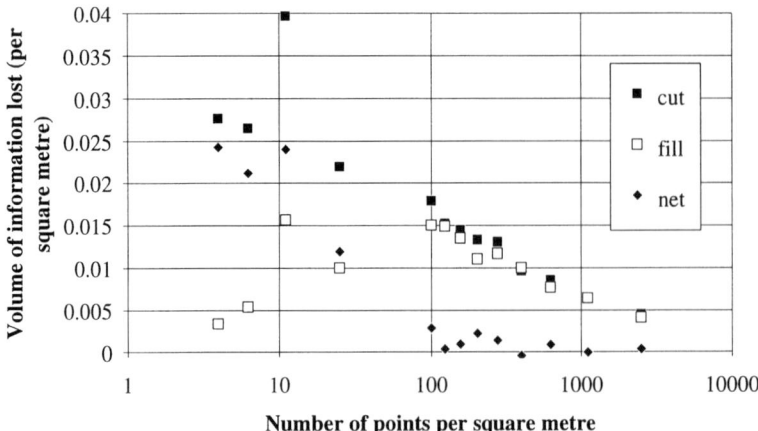

Figure 14.2 The effects of a progressive reduction in grid-size spacing (expressed as number of points per square metre) upon the volume of information recorded in the surface (expressed as the net, cut and fill differences per unit area obtained by comparing the thinned surface from the full density surface)

monitoring using total-station survey, suggested that there would be a rapid reduction in surface quality with point densities between 4.0 and 2.0 points per square metre, corresponding to average point spacings of between 0.5 and 0.7 m. At spacings of less than *c.* 0.3 m (greater than 3.5 points per square metre) increasing point density resulted in only small improvements in surface quality. It was argued that further significant reductions in error could not be expected until the micro-scale bedform elements are recorded, requiring a spacing of 0.01 to 0.05 m (or 4000 to 10 000 points per square metre) for the grain size in this stream.

The problem with this method was its dependence upon the amount of infor-mation contained in the original surface: calculations were based upon progressive thinning of the original surface and calculation of the volumes of cut and fill per unit area, obtained by comparing the original surface with the thinned surface. To overcome this problem in part, a DEM of much higher resolution was obtained using automated extraction methods, based upon small-scale, scanned, overlapping photography (Butler *et al.*, in review) acquired using semi-metric cameras mounted on a gantry 2 m above exposed river-bed gravels. The original DEM was thinned using a progressive reduction in grid spacing. As in Lane *et al.* (1994), the thinned DEM was compared to the original DEM to determine volumes of cut, fill and net change per unit area. The results are shown in Figure 14.2.

The results suggest that with point densities of around 100 per square metre or a point spacing of 0.1 m, cut and fill patterns will on average cancel each other out, resulting in zero net change. However, as point densities are reduced further, a systematic bias is introduced, with greater cut than fill in this situation. In the absence of systematic surface structure, the nature of this bias will be dependent upon the location of the grid origin. If systematic structure is present, the bias will reflect the interaction of grid origin with the structure, for each spacing. These results suggest that volumes of information loss are significantly greater than detailed in Lane *et al.* (1994), and this almost certainly reflects: (i) the greater volume of topographic information contained in the original surface used in this study; and (ii) the fact that the thinning process in the Lane *et al.* (1994) study was stratified, retaining critical points during the thinning process (e.g. topographic highs and lows). This last point is a salutary reminder that field sampling design should be defined by the surface being sampled.

These results have been based upon exposed gravels obtained for a particular environment. Research is needed to assess how these results may differ for sub-aqueous gravels, and how the results vary with different grain sizes and bedform type structures. Ultimately, experiments such as these need to be judged in terms of the use to which the acquired information will be put. For instance, they can be used to obtain an estimate of surface "noise" and should be judged with respect to observed "signals" such as observed volumes of erosion and deposition. For lumped estimates the noise can be scaled up to the area of observed change and compared with erosion and deposition estimates. For distributed estimates, it may be used to specify a minimum depth of cut or fill which must be exceeded on a node-by-node basis for the observed change to be deemed significant.

Point Distribution and Surface Quality

Implicit to the above argument was the importance of point distribution for surface quality and this may be central to reducing requisite point densities to levels that are feasible for field data collection. It follows from Table 14.2 that the inclusion of breaks of slope during data collection, as well as during surface reconstruction, will be critical to adequate representation of river bed topography. In the case of automated methods, which tend to be dominated by grid-based sampling, point distribution effects will be related to positioning and orientation of the grid. Provided that data collection is of sufficiently high density, this is unlikely to have a significant effect on surface quality. However, in non-automated methods, whether involving manually digitised photogrammetry or bed survey, random point location, whether in terms of choice of which clast to digitise or where to locate a survey pole on the bed, could have a major effect on surface quality.

This can be assessed by measuring the same area of an epoch to the same density on two separate occasions. This is possible for manually digitised photogrammetric methods but not for bed survey methods, as with the latter there may have been significant erosion or deposition between measurement periods. Repeat measurement of this kind will embody both random and systematic sampling effects (see Cooper, Chapter 2). Random choices of which clast to digitise will be countered partly by systematic operator choice of those clasts which are most likely to improve surface representation.

Figure 14.3 shows the two triangulated surfaces obtained in such an assessment superimposed on one another. There is a clear difference in the planform position of points, although this is difficult to quantify. The two surfaces were compared to obtain a surface of difference (Figure 14.4). Volume comparison of the two surfaces suggested an erosion per unit area of 0.0025 m^3/m^2 and a deposition per unit area

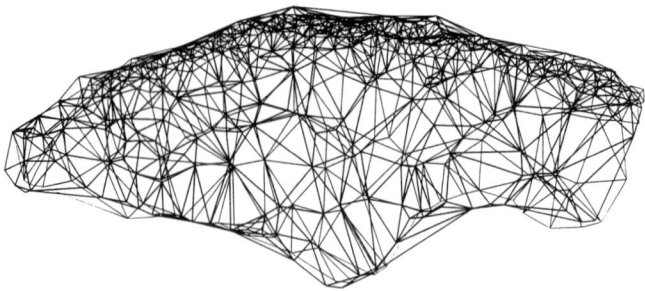

Figure 14.3 The two surfaces obtained from repeat digitisation of the same area containing the same number of points, on two separate occasions, to illustrate the different point distributions

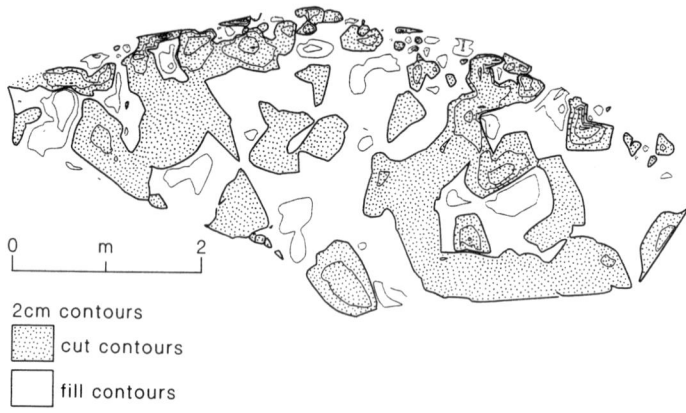

Figure 14.4 The surface of difference used to illustrate the effects of different point densities on surface representation. This was obtained by subtracting two surfaces, covering the same area and with the same number of points, but collected on two separate occasions

of 0.0027 m^3/m^2 which in net terms is 0.0002 m^3/m^2. This was spatially variable and clearly dependent on whether or not a clast digitised in one epoch was detected in the second. This is reflected in Figure 14.4 in localised differences of up to 4 cm. This analysis implies that the amount of surface error might be related to both the sedimentological heterogeneity of a particular area and the typical clast diameter, which in combination might be labelled the surface "micro-topography". With homogeneous small clasts (sand or gravel), the effects of surface error due to point distribution effects are likely to be reduced as compared with more heterogeneous surfaces or surfaces with coarse particle sizes, in which a higher local rate of change of micro-elevation makes point distribution more important. As with point density effects, the magnitude of these errors needs to be evaluated with respect to the use to which the terrain surfaces will be put.

VISUALISATION AND QUANTIFICATION OF VOLUMES OF CHANGE

One of the most useful applications of DEMs involves calculation of DEMs of difference by comparing DEMs obtained at consecutive time periods, and calculation of the associated spatially distributed and total volumes of erosion and deposition. This section will: (i) detail some of the methodological issues associated with deriving DEMs of difference and associated volumes of change; (ii) illustrate how these may be used to understand and interpret river channel dynamics; and (iii) detail how they may be used for the determination of estimates of both one- and two-dimensional bed material transport rates.

Determination of Volumes of Erosion and Deposition from DEMs

DEM comparison, to calculate both DEMs of difference and volumes of change, needs careful consideration, as the manner in which it is undertaken may significantly affect the results obtained. Traditionally, most estimates of erosion and deposition in dynamic river channels have been obtained using cross-section comparison (e.g. Carson and Griffiths, 1989; Ferguson et al., 1992; Goff and Ashmore, 1994) using fixed cross-sections established in the field and repeatedly surveyed through time. The calculation of volumes of erosion and deposition is undertaken using either the trapezoidal or the prismoidal rule for volumes (Bannister et al., 1992). The trapezoidal rule for volumes is often labelled the mean end area method and is based upon the following:

$$V_e = \sum_{i=1}^{n} \frac{\delta_{i-1,i}(E_{i-1} + E_i)}{2}$$

$$V_d = \sum_{i=1}^{n} \frac{\delta_{i-1,i}(D_{i-1} + D_i)}{2} \tag{14.1}$$

$$V_{net} = V_d - V_e$$

where V_e = volume of erosion; V_d = volume of deposition; V_{net} = net change in volume; D = area of cross-section associated with deposition; E = area of cross-section associated with erosion; n = number of cross-sections $\delta_{i,i-1}$ = distance between consecutive cross-sections i and $i-1$.

Equation (14.1) is correct provided the area (of erosion or deposition) of the cross-section midway between cross-section i and cross-section $i-1$ is the mean of the two. A more precise formula makes use of the prismoidal rule for volumes. The volume of a prismoid is given by:

$$V = \frac{D}{6} (A_1 + 4M + A_2) \tag{14.2}$$

where V = volume; D = distance between end cross-sections 1 and 2 with areas A_1 and A_2; M = area of cross-section midway between 1 and 2.

Application of the more precise prismoidal volume formula is, however, not straightforward. The main problem with this formula is that if the end section areas are simply averaged, the formula simplifies to the trapezoidal rule for volumes. Determination of M is a major problem (Bannister *et al.*, 1992). In river channel studies (e.g. Ferguson and Ashworth, 1992), this problem has been overcome through treating each end cross-section as the end area of a prismoid of length D. The dimensions of the cross-section midway between these cross-sections are estimated as the average of the end sections (i.e. averaging the dimensions and calculating the mid-area rather than calculating the mid-area as the average of the end section areas). To apply this formula, Ferguson and Ashworth (1992) determined the average depth of erosion or deposition (d_{i-1}, d_i) in the end cross-sections and the width of each end cross-section (w_{i-1}, w_i) and used the following formula:

$$V = \frac{D}{6} (2w_i d_i + w_i d_{i-1} + w_{i-1} d_i + 2w_{i-1} d_{i-1}) \tag{14.3}$$

The problem with the prismoidal rule for volumes is in its application to irregular sections and the following issues need mention: (i) in equation (14.3) the average channel change (depth of erosion or deposition) across a cross-section is determined in what will be, by definition, an irregular pattern of erosion and deposition; (ii) this problem may be compounded by variability in erosion and deposition *between* cross-sections, the associated error being dependent upon cross-section spacing; and (iii) although there are alternatives to averaging dimensions, such as levelling an additional intermediate cross-section or using consecutive cross-sections as alternate middle and end cross-sections, these are not feasible in dynamic river channels. Additional levelling is limited by time constraints on the field acquisition of levelling data, and using intervening cross-sections would result in a significant increase in the possibility of unmeasured intervening erosion or deposition between cross-sections.

Determination of volumes of change from DEMs avoids many of these problems because the calculation is fully two-dimensional, so reducing the magnitude of the errors that come from using equations (13.2) and (13.3), as well as providing spatially distributed information on erosion and deposition patterns. In the case of two surfaces represented by gridded data with the same grid origin and spacing, surface comparison simply involves subtracting corresponding grid nodes. Triangulation of the resulting surface of difference provides all that is necessary for the visualisation of patterns of erosion and deposition, and the volumes of erosion and deposition can be calculated using simple geometrical formulae.

However, in situations where the data set is irregular, or where the grid origin or spacing are different, such a simple comparison is not possible. There are two alternatives in this situation. The first is to superimpose a grid over each surface and sample the elevations at each grid node. This has the advantage of being a straightforward calculation but it ignores breaks of slope and, if the grid is insufficiently dense, significant errors may be introduced. With increasing grid density, there is a rapid increase in computational time (Lane *et al.*, 1994). The second alternative undertakes an exact comparison of two surfaces, using the triangle geometry associated with both. Each triangle from one surface is projected onto the second surface, the intersection with the surface defining new nodes on the surface of difference. The process is then repeated for the projection of points in the second surface onto the first surface, to provide a second set of nodes of erosion (or deposition). Both sets of nodes are then incorporated into a surface of difference, which contains a number of nodes equal to the number in the first surface plus the number in the second surface. These can be triangulated to allow visualisation of the surface of difference and the calculation of volumes of erosion and deposition during the two periods.

In practice, the procedure is simplified if visualisation of the surface of difference is not required. The total erosion volume and the total deposition volume may be obtained by calculating the volume of each prismoid formed by the triangle on the first surface and the intersection of its vertices with the second surface. This is an exact estimate which, when repeated for all triangles on the first surface, results in total volumes of erosion and deposition and a net volume of change (erosion volume minus deposition volume).

It is possible to assess the effectiveness of a cross-section-based approach by reference to volume calculations based on triangle volumes. For this purpose, volume estimates were calculated from the same area using the mean end area method, with parallel cross-sections located along the river channel. The reach length was divided by the number of cross-sections used in the calculation, minus one, to define the cross-section spacing, and estimates of erosion and deposition were calculated. This was repeated for a number of cross-section spacings. To consider the quality of these calculations, they were compared with the exact amounts of erosion and deposition revealed by the triangle-based volume calculation. The comparison has the advantage of cancelling out the width and length effects (both of which will increase the magnitude of the error in volume calculation), such that no further standardisation is necessary and the amount of information retained can be directly related to cross-section spacing (see Lane (1994) for justification).

The analysis was undertaken for two separate periods to assess the consistency of the results. Figure 14.5 shows the percentage of information retained for each cross-section spacing (the volume of cut, fill or net change as a percentage of the volume of cut, fill or net change revealed by the triangle comparison). This suggests that a cross-section spacing of less than 2 m is required for an estimate of cut or fill to be within 20% of the correct value. This can be compared with previous attempts to use cross-section-based approaches to quantify spatial patterns of erosion and deposition in braided streams of similar size. These have used cross-section spacings as great as 10 m (Table 14.1) in streams of comparable width to that in this study. Such a comparison, however, is not entirely valid as the probability of intervening cut and fill between "end areas" will vary between environments. Similarly, effective design of a sampling strategy (e.g. Ferguson and Ashworth, 1992) may reduce this problem, but this is only feasible if the approach is terrain-based (as opposed to cross-section-based). Further, the choice between cross-section-based and DEM-based means of surface comparison is determined by the form in which the data

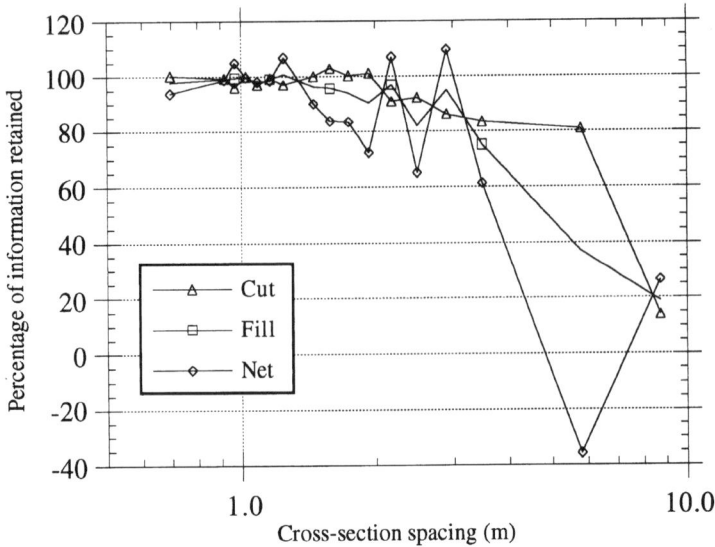

Figure 14.5 The effects of different cross-section spacings on the volume of information retained

have been collected. With cross-sections that have a large spacing, the effects of the uncertainties on volume calculations due to low densities of information will almost certainly be greater than the uncertainties that arise from use of equations (14.2) or (14.3) rather than DEM comparison.

In addition to the manner in which surfaces are compared, it is also important to assess the temporal resolution of the DEMs used in the calculation, and the manner in which this may affect estimates of volumes of change. Only when the topographical information is obtained at closely spaced points in time is it possible to detect the real nature of the river dynamics, and use information obtained from comparing DEMs to estimate volumes of erosion and deposition. Figure 14.6, taken from Lane *et al.* (1994), shows the cumulative estimates of erosion and deposition on the basis of a full data set, with the cumulative amounts of cut and fill obtained by comparing surfaces at a much coarser temporal resolution. The figure suggests that the cumulative amounts of cut and fill are progressively underestimated as the length of time between surfaces is increased.

One of the key reasons for the patterns in Figure 14.6 is that the environment is dominated by a diurnal fluctuation of discharge. The full data set is based upon DEM acquisition at consecutive discharge peaks and troughs, and the improved temporal resolution obtained in this study allowed quantification for the first time of the effects of diurnal discharge fluctuation on erosion and deposition processes in a braided stream (Lane *et al.*, 1996). This is an important point. Figure 14.6 does not imply that it is necessary to *continuously* sample river channels at close intervals. Rather, what is required is design of a fieldwork strategy that allows

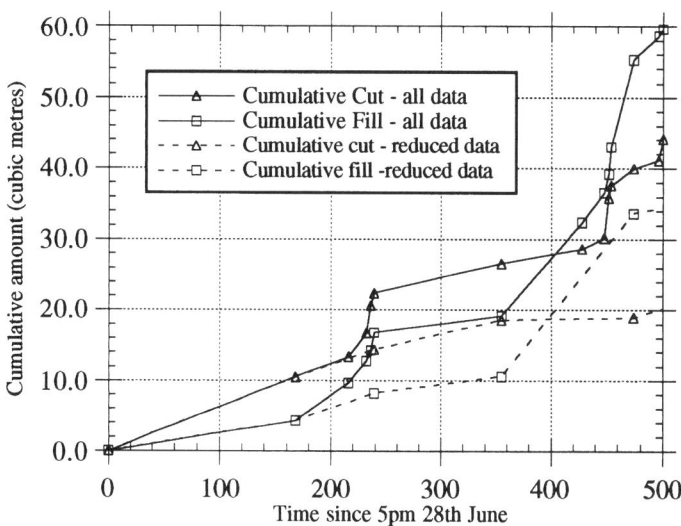

Figure 14.6 Cumulative volumes of erosion and deposition for all epochs and for only a small selection of those epochs. From Lane *et al.* (1994)

flexible data collection in terms of both time (the frequency of resurvey) and space (stratification of sampling to zones of maximum activity), and some observation of when it is necessary to resurvey. With the use of historical data, where one has no observations of river channel behaviour, some attempt to evaluate the representativeness of the acquired data will be required. For instance, this may be by considering the history of process events with respect to erosion thresholds within the study reach and the upstream catchment.

Visualisation of DEMs of Difference and its Use to Further Understand River Channel Dynamics

Lane *et al.* (1996) described the use of DEMs of difference of a sub-reach of river channel in the understanding of channel dynamics. This section will describe the use of DEMs obtained at a lower frequency than this study but from over a larger area for a short period between 5 and 24 July 1992. The 5 July DEM shows a clearly defined divided reach (Figure 14.7a). Three medial bars are apparent (A, B and C on Figure 14.7a), with a narrow channel separating them, and a major and well scoured confluence downstream end of the reach (D on Figure 14.7a). Between the 5 and 13 July, the reach maintained its overall morphology, despite numerous small channel changes. The main true right channel scoured a little way upstream and towards the true right (A on Figure 14.7b) which was concurrent with a small upstream extension of the major medial bar (B on Figure 14.7b). Similar changes have been observed in other glacial streams (e.g. Ashworth *et al.*, 1992), and

(a)

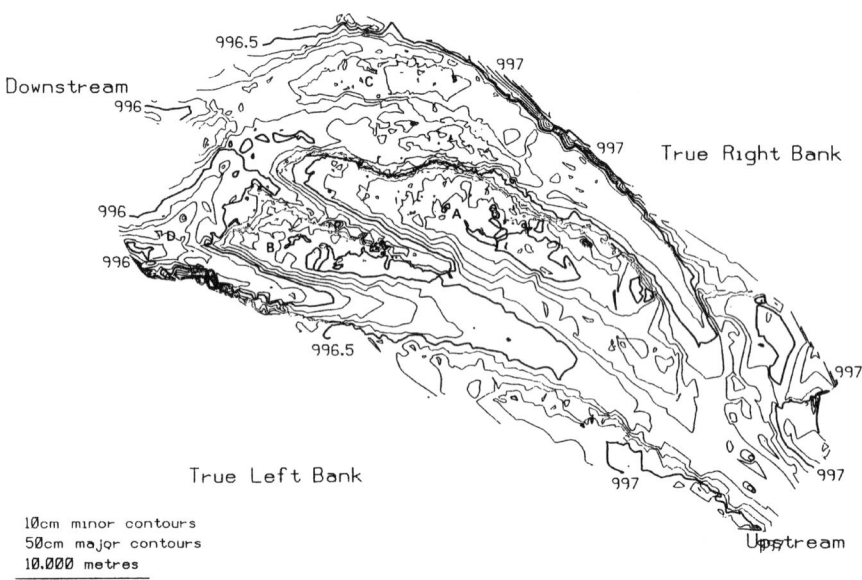

Figure 14.7 A sequence of contour plots of river channel morphology and river channel change for the full reach

Date	Digital terrain model	Surface of difference between date and next date
5 July	a	b
13 July	c	d
18 July	e	f
24 July	g	

associated with trapping of coarse material on the bar head. In this situation, it resulted in flow concentration and exacerbated scour to the true right of this medial bar. Further downstream in the true right channel a new medial bar had formed, with scour on either side, resulting in some bank recession on the main medial bar (C on Figure 14.7b). There was some scour of the small channel to the true left of the main medial bar, notably at its upstream end (D on Figure 14.7b), and fill in the major confluence at the downstream end of the reach (E on Figure 14.7b).

Due to the spatially limited nature of change during this period, the topography of 13 July (Figure 14.7c) was largely unchanged as compared to that of 5 July (Figure 14.7a) except for the shift in position of the minor medial bar in the true right channel. However, between the 13 and 18 July, and accompanying a rise in discharge, major changes in the reach-scale morphology are evident. The true right of the channel filled considerably (A on Figure 14.7d). There was some minor scour on the true left of this channel, with bank recession along the margin of the major

(b)

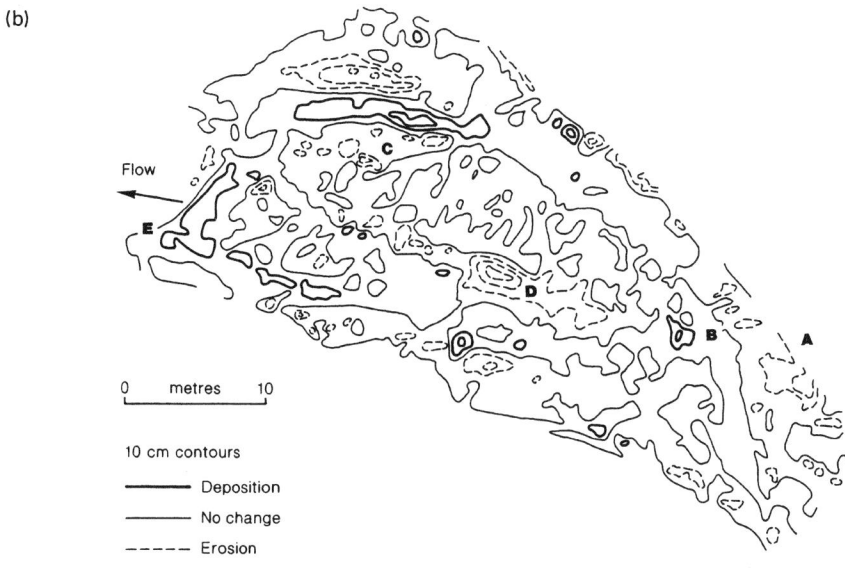

Flow

0   metres   10

10 cm contours

———— Deposition

———— No change

----- Erosion

(c)

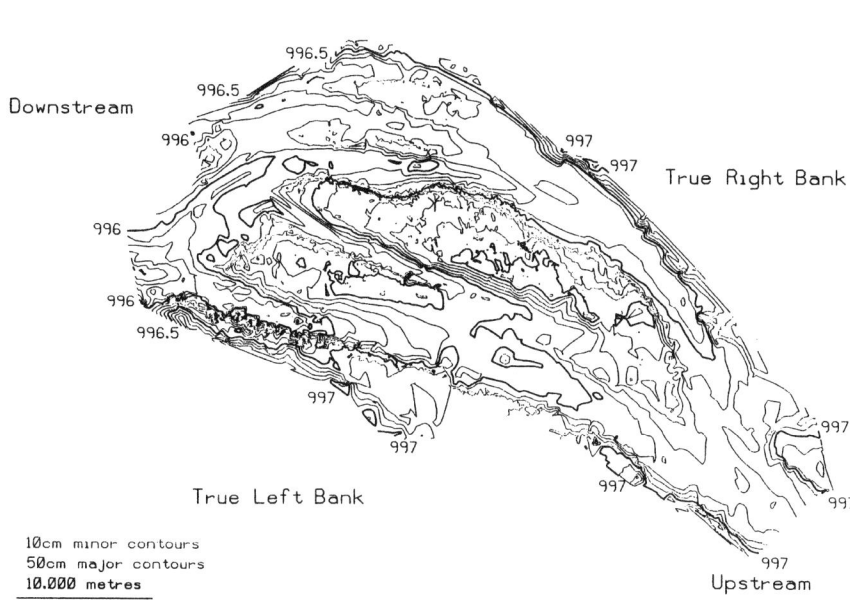

Downstream

996.5

996.5

996

996

996

996.5

997

997

True Left Bank

997

True Right Bank

997

997

997

997

997

Upstream

10cm minor contours
50cm major contours
10.000 metres

Figure 14.7   *(continued)*

(d)

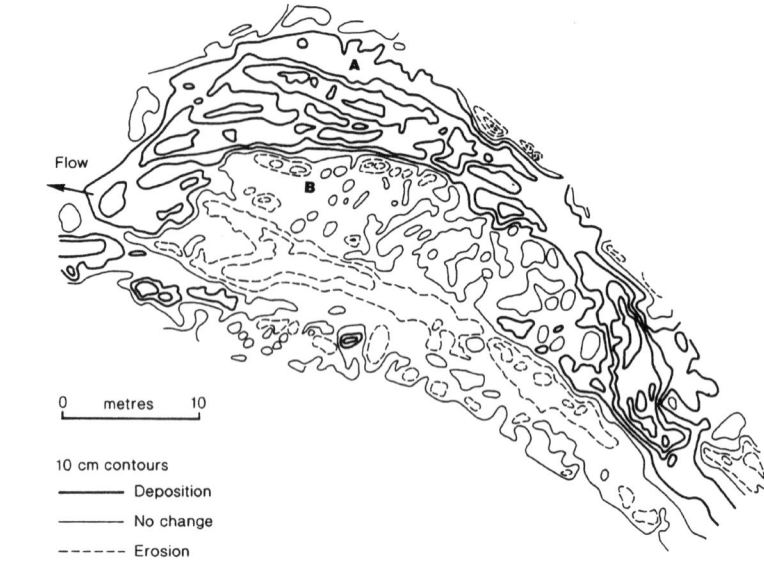

0   metres   10

10 cm contours
———————— Deposition
———————— No change
– – – – – Erosion

(e)

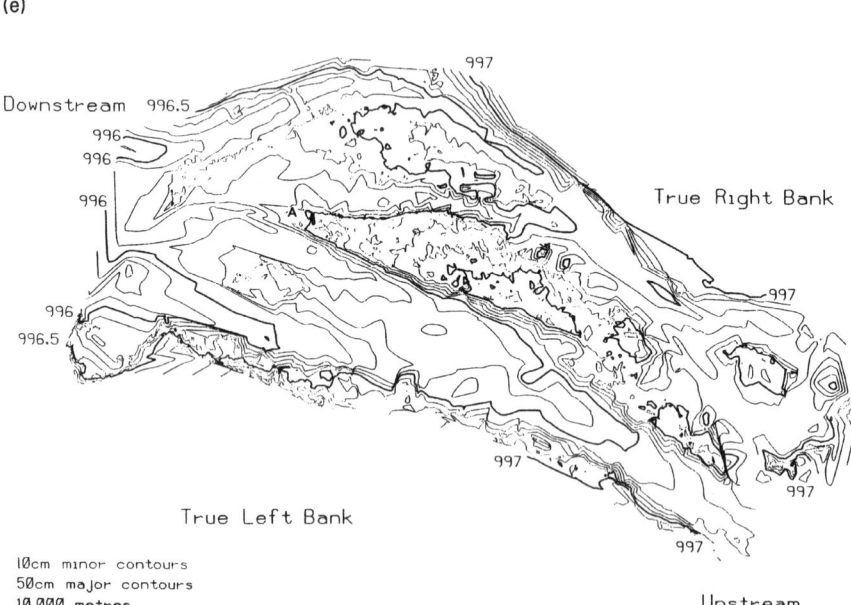

Downstream   996.5

997

996
996

996

True Right Bank

997

996
996.5

997

997

True Left Bank

997

10cm minor contours
50cm major contours
10.000 metres

Upstream

Figure 14.7   (*continued*)

(f)

Flow

0   metres   10

10 cm contours

——————— Deposition
——————— No change
– – – – – Erosion

(g)

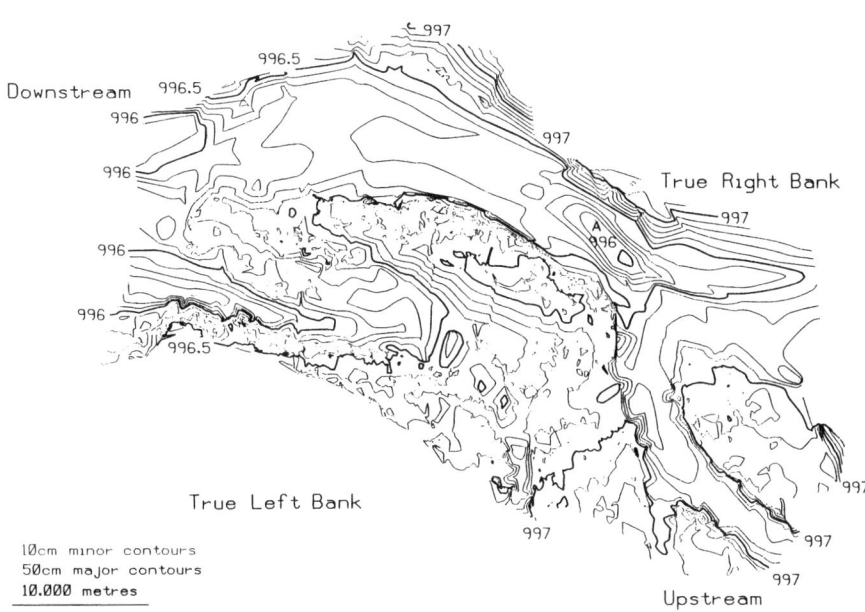

Downstream

996.5

996.5

996

996

996

996

996.5

997

997

True Right Bank

997

997

A
996

997

True Left Bank

997

997

997

Upstream

10cm minor contours
50cm major contours
10.000 metres

Figure 14.7   (continued)

medial bar (B on Figure 14.7d), but most of the scour was in the true left of the reach, and particularly on the true left side of the medial bar. These changes are reflected in the contour plot of the 18 July DEM (Figure 14.7e) with upstream extension of the 996.5 m contour in the true left channel and downstream migration of the 996.0 and 996.5m contours in the true right channel. The erosion of the downstream end of the major medial bar is also apparent (A on Figure 14.7e).

The changes between 18 July and 24 July were even more dramatic (Figure 14.7f). Approximately 50 m upstream of the reach, there was a major avulsion into a previously dry relict channel that entered the reach at A on Figure 14.7f. This was at a much lower elevation than the reach in general, as a result of continued buried ice melt-out, and this resulted in substantial erosion into the true right channel. The lower elevation also resulted in headward extension of a new channel (B on Figure 14.7f) which captured most of the flow from the true left channel, where there was substantial deposition, except at the downstream end (C on Figure 14.7f). The main channel now left the reach further towards the true right, and this shift was observed to be having important effects further downstream. Figure 14.7g shows the reach topography on 24 July, illustrating upstream extension of the 996.5 m contour on the true right and downstream extension on the true left. Most notable is a marked confluence scour (A on Figure 14.7g) of 50 to 60 cm in the true right channel.

In summary, the reach-scale DEMs of difference allowed the visualisation of major changes in river channel pattern, which allows a number of conclusions to be made about the nature of channel changes in this sort of system. First, there was evidence of the way in which the gross dynamics of this reach were controlled by processes exogenous to the reach. For instance, there was evidence of topographic adjustment and alignment to the prevailing discharge and sediment regimes, reflected: (i) in the progressive increase in channel dynamism as discharge, the magnitude of discharge fluctuation and, by implication, sediment supply from upstream all increased; and (ii) by repositioning of zones of maximum channel activity as the main channel repositioned itself (e.g. topographic orientation in response to an upstream avulsion). This emphasises the need to consider the reach, and its morphological change, as only one part of the full proglacial river channel system within this drainage basin. Just as the dynamics of this reach were modified substantially by an upstream avulsion, so the shift of the main exit channel to the true right during the study period had substantial downstream implications for morphological adjustment. Second, within the reach, some evidence of form–process feedback was clearly evident (e.g. the upstream bar and scour extension between 13 and 18 July). This was spatially distributed, and just as the reach itself moderated the effects of processes imposed upon it from upstream on downstream zones, so there was a coupling of zones within the reach (e.g. erosion in one zone resulted in the necessary topographic gradients for there to be headward extension of a new channel into a second zone). The reach could be viewed as a palimpsest of dynamic sources and sinks (Lane *et al.*, 1996), where the dynamism itself was maintained by the continual creation and recreation of those sinks. Third, there was a clear coupling of channel changes across different temporal and spatial scales. Although events external to this reach were critical in determining the nature of channel change, smaller scale processes within the reach were also important. For example, the stability of the main medial bar (A in Figure 14.7a), apart from bank erosion and upstream accretion, suggests that the overall structure of the reach reflects modification of a "fossil geomorphology" formed during a previous major process event or perhaps associated with the cumulative effects of meltwater processes towards the end of the previous period of channel activity (during the summer of 1991). Within this structure, however, some aspects of different "styles of channel change" were noted and these progressively modify the relict landform (e.g. the upstream accretion and the mid-channel stalling of bedload). There is a coupling of interactions between form and process across temporal and spatial scales, with form–process interactions at smaller scales being both conditioned by, and involved in the conditioning of, larger topographic scales.

Estimation of Bed Material Transport Rates from DEMs

In addition to visualising channel changes, DEMs and DEMs of difference may provide an alternative means for the estimation of bedload, or strictly speaking bed material, transport rates. Research in the 1960s and 1970s (Hubbell, 1964; Neill, 1969, 1971) suggested the potential for estimating bed material transport rates from morphological information. Recent research in dynamic gravel-bed river streams (e.g. Ferguson and Ashworth, 1992; Goff and Ashmore, 1994; Lane *et al.*, 1995b; Martin and Church, 1996) has illustrated the potential of volume change estimates for the estimation of bed material transport rates, and this has been thoroughly reviewed by Ashmore and Church (1995). The method is based upon the continuity equation for sediment transport, expressed in either one or two dimensions. In one-dimensional, steady-state, finite difference form, this is given by:

$$-\frac{\Delta i_{by}}{\Delta y} = \frac{\rho(1-\varepsilon)\Delta z}{\Delta t} \qquad (14.4)$$

where Δi_{by} = mass transport rate (kg/m/s) in the down-stream direction; Δz = change in elevation; Δt = change in time over which calculation is being made; ρ = sediment density; ε = sediment porosity; and y = downstream direction. In two-dimensional form, this is given by:

$$-\frac{\Delta i_{bx}}{\Delta x} - \frac{\Delta i_{by}}{\Delta y} = \frac{\rho(1-\varepsilon)\Delta z}{\Delta t} \qquad (14.5)$$

where x = cross-stream direction. One-dimensional application allows the estimation of temporal fluctuations in bed material transport rate in a series of river reaches (e.g. Goff and Ashmore, 1994; Ashmore and Church, 1995; Martin and Church, 1996). Two-dimensional application allows estimation of spatial variation in bedload transport rates from within a river reach (e.g. Lane *et al.*, 1995b).

A DEM of difference provides the essential morphological information required for both equations (14.4) and (14.5). One-dimensional application (Figure 14.8)

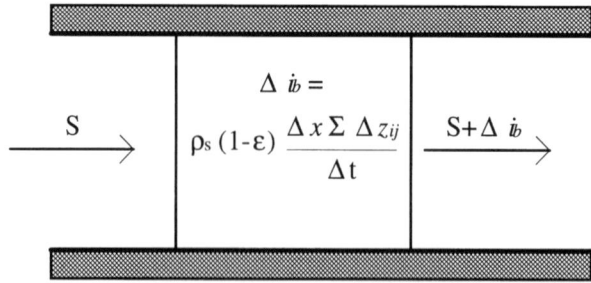

The upstream sediment supply is corrected for the effects of observed
erosion and deposition in the reach under consideration to predict
the downstream transport rate. The change in elevation in each
i,j grid cell of width and length Δx is summed to give the net elevation
change, indicating a reduction in transport rate if positive and an
addition to transport if negative.

Figure 14.8   The one-dimensional method used to estimate bed material transport
rate

makes use of the net volume of change obtained from comparing the two river bed
surfaces, dividing by the area of surface common to both DEMs during calculation
of the DEM of difference to obtain Δz. Two-dimensional applications (Figure 14.9)
use the value of Δz obtained at every grid node on the DEM of difference to
determine the net contribution of the grid cell to the transport rate (positive or
negative) in the time between DEM acquisition. For a regular grid ($\Delta x = \Delta y$), the
change in bedload transport rate is given by

$$\Delta i_b = \frac{\rho(1 - \varepsilon)\Delta z_{ij}\Delta x}{\Delta t} \tag{14.6}$$

The computational grid was oriented such that the dominant direction of sediment movement was in the i direction and a flow-driven sediment routing model
was developed. Each (i, j) cell not touching either a sidewall or the line of input or
output cells (Figure 14.9b) has supply from up to three upstream cells: $(i–1, j–1)$,
$(i–1, j)$ and $(i–1, j+1)$. This defines the total transport rate to the cell which is then
modified by erosion or deposition within the cell. The output from grid cell (i,j) is
then:

$$i_{b(i,j)} = i_{b(i-1,j+1)} + i_{b(i-1,j)} + i_{b(i-1,j-1)} - \frac{\rho_s(1 - \varepsilon)z_{ij}x}{t} \tag{14.7}$$

If it is assumed that the direction of the transport of material moved as bedload
matches the direction of flow movement, then $i_{b(i,j)}$ can be divided between the next
row of j-cells by:

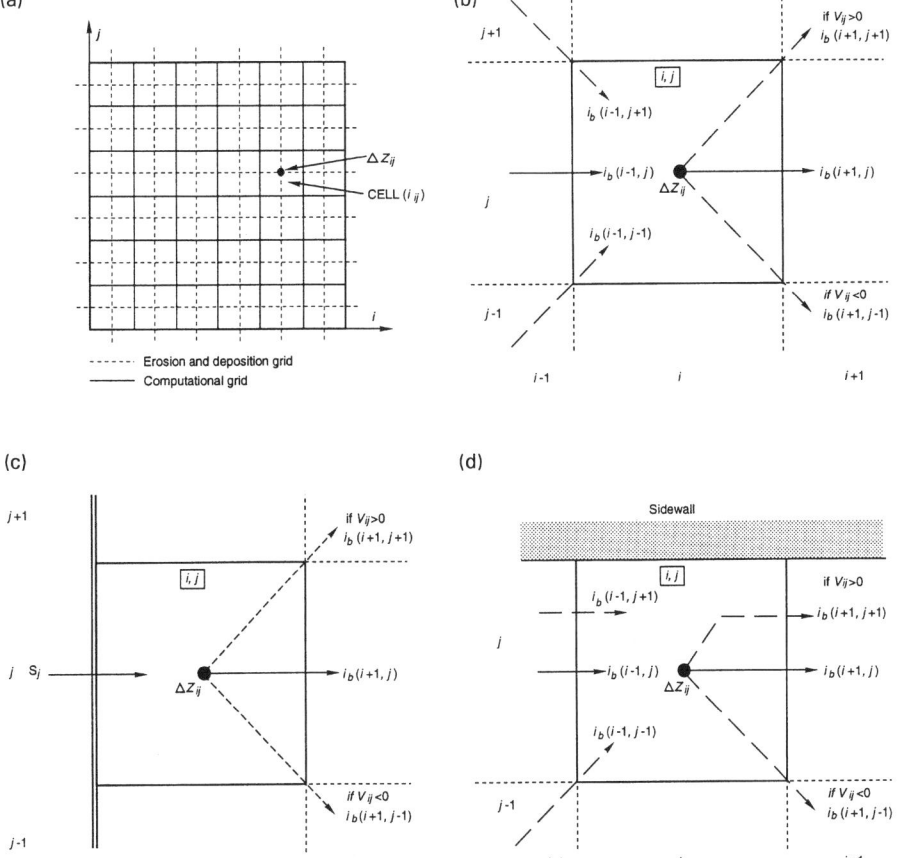

Figure 14.9 The two-dimensional method used to estimate spatially distributed bed material transport rate. (a) Erosion and deposition grid located over the computational (i,j) grid; (b)–(d) the three types of grid cell for which routing is required

$$i_{b(i+1,j)} = i_{b(i,j)} \left(\frac{U^2}{U^2 + V^2} \right)$$

if $V > 0$ then $\quad i_{b(i+1,j+1)} = i_{b(i,j)} \left(\frac{V^2}{U^2 + V^2} \right) \quad$ and $\quad i_{b(i+1,j-1)} = 0 \qquad$ (14.8)

if $V < 0$ then $\quad i_{b(i+1,j-1)} = i_{b(i,j)} \left(\frac{V^2}{U^2 + V^2} \right) \quad$ and $\quad i_{b(i+1,j+1)} = 0$

where U = flow velocity in the i direction; and V = flow velocity in the j direction. This needs modification for two special cases. First, for a grid cell (i,j) adjacent to the upstream supply condition (Figure 14.9c), equation (14.7) becomes:

$$i_{b(i,j)} = S_j - \frac{\rho_s(1-\varepsilon)\Delta z_{ij}\Delta x}{\Delta t} \tag{14.9}$$

where S_j = sediment supplied to boundary cell j from upstream. Second, for grid cells adjacent to sidewalls with maximum and minimum values of j (Figure 14.9d), equation (14.8) becomes:

For minimum j

if $V < 0$ then $i_{b(i+1,j\min)} = i_{b(i,j)}$

if $V > 0$ then $i_{b(i+1,j\min)} = i_{b(i,j\min)}\left(\dfrac{U^2}{U^2+V^2}\right)$ and

$$i_{b(i+1,j\min+1)} = i_{b(i,j)}\left(\frac{V^2}{U^2+V^2}\right)$$

For maximum j

if $V > 0$ then $i_{b(i+1,j\max)} = i_{b(i,j)}$

if $V < 0$ then $i_{b(i+1,j\max)} = i_{b(i,j\max)}\left(\dfrac{U^2}{U^2+V^2}\right)$ and

$$i_{b(i+1,j\max-1)} = i_{b(i,j)}\left(\frac{V^2}{U^2+V^2}\right)$$

Application of this model proceeds in a downstream manner, routing sediment into the first set of i cells, correcting for observed changes in sediment transport due to erosion and deposition and then splitting the output between the $j-1$, j and $j+1$ of the next set of i cells according to flow direction. This type of routing is computationally efficient and acceptable if the grid spacing is small.

The routing model requires knowledge of the flow direction in each grid cell. Two alternatives were available for specification of the flow direction. The first could make use of the information obtained from rapid survey of the spatial patterns of flow velocity obtained from a two-dimensional current meter (e.g. an electromagnetic current meter). Although this was undertaken during the 1992 field season, this could only provide a reasonable estimation of the magnitude of flow velocity. Establishing flow direction was more difficult owing to the problems of correctly establishing the sensor head orientation, particularly because high suspended sediment concentrations obscured the stream bed, although redesign of the mapping methodology for the 1995 season has reduced this problem (Lane *et al.*, in review). Further, a numerical interpolation procedure would have been required to assign a flow model prediction to each grid cell given that the grid spacing used in this problem (0.25 m) was much finer than the density of EMCM measurements.

The second alternative used a depth-averaged flow model to provide estimates of flow direction (Lane *et al.*, 1995a). This used the DEM acquired at the start of the

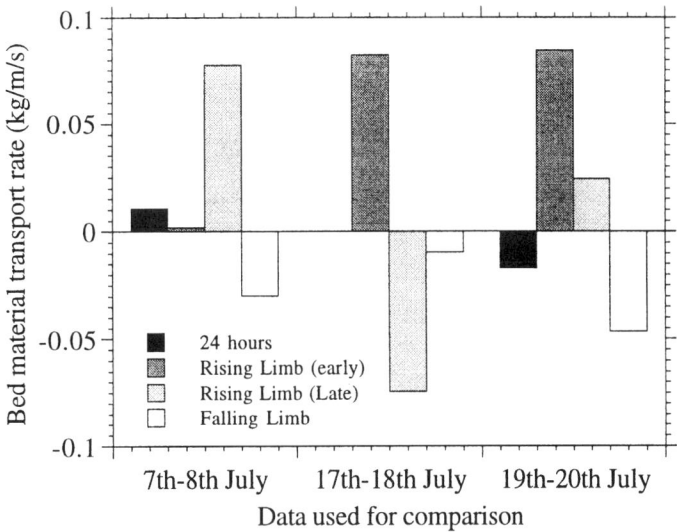

Figure 14.10 Simulation of the effects of different frequencies of sampling bed morphology on estimation of bed material transport rates

period over which the DEM of difference was calculated as the topographic boundary condition, combined with field measurements of bed roughness and a distributed upstream discharge condition. Flow predictions were obtained for an irregular grid with much smaller spacings than the computational grid used for the routing model, and this allowed estimation of flow direction for each grid cell.

In the one-dimensional application, the change in bed material transport from within a river reach was added to an estimate of sediment supply from upstream of the reach obtained from continual Helley-Smith sampling, and compared with the results of continual Helley-Smith sampling at the downstream end of the reach. The predicted and observed estimates of bedload transport agreed to within an order of magnitude (Lane *et al.*, 1995b). In the two-dimensional application, distributed estimates of erosion and deposition depth were used to estimate distributed volumes of bed material change. Helley-Smith bedload samples distributed across the upstream end of the reach were routed downstream according to flow direction, estimated using a depth-averaged two-dimensional model of open channel flow, adding and subtracting on a grid cell basis the estimates of bed material change. Predictions were compared with spatially distributed estimates of bedload transport rate obtained from point Helley-Smith samples, with a correlation of 0.712.

For the estimation of spatially distributed but time-integrated transport rates, these results are encouraging. However, it is worth stressing their time-integrated nature. As the time between morphological surveys increases, so the probability of undetected erosion or deposition increases (Lane *et al.*, 1994). It is possible to simulate the effects of different morphological sampling frequencies upon estimates of bedload transport rate, and Figure 14.10 illustrates the effects of different frequencies of morphological survey upon estimates of bedload transport rate using a one-dimensional model. The transport rates were determined from comparison of DEMs obtained at low flow on consecutive days in the case of the 24 h data, and *within* 24 h periods in the case of other data. The latter were spaced in time such that a DEM was collected at both the top of the rising limb and the bottom of the falling limb, and that there was unidirectional increase or decrease in discharge during terrain comparison periods (Lane *et al.*, 1996). The results illustrate how there can be substantial change in the magnitude and sign of the bed material transport estimate as the time scale of comparison is altered. Most striking is the 17–18 July comparison, when early rising limb erosion was nearly exactly countered by late rising limb and falling limb deposition. The apparent effect is a zero bed material transport rate. This should not caution against the use of morphological information in the estimation of bed material transport rates, but does emphasise their time-integrated nature: as the integration period changes so the estimate of transport rate will change. This does become a problem if calculated transport rates are to be linked to process. In addition to the problem of to which aspect of discharge to daily transport rate estimates relate (e.g. peak, difference between peak and trough, volume of water passing through during a diurnal cycle), any derived relationships will alter with the integration period.

In addition to the frequency of survey problem, these methods only provide an estimate of the *change* in bed material transport due to erosion or deposition in the reaches under consideration. Further information is required to estimate actual bed material transport rates, due to the possible complicating effect of sediment supply from upstream. The importance of this is implicit to both of the methods described above: they required specification of an upstream sediment supply condition. This may be dealt with by means other than measuring upstream supply. First, in some situations (e.g. McLean, 1990), this may be a zero supply condition at either the upstream or the downstream end of the reach. Second, it may be possible to introduce a step length specification (e.g. Ferguson and Ashworth, 1992) which will also deal with this problem (see Lane (in press) for explanation). Third, minimum transport rates should not become negative, due to there being more sediment deposited in a reach or grid cell than is supplied from upstream, and this may be used to determine minimum upstream sediment supply rates. These issues all merit further and closer investigation.

CONCLUSION

The use of DEMs in fluvial geomorphology is a recent development, but one which has much potential. Provided DEMs of appropriate quality and spatial and temporal resolution can be acquired, they can provide a useful means of both visualising and understanding river channel change, and act as critical boundary conditions for application to numerical modelling. Further research is needed into the quality and density of data that the various terrain acquisition methods can provide, and the relationship between data quality issues and the required use of the acquired information. This will require both downscaling and upscaling. Further understanding of the effects of river bed structure can only come from a better appreciation of the nature of bed micro-topography, which has both methodological and substantive importance. In methodological terms, this is particularly the case for gravel-bed rivers, where spatial variation in both grain size and bed structure will mean that low density sampling could have a significant effect upon the representation of topography at single epochs, and hence on estimates of erosion or deposition. This takes on substantive importance with the development of distributed flow and sediment transport modelling for gravel-bed rivers, owing to the importance of surface topography for the formation of flow structures, for sediment entrainment and for sediment transport. Recent research using automated digital photogrammetry of exposed gravel bars (Butler *et al.*, in review) has suggested that the relationship between surface grain size and topographic expression remains unclear, particularly where the surface is well imbricated. This has clear implications for roughness determination and for the treatment of surface topography in such models. It emphasises the need to see surface roughness as a continuum of topographic scales, to assess the smallest scale of topography measured by the data collection methods being used, and to use this to inform the design of computational grids, where "roughness" is defined as a sub-grid-scale phenomenon. There is a clear need for intensive studies of topography at this scale to determine the surface roughness associated with particular grain size distributions.

In addition to this downscaling, upscaling is equally important. Developments in both remote sensing and photogrammetry will allow data to be acquired more rapidly and from larger areas. The automation of both photogrammetric and survey data acquisition is one step towards such improvements, particularly if photogrammetric methods can be automated for the special case of terrestrial oblique imagery. The new series of high resolution satellite imagery will accelerate this process (e.g. Gilvear *et al.*, Chapter 9). However, there is equally exciting potential in the coupling of micro-topographic scales to larger morphological scales through the development of Geographical Information System-based methods of predicting micro-topography on the basis of macro-topographic information, perhaps coupling this with process models. For instance, topographic analysis may be used to identify typical channel features (e.g. bar forms, confluences, diffluences) which may be used to estimate smaller scale topographic properties that may be required as input to process models. Similarly, one of the easiest means of distributing surface roughness might be to combine a spatially distributed, depth-averaged flow model with a simple entrainment criterion for a stable channel, to estimate the equivalent grain size required for the channel to be stable. Such a means of distributing equivalent grain size may then be used iteratively to improve flow model estimates, and to provide the distributed boundary condition information needed for sediment transport modelling when the channel becomes unstable.

ACKNOWLEDGEMENTS

This research has been supported by the NERC, the Royal Society, Fitzwilliam College, University of Cambridge, and City University, London. A significant proportion was completed whilst the author was under the supervision of Professor K. S. Richards and Dr J. H. Chandler, and their advice and encouragement are gratefully acknowledged. Field data collection in 1992 was made possible by the extensive support of Cambridge University undergraduates, and Nick Spedding (University of Aberdeen), Graham Wood (Oxford Brookes University) and Mark Roberts (University of Bristol) in particular.

REFERENCES

Ackermann, F., 1978. Experimental investigation into the accuracy of contouring through digital terrain modelling. *Proceedings of the Digital Terrain Modelling Symposium*, St. Louis, 165–92.

Andrews, E. D. and Nelson, J. M., 1989. Topographic response of a bar in the Green River, Utah, to variation in discharge. In *River Meandering*, eds S. Ikeda and G. Parker, American Geophysical Union Monograph, 12, 463–485.

Ashmore, P. E., 1991. How do gravel-bed rivers braid? *Canadian Journal of Earth Sciences*, **28**, 326–341.

Ashmore, P. E. and Church, M., 1995. Sediment transport and river morphology: a paradigm for study. Paper presented to the 4th Gravel-bed Rivers Conference, *Gravel-bed Rivers and the Environment*, Gold Bar, Washington State, USA, 20–26 August 1995.

Ashmore, P. E., Ferguson, R. I., Prestegaard, K. L., Ashworth, P. J. and Paola, C., 1992. Secondary flow in anabranch confluences of a braided gravel-bed stream. *Earth Surface Processes and Landforms*, **17**, 299–311.

Ashworth, P. J. and Ferguson, R. I., 1986. Interrelationships of channel processes, changes and sediments in a proglacial braided river. *Geografiska Annaler*, **68A**, 361–371.

Ashworth, P. J., Ferguson, R. I. and Powell, M. D., 1992. Bedload transport and sorting in braided channels. In Billi, P., Hey, R. D., Thorne, C. R. and Tacconni, P. (eds) *Dynamics of Gravel-Bed Rivers*, Wiley, Chichester, 497–515.

Bannister, A., Raymond, S. and Baker, R., 1992. *Surveying*. Longman, Harlow, Essex, 510pp.

Bridge, J. S., 1993. The interaction between channel geometry, water flow, sediment transport and deposition in braided rivers. In *Braided Rivers*, eds J. L. Best and C. S. Bristow, Geological Society Special Publication, 75, 13–71.

Butler, J. B., Lane, S. N. and Chandler, J. H., in review. Application and assessment of automated close range photogrammetry for the determination of the microtopographic structure of gravel-bed bar surfaces. Submitted to *Photogrammetric Record*.

Carson, M. A. and Griffiths, G. A., 1989. Gravel transport in the braided Waimakariri River: mechanisms, measurements and predictions. *Journal of Hydrology*, **109**, 201–220.

Clifford, N. J., Robert, A. and Richards, K. S., 1992. Estimation of flow resistance in gravel-bedded rivers: a physical explanation of the multiplier of roughness length. *Earth Surface Processes and Landforms*, **17**, 111–126.

Davies, T. H. R., 1987. Problems of bedload transport in braided gravel-bed rivers. In *Sediment Transport in Gravel-Bed Rivers*, eds C. R. Thorne, J. C. Bathurst and R. D. Hey, Wiley, Chichester, 793–828.

Dietrich, W. E. and Smith, J. D., 1984. Bedload transport in a river meander. *Water Resources Research*, **20**, 1355–1380.

Dietrich, W. E. and Whiting, P. J., 1989. Boundary shear stress and sediment transport in river meanders of sand and gravel. In *River Meandering*, eds S. Ikeda and G. Parker, American Geophysical Union Monograph, 12, 1–50.

Ferguson, R. I., 1993. Understanding braiding processes in gravel-bed rivers: progress and unresolved problems. In *Braided Rivers*, eds J. L. Best and C. S. Bristow, Geological Society Special Publication, 75, 13–71.

Ferguson, R. I. and Ashworth, P. J., 1992. Spatial patterns of bedload transport and channel change in braided and near-braided rivers. In *Dynamics of Gravel-bed Rivers*, eds P. Billi, R. D. Hey, C. R. Thorne and P. Tacconi, Wiley, Chichester, 477–492.

Ferguson, R. I., Prestegaard, K. and Ashworth, P. J., 1989. Influence of sand on hydraulics and gravel transport in a braided gravel-bed river. *Water Resources Research*, **25**, 635–643.

Ferguson, R. I., Ashmore, P. E., Ashworth, P. J., Paola, C. and Prestegaard, K. L., 1992. Measurements in a braided river chute and lobe: I Flow pattern, sediment transport and channel change. *Water Resources Research*, **28**, 1877–1886.

Goff, J. R. and Ashmore, P. E. 1994. Gravel transport and morphological change in braided Sunwapta River, Alberta, Canada. *Earth Surface Processes and Landforms*, **19**, 195–212.

Gomez, B., Hubbell, D. W. and Stevens, H. H., 1990. At-a-point bedload sampling in the presence of dunes. *Water Resources Research*, **26**, 2717–2731.

Hassan, M. and Reid, I., 1991. The influence of microform bed roughness elements on flow and sediment transport in gravel-bed rivers. *Earth Surface Processes and Landforms*, **15**, 739–750.

Hubbell, D. W., 1964. *Apparatus and techniques for measuring bedload*. US Geological Survey Water-Supply Paper, 1748.

Lane, S. N., 1994. *Monitoring and modelling morphology, flow and sediment transport in a gravel-bed stream*. Unpublished PhD Thesis, University of Cambridge.

Lane, S. N., in press. The reconstruction of sediment supply rates from morphological information. Forthcoming in *Catena*.

Lane, S. N. and Richards, K. S., 1997. Linking river channel form and process: time, space and causality revisited. *Earth Surface Processes and Landforms*, **22**, 249–60.

Lane, S. N., Chandler, J. H. and Richards, K. S., 1994. Developments in monitoring and terrain modelling small-scale river-bed topography. *Earth Surface Processes and Landforms*, **19**, 349–368.

Lane, S. N., Richards, K. S. and Chandler, J. H., 1995a. Within reach spatial patterns of process and channel adjustment. *Rivers*, ed. E. J. Hickin, Wiley, Chichester, 105–130.

Lane, S. N., Richards, K. S. and Chandler, J. H., 1995b. Morphological estimation of the time-integrated bedload transport rate. *Water Resources Research*, **31**, 761–772.

Lane, S. N., Richards, K. S. and Chandler, J. H., 1996. Discharge and sediment supply controls on erosion and deposition in a dynamic alluvial channel. *Geomorphology*, **15**, 1–15.

Lane, S. N., Biron, P. A., Bradbrook, K. F., Butler, J. B., Chandler, J. H., Crowell, M. D., McLelland, S. J., Richards, K. S. and Roy, A. G., in review. Integrated three-dimensional measurement of river channel topography and flow processes using acoustic doppler velocimetry. To be submitted to *Earth Surface Processes and Landforms*.

Laronne, J. B. and Duncan, M. J., 1992. Bedload transport paths and gravel-bar formation. In Billi, P., Hey, R. D., Thorne, C. R. and Tacconni, P. (eds) *Dynamics of Gravel-Bed Rivers*, Wiley, Chichester, 177–200.

Leopold, L. B. and Maddock, T. Jr, 1953. *The hydraulic geometry of stream channels and some physiographic implications*. Professional Paper of the US Geological Survey, 252, 1–57.

Li, Z., 1992. Variation of the accuracy of Digital Terrain Models with sampling interval. *Photogrammetric Record*, **14**, 113–28.

Martin, Y. and Church, M., 1996. Bed-material transport estimated from channel surveys – Vedder River, British Colombia. *Earth Surface Processes and Landforms*, **20**, 247–261.

McLean, D. G., 1990. *The relation between channel instability and sediment transport on Lower Fraser River*. Unpublished PhD Thesis, University of British Colombia.

Murray, A. B. and Paola, C., 1994. A cellular model of braided rivers. *Nature*, **371**, 54–57.

Naden, P. and Brayshaw, A. C., 1987. Bedforms in gravel-bed rivers. In *River Channels: Environment and Process*, ed. K. S. Richards, IBG Special Publication, 18, Blackwell, Oxford, 249–271.

Neill, C. R., 1969. Bedforms in the Lower Red Deer River, Alberta. *Journal of Hydrology*, **7**, 58–85.

Neill, C. R., 1971. River bed transport related to meander migration rates. *ASCE, Journal of the Waterways and Harbours Division*, **97**, 783–786.

Nelson, J. M. and Smith, J. D., 1989. Flow in meandering channels with natural topography.

In *River Meandering*, eds S. Ikeda and G. Parker, American Geophysical Union Monograph, 12, 69–102.

Prestegaard, K. L., 1983. Bar resistance in gravel-bed streams at bankfull stage. *Water Resources Research*, **19**, 472–476.

Richards, K. S., Sharp, M. J., Arnold, N. S., Lawson, W., Nienow, P., Willis, I. C., Gurnell, A., Clark, M., Tranter, M., Hill, C. and Brown, G., 1992. *Integrated approaches to modelling hydrology and water quality in glacierised catchments.* NERC Final Report to Grant GR3/7004a.

Shannon, C. E., 1949. Communication in the presence of noise. Proceedings of the Institute of Radio Engineers, **37**, 10–21.

Shearer, J. W., 1990. The accuracy of Digital Terrain Models. In *Terrain Modelling in Surveying and Civil Engineering*, eds G. Petrie and T. J. M. Kennie, Whittles, London, 315–336.

15 Mass Balance and Flow Variations of Haut Glacier d'Arolla, Switzerland, Calculated using Digital Terrain Modelling Techniques

I. C. WILLIS[1], N. S. ARNOLD[1], M. J. SHARP[2], J-M. BONVIN[3] and B. P. HUBBARD[4]

[1] Department of Geography, University of Cambridge, UK
[2] Department of Earth and Atmospheric Sciences, University of Alberta, Canada
[3] Grand Dixence SA, Valais, Switzerland
[4] Centre for Glaciology, University of Wales, UK

ABSTRACT

This study uses digital terrain modelling techniques to examine spatial variations in mass balance and flow of a valley glacier between September 1992 and September 1993. The spatial pattern of glacier surface height change (thickening and thinning) was calculated from digital terrain models (DTMs) constructed from aerial photographs using analytical photogrammetry. Most of the glacier experienced net lowering, with a mean of 0.80 m of water equivalent (w.e.) and a maximum of more than 4 m w.e. The spatial pattern of glacier surface mass balance (accumulation and ablation) was calculated with a distributed energy balance melt model using measurements of winter snow depth and calculations of summer ablation. Most of the glacier experienced net ablation, with a mean of 1.25 m w.e. and a maximum of about 5.0 m w.e. The relationships between spatial variations in surface lowering, ablation and terrain characteristics were examined using statistical techniques. Nearly 20% of the pattern of surface lowering is explained by ablation variations, and about 15% of the surface lowering pattern is explained by variations in altitude, slope angle, slope aspect and shading. Most of the remaining variation in glacier thinning is attributed to spatial variations in the longitudinal flux divergence. The spatial pattern of the flux divergence was computed from the difference between the spatial patterns of glacier thickening and thinning and surface accumulation and ablation, using the continuity equation for ice thickness. These computations of flux divergence together with knowledge of spatial variations of glacier thickness are used to compute the spatial pattern of the gradient in depth-averaged velocity. Flow over most of the glacier is compressive, with strong compression near the glacier snout and margins, but there are a few areas of extending flow in regions where crevasses are found. Integration of the depth-averaged velocity gradient along the glacier centreline produces velocities which are generally less than the measured surface velocities. The ratio of surface to depth-averaged velocity appears to be dependent on ice thickness.

Landform Monitoring, Modelling and Analysis. Edited by S. N. Lane, K. S. Richards and J. H. Chandler.
© 1998 John Wiley & Sons Ltd.

INTRODUCTION

Digital terrain modelling (DTM) techniques are now used widely in hydrology, geomorphology and biogeography (Moore *et al.*, 1991) and are increasingly being used in glaciology (Brunner, 1987). Within glaciology, DTM techniques have been used to: (i) aid the identification of snow, firn and ice fields from satellite images (Parrot *et al.*, 1993); (ii) model snow distribution across glaciers (Copland, Chapter 17; Turpin *et al.*, Chapter 16); (iii) reconstruct glacier hydrological pathways (e.g. Björnsson, 1988; Holmlund, 1988; Sharp *et al.*, 1993; Fountain and Vaughn, 1995); (iv) model glacier surface energy balance (Munro and Young, 1982; Escher-Vetter, 1985; Gratton *et al.*, 1993; Arnold *et al.*, 1996); (v) determine glacier mass balance (Haakensen, 1986; Knudsen, 1986; Reinhardt and Rentsch, 1986; Krimmel, 1989; Etzemüller *et al.*, 1993); and (vi) study glacier flow (Rentsch *et al.*, 1990; Etzelmüller *et al.*, 1993). The present study uses DTM techniques to study spatial variations in the mass balance and flow of Haut Glacier d'Arolla in Switzerland.

DTM techniques have previously been used to calculate glacier mass balance by the "geodetic" method which involves surveying changes in glacier surface topography over time to calculate glacier volumetric change (e.g. Haakensen, 1986; Knudsen, 1986; Reinhardt and Rentsch, 1986; Krimmel, 1989; Etzemüller *et al.*, 1993). Topographic data were traditionally obtained by terrestrial surveys which are time-consuming and are limited to the accessible parts of the glacier (e.g. Østrem and Brugman, 1991). Thus, photogrammetric surveys are now increasingly being used since they allow more extensive topographic data to be obtained more quickly and accurately. Early DTM-based studies of glacier mass balance used DTMs that were constructed manually from existing contour maps (e.g. Haakensen, 1986; Knudsen, 1986; Krimmel, 1989). The resulting DTMs had a coarse resolution (typically 100 × 100 m at best) and the mean error in the height determination was of the order of several metres. More recently, DTM studies of glacier mass balance have used more advanced photogrammetric techniques, where DTMs are digitised directly from aerial photographs (e.g. Reinhardt and Rentsch, 1986; Etzemüller *et al.*, 1993). These DTMs can be produced to a high grid-cell resolution (typically 5 × 5 m) and have a mean height error of a few centimetres. This technique is limited to the determination of the average (or specific) net mass balance of the whole glacier. The average (or specific) net mass balance of individual elevation intervals cannot be determined by this method because spatial patterns of glacier surface elevation change are also influenced by patterns of glacier flux divergence (i.e. spatial variations in the gradients of longitudinal and transverse glacier flux).

DTM techniques have previously been used to study glacier flow by mapping the movement of crevasses, boulders, ogives and other glacier surface features that are visible on aerial photographs (e.g. Rentsch *et al.*, 1990; Etzelmüller *et al.*, 1993). This approach is limited to the determination of surface velocities. Furthermore, over long time intervals between consecutive surveys, errors in the determination of glacier flow may result if crevasses are mapped, because crevasses may close and new ones open up so that their apparent rate of movement will be less than the glacier flow velocity.

The present study uses a combination of existing and new DTM techniques to study the mass balance and flow of Haut Glacier d'Arolla in Switzerland, and avoids some of the limitations encountered in previous work described above.

AIMS AND APPROACH

The aims and approach of the present study are as follows:

1. to calculate the spatial pattern of glacier surface height change (thickening and thinning) between September 1992 and September 1993 from the differences between two DTMs constructed from aerial photographs using analytical photogrammetry (see Dixon *et al.*, Chapter 4);
2. to calculate the spatial pattern of glacier surface mass balance (accumulation and ablation) between September 1992 and September 1993 using measurements of winter snow water equivalent depth and calculations of summer ablation derived from a distributed energy balance melt model;
3. to examine the degree to which the spatial pattern of surface lowering may be explained in terms of patterns of ablation, and in terms of terrain characteristics (i.e. altitude, slope angle, slope aspect and shading) using statistical methods;
4. to compute the spatial pattern of the longitudinal glacier flux divergence using knowledge of the spatial pattern of glacier thickening and thinning calculated in 1 and the spatial pattern of surface accumulation and ablation calculated in 2, together with the continuity equation for ice thickness; and
5. to compute the spatial pattern of the glacier velocity gradient using the computations of longitudinal flux divergence calculated in 4, and knowledge of spatial variations in glacier thickness.

STUDY SITE

Haut Glacier d'Arolla, Valais, Switzerland, covers an area of 6.3 km^2 and is situated at the head of the Val d'Hérens, a southern tributary of the Rhône Valley (Figure 15.1). The glacier tongue flows approximately south to north and is fed by a large upper eastern basin and a smaller western one (Figure 15.1). Over most of its length, the glacier surface slopes gently from *c.* 3000 m above sea level (a.s.l.) near the head to *c.* 2560 m a.s.l. at the snout. The upper eastern basin, however, is fed by a series of ice falls off the north face of Mont Brûlé which extend to *c.* 3500 m a.s.l. In addition to Mont Brûlé, the glacier is surrounded by the Bouquetins Ridge to the east and the mountains of L'Évêque and Mont Collon to the west (Figure 15.1). The glacier has been the focus of an integrated study of glacier hydrology, water quality and ice dynamics since 1989 (Richards *et al.*, 1996). Since 1989, the glacier has had a strongly negative mass balance and the snout has retreated by over 100 m (Willis *et al.*, unpublished data).

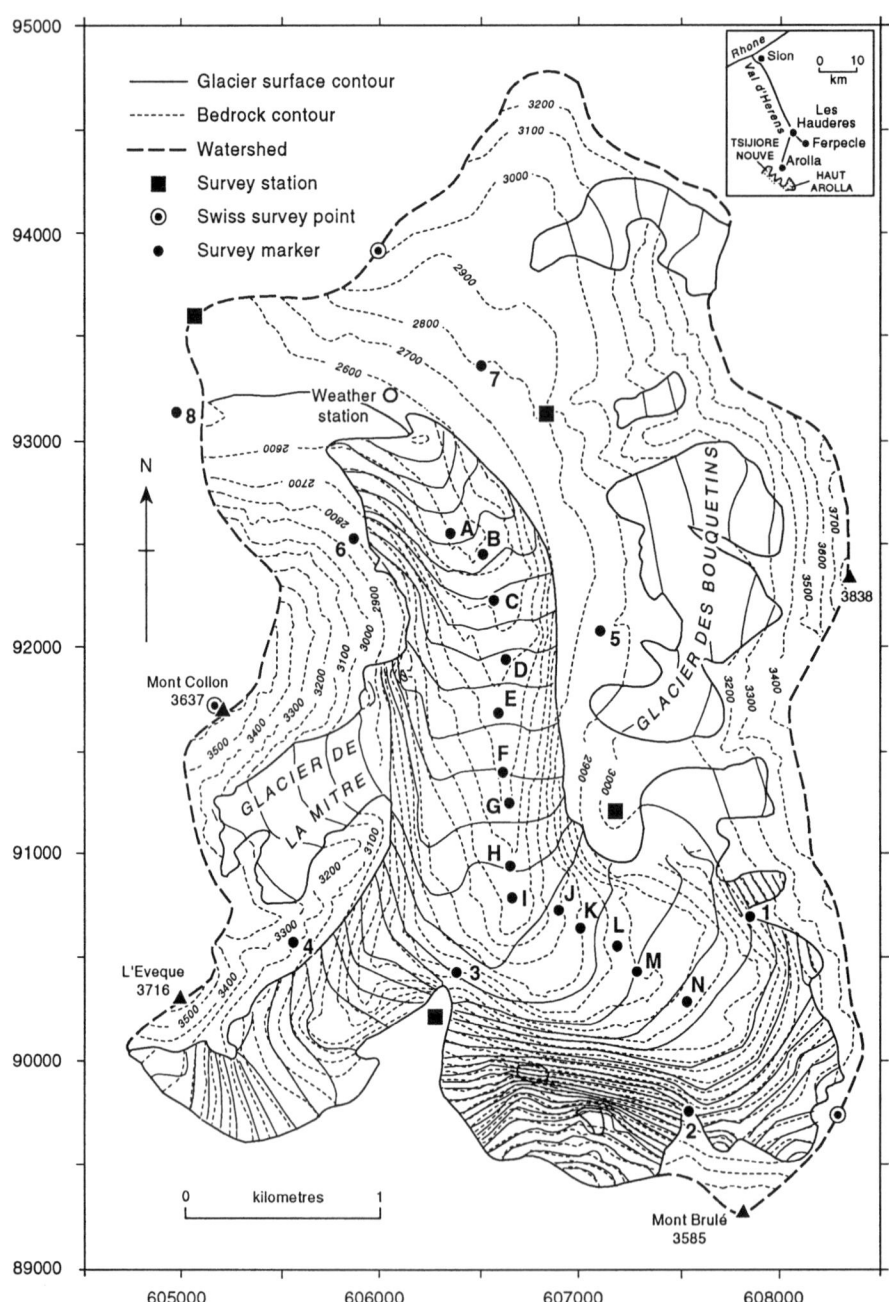

Figure 15.1 Map of Haut Glacier d'Arolla showing survey stations, Swiss Survey Trigonometric Points, glacier and bedrock survey markers, and meteorological station. Ablation stakes are also located at the glacier centreline survey markers labelled A–N. Surface contours are taken from the Swiss Survey 1:25 000 map (extraglacial area) and 1989 and 1990 ground survey (glacier). Bedrock contours are based on radio echo-sounding data collected in 1989 and 1990

METHODS AND RESULTS

Spatial Pattern of Glacier Surface Height Change

DTMs were constructed using analytical photogrammetry from vertical aerial photos taken *c*. 1000 m above the glacier on 17 September 1992 and 20 September 1993. Ground survey on the days that the photographs were taken consisted of angle and distance measurements between four survey stations located around the glacier (Figure 15.1). These were fixed into the Swiss Grid Coordinate System by surveying to three Swiss Survey Trigonometric Points (Figure 15.1). Finally, markers which were visible on the photographs (15 on the glacier and eight off the glacier) were surveyed from as many survey stations as possible (Figure 15.1). The coordinates of the survey stations and the markers were then determined using a least-squares adjustment procedure. A Leica DSR14 Analytical Plotter was used to produce breakline and vector data representing the glacier surface. Intergraph/Bentleys terrain modelling software was used to generate high resolution contour data at 1 m intervals across the glacier. These were interpolated onto a regular 20 × 20 m grid using the bilinear interpolation routine within UNIMAP (UNIRAS, 1990). Digital terrain information for parts of the upper glacier could not be produced owing to the difficulties of obtaining elevation data from the snow-covered, "textureless" surface. Therefore the resulting DTMs are largely for ice- and firn-covered areas of the glacier and each contain 7212 grid cells. Errors in the elevation calculations are typically less than a few centimetres but are of the order of 10 cm over parts of the upper glacier. The spatial pattern of surface height change between September 1992 and 1993 was obtained by subtracting the 1992 DTM from the 1993 DTM (Figure 15.2). Thus a decrease in surface elevation is shown by negative values and an increase by positive values. Units were converted to water equivalent (w.e.) by multiplying by the density of ice ($900 \ kg \ m^{-3}$).

Thinning occurred over most of the glacier with a maximum of over 4.0 m w.e. on the eastern side of the lower glacier tongue and near the top of the upper eastern basin. Some areas of the glacier experienced thickening of a few decimetres, including some areas near the glacier head but also the sides of the main glacier tongue. Some of the thickening towards the glacier sides may be due to the effects of shading by surrounding topography and moraine cover retarding ablation. The average surface height change across the glacier was −0.80 m w.e. and the net surface volume change was $-2.31 \times 10^6 \ m^3$ w.e.

Spatial Pattern of Glacier Surface Mass Balance

Air temperature data collected at a site located 100 m in front of the glacier (Figure 15.1) were used to identify when the mean daily temperature first rose above 0°C. On this basis, the 1992/1993 mass balance year was split into an accumulation season (17 September 1992 to 19 May 1993) and an ablation season (20 May to 20 September 1993).

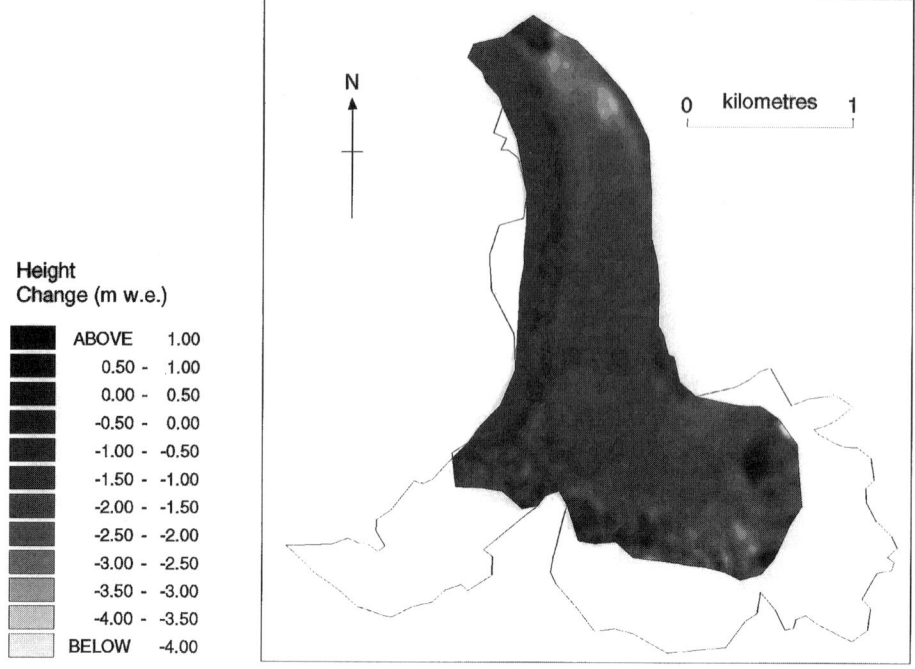

Height Change (m w.e.)

ABOVE	1.00
0.50 -	1.00
0.00 -	0.50
-0.50 -	0.00
-1.00 -	-0.50
-1.50 -	-1.00
-2.00 -	-1.50
-2.50 -	-2.00
-3.00 -	-2.50
-3.50 -	-3.00
-4.00 -	-3.50
BELOW	-4.00

Figure 15.2 Map of surface height change between September 1992 and September 1993 calculated from the difference between the 1992 and 1993 DTMs. Positive values indicate glacier thickening and negative values indicate glacier thinning

The spatial pattern of winter accumulation was calculated as follows. Snow depth was measured between 19 and 22 June 1993 at a network of 87 points distributed widely over the glacier (Figure 15.3). The points were divided into four areas representing the lower tongue, the upper tongue, the upper eastern basin and the upper western basin. Within each area, a snowpit was dug in which snow density was measured at 20 cm intervals. Within each area, the vertical variation of snow density in the snowpit was small and so the mean density was used to represent that of the snow measured at all the points in the area. In this way, the snow water equivalent depth was calculated at the 87 points (Copland, Chapter 17). The calculations of snow water equivalent depth were interpolated onto the 20 × 20 m grid of the DTM using the bilinear interpolation routine in UNIMAP. To compute the spatial distribution of snow water equivalent depth at the end of the winter (i.e. on 19 May 1993), a distributed surface energy balance melt model developed for Haut Glacier d'Arolla by Arnold *et al.* (1996) was used to calculate the pattern of surface melt between 20 May and 20 June (see below for a description of the surface energy balance melt model). The model uses a parameterisation scheme developed by Oerlemans (1993) in which albedo is a function of snow depth. As the pattern of snow depth between 20 May and 20 June is not known, the model used the snow depth distribution measured in June to represent the distribution between

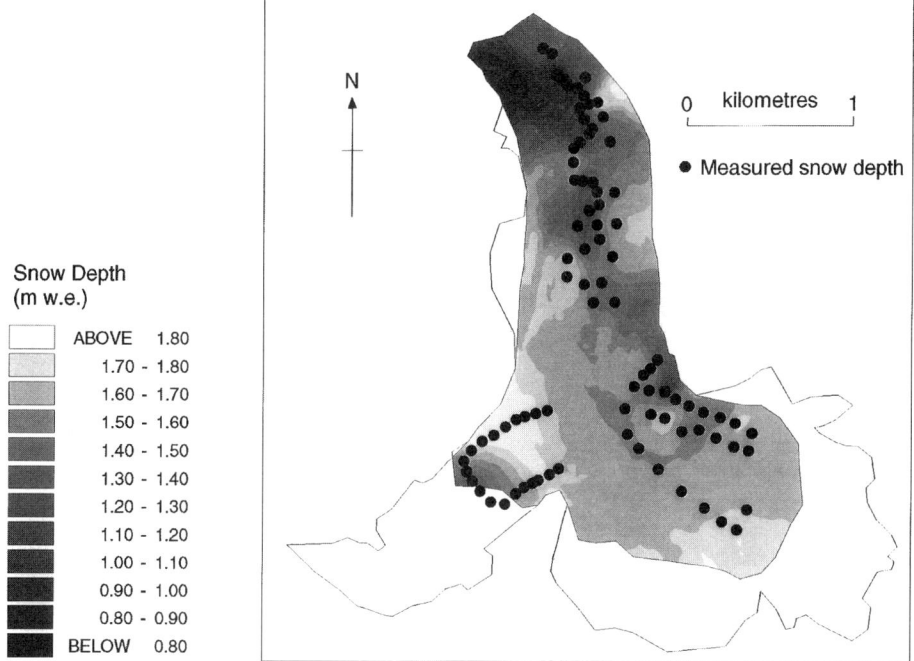

Figure 15.3 Map of 87 locations at which snow depth was measured in mid-June 1993 and the snow distribution map for mid-May 1993 produced from these measurements

20 May and 20 June for the purposes of the albedo parameterisation scheme. This should not have significantly affected the results as albedo is insensitively related to snow depth for depths over 1 m (Oerlemans, 1993). For each grid cell, the calculated melt between 20 May and 20 June was added to the snow water equivalent depth measured between 19 and 22 June to give an estimate of the total winter accumulation. The spatial pattern of winter accumulation is shown in Figure 15.3 and varies from less than 0.8 m at the glacier snout to more than 1.8 m near the glacier head.

The spatial pattern of summer ablation was calculated using the distributed surface energy balance model referred to above. The 1993 DTM of the glacier surface was combined with a 20 × 20 m DTM of the surrounding topography up to and including the drainage divides. This DTM was obtained from Swiss Survey 1:25 000 maps using contour-following software developed by Mayo (1993) and the bilinear interpolation procedure within UNIMAP. This "Catchment DTM" was used in conjunction with standard solar altitude and azimuth theory (e.g. Walraven, 1978) to compute hourly patterns of shading across the glacier for every third day of the summer. For each of these days, the computed hourly shading pattern was also used as the pattern for the two subsequent days. Hourly measurements of incoming short-wave radiation, air temperature, relative humidity and wind speed

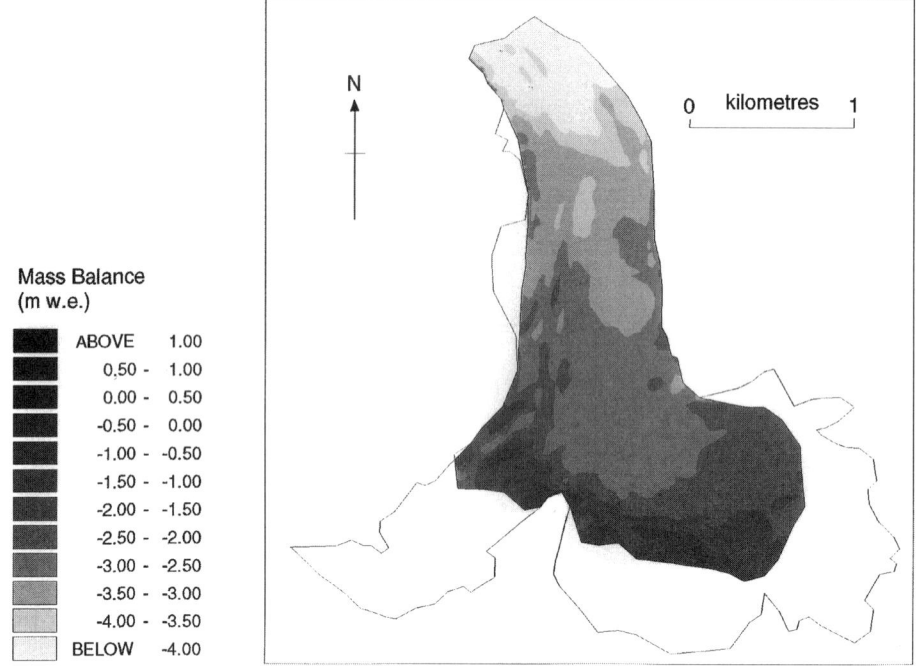

Figure 15.4 Map of net mass balance between September 1992 and September 1993. Positive values are net accumulation and negative values are net ablation

collected at an elevation of 2547 m a.s.l. at a site in front of the glacier (Figure 15.1) were used in conjunction with the DTM information (elevation, slope angle and slope aspect) and the shading information to compute the energy receipt (short-wave and long-wave radiation and sensible and latent heat) (W m^{-2}) for each DTM grid cell for each hour of the summer. Units were converted to water equivalent by dividing by the latent heat of fusion of water (3.34×10^5 J kg^{-1}) and the density of water (1000 kg m^{-3}). For full details of the energy balance model, see Arnold *et al.* (1996). Knowing the snow cover distribution at the start of the melt season, the model was used to determine the total amount of snow and ice melt in each grid cell over the period 20 May to 20 September 1993.

The spatial pattern of annual surface mass balance was calculated by subtracting the summer ablation values from the winter accumulation values for each grid cell (Figure 15.4). Thus, positive values represent accumulation and negative values represent ablation. Annual surface mass balance increases upglacier from about – 5.0 m w.e. ablation at the snout to about 0 m w.e. in the upper glacier. There is also some cross-glacier variation of mass balance particularly on the glacier tongue, where ablation tends to be higher towards the centre than at the sides. The average mass balance across the glacier is –1.25 m w.e. and the net surface mass balance across the glacier was -3.59×10^6 m^3 w.e. The glacier's equilibrium line altitude (ELA) was obtained by regressing net mass balance against elevation for all grid cells. The result is a 1992–1993 ELA (i.e. where net mass balance = 0) of 3012 m a.s.l.

To check the accuracy of the surface mass balance calculations, field measurements of mass balance were made at a line of wooden stakes drilled into the glacier surface along its centreline, extending from c. 2630 m a.s.l. near the snout to c. 2988 m a.s.l. in the upper eastern basin (at the centreline markers A–N shown in Figure 15.1). All the stakes were below the position of the 1992–1993 ELA and were therefore within the glacier's ablation area. Unfortunately, we have no stake or snowpit measurements from the glacier's accumulation areas as these are not readily accessible owing to very steep slopes and crevasses. The net ablation at each stake was measured between 18 September 1992 and 12 September 1993. These measurements are shown in Figure 15.5 together with the net ablation in the corresponding grid cells of the DTM calculated by the energy balance model between 17 September 1992 and 20 September 1993. It shows a very good correspondence ($r = 0.99$). The mean and standard deviation of the differences between the measured and calculated values (measured minus calculated) are –24 cm and 20 cm respectively, showing that the model slightly overestimates melt. The overestimation is greatest at high ablation rates on the lower glacier (Figure 15.5). The mean and standard deviation of the absolute differences between measured and calculated melt are 26 cm and 17 cm respectively, confirming that the model seems to reproduce accurately the pattern of ablation across the glacier.

Relationships Between Surface Lowering, Ablation and Terrain Characteristics

There is some similarity between the spatial pattern of surface elevation change between September 1992 and September 1993 (Figure 15.2) and the spatial pattern of surface mass balance over approximately the same time period (Figure 15.4). Statistical techniques were used to identify the degree to which the spatial pattern of glacier thinning can be explained: first, by spatial variations in net ablation, and second, by spatial variations in terrain characteristics. Those parts of the glacier which experienced a net thickening are ignored, as an increase in mass is controlled by factors different from those that control mass decrease. Results from 5734 grid cells were therefore used in the analysis. The correlation coefficient between thinning and ablation is 0.43 and the regression equation of thinning on ablation is shown in Table 15.1. The slope of the regression line suggests that surface lowering is, on average, only 65% of ablation. Nearly 20% of the variation in surface lowering may be explained in terms of ablation variations.

As ablation variations across Haut Glacier d'Arolla are influenced by terrain characteristics (Arnold et al., 1996), the relationships between patterns of surface thinning and terrain attributes were investigated. Four terrain attributes were identified as follows:

(i) *Altitude*. Altitude is likely to influence net surface ablation and hence lowering via two main effects. First, altitude influences the distribution of winter accumulation and therefore the initial snow depth at the start of the summer (Figure

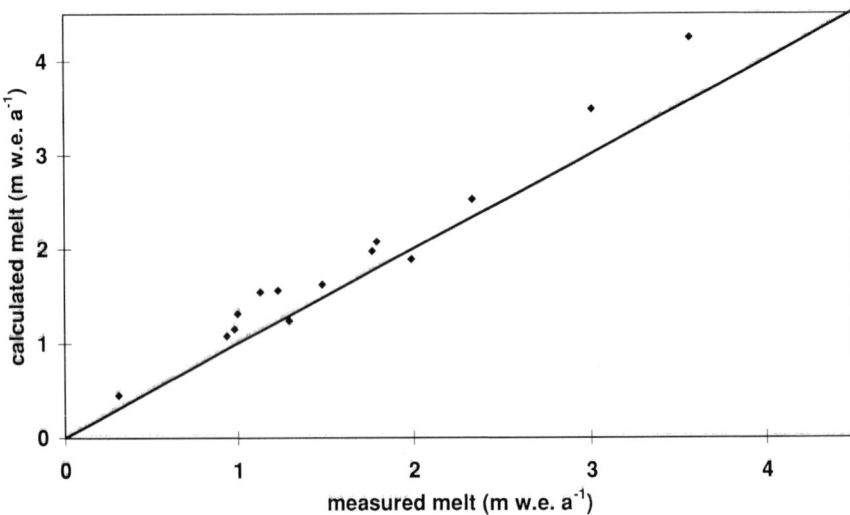

Figure 15.5 Relationship between net ablation measured at centreline stakes A–N between September 1992 and September 1993 and net ablation calculated from the energy balance model at corresponding grid cells over approximately the same time period

Table 15.1 Results of the regression analysis. Figures in parentheses are respectively the standard error and the t-ratio of the coefficients

Relationship between surface lowering and ablation

Lowering^a = 0.65 (ablation)
          (0.0062)
          (104.59)

R^2 = 0.19

Relationship between surface lowering and terrain factors

Lowering^a = 0.0027 (altitude) + 0.039 (slope angle) + 5.30 (cos aspect) − 0.42 (potential
          (0.000086)   (0.0018)   (0.20)   (0.020) short-wave
          (31.07)   (21.90)   (27.13)   (−21.40) radiation)

R^2 = 0.15

^{a} Values of lowering are negative so higher values indicate less lowering.

15.3). This controls the length of time that the glacier surface is covered by snow during the summer. Thicker winter snow cover will be associated with a longer time over the summer during which the glacier is snow-covered, lower net ablation, and therefore reduced surface lowering. Second, altitude influences the air temperature and vapour pressure during the summer and therefore the sensible and latent heat fluxes. Higher elevations will be associated with lower air temperatures and vapour pressures, reduced turbulent energy fluxes, lower ablation and therefore reduced surface lowering.

(ii) *Slope angle.* Slope angle influences the incoming short-wave radiation receipt. Higher slope angles will be associated with either lower or higher incoming radiation, depending on the sun angle. Ablation and lowering will be affected accordingly.

(iii) *Slope aspect.* Slope aspect also influences the incoming short-wave radiation receipt. More northerly facing slopes will be associated with lower incoming radiation, lower ablation and therefore reduced lowering.

(iv) *Shading.* Shading also influences the incoming short-wave radiation receipt. Parts of the glacier that experience extensive shading during the summer will be associated with lower incoming radiation, lower ablation and therefore reduced lowering.

These four factors, which vary over the glacier, might be expected to account for some of the variation in surface lowering observed in Figure 15.2. Values for all four factors were obtained for each grid cell of the DTM in which surface lowering between September 1992 and September 1993 was observed. The altitude of each cell (in metres a.s.l.) was obtained directly from the DTM data. The slope angle of a given cell (in degrees) was obtained by fitting a quadratic surface to the cell and its eight neighbouring cells. The maximum slope angle from the centre of the given cell across this surface was then used (see Zevenbergen and Thorne, 1987). The slope aspect of a given cell (in degrees from north) was obtained in a similar way. For the statistical analysis, the cosine of the aspect was taken to produce a "north/ south scalar". This produced values ranging from -1 for true south-facing slopes to $+1$ for true north-facing slopes, with values of 0 for true east- or west-facing slopes (Copland, Chapter 17). The effects of shading were analysed by determining the value of potential direct short-wave radiation for each grid cell (in metres w.e.). For each hourly time step of the model, the incoming short-wave radiation to each grid cell was predicted from theory, taking into account the effects of slope angle and aspect. It was assumed that diffuse radiation is 20% of the total radiation and so the predicted incoming short-wave radiation was reduced by 80% if the grid cell was in shade (Oerlemans, 1993; Arnold *et al.*, 1996). The pattern of melt due to total potential direct short-wave radiation over the whole summer is shown in Figure 15.6. The highest potential direct short-wave radiation (i.e. lowest shading) is found on the eastern part of the lower glacier tongue and in the central part of the upper eastern basin. Figure 15.6 also demonstrates the shading effects of Mont Brûlé, L'Évêque and Mont Collon (compare with Figure 15.1).

The correlation coefficients between thinning and altitude, slope angle, slope aspect and potential short-wave radiation are 0.281, 0.094, 0.046 and -0.014 respectively. Only the correlation between thinning and altitude is significant ($p > 0.001$) showing that thinning declines (i.e. becomes less negative) with increasing altitude. The lack of correlation between thinning and the other terrain

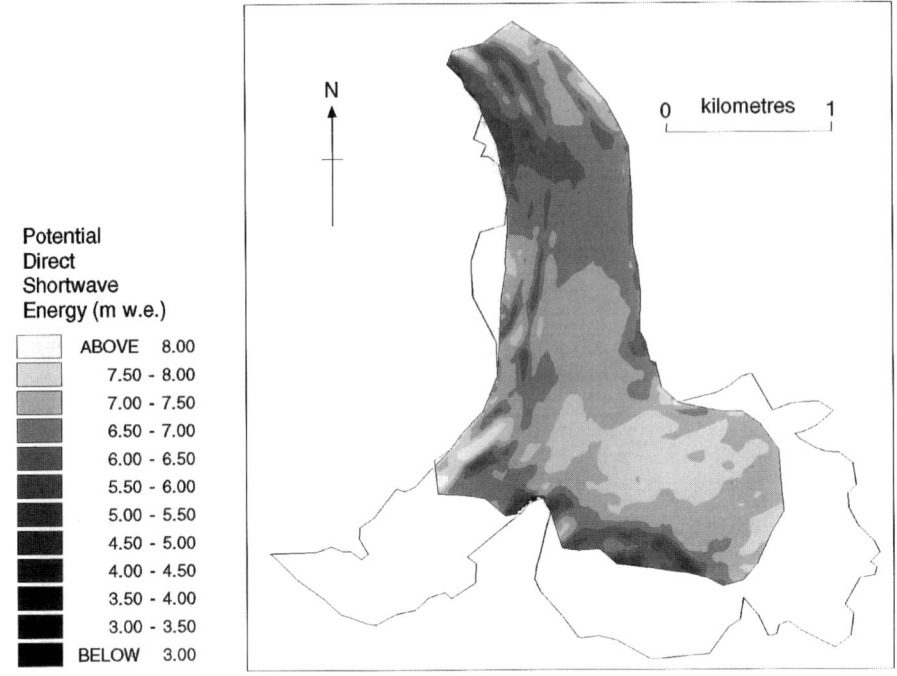

Potential
Direct
Shortwave
Energy (m w.e.)

☐	ABOVE 8.00
	7.50 - 8.00
	7.00 - 7.50
	6.50 - 7.00
	6.00 - 6.50
	5.50 - 6.00
	5.00 - 5.50
	4.50 - 5.00
	4.00 - 4.50
	3.50 - 4.00
	3.00 - 3.50
	BELOW 3.00

Figure 15.6 Map of potential direct short-wave radiation across Haut Glacier d'Arolla between 20 May and 20 September 1993

characteristics may be due to the dominant influence of altitude. To investigate this, a stepwise multiple regression analysis was performed with thinning as the dependent variable and the four terrain attributes as potential independent variables. Taken together, variations in all four terrain attributes contribute significantly ($p > 0.001$) to the variation in thinning. The resulting regression equation is shown in Table 15.1. The signs of the model coefficients show that thinning declines (i.e. becomes less negative) with increasing altitude, increasing slope angle, more northerly aspect, and decreasing potential short-wave radiation (increasing shading). Fifteen per cent of the variation in surface lowering may be explained in terms of these terrain variations.

The statistical analysis demonstrates that about 20% of the spatial variability in surface lowering is due to variation in ablation, and about three-quarters of this may be accounted for by variation in altitude, slope angle, slope aspect and shading, which themselves will influence melt. This highlights the fact that annual surface lowering is not the same as annual ablation. Surface lowering of a grid cell is not only a function of ablation but also of the flux divergence across the cell. In the next section, we use the measurements of annual surface height change (i.e. thickening and thinning) determined from the difference between the 1992 and 1993 DTM surfaces, together with the calculations of annual net surface mass balance (i.e. accumulation and ablation) from the measurements of winter accumulation and calculations of summer ablation, to compute the spatial pattern of the annual flux divergence across the glacier. We then use these to estimate the spatial pattern of annual velocity gradient.

Spatial Pattern of Glacier Flux Divergence and Velocity Gradient

It was shown above that the net volume change across the whole glacierised area was -2.31×10^6 m^3 w.e. whereas the net surface mass balance was -3.59×10^6 m^3 w.e. This gives a difference of -1.28×10^6 m^3 w.e. As the area for which the surface height change and mass balance were calculated is predominantly the ice- and firn-covered parts of the glacier, this implies that an ice flux of about 1.28×10^6 m^3 w.e. crossed the 1992–1993 ELA and flowed from the accumulation area into the ablation area between September 1992 and September 1993. At the 1992–1993 ELA (i.e. 3012 m a.s.l.), the glacier has an average thickness of approximately 70 m and is approximately 2.5 km wide (Figure 15.1) giving a width- and depth-averaged ice velocity across the 1992–1993 ELA of about 7.3 m a^{-1}. This compares favourably with a measured centreline surface velocity at the 1992–1993 ELA of about 8 m a^{-1} (see below).

The annual surface height change of a block of glacier is a function of the annual surface mass balance, the annual basal mass balance, and the depth-averaged longitudinal and transverse divergence of ice flux across the ice block (Paterson, 1994). For each 20×20 m grid cell of the DTM, this may be expressed in terms of the continuity equation for ice thickness as:

$$\frac{dh}{dt} = b + b' - \frac{dQ_x}{dx} - \frac{dQ_y}{dy} \tag{15.1}$$

where h is the ice thickness; b is the ice mass per unit area and time added or removed at the surface; b' is the ice mass per unit area and time added or removed at the bed; dQ_x is the mass flux in the ice flow direction, x; and dQ_y is the mass flux orthogonal to the ice flow direction, y.

On typical valley glaciers, b' is small compared with b and we can assume that $b' = 0$. Furthermore, if we assume that the ice deforms in plain strain (i.e. $dQ_y = 0$), equation (15.1) can be reduced and rearranged to:

$$\frac{dQ_x}{dx} = b - \frac{dh}{dt} \tag{15.2}$$

Clearly, the assumption that ice deforms in plain strain is not necessarily valid, especially where flowlines converge or diverge such as near the glacier snout, towards the glacier sides, or where ice flow from the upper basins converges on the glacier tongue. Nevertheless, for each grid cell of the DTM, the ice flux divergence in the direction of flow was approximated by subtracting the measured annual ice thickness change from the calculated annual surface mass balance.

The ice flux divergence can be related to the glacier velocity gradient as follows:

$$\frac{\mathrm{d}Q_x}{\mathrm{d}x} = h\frac{\mathrm{d}\bar{u}}{\mathrm{d}x} + \bar{u}\frac{\mathrm{d}h}{\mathrm{d}x} \tag{15.3}$$

where $\bar{u}$ is the depth-averaged ice velocity. This equation has two unknowns ($\mathrm{d}\bar{u}/\mathrm{d}x$ and $\bar{u}$) and so it has no analytical solution. But if we assume that the change in ice thickness in the ice flow direction is negligible (i.e. $\mathrm{d}h/\mathrm{d}x = 0$), which is not unreasonable over distances of between 20 and 28.28 m, then equation (15.3) becomes:

$$\frac{\mathrm{d}\bar{u}}{\mathrm{d}x} = \frac{1}{h}\left(\frac{\mathrm{d}Q_x}{\mathrm{d}x}\right) \tag{15.4}$$

For each grid cell of the DTM, the velocity gradient in the direction of ice flow was approximated by dividing the flux divergence by the ice thickness.

Ice thickness was determined for each grid cell of the DTM by subtracting the 1993 glacier surface elevation from the glacier bed elevation. Gridded values of glacier bed elevation were obtained by linear interpolation of 242 measurements of bed elevation which had been obtained by ground survey and radio echo-sounding techniques in 1989 and 1990 (Sharp et al., 1993).

Over most of the glacier $\mathrm{d}\bar{u}/\mathrm{d}x$ is negative and therefore flow is compressive (Figure 15.7). Rates of compression vary across the glacier, with increasing compression towards the lower part of the glacier tongue and towards the glacier sides, although some of the increase in compression in these areas is probably due to the assumption of zero ice flux orthogonal to the ice flow direction. There are some areas where $\mathrm{d}\bar{u}/\mathrm{d}x$ approaches zero and a few where it becomes positive such that flow is extending. Small areas of extending flow occur near the head of the glacier in the upper eastern and western basins, on the inside of the bend where the eastern basin meets the glacier tongue, towards the central part of the upper tongue, and on the eastern part of the lower tongue. These are all areas where surface crevasses are found, which gives us confidence in the accuracy of the estimates of glacier velocity gradients.

The calculated depth-averaged velocity gradients shown in Figure 15.7 should be most accurate for the centreline where the assumption of zero ice flux orthogonal to

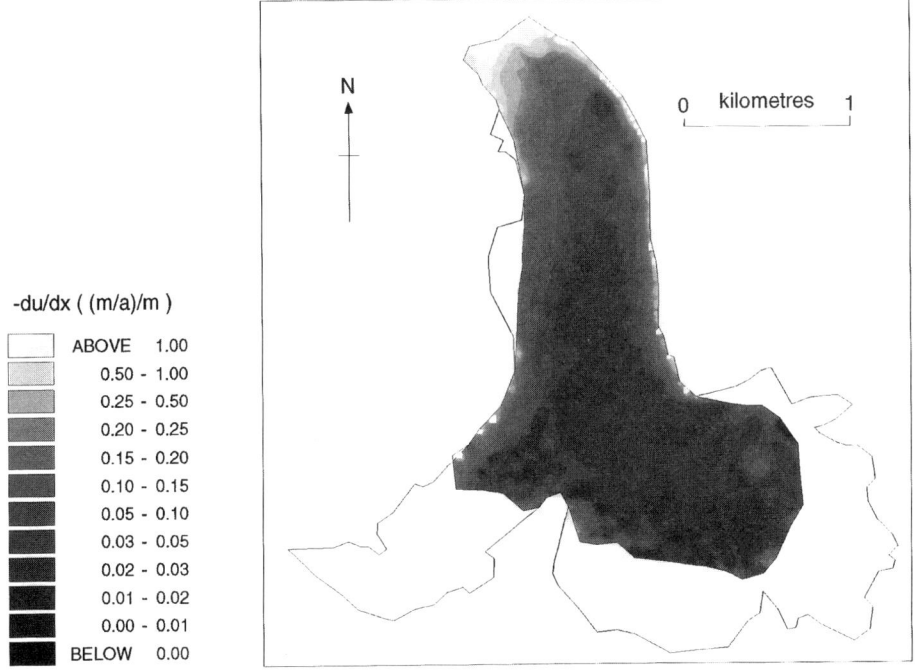

-du/dx ((m/a)/m)

ABOVE	1.00
0.50 -	1.00
0.25 -	0.50
0.20 -	0.25
0.15 -	0.20
0.10 -	0.15
0.05 -	0.10
0.03 -	0.05
0.02 -	0.03
0.01 -	0.02
0.00 -	0.01
BELOW	0.00

N

0   kilometres   1

Figure 15.7 Map of velocity gradients in the direction of glacier flow. The map shows $-(d\bar{u}/dx)$ so that deceleration downglacier is indicated by positive values. Thus, most of the glacier is undergoing compressive flow

the ice flow direction is probably most valid. The depth-averaged velocities along the glacier centreline (the line of stakes in Figure 15.1) were calculated by integrating the values of $d\bar{u}/dx$ downglacier from Stake N just below the 1992–1993 ELA where the velocity was set to 8.0 m a^{-1} (the annual surface velocity measured here between 30 August 1994 and 24 August 1995). Calculated depth-averaged velocities increase from 6.5 m a^{-1} at Stake C to 8 m a^{-1} at Stake N (Figure 15.8). The annual surface velocities measured along the glacier centreline between 30 August 1994 and 24 August 1995 are also shown on Figure 15.8. The observed velocities increase more steeply on the glacier tongue than the calculated velocities, ranging from 6.1 m a^{-1} at Stake C to 11.7 m a^{-1} at stake H before declining in the upper eastern basin to 8 m a^{-1} at Stake N. The discrepancy is presumably because the calculated velocities are averaged over the glacier thickness, whereas the observed velocities are measured at the surface. As the glacier thickness increases between Stake C and Stake H and then decreases to Stake N (Figure 15.1), the results suggest that the ratio of surface to depth-averaged velocity is dependent on

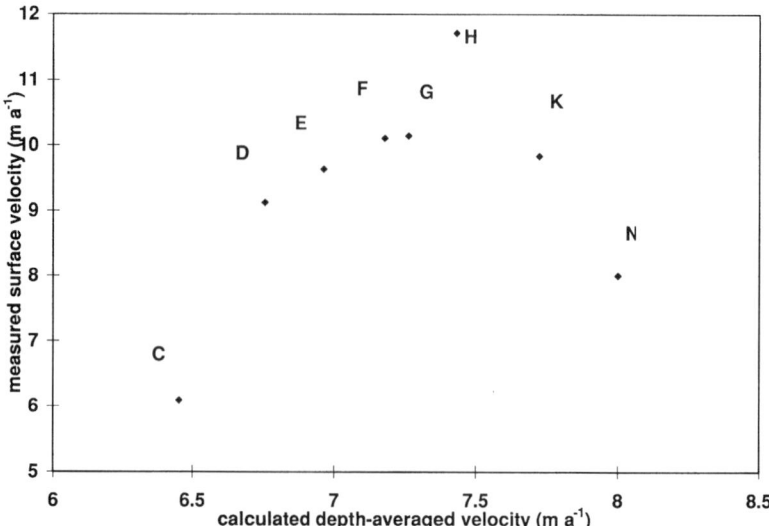

Figure 15.8 Plot of calculated depth-averaged velocities (September 1992 to September 1993) against measured surface velocities (August 1994 to August 1995). The letters refer to the centreline stakes shown in Figure 15.1

ice thickness. This implies that the relative contribution of ice deformation to surface motion increases with glacier thickness. Thus, where ice is thin (e.g. at Stakes C and N), a relatively high proportion of surface movement is due to basal motion, but where ice is thickest (i.e. at Stake H), a higher proportion of surface movement is due to ice deformation.

CONCLUSIONS

Digital terrain models are useful for studying spatial variations in the mass balance and flow of valley glaciers. Our DTM data show that most of Haut Glacier d'Arolla experienced net thinning over the mass balance year 1992–1993, presumably due to net ablation. A few areas near the head of the glacier experienced net thickening presumably due to net accumulation. Some areas near the margins of the glacier tongue also experienced a net thickening, probably due to low ablation rates (due to shading by surrounding mountains and a high surface debris cover supplied from the valley sides) and high rates of longitudinal compression. Our energy balance model calculations confirm that most of Haut Glacier d'Arolla experienced net ablation over the mass balance year 1992–1993. There was a strong longitudinal gradient of net ablation due to the effects of altitude on: (i) winter accumulation and the length of time that ice is exposed during the summer; and (ii) air temperature and vapour pressure and therefore the turbulent energy fluxes.

About 20% of the spatial variation in glacier thinning over the mass balance year 1992–1993 is accounted for by spatial variation in ablation. About 15% of the variability in thinning may be explained by variations in the four terrain characteristics of altitude, slope angle, slope aspect and shading, presumably due to the effects of these attributes on surface melting. Some of the variation in glacier thinning not accounted for by ablation may be explained by errors in the measurements of thinning or, more likely, in the calculations of the energy balance components. Most of the remaining variation in glacier thinning will be due to spatial variations in the longitudinal and transverse flux divergence.

Our data enable us to estimate the pattern of the gradient of annual depth-averaged velocity across the glacier using a technique that has not been used hitherto. The technique assumes that the variation in ice depth across the DTM grid cells in the direction of glacier flow is negligible, so that an analytical solution to the continuity equation for ice thickness can be obtained. A numerical solution could be obtained using a finite-difference iteration scheme, although the result would probably not be substantially different from our approximation. The results suggest that the flow of Haut Glacier d'Arolla is compressive along most of its length, with strong rates of compression towards the snout and near the glacier margins. There are a few areas of extending flow which correspond with the position of crevasses. The pattern of annual depth-averaged velocity along the glacier centreline calculated from the pattern of velocity gradient is supported by recent field measurements of annual surface velocity. The results suggest, however, that the ratio of surface to depth-averaged velocity is dependent on ice thickness and that basal motion contributes relatively more to surface motion where ice is thin.

ACKNOWLEDGEMENTS

We gratefully acknowledge financial support from NERC (Grants GR3/8971 and GR3/8114) and Grande Dixence SA. The Weather Station was borrowed from the NERC equipment pool. We thank Bureau Elzingre for organising the photogrammetric work and production of the DTMs. We also thank all members of the Arolla Glaciology Project who helped collect the data, especially Luke Copland (early summer snow survey), Pete Nienow and Michael Nielsen (ground survey for photogrammetry) and Ben Brock (ablation survey). The annual surface velocities along the glacier centreline were measured by Pete Nienow and Doug Mair. Finally, we thank Patricia and Basile Bournissen and Yvonne Bams for logistical help in Arolla.

REFERENCES

Arnold, N. S., Willis, I. C., Sharp, M. J., Richards, K. S. and Lawson, W. J., 1996. A distributed surface energy balance model for a small valley glacier. I. Development and testing for Haut Glacier d'Arolla, Valais, Switzerland. *Journal of Glaciology*, **42**, 77–89.

Björnsson, H., 1988. *Hydrology of Ice Caps in Volcanic Regions*. Viisindafélag Íslendinga, Societas Scientarium Islandica, 45, Reykjavík.

Brunner, K., 1987. Glacier mapping in the Alps (with three map sheets). *Mountain Research and Development*, **7**, 375–385.

Escher-Vetter, H., 1985. Energy balance calculations for the ablation period 1982 at Vernagtferner, Oetztal Alps. *Annals of Glaciology*, **6**, 158–160.

Etzelmüller, B., Vatne, G., Ødegård, R. S. and Sollid, J. L., 1993. Mass balance changes of surface slope, crevasse and flow pattern of Erikbreen, northern Spitsbergen: an application of a geographical information system. *Polar Research*, **12**, 131–146.

Fountain, A. G. and Vaughn, B. H., 1995. *Changing drainage patterns within South Cascade Glacier, Washington, USA, 1964–1992*. International Association of Hydrological Sciences Publication, 228, 379–386.

Gratton, D. J., Howarth, P. J. and Marceau, D. J., 1993. Using Landsat-5 Thematic Mapper and digital elevation data to determine the net radiation field of a mountain glacier. *Remote Sensing of Environment*, **43**, 315–331.

Haakensen, N., 1986. Glacier mapping to confirm results from mass-balance measurements. *Annals of Glaciology*, **8**, 73–77.

Holmlund, P., 1988. An application of two theoretical melt water drainage models on Storglaciären and Mikkaglaciären, northern Sweden. *Geografiska Annaler*, **70A**, 1–7.

Knudsen, N. T., 1986. Recent changes of Nordbogletscher and Nordgletscher, Johan Dahl Land, South Greenland. *Annals of Glaciology*, **8**, 106–110.

Krimmel, R., 1989. Mass balance and volume of South Cascade Glacier, Washington 1958–1985. In *Glacier Fluctuations and Climatic Change*, ed. J. Oerlemans, Kluwer, The Netherlands.

Mayo, T. R., 1993. *Intelligent systems for cartographic data capture*. Unpublished PhD Thesis, University of Cambridge.

Moore, I. D., Grayson, R. B. and Ladson, A. R., 1991. Digital terrain modelling: a review of hydrological, geomorphological, and biological applications. *Hydrological Processes*, **5**, 3–30.

Munro, D. S. and Young, G. J., 1982. An operational net shortwave radiation model for glacier basins. *Water Resources Research*, **18**, 220–230.

Oerlemans, J., 1993. A model for the surface balance of ice masses: Part 1. Alpine glaciers. *Zeitschrift für Gletscherkunde und Glazialgeologie*, **27/28**, 63–83.

Østrem, G. and Brugman, M., 1991. *Glacier Mass Balance Measurements, A Manual for Field and Office Work*. NHRI Science Report No 4.

Parrot, J. F., Lyberis, N., Lefauconnier, B. and Manby, G., 1993. SPOT multispectral data and digital terrain model for the analysis of ice-snow fields on arctic glaciers. *International Journal of Remote Sensing*, **14**, 425–440.

Paterson, W. S. B., 1994. *The Physics of Glaciers*, 3rd edition. Pergamon, Oxford.

Reinhardt, W. and Rentsch, H., 1986. Determination of changes in volume and elevation of glaciers using digital elevation models from the Veragtferner, Ötztal Alps, Austria. *Annals of Glaciology*, **8**, 151–155.

Rentsch, H., Welsch, W., Heipke, C. and Miller, M. M., 1990. Digital terrain models as tools for glacier studies. *Journal of Glaciology*, **36**, 273–278.

Richards, K., Sharp, M., Arnold, N., Gurnell, A., Clarke, M., Tranter, M., Nienow, P., Brown, G., Willis, I. and Lawson, W., 1996. An integrated approach to modelling hydrology and water quality in glacierised catchments. *Hydrological Processes*, **10**, 479–508.

Sharp, M., Richards, K., Willis, I., Arnold, N., Nienow, P., Lawson, W. and Tison, J.-L., 1993. Geometry, bed topography and drainage system structure of Haut Glacier d'Arolla, Switzerland. *Earth Surface Processes and Landforms*, **18**, 557–571.

UNIRAS, 1990. *Unimap 2000 Users Manual*. Version 6. UNIRAS Ltd, Soborg, 255pp.

Walraven, R., 1978. Calculating the position of the sun. *Solar Energy*, **20**, 393–397.

Zevenbergen, L. W. and Thorne, C. R., 1987. Quantitative analysis of land surface topography. *Earth Surface Processes and Landforms*, **12**, 47–56.

16 The Transient Snowline on Glaciers: Topographic Controls and Implications for Melt Predictions

O. C. TURPIN[1,2], R. I. FERGUSON[1] and C. D. CLARK[2]
[1] Department of Geography, University of Sheffield, UK
[2] Sheffield Centre for Earth Observation Science, University of Sheffield, UK

ABSTRACT

The transient snowline (TSL) on glaciers is important in meltwater modelling because ice absorbs more radiant heat energy than snow. The TSL may be irregular and patchy because of local spatial variability in winter snow accumulation and subsequent snow-melt, but in simple glacier hydrology models for use in data-poor situations, the TSL is often approximated by a line parallel to the contours that rises in response to snow-melt. This paper explores the implications of this assumption. A digital elevation model for a Swiss glacier and a Landsat image part way through the melt season show that the "snowline" is actually a mosaic of snow and ice extending over a 600 m vertical range. Elevation is the best predictor of the snow/ice distribution within this region and slope angle is the only other topographic variable that assists explanation. A simple mathematical analysis suggests that approximating the TSL by a horizontal line with the same snow-covered area will lead to overestimation of meltwater production, but only by a few per cent. This is confirmed by applying a glacier melt model to the snow/ice mosaic derived from the Landsat image and to the same areas of snow and ice divided by a TSL at constant elevation. The assumption of a simple horizontal TSL therefore appears to be unrealistic in detail, but acceptable for hydrological modelling in data-poor situations.

INTRODUCTION

Meltwater from alpine glaciers and seasonal snowpacks forms the majority of annual runoff in rivers draining major mountainous regions such as the European Alps, the Rocky Mountains and the greater Himalaya. It is a vital resource for drinking water and crop irrigation in dryland environments near high mountains (e.g. Butz, 1989) and also for hydroelectric power in many countries (e.g. Bezinge, 1987). Accurate daily and seasonal predictions of glacier runoff are of great value to applied hydrologists seeking to maximise the practical and economic value of the resource. There is also growing concern at the possible sensitivity of high-mountain water

Landform Monitoring, Modelling and Analysis. Edited by S. N. Lane, K. S. Richards and J. H. Chandler.
© 1998 John Wiley & Sons Ltd.

resources to global warming (Oerlemans and Fortuin, 1992; Willis and Bonvin, 1995).

This paper reports one aspect of a project seeking to develop a relatively simple physically based model for daily and seasonal runoff from alpine glaciers. The model is designed to require minimal data input and to be readily transferable to glaciers in remote or underdeveloped areas. When melt is predicted, or snow falls, the model recalculates the elevation of the transient snowline (TSL) which separates snow from bare ice. The seasonal rise in TSL up alpine glaciers is important because radiative energy usually accounts for the majority of melt (Megahan et al., 1967; Paterson, 1994) and glacier ice absorbs far more radiation than snow, so TSL rise controls total runoff and its distribution through the melt season (Munro and Young, 1982).

Glacier snowlines can be irregular and patchy because of spatial variations in initial snow cover and subsequent melting. In simple models, however, the TSL is often assumed to rise parallel to the surface contours and to depend only on elevation. In this paper we consider (1) whether the distribution of snow and ice in the vicinity of the TSL can be better predicted if topographic variables additional to elevation are considered; and (2) whether approximating an irregular snowline by a horizontal line makes much difference to predictions of snow and ice melt. In order to achieve these aims we couple a classified melt season satellite image of a Swiss glacier to a digital elevation model (DEM), perform statistical analyses of snow/ice cover within the region of the TSL in relation to a variety of topographic variables derived from the DEM, analyse mathematically the implications for meltwater generation, and support this analysis by running a melt model first with the classified distribution of snow and ice and then with the same total snow and ice areas separated at a fixed elevation.

BACKGROUND

The Physical Basis for Modelling Glacier Runoff

A conceptually based approach to modelling runoff from high mountain basins containing glaciers must take account of the main hydrological processes and the initial and boundary conditions for their operation. Runoff in such basins is a combination of off-glacier snowmelt, glacier snowmelt, glacier icemelt, and rainfall. At the end of winter, the snow-covered area (SCA) is extensive and normally no glacier ice is exposed. Snow water equivalent (SWE) generally increases with elevation, but may also depend on other topographic properties such as slope and aspect due to the influence of wind direction and avalanching during and after snowfall. In alpine basins the snowpack is usually isothermal at the melting point from early in the melt season, so heat energy inputs cause surface melting. In early season there may be significant meltwater storage in the snowpack and sub-glacial cavities (e.g. Östling and Hooke, 1986), but later the englacial and sub-glacial drainage system typically drains surface melt to the proglacial river within hours or at most days (e.g. Lang, 1973). As melting proceeds, snow is progressively removed from off-glacier areas and from lower parts of the glacier, resulting in the exposure of ice. The TSL separating ice from snow is important because ice has a lower albedo, hence absorbs more solar radiation and melts faster. The seasonal regime of runoff thus depends not just on heat supply but also on TSL rise, and generally peaks one to two months after the midsummer maximum of potential radiation (e.g. Elliston, 1973).

In principle all this can be modelled (e.g. Baker *et al.*, 1982; Blöschl *et al.*, 1991; Arnold *et al.*, 1996), but the data requirements are high. Energy balance computations require meteorological data extrapolated to the relevant elevations and corrected for shading by cloud and topography; values are also required for surface parameters such as albedo and aerodynamic roughness that can be temporally and spatially variable beyond the basic difference between snow and ice (US Army Corps of Engineers, 1956; Brugman, 1991; Van de Wal *et al.*, 1992; Braithwaite, 1995). The initial spatial distribution of SWE must be known and account must be taken of any summer snowfall. Meltwater must be routed through the snowpack and the englacial and sub-glacial conduit system, which probably evolves through the melt season (e.g. Gurnell *et al.*, 1992). Great progress has been made in spatially distributed modelling of these coupled processes in research catchments by using automatic weather stations, DEMs, intensive ground surveys of SWE, and borehole and tracer studies of meltwater routing (e.g. Richards *et al.*, 1996). But this fully distributed approach requires initial conditions and subsequent input data for what will typically be a minimum of the order of 10^4 grid cells. At present these data requirements are very demanding and expensive.

Semi-distributed Models for Glacier Runoff

Whilst fully distributed models are required to obtain maximum understanding of glacier hydrology, there is a clear complementary requirement for simpler models that can be used to simulate seasonal runoff regimes and totals in data-poor situations, for use in water resource planning and in exploring scenarios of climate change. In rainfall runoff hydrology the natural approach for data-poor situations is a spatially lumped model. However, the fundamental importance of TSL rise for glacier melting suggests that a semi-distributed approach which differentiates between snow and ice is preferable, even when data are sparse. The present paper reports on aspects of the building and testing of a semi-distributed model.

The importance of TSL rise for glacier runoff has much in common with the importance of snowpack depletion for nival runoff. In both cases the decline in SCA over time must be modelled or otherwise prescribed, even though in the nival case runoff declines as bare ground is exposed whereas in the glacial case runoff rises as faster melting ice is exposed. SCA depletion (or TSL rise) can be handled in several ways. The first is to impose it exogenously through a function relating SCA to elapsed time (e.g. Martinec, 1976) or cumulative degree-days (Rango and Martinec, 1982), or on the evidence of repeated ground observation or remotely sensed data (e.g. Baker *et al.*, 1982; Munro and Young, 1982; Baumgartner *et al.*, 1986). Alternatively, SCA can be made endogenous using parametric assumptions about the spatial distribution of initial SWE within the basin and thus the relationship between loss of SWE and reduction in SCA, as done by Gottlieb (1980) for a glacier and Ferguson (1984, 1986) for transient snowpacks. Some models (e.g. Ferguson, 1984) treat the entire snow-covered area as a unit for melt calculations; others (e.g. Rango and Martinec, 1982) divide the catchment into several elevation zones within each of which SCA changes over time. In this paper, TSL rise is modelled endogenously within elevation zones. In keeping with the simplified nature of such models, melt calculations are invariably done using either a simple temperature index or a parametric approximation of the energy balance, possibly with allowance for seasonal change in surface properties such as albedo but not for spatial variability within them.

FIELD SITE, DEM AND IMAGE ANALYSIS

In this section we describe the Swiss glacier used in this study and explain the techniques used to create the DEM, couple a Landsat image to it, and derive a classification of snow and ice from the digital image. These provide a site-specific data base for subsequent analyses of topographic controls on the distribution of snow and ice, and of the effect of snow/ice distribution on melt production.

The Findelengletscher Basin

The results reported in this paper relate to the Findelengletscher, Valais, Switzerland. The gauged basin has an area of 24.6 km^2 of which 76% is glacierised (Collins, 1995). The glacier flows approximately east–west, extending from over 4100 m above sea level (a.s.l.) to about 2500 m at the snout. Discharge in the proglacial stream is measured hourly during the melt season by the hydroelectric company Grande Dixence SA who also maintain a meteorological station near the glacier snout that records hourly temperature, precipitation and incoming radiation.

Digital Elevation Model

A DEM of the glacier was created using Arc/Info. The contours, spot heights and boundary of the Findelengletscher taken from the 1988 edition 1:25 000 Landes-karte der Schweiz topographic map were digitised to create line and point cover-ages. These were used to create a triangulated irregular network (TIN). The original version of the TIN contained a number of erroneous flat triangles caused by the selection of three points of the same elevation. Extra interpolated points were digitised to resolve these problems. The TIN was interpolated to form a 30 m raster grid, using quintic interpolation which yields continuous first-order derivatives such as slope (ESRI, 1991).

   Without additional data it is not possible to test the accuracy of the DEM, but an indication can be obtained by seeing how well the map spot heights are predicted by a DEM created without them, i.e. from contours alone. Spot heights representing peaks or closed hollows were excluded, reducing the number used to 22, as it would be unrealistic to expect a contour-based DEM to interpolate these features accurately.

The root mean square error (RMSE) was 3.7 m, and we presume that the DEM using spot heights as well as contours is slightly more accurate than this.

Satellite Image Analysis

The extent of the snow- and ice-covered areas on the glacier was estimated from a Landsat Thematic Mapper (TM) image (path 195/row 28) acquired on 7 July 1984 with solar elevation 58° and azimuth 127°. Processing of the image consisted of seven steps: geometric correction, registration of the image to the DEM, masking of non-snow/ice pixels, unsupervised classification of remaining pixels into spectral clusters, grouping of spectral clusters into homogeneous land cover classes, post-classification, and masking of pixels outside the glacier basin.

The few available ground control points were used for geometrical correction of the image to the Swiss national grid system; the RMSE was 1.7 pixels (51 m). The corrected image and the DEM possess the same map coordinates making it possible to overlay them. The 30 m resolution of the DEM allows direct comparison of topography with pixel values in the image. In this investigation we were only concerned with the area of the Findelengletscher, so the rest of the image was discarded to save subsequent processing time. As snow/ice pixels have a very low reflectance in TM band 5 (mid-infrared) non-snow/ice pixels were masked out using a threshold value to reduce the spectral variability within the image to just the snow and ice facets. The remaining pixels were grouped into 20 spectral clusters using an ISODATA unsupervised clustering algorithm with all seven TM bands as input, then allocated to one of four classes: "snow", "ice", "intermediate" or "other" (Figure 16.1). No ground truth data simultaneous with the satellite overpass were available so the spectral clusters were assigned to surface types using a combination of visual examination of the image, personal knowledge of the field area, and published information on the spectral response of snow and ice (e.g. O'Brien and Munis, 1975; Zheng et al., 1984; Hall et al., 1989).

The spectral variability of the snow-covered area was spread over 10 spectral clusters, that of bare ice over five. Two small clusters that occurred along the boundaries between the ice and snow classes, like the "transition zone" of Baumgartner et al. (1987), were labelled intermediate. Their spectral response may be due to the occurrence of mixels and/or related to differential shading across the glacier. The "other" class contain the non-snow/ice pixels that had too low a reflectance in band 5 to be masked out in the earlier processing step. A few pixels classified as ice or intermediate were so high in the accumulation area that they were interpreted as snow in shadow, which has a similar spectral signature to ice. All such pixels above 3400 m were reclassified as snow. The area outside the glacier basin was removed by multiplying the image by a mask of the basin.

TOPOGRAPHIC CONTROLS OF SNOW AND ICE DISTRIBUTION

Over the full altitudinal range of an alpine glacier, elevation has an overriding control on the distribution of snow and ice because winter accumulation of snow

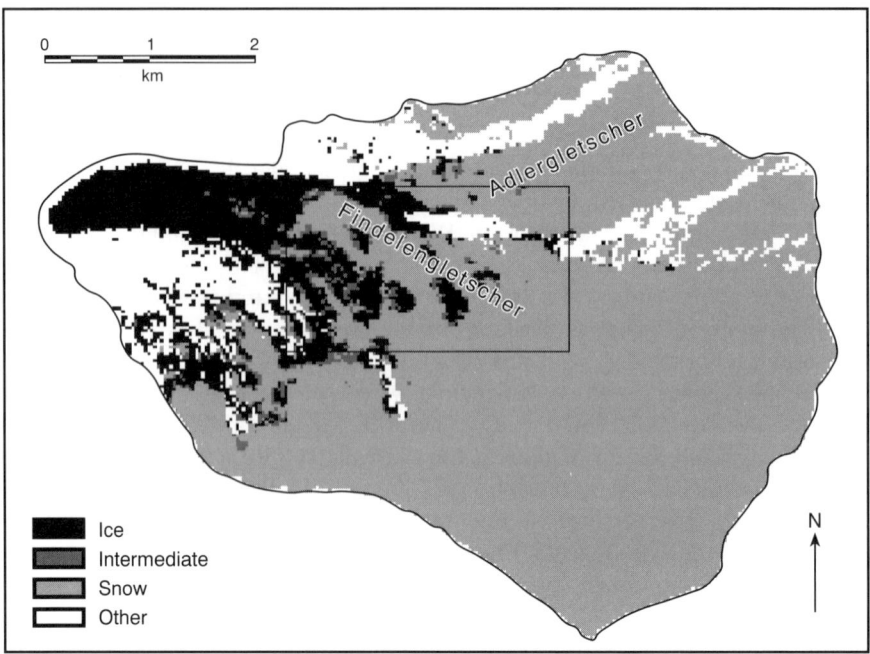

Figure 16.1   Classified Landsat TM scene of the Findelengletscher on 7 July 1984

tends to be higher, and summer melting of snow lower, at higher elevations. A TSL at a single elevation is therefore a reasonable first approximation for modelling purposes. However, other topographic variables influence snow accumulation and melting and may therefore affect the local distribution of snow and ice close to the transient snowline. This suggests the possibility of more refined modelling of TSL rise. In this section we examine the extent to which the division between snow and ice can be better understood if topographic variables additional to elevation are considered.

Derivation of Topographic Variables

In view of the overriding influence of elevation at the glacier basin scale, we restricted the topographic analysis to a rectangular sub-scene, shown in Figure 16.1, which excluded the entirely snow-free glacier snout below 2793 m a.s.l. and the completely snow-covered part above 3385 m. This subscene comprised 2.6 km^2 snow, 1.0 km^2 ice, and 0.2 km^2 other terrain. The glacier flows westwards through it with hummocky ice-cored lateral moraine in the southwest corner and the steep slopes of the tributary Adlergletscher icefall in the northeast corner.

The topographic properties analysed were elevation, slope, aspect, plan curvature, profile curvature and overall curvature. Raster grids of these variables were created from the DEM using the "curvature" function in Arc/Info Grid which fits a nine-point polynomial surface and computes its derivatives, as in Zevenbergen and

Thorne (1987). Overall curvature is calculated as plan curvature minus profile curvature (Moore *et al.*, 1991). Statistical analysis was performed to see to what extent each variable differed between snow and ice areas, and to what extent snow and ice could be discriminated using combinations of variables. Aspect, which is measured on a circular scale, was analysed using directional statistics and on a linear scale using its sine and cosine (east–west and north–south scalars respectively).

Analysis of Topographic Controls Determining Position of TSL

Figure 16.2 illustrates the differences in the statistical distributions of values for snow-covered and bare-ice parts of the sub-scene for each topographic variable. Table 16.1 lists summary statistics and the results of two-sample t-tests for difference in means between snow and ice. It is worth noting that with such large sample sizes (snow 2844, ice 1084), a very small difference in means can be statistically significant even when there is great overlap between the two samples, so that knowing the value of the variable is of very little help in predicting whether the surface is snow or ice. Spatial autocorrelation is also neglected in such tests.

The clearest differences are in elevation and slope angle. Ice pixels in the sub-scene are concentrated at lower elevations than snow, with very little bare ice above 3150 m. Ice pixels are also generally on steeper slopes than snow, partly due to the inclusion of the end of the steeply sloping snow-free Adlergletscher tongue. This may relate to redistribution of new snow from steeper to gentler slopes by wind or avalanches.

The orientation of ice pixels ranges from southwest to north with no prominent mode, but that of snow pixels is more bell-shaped distribution with a mode just north of west. The north–south scalar showed a wider spread of values than the east–west scalar owing to the predominant flow direction of the glacier. Planform, profile and overall curvature are all concentrated close to zero but with very long tails. Although some of these variables show a significant statistical difference between snow and ice, the samples are almost completely overlapping and it seems doubtful whether there is any real physical difference.

The intercorrelations amongst the topographic variables are shown in Table 16.2. Although several correlations are statistically significant, very few show appreciable predictive value. The only correlations with $r^2 > 0.25$ are the inevitable ones between overall curvature and its two components, profile and plan curvature. The correlation between profile and plan curvature is moderately strong. The only other significant correlations are those of the north–south scalar (cosine of aspect) with elevation and slope (both negative), and with the east–west scalar (positive). One unexpected result was the near independence of elevation and slope, which at the scale of the entire basin are positively correlated because of the overall concavity of landform.

Table 16.2 also shows the correlation of each topographic variable with a snow/ice dummy variable (snow = 1, ice = 2). These are statistically equivalent to the t-tests in Table 16.1 and show the same pattern, with the elevation and slope variables correlated most strongly with snow/ice distribution. In an attempt to see whether combinations of topographic variables gave greater explanation of the

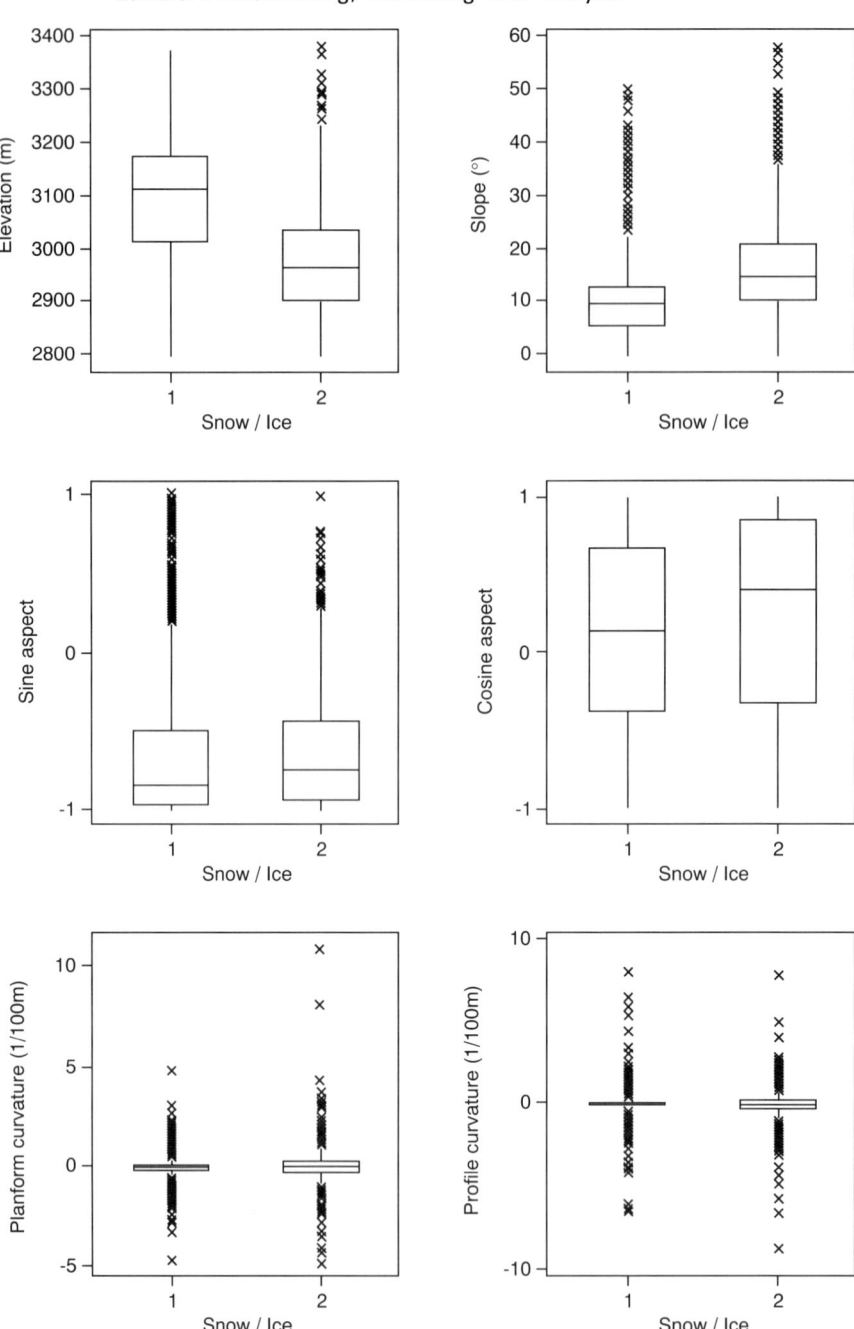

Figure 16.2 Boxplots of topographic variables within the sub-scene, for snow (1) and ice (2) separately. Boxes show medians and quartiles, whiskers extend $1.5(Q_3-Q_1)$ beyond the quartiles, crosses mark outliers. The main differences are in elevation and slope

Table 16.1 Comparison of topographic means, standard deviations and statistical difference for snow and ice classes

Topographic variable	Snow (n=2844) (mean ± s.d.)	Ice (n=1084) (mean ± s.d.)	Significant difference? (p = 0.01)
Elevation (m)	3095 ± 124	2974 ± 102	yes
Slope angle (°)	10.5 ± 6.6	17.3 ± 9.9	yes
Vector orientation (°)[a]	280 and 0.32	291 and 0.32	no[b]
East–west scalar (sine aspect)	−0.67 ± 0.41	−0.64 ± 0.37	no
North–south scalar (cos aspect)	0.12 ± 0.61	0.25 ± 0.63	yes
Overall curvature (1/100 m)	−0.05 ± 0.85	0.10 ± 1.71	yes
Profile curvature (1/100 m)	−0.01 ± 0.53	0.02 ± 0.90	no
Planform curvature (1/100 m)	0.09 ± 0.44	−0.03 ± 1.06	yes

[a] Vector mean direction and circular variance. Note these are not additive
[b] Two-sample test statistic for mean direction (Davis, 1986)

Table 16.2 Correlation matrix between topographic variables and the snow/ice dummy variable. Correlations with an absolute value of 0.06 or greater, shown in bold type, are significant at the p = 0.01 level

	Elevation	Slope	E–W scalar	N–S scalar	Overall curvature	Planform curvature	Profile curvature
Slope	0.02						
E–W scalar	−0.01	0.01					
N–S scalar	**−0.32**	**−0.12**	**0.24**				
Overall curvature	−0.02	0.01	0.00	0.03			
Planform curvature	−0.01	0.03	0.00	0.01	**0.80**		
Profile curvature	0.01	0.00	0.00	−0.04	**−0.87**	**−0.50**	
Snow/ice class	**0.42**	**−0.37**	−0.04	**−0.09**	**−0.06**	**−0.08**	0.02

pattern of snow and ice, multiple linear regression was used to predict the snow/ice dummy variable. This is a form of linear discriminant analysis. Table 16.3 lists the R^2 values obtained with different combinations of topographic predictors. Even when all seven variables were used, only 32% of the variance of the snow/ice dummy variable was explained. Elevation and slope explained 17% and 14% of the variance respectively, and 32% in combination; some of the orientation and curvature variables were statistically significant but they accounted for less than 1% of the snow/ice variance either alone or in combination with elevation and/or slope. The multiple regression equation relating the snow/ice dummy variable to elevation and slope showed that snow cover is more likely at higher elevations, and on lower gradients at a given elevation. A multivariate linear discriminant analysis using this equation correctly classified 72% of the pixels.

Although only 32% of the variability of the snow/ice dummy variable was explained by topographic variables, it still seems likely that topography has a major influence on the occurrence or absence of snow in the region of the TSL. One possible explanation for this paradox is that micro-scale topography at a finer

Table 16.3 Results of simple and multiple regression of snow/ice dummy variable on topographic variables listed in Table 16.1

Predictor(s)	R^2
All variables	0.32
Elevation, slope and planform curvature (best)	0.32
Elevation and slope	0.32
Elevation alone	0.17
Slope alone	0.14
All other variables (alone)	<0.01

resolution than the DEM is important. Alternatively, the interaction of accumulation and ablation at the local scale may result in a snow/ice distribution that is not linearly associated with any of the variables analysed.

A similarly low level of statistical explanation was found by Elder *et al.* (1989) in the much smaller, nival, Emerald Lake Basin in the Californian Sierra Nevada. Only 40% of variance in measured SWE was explained by elevation, slope and net radiation, with the last of these the most important. In the glacial context of the present study, net radiation is itself a function of whether the surface is snow or ice and cannot be used as a predictor. Global radiation may explain some of the observed snow/ice variance, but it is essentially a function of slope and aspect which are already in our predictor set. We calculated direct solar radiation for each pixel (taking account of slope orientation and shading by surrounding relief, but not considering reflected diffuse radiation from neighbouring pixels or shading by clouds) at three-hourly intervals from the beginning of the melt season (assumed 1 May) to the date of the image. The cumulative total radiation on its own explained only 6% of variance in snow/ice cover; when added to the multiple regression on elevation and slope, radiation improved the R^2 by only 4%. These are statistically significant results because of the large sample size, but show very limited explanatory power. This could be because radiation can only explain differences in snow cover produced by differential melting, not differential accumulation. Moreover, the radiation calculations assumed a clear sky with mostly direct radiation, highly sensitive to slope and shadow, disregarding the effect of cloudy days when radiation is far more uniform (Munro and Young, 1982).

An important factor during accumulation is the sloughing or avalanching of snow from steep slopes to less steep slopes below. In the sub-scene we analysed there may have been some redistribution of snow from the tongue of the Adlergletscher onto the Findelengletscher. Gradient is a good indicator of where avalanches are likely to be triggered, while profile curvature indicates where redistributed snow is likely to accumulate, but neither of these variables shows how much redistributed snow a pixel is likely to accumulate. This probably depends on uphill slope, length or area. The variability of snow/ice cover on a glacier may therefore depend on topographic properties not only of individual pixels but also of their neighbours.

EFFECT OF TSL ASSUMPTIONS ON MELTWATER MODELLING

The foregoing image analysis shows that, at least in the case studied, the TSL on a gently sloping alpine glacier in the melt season is an extensive mosaic rather than a simple line at constant elevation. It also shows that the distribution of snow and ice around the TSL is not well predicted from topographic variables derived from a 30 m resolution DEM. These findings are potentially worrying for a simplified modelling approach which approximates the TSL as a line of constant elevation. However, it is important to examine whether the discrepancy between the actual TSL mosaic and the simplifying approximation of a TSL parallel to the contours makes much difference to predicted volumes of meltwater production. We examine this in two stages. The first is a simple but general mathematical analysis. We then look at the specific case of the Findelengletscher in July 1984 using meteorological data and a melt model to compare meltwater production from the TSL mosaic inferred from the Landsat image with that from a horizontal TSL with the same snow-covered area.

Mathematical Analysis of Implications of TSL Pattern for Meltwater Modelling

If radiant heat is a significant contribution to melting, and the albedo of ice is substantially lower than that of snow, it is obvious that predictions of meltwater production are sensitive to errors in the specification or simulation of the SCA. It is less obvious whether the precise nature and position of the TSL matters if the SCA is correct. A simple but general mathematical analysis can be used to show that approximating an irregular TSL by a horizontal TSL with the same SCA normally leads to overestimation of daily meltwater production. Representative parameter values can then be used to assess the likely magnitude of the overestimation.

The analysis compares an assumed horizontal TSL at elevation z with an irregular TSL defining the same snow-covered area (Figure 16.3). The lower diagram has equal areas A_1 of ice at elevations above z and A_2 of snow at elevations below z. Assume that snow at elevations just above z melts at some rate s, ice at the same elevation melts at the higher rate $s(1+r)$, and snow or ice at elevations just below z melts at $(1+k)$ times the rate for the same surface type just above z. By adding up the meltwater production from areas A_1 and A_2 and writing $A_1 = A_2 = A$ it can be seen that the irregular TSL yields $As(2+k+r)$ whereas the horizontal TSL yields $As(2+k+r+kr)$. This is an increase of $Askr$ or in percentage terms $E=100kr/(2+k+r)$. Because of the vertical gradient of melt rate ($k>0$), more extra melt occurs in A_2 when ice replaces snow here than is lost from A_1 when snow replaces ice there. Only if melting was entirely through radiation with no vertical gradient ($k=0$), or entirely through sensible heat with no accentuated melting of ice compared to snow ($r=0$), would the two scenarios produce the same amount of melt ($E=0$). With positive values of k and r there is always a difference, with $E>0$, but calculations with plausible values suggest it will be small because it is unlikely that both k and r will have high values on the same glacier. If melting is mainly by radiation, and the albedos of snow and ice are 0.66–0.88 and 0.21–0.42 respec-

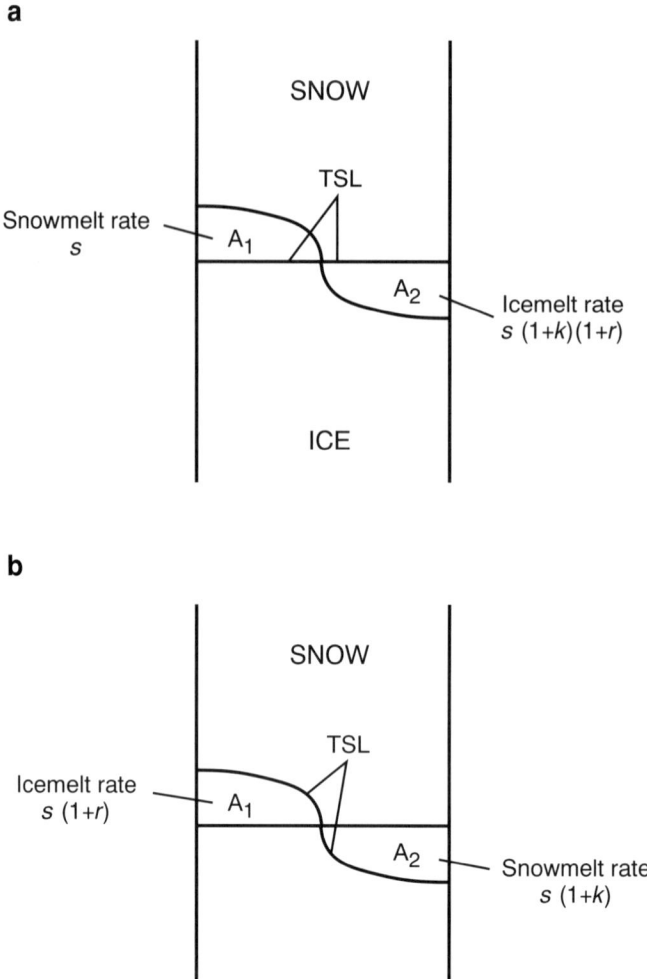

Figure 16.3 Schematic diagram showing theoretical melt for regular (a) and irregular (b) TSL. See text for mathematical analysis and explanation of terms

tively, as quoted by Röthlisberger and Lang (1987) for summer measurements on Aletschgletscher, r could have a value of 1 to 2. The vertical gradient in radiant melting is likely to be small, say $k < 0.1$. In a maritime scenario, with melting mainly by sensible heat, k would be higher, possibly as high as 0.3, but this would be offset by a lower r value, probably less than 0.5. Both scenarios give $E \approx 5\%$ from $A_1 + A_2$, and presumably more like 1 to 2% from the whole glacier since the high melt contribution from the snow-free snout is unchanged. It seems, therefore, that whilst approximating an irregular TSL by a horizontal TSL with the same SCA leads to a systematic error in simulated meltwater production, the error is very

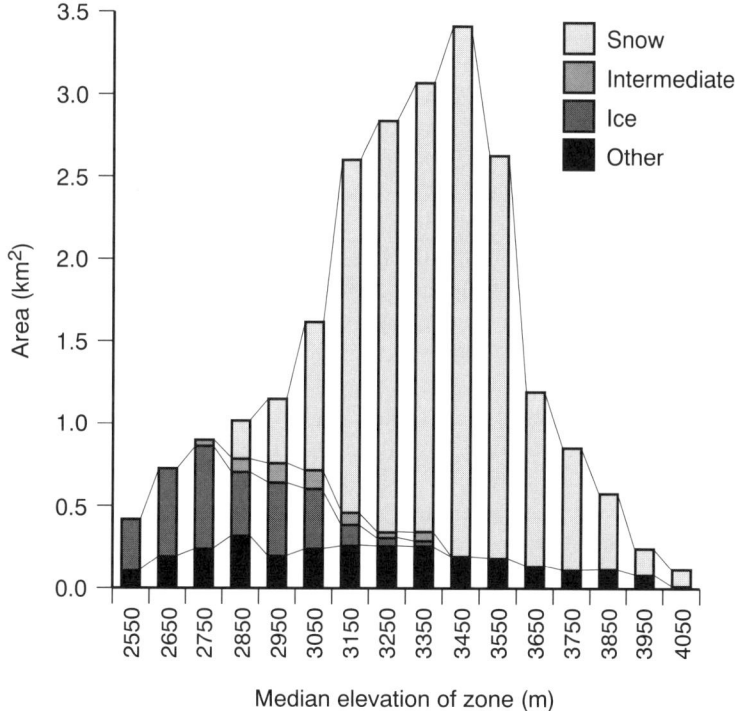

Figure 16.4   Distribution of classes classified from the image by elevation zone. Note the presence of both snow and ice, and of intermediate mixels, between 2700 m and 3400 m

small for all likely circumstances. More substantial errors in predicted meltwater production could of course result from errors in SCA, particularly if melting is dominated by radiation so that the melt rates of ice and snow differ substantially.

Calculation of Difference in Melt Production for TSL Scenarios

The classified image of Figure 16.1 suggests a very irregular TSL on 7 July 1984, distributed over a considerable vertical range. The areas within the snow, ice and intermediate classes in each elevation zone were obtained using the DEM. Exposed ice occupies only 12% of the basin, predominantly in the bottom six of the sixteen 100 m elevation zones (below 3100 m a.s.l.), with small amounts in the next three zones (Figure 16.4). Snow occupies 73% of the basin, mainly above 3000 m but with a few pixels down to 2700 m. Between 2700 and 3400 m there is a progressive increase in the ratio of snow to ice. The intermediate class constitutes only 2% of the whole glacier basin, and is concentrated in the zones where snow and ice coexist, with a maximum in the zone with the closest to equal proportions of snow and ice. This supports the interpretation that "intermediate" pixels are mixels of snow and ice. "Other" pixels make up 12% of the surface area and are distributed throughout the elevation range; they represent snow-free mountain sides, mainly lateral moraines low in the basin and cliffs higher up.

Before calculating meltwater production from this configuration of snow and ice, the "intermediate" area in each elevation zone was divided equally into snow and ice, in line with the interpretation that the intermediate area comprised mixels, and new total snow and ice areas were calculated. An alternative scenario was then constructed in which the same total snow-covered area was defined by a simple snowline parallel to the contours; this fell at an elevation of 2995 m so that the 2900–3000 m zone was predominantly ice, the four lower zones all ice, and the zones above 3000 m all snow. Calculation of the areas A_1 and A_2 (see Figure 16.3) indicates that although the snow/ice mosaic stretches over seven elevation zones, the area of misplaced snow and ice under this assumption is relatively small ($A_1 = A_2 = 0.64$ km^2). A melt model was used to predict daily snow and ice melt rates in each elevation zone using meteorological data for 1–14 July 1984. Predicted total volumes of meltwater generated by the alternative TSL scenarios were then obtained by multiplying the area of snow and/or ice in each zone by the relevant melt rate.

Melting in each zone was computed at hourly intervals by a simplified energy balance model (based on Boyce, 1988). Only the energy resulting from the radiant and sensible heat terms is calculated, as these normally account for the majority of melt occurring on an alpine glacier surface (Paterson, 1994). Net radiation was calculated assuming an ice albedo of 0.4, a snow albedo that declines linearly from 0.7 on 1 May to 0.4 on 30 September, and atmospheric emissivity (used to calculate net long-wave radiation) estimated from diurnal temperature range. Sensible heat supply was estimated from air temperature at the meteorological station near the glacier snout using a lapse rate of 9°C km^{-1}, a wind speed of 10 m s^{-1} in the absence of data, a surface roughness height optimised by Boyce at 1.0 mm, an air density of 1.0 kg m^{-3}, and a glacier surface temperature of 0°C. Diurnal refreezing of meltwater was accounted for by requiring that any negative degree-hours over-night must be eliminated before melt occurs, and any negative overnight radiation balance has to be eliminated before radiant melt occurs.

Since it is the *difference* in predicted meltwater volume that is of interest here, absolute accuracy of the melt model is not essential. The model did in fact give fairly good simulations of measured daily runoff (R^2-type goodness of fit 79% for 1984 using default parameter values, not optimised), and of ice ablation (field measurements in 1994); details of these tests and subsequent development of the model will be reported elsewhere.

The Effect of TSL Pattern on Melt Production from Findelengletscher

The meteorological conditions, and therefore the volume of melt predicted, varied markedly during the period 1–14 July 1984. Mean daily temperature measured at the glacier snout ranged from 2°C to 15°C (mean 9°C). Figure 16.5 shows the sensible and radiative melt predicted by the model for 7 July, when conditions were close to the 14 day average. Sensible energy creates most melt in the lowest four zones, but the depth of sensible melt is reduced up-glacier because of lower temperatures. The up-glacier reduction in radiative melt is more gradual. It occurs

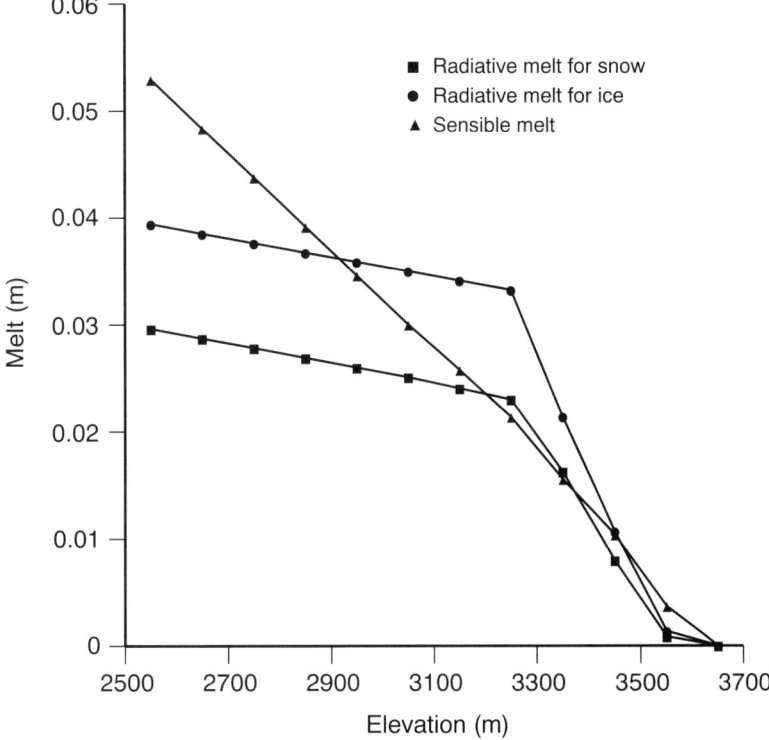

Figure 16.5 Simulated sensible and radiant melt rates in each elevation zone using meteorological conditions on 7 July 1984

because the fixed total radiative melt in each zone has to overcome an overnight cold content (calculated in the way outlined in the last section) which increases with elevation. Radiative melt is greater on ice than snow because of the difference in albedo. Above 3300 m both radiative and sensible melt fall away more rapidly as the negative cumulative daily temperature (simulating nightly snowpack cold content) restricts melt production; no melt occurs above 3700 m.

The total volume of melt over the whole glacier on 7 July was 0.03% greater when a TSL at a constant elevation was assumed than when the classified distribution of snow and ice was used. The assumption of a snowline at constant elevation results in slightly less snowmelt and slightly more ice melt than for the classified distribution. More ice melt occurs because exposed ice has a lower mean elevation and therefore long-wave radiant energy emission is lower. Snow has a higher elevation and therefore melts more slowly, but the difference in albedo means that ice melts faster than snow at any elevation where radiant energy is supplied. Melting due to sensible heat input remains the same under both scenarios. The distribution of melt up the glacier also varies (Figure 16.6). Melt produced by

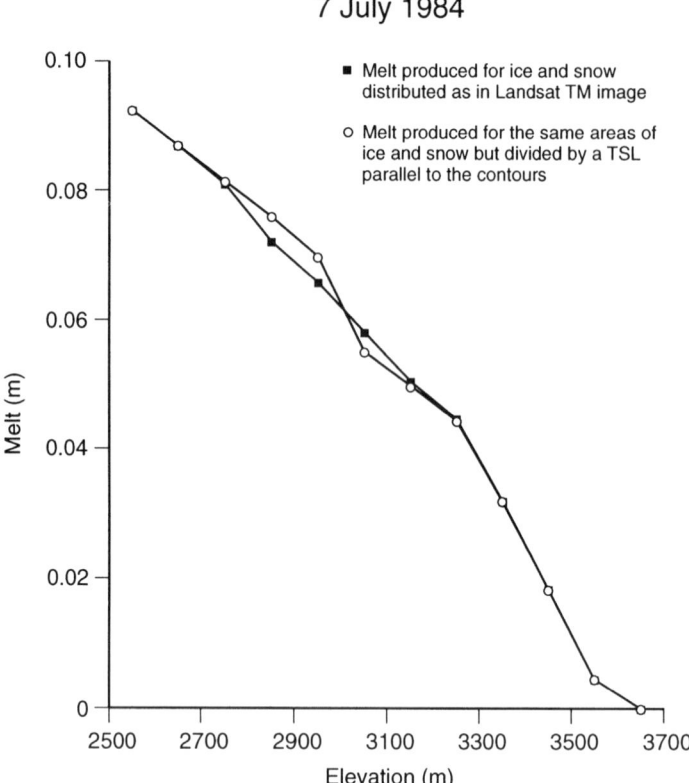

Figure 16.6 Simulated total melt rate in each elevation zone on 7 July 1984 (meteorological conditions typical of the study period) under alternative TSL scenarios. Total melt is 0.03% higher when a TSL parallel to the contours is assumed than when using the patchy snow/ice distribution derived from the Landsat image

the classified distribution of snow and ice reduces more linearly with elevation, whereas there is a prominent drop in melt between 2950 m and 3050 m when a TSL at constant elevation is assumed; this is due to the sharp transition from ice to snow at 2995 m. Above the mean TSL the classified distribution produces more melt up to the maximum ice elevation (3400 m); below the mean TSL it generates less melt down to the minimum snow elevation (2700 m).

The maximum difference in melt between the two scenarios occurred when meteorological conditions allowed melt to occur only over part of the snow/ice mosaic. For example, using the meteorological conditions on 4 July, melt is only predicted by the model up to an elevation of 3000 m (Figure 16.7). On this day radiative melt provides the majority of melt and the difference in melt for ice and snow is proportionately large. The volume of melt for the glacier as a whole is 1.53% higher if a TSL at constant elevation is assumed rather than the distributed

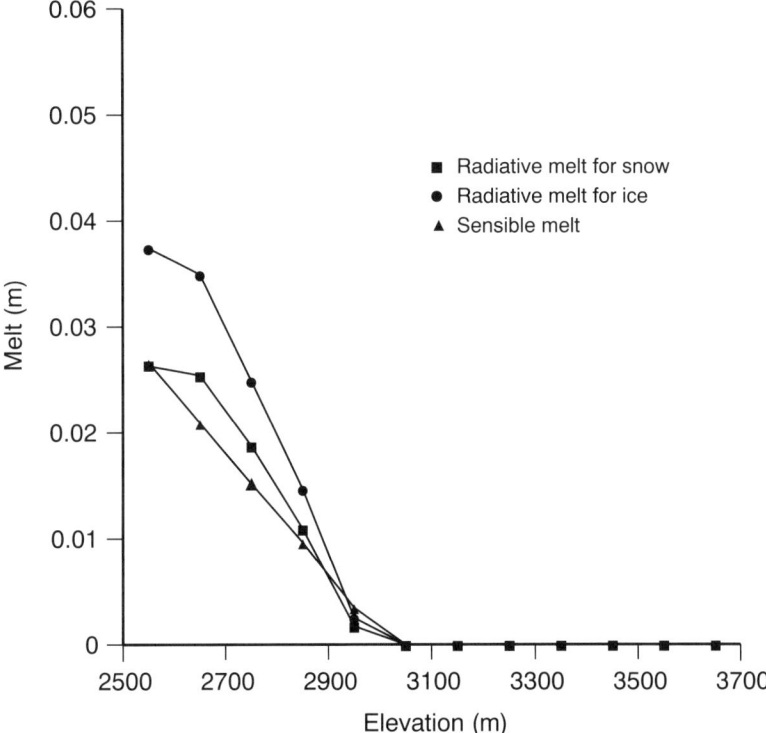

Figure 16.7 Simulated sensible and radiant melt rates in each elevation zone using meteorological conditions on 4 July 1984 (cold but sunny weather). The maximum difference in melt between the two scenarios occurred on this day: TSL parallel to the contours produced 1.53% more melt than the classified snow/ice distribution

snow/ice mosaic. This is not only because the ice now has a lower mean elevation and therefore melts faster, while the snow has a higher mean elevation and melts more slowly, but also because the proportion of ice exposed to melting is higher. This extra melt can be explained mathematically using the notation of Figure 16.3. Instead of assuming that melt occurs across the whole of the snow/ice mosaic, if we assume it only occurs below the level of the mean TSL (i.e. not in area A_1) then it can be seen that the irregular TSL yields $As(1+k)$ while the horizontal TSL generates $As(1+k+r+kr)$, an increase of $Asr(1+k)$.

Very small differences in melt occurred on days with strong melting across the whole glacier. For example, predicted melt on 10 July was dominated by exceptionally high sensible melt up to an elevation of 3500 m. The smaller radiative melt component, and the smaller proportional difference between snow and ice melt, led to a negligible difference between the two scenarios. On 2 July there was insufficient energy to cause predicted melt above 2700 m. The area below this height was entirely snow-free according to the classified image, hence imposing a TSL at constant elevation makes no difference to the volume of melt.

Total melt predicted by the model over the two week period using the snow/ice distribution as in the classified image was 8.32×10^6 m³. Melt for the same period, assuming a TSL at constant elevation, was 0.1% more. Over the two week period, assuming a TSL parallel to contours resulted in an average of 640 m³ more melt per day.

By the beginning of July the albedo of snow that fell during the previous winter is much reduced from its initial high value. Hence the albedo of snow on the days analysed does not differ greatly from the ice albedo. A final analysis was therefore performed with a snow albedo of 0.8 to represent new snow and an ice albedo of 0.2 to determine what effect a simple TSL might have under extreme albedo conditions. When melting occurred across all of the snow/ice mosaic, melt generated using the TSL at constant elevation produced approximately 0.3% more melt than the irregular TSL. This figure varied depending on the proportions of radiative and sensible melt and the intensity of melting. When melting was predicted to occur only in the lower part of the snow/ice mosaic, the difference in melt between the two scenarios was much greater (4.9% using meteorological conditions on 4 July).

CONCLUSIONS

The combination of a Landsat TM scene of the Findelengletscher and a DEM showed that the TSL is not a simple line at a single elevation separating ice from snow, but a mosaic of snow and ice patches grading from completely bare ice to total snow cover. Although this analysis was performed using satellite data from only one date, it is probable that the TSL is hardly ever a simple line dividing ice from snow, because of the combined effect of local variability in both accumulation and ablation. The snow/ice distribution on 7 July 1984 had very little detectable relationship to any topographic variable apart from elevation, and to a lesser extent, slope angle. This may be because accumulation and ablation depend on local topography in different ways, and accumulation on upslope as well as local topography. It should be possible to relate data on the snow/ice distribution from remotely sensed sources to localised differences in ablation if the variation of snow accumulation with topography is already known.

Using a simple but general mathematical analysis we have shown that the melt produced by a snow/ice distribution separated by a TSL parallel to the contours will always produce slightly more melt than a more scattered mosaic of snow and ice when radiant and sensible energy both contribute towards melting. This analysis has been substantiated by comparing the volume of melt generated from the snow/ice distribution inferred from the Landsat image with that produced by equal areas of snow and ice divided by a horizontal TSL. The extra volume of melt was found to be less than 1% for the whole glacier on all days tested. On the basis of this evidence, modelling TSL movement as being solely dependent on elevation creates only a minor error in melt predictions, in a known direction. Errors from other sources, such as the uncertainties involved in setting winter and summer precipitation gradients, are likely to have a bigger effect on predicted melt. If glacier runoff is required then the problems associated with routing melt through the snowpack and glacier will provide large additional sources of error.

Satellite image analysis of snow cover distribution has been widely used to provide input to snowmelt runoff models for large areas (e.g. Baumgartner *et al.*, 1986). However, it has not previously been combined with terrain information from a DEM to provide input for the development and testing of glacier melt and runoff models. Satellite imagery has the potential to be especially useful for the development and testing of the snow cover component of grid-based fully distributed models along similar lines to those employed by Blöschl *et al.* (1991) using oblique air photos.

ACKNOWLEDGEMENTS

We thank Grande Dixence SA for provision of meteorological data, NERC for supplying the Landsat TM image to Chris Clark, Graham Boyce for making available the original model source code, Steve Wise for comments and assistance with the GIS aspects of the paper, Dave Collins for general assistance, and Ian Willis for constructive criticisms of the first version of the paper.

REFERENCES

Arnold, N. S., Willis, I. C., Sharp, M. J., Richards, K. S. and Lawson, W. J., 1996. A distributed surface energy balance model for a small valley glacier. I. Development and testing for the Haut Glacier d'Arolla, Valais, Switzerland. *Journal of Glaciology* **42**, 77–89.

Baker, D., Escher-Vetter, H., Moser, H., Oerter, H. and Reinwarth, O., 1982. A glacier discharge model based on results from field studies of energy balance, water storage and flow. *International Association of Hydrological Sciences Publication*, **138**, 103–112.

Baumgartner, M. F., Martinec, J. and Seidel, K., 1986. Large area deterministic simulation of natural runoff from snowmelt based on Landsat MSS data. *IEEE Transactions on Geoscience and Remote Sensing*, **GE-24**, 1013–1017.

Baumgartner, M. F., Seidel, K. and Martinec, J., 1987. Toward snowmelt runoff forecast based on multisensor remote sensing information. *IEEE Transactions on Geoscience and Remote Sensing*, **GE-25**, 746–750.

Bezinge, A., 1987. Glacial meltwater streams, hydrology and sediment transport: the case of the Grande Dixence hydroelectricity scheme. In *Glacio-fluvial Sediment Transfer: an Alpine Perspective*, eds A. M. Gurnell and M. J. Clark, Wiley, Chichester, 473–498.

Blöschl, G., Kirnbauer, R. and Gutknecht, D., 1991. A spatially distributed snowmelt model for application in alpine terrain. *International Association of Hydrological Sciences Publication*, **205**, 51–60.

Boyce, G. A. J., 1988. *A model for glacier derived runoff from Findelengletscher, Switzerland*. MSc Thesis, Department of Civil Engineering, University of Newcastle, 97pp.

Braithwaite, R. J., 1995. Aerodynamic stability and turbulent sensible-heat flux over a melting ice surface, the Greenland ice sheet. *Journal of Glaciology*, **41**, 562–571.

Brugman, M. M., 1991. Scale dependent albedo variations and runoff from a glacierized alpine basin. *International Association of Hydrological Sciences Publication*, **205**, 61–77.

Butz, D., 1989. The agricultural use of melt water in Hopar settlement, Pakistan. *Annals of Glaciology*, **13**, 35–39.

Collins, D. N., 1995. Dissolution kinetics, transit times through sub-glacial hydrological pathways and diurnal variations of solute content of meltwaters draining from an alpine glacier. *Hydrological Processes*, **9**, 897–910.

Davis, J. C., 1986. *Statistics and Data Analysis in Geology*, 2nd edn, Wiley, Chichester, 646pp.

Elder, K., Dozier, J. and Michaelsen, J., 1989. Spatial and temporal variation of net snow accumulation in a small alpine watershed, Emerald Lake basin, Sierra Nevada, California, U.S.A. *Annals of Glaciology*, **13**, 56–63.

Elliston, G. R., 1973. Water movement through the Gornergletscher. *International Association of Hydrological Sciences Publication*, **95**, 79–84.

ESRI, 1991. *Surface Modelling with TIN*. Environmental Systems Research Institute Inc., Redlands, California, US.

Ferguson, R. I., 1984. Magnitude and modelling of snowmelt runoff in the Cairngorm Mountains, Scotland. *Hydrological Sciences Journal*, **29**, 49–62.

Ferguson, R. I., 1986. Parametric modelling of daily and seasonal snowmelt using snowpack water equivalent as well as snow covered area. *International Association of Hydrological Sciences Publication*, **155**, 151–161.

Gottlieb, L., 1980. Development and applications of a runoff model for snowcovered and glacierized basins. *Nordic Hydrology*, **11**, 255–272.

Gurnell, A. M., Clark, M. J. and Hill, C. T., 1992. Analysis and interpretation of patterns within and between hydroclimatological time series in an alpine glacier basin. *Earth Surface Processes and Landforms*, **17**, 821–839.

Hall, D. K., Chang, A. T. C., Foster, J. L., Benson, C. S. and Kovalick, W. M., 1989. Comparison of in-situ and Landsat derived reflectance of Alaskan glaciers. *Remote Sensing of the Environment*, **28**, 23–31.

Lang, H., 1973. Variations in the relation between glacier discharge and meteorological elements. International Association of Hydrological Sciences Publication, **95**, 85–94.

Martinec, J., 1976. Snow and ice. In *Facets of Hydrology*, ed. J. C. Rodda, Wiley, Chichester, 85–118.

Megahan, W. F., Meiman, J. R. and Goodell, B. C., 1967. Net allwave radiation as an index of natural snowmelt and snowmelt accelerated with albedo reducing materials. *Proceedings of an International Hydrology Symposium*, Fort Collins, Colorado, 149–156.

Moore, I. D., Grayson, R. B. and Landson, A. R., 1991. Digital terrain modelling: a review of hydrological, geomorphological, and biological applications. *Hydrological Processes*, **5**, 3–30.

Munro, D. S. and Young, G. J., 1982. An operational net shortwave radiation model for glacier basins. *Water Resources Research*, **18**, 220–230.

O'Brien, H. W. and Munis, R. H., 1975. Red and near infrared spectral reflectance of snow. In *Operational Applications of Satellite Snowcover Observations*, ed. A. Rango, Science and Technology Information Office, NASA, Washington, DC, NASA SP-391.

Oerlemans, J. and Fortuin, J. P. F., 1992. Sensitivity of glaciers and small ice caps to greenhouse warming. *Science*, **258**, 115–117.

Östling, M and Hooke, R le B., 1986. Water storage in Storglaciären, Kebnekaise, Sweden. *Geografiska Annaler*, **68A**, 279–290.

Paterson, W. S. B., 1994. *The Physics of Glaciers*, 3rd edn, Permagon Press, Oxford, 480pp.

Rango, A. and Martinec, J., 1982. Snow accumulation derived from modified depletion curves of snow coverage. *International Association of Hydrological Sciences Publication*, **138**, 83–90.

Richards, K., Sharp, M., Arnold, N., Gurnell, A., Clark, M., Tranter, M., Nienow, P., Brown, G., Willis, I. and Lawson, W., 1996. An integrated approach to modelling hydrology and water quality in glacierized catchments. *Hyrdological Processes*, **10**, 479–508.

Röthlisberger, H. and Lang, H., 1987. Glacial hydrology. In *Glacio-fluvial Sediment Transfer: an Alpine Perspective*, eds A. M. Gurnell and M. J. Clark, Wiley, Chichester, 207–284.

US Army Corps of Engineers, 1956. *Snow Hydrology*. US Army Corps of Engineers

Summary Report of Snow Investigations, North Pacific Corps of Engineers, Portland, Oregon.

Van de Wal, R. S. W., Oerlemans, J. and Van der Hage, J. C., 1992. A study of ablation variations on the tongue of Hintereisferner, Austrian Alps. *Journal of Glaciology*, **38**, 319–324.

Willis, I. and Bonvin, J-M., 1995. Climate change in mountain environments: hydrological and water resource implications. *Geography*, **80**, 247–261.

Zevenbergen, L. W. and Thorne, C. R., 1987. Quantitative analysis of land surface topography. *Earth Surface Processes and Landforms*, **12**, 47–56.

Zheng, Q., Cao, C. M., Feng, X., Liang, F., Chen, X. and Sheng, W., 1984. Study on spectral reflectance characteristics of snow, ice and water of northwest China. *Scientia Sinica (Ser. B)*, **27**, 647–656.

17 The Use of Terrain Analysis in the Evaluation of Snow Cover Over an Alpine Glacier

L. COPLAND
Department of Earth and Atmospheric Sciences, University of Alberta, Canada

ABSTRACT

The terrain parameters of elevation, slope angle, aspect, profile curvature and planform curvature were calculated from a digital elevation model of Haut Glacier d'Arolla, Switzerland. Principal components and cluster analysis were used to divide the glacier into similar zones based on these terrain parameters. Snow water equivalent (SWE) measurements made at two-week intervals during the summer melt season indicate that elevation is the primary control on variations in SWE. Regression of SWE with elevation provides the best estimates of SWE, although terrain zonation also produces an effective partitioning of the glacier into areas of similar SWE. Improved evaluation of snowpack conditions may therefore be possible by combining terrain zonation with elevation-based regression predictions. Determination of terrain curvature may also help in accounting for the redistribution of snow by processes such as wind drift and avalanching. This should improve the areal extrapolation of point measurements of SWE, and stratification of terrain before sampling should increase the accuracy and decrease the labour requirements of snow surveys.

INTRODUCTION

Increased demand for water in mountain regions due to population growth, coupled with resource development, has made accurate assessment of snow cover distributions essential for resource managers. Most runoff in alpine areas is from the seasonal snowpack (Elder *et al.*, 1991), and more water may be produced per unit area than in non-alpine regions (Alford, 1980). Knowledge of snow cover distribution is also required for glacier energy and mass balance models (Arnold *et al.*, 1996; Willis *et al.*, Chapter 15). In addition, there is a need to improve the calibration and testing of snowmelt models as the use of runoff data alone provides little information about the spatial distribution of melt (Blöschl *et al.*, 1991).

To meet the need for snow cover data, traditional methods of evaluation have been based on areal extrapolation of the product of point measures of snow depth

Landform Monitoring, Modelling and Analysis. Edited by S. N. Lane, K. S. Richards and J. H. Chandler.
© 1998 John Wiley & Sons Ltd.

and density to provide estimates of snow water equivalence (SWE). An important question is the optimum number of measurements needed to effectively characterise the SWE of a snowpack. Given the wide range of factors (e.g. elevation, slope angle, aspect, curvature, energy availability) that interact to determine the final snow cover distribution, linear extrapolation of point values is not always effective. These distribution factors are exaggerated in high relief basins because of the rapidly varying topography, resulting in a heterogeneous snowpack that changes markedly over space and time (Elder *et al.*, 1991).

Past work has provided a relatively good understanding of snowpack variation within regions of mild relief (e.g. Steppuhn and Dyck, 1974; Adams, 1976; Rawls *et al.*, 1980). Direct extrapolation of snowpack distribution relationships from low relief areas to high relief alpine regions is problematic, however, owing to the complexity of the terrain. Rychetnik (1987) related the date of disappearance of snow cover to snow depths, elevations, slope angles and aspects. Elder *et al.* (1989, 1991) classified snow distribution as a function of solar radiation, slope angle and elevation. Blöschl and Kirnbauer (1992) used aerial photographs to detect snow presence, but ignored SWE and only considered elevation and slope angle. A major limitation in these and other studies has been the effective determination of snow redistribution by processes such as wind drift and avalanching. Redistribution does not change the total SWE in a watershed, but can be hydrologically important because it causes rapid variations in SWE over short distances. For example, Woo *et al.* (1983) found SWE to vary from 30% on hilltops to 300% in gullies as compared to flat areas in the Canadian Arctic, and Golding (1974) reported mean water equivalents of 70% on ridge tops and 170% at valley bottoms for an Albertan basin.

The aim of the study reported here was to investigate the relationship between snow cover and terrain in an alpine area. Cluster analysis was used to determine whether terrain-based stratification of a glacier can improve estimation of the snow cover distribution. This involved dividing the study area into similar terrain zones based on elevation, slope angle, aspect, profile (downslope) curvature and planform (across-slope) curvature. Field measurements of SWE were used to indicate whether this zoning provided an effective partitioning of the snow cover over the study area. SWE predictions based on terrain zonation of the study area were also compared with SWE predictions based on regression of elevation with SWE.

STUDY SITE

Haut Glacier d'Arolla is a temperate valley glacier located in Valais, Switzerland, at 45°58'N, 7°32'E (Figure 17.1). The glacier covers an area of 6.3 km², within a basin of 11.7 km². The glacier is located above the tree line, and vegetation is virtually non-existent in the surrounding area. The glacier ranges in elevation from 2550 to almost 3500 m, with a mean of 2952 m, while the entire basin ranges from 2460 to 3838 m. The glacier has a mean surface slope angle of 15.8°, faces predominantly north, and is surrounded by steep cliffs.

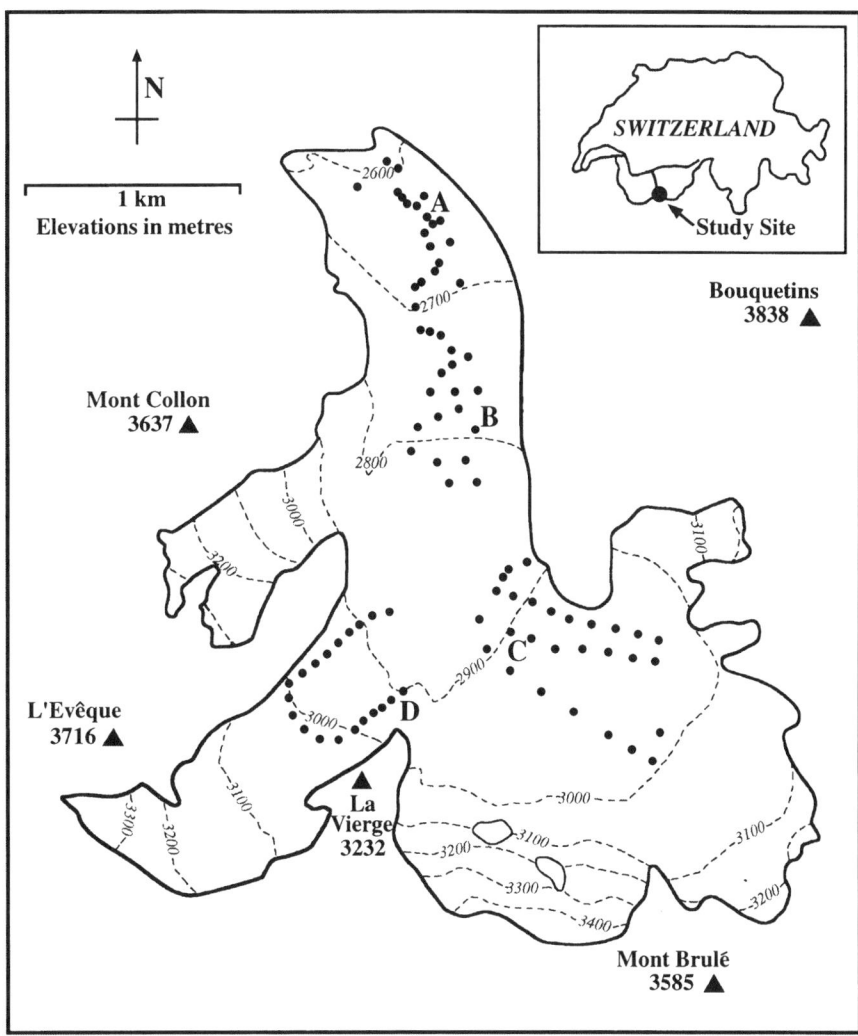

Figure 17.1 Map of Haut Glacier d'Arolla, Switzerland. Black dots show the location of the main sample point network. A, B, C and D are the locations of the snow density pits

METHODOLOGY

Terrain Analysis

Before field sampling, a DEM of Haut Glacier d'Arolla with a nodal spacing of 20 m was constructed from a combination of field survey and data capture from published maps (Figure 17.2). Using equations given in Zevenbergen and Thorne (1987), a

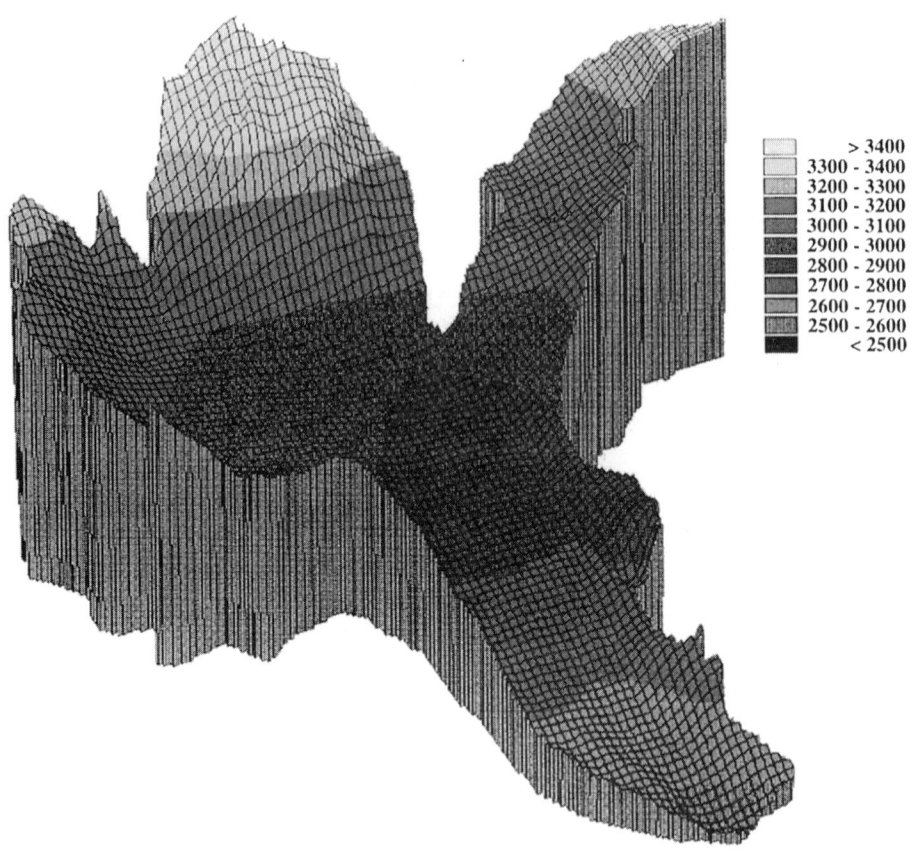

	> 3400
	3300 - 3400
	3200 - 3300
	3100 - 3200
	3000 - 3100
	2900 - 3000
	2800 - 2900
	2700 - 2800
	2600 - 2700
	2500 - 2600
	< 2500

Figure 17.2 Three-dimensional representation of the digital elevation model of Haut Glacier d'Arolla

computer program was written to calculate the terrain parameters of slope angle, aspect, profile curvature, and planform curvature at each node from a regular 3×3 sub-matrix as it passed over the digital elevation model (DEM). To make the results more easily interpretable in statistical calculations, a modification was made to the aspect data. This is because two slopes with aspects of 3° and 358° are physically very similar, yet mathematically near the extremes of a continuum from 0 to 360°. To resolve this problem, a north–south scalar determined by the cosine of the aspect, and an east–west scalar determined by the sine of the aspect, were used. For the north–south scalar, 0° is represented by a value of 1, and 180° is represented by –1; for the east–west scalar, 90° is represented by 1, and 270° is represented by –1.

It was anticipated that terrain parameters would control the snow cover distribution because: (i) elevation relates to increases in precipitation with altitude, and processes controlled by air temperature such as the transition from rain to snow; (ii) slope angle relates to variations in snow depth induced by wind drift and avalanching, as well as differences in direct beam and diffuse radiation; (iii) aspect relates to variations in snow cover caused by variations in solar radiation and prevailing wind direction, and also defines the slope direction and hence the direction of flow; and (iv) curvature relates to snow redistribution. This is due to the relatively low SWE in convex areas where flow (e.g. from wind or avalanching) accelerates and diverges, and the relatively high SWE in concave areas where flow decelerates and converges. Concave areas are also shadowed for longer than convex areas.

In data analysis, a routine was included to remove the problem of erroneous results at the edge of the study site, by ignoring a point if it, or any of the surrounding eight points, had a missing value. This resulted in the loss of a 20 m wide strip of values around the edge of the study area, although this was not significant because field sampling was undertaken away from the glacier margin. Side slopes were not included because this study was only concerned with the distribution of snow over the glacier. The position in map coordinates of each node was also calculated to allow location of points in the field. The final computer program produced a data set with the coordinates, elevation, slope angle, north–south scalar aspect, east–west scalar aspect, profile curvature and planform curvature evaluated at every 20 m interval over the surface of Haut Glacier d'Arolla (Figures 17.3 to 17.7).

Terrain Clustering

Correlation of the terrain parameters showed collinearity between some factors. For example, elevation was positively correlated with slope angle because higher parts of the glacier tend to have steeper slopes. In order to remove such statistical problems, principal components analysis (PCA) was performed on the terrain values. PCA involves rewriting a data set such that the new variables are weighted representations of the original values and uncorrelated with one another (Johnston, 1980). The correlation matrix was used for the PCA process so that the component scores were calculated from standardised variables. The correlation between each variable and component (the loadings matrix) is given in Table 17.1. A component was deemed significant if the eigenvalue was greater than 1, because this represents the variance of the original variables. The first three components are significant in Table 17.1, accounting for 66.54% of the total variance. Table 17.2 shows that the significant components represent the original data set quite well, except for east–west scalar aspect. This is likely to be of minimal significance, however, as the greatest contrast in radiation receipts is between north- and south-facing slopes. In general, component 1 is an elevation/slope angle factor, component 2 is a curvature factor, and component 3 is a north–south scalar aspect factor.

The three significant, standardised components identified by PCA were then clustered using nearest centroid sorting. In this sorting routine a site is assigned to the cluster for which the value between the site and the cluster centre is smallest. Cluster analysis involves the placing of objects into more or less homogeneous groups, although formal tests of significance do not yet exist (Norusis, 1990).

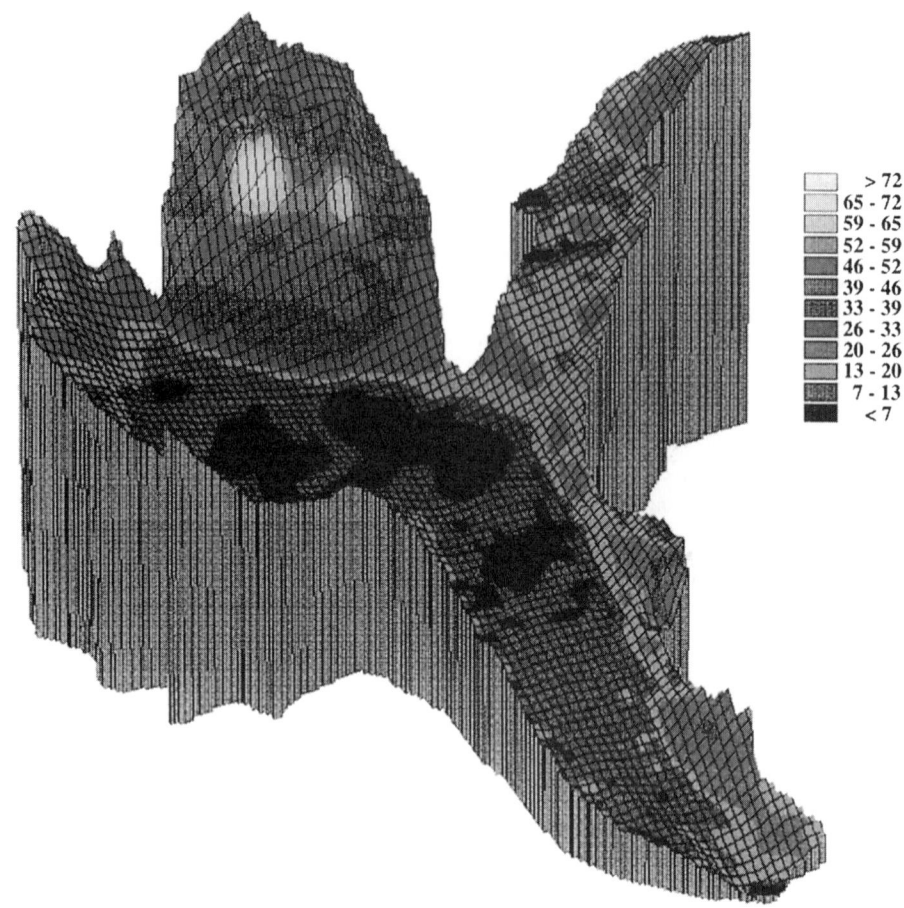

Figure 17.3 Distribution of surface slope angle across Haut Glacier d'Arolla

Zoning with six to 16 clusters was completed, with semi-qualitative assessment indicating which zonation best represented the division of the glacier. This assessment involved visual analysis of maps of the clusters, and of the distribution of points between zones. In general, a few zones tended to contain a high proportion of the data, while many zones contained a few points. Clustering with 10 zones was chosen as this provided a reasonably even spread of points between zones, and also divided the terrain into regions that match quite well with personal knowledge of the glacier (Table 17.3; Figure 17.8). When Tables 17.2 and 17.3 are viewed together it is apparent which terrain features were important in defining the different zones. For example, zone 1 was generally defined by just below average slope angles and elevations (component 1 = 0.482), minimal curvature (component 2 = 0.028), and was predominantly north-facing (component 3 = 0.626). Zone 6 was defined by slightly lower slope angles and elevations (component 1 = 0.122), greater curvature (component 2 = 0.224), and was more south-facing (component 3 = −0.753).

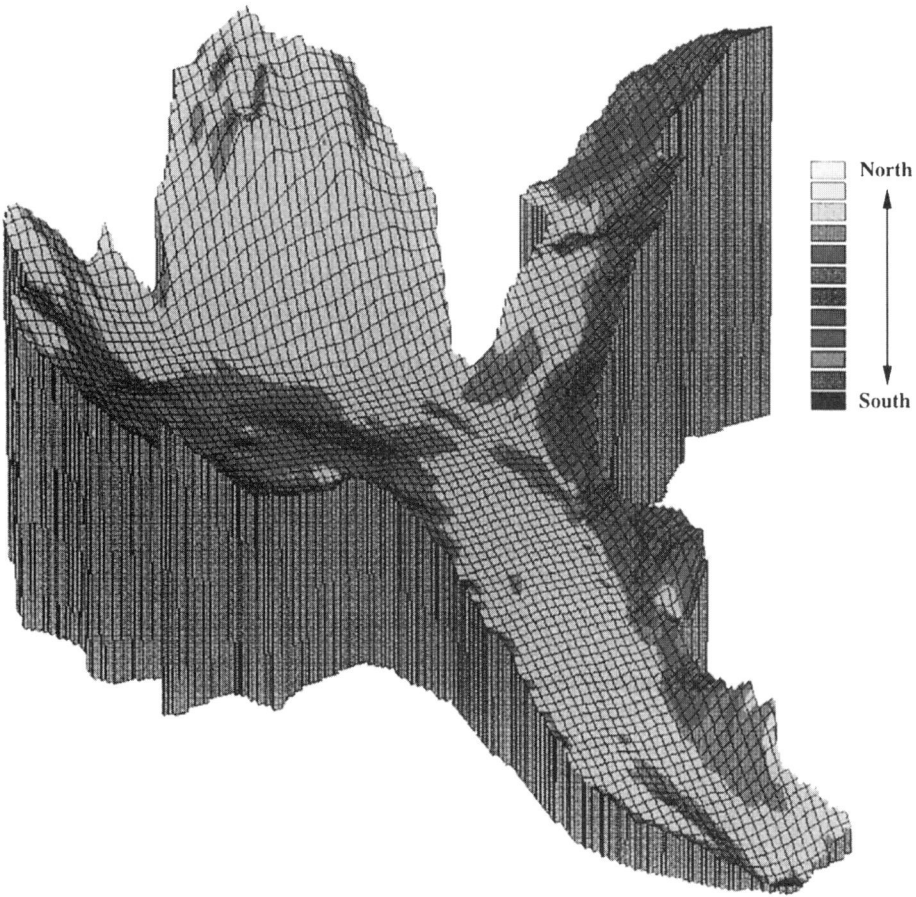

Figure 17.4 Distribution of north–south scalar aspect across Haut Glacier d'Arolla

Field Methods

Following terrain analysis, an intensive field programme was completed at Haut Glacier d'Arolla between 19 June and 2 August 1993. SWE was recorded every two weeks at a network of 87 points (marked with wooden poles) over the glacier surface. As one of the aims was to test the validity of terrain clustering, measurement sites were not selected on the basis of the cluster zones. Classifying the sample network before data collection would have implied pre-existing knowledge of the snow distribution and may have biased the results. Figure 17.1 shows that the sample points were grouped into three zones due to the logistical need to complete the first two surveys over three days. Some melting probably occurred during survey periods, although the time taken to complete each survey was relatively short in comparison to the time between surveys.

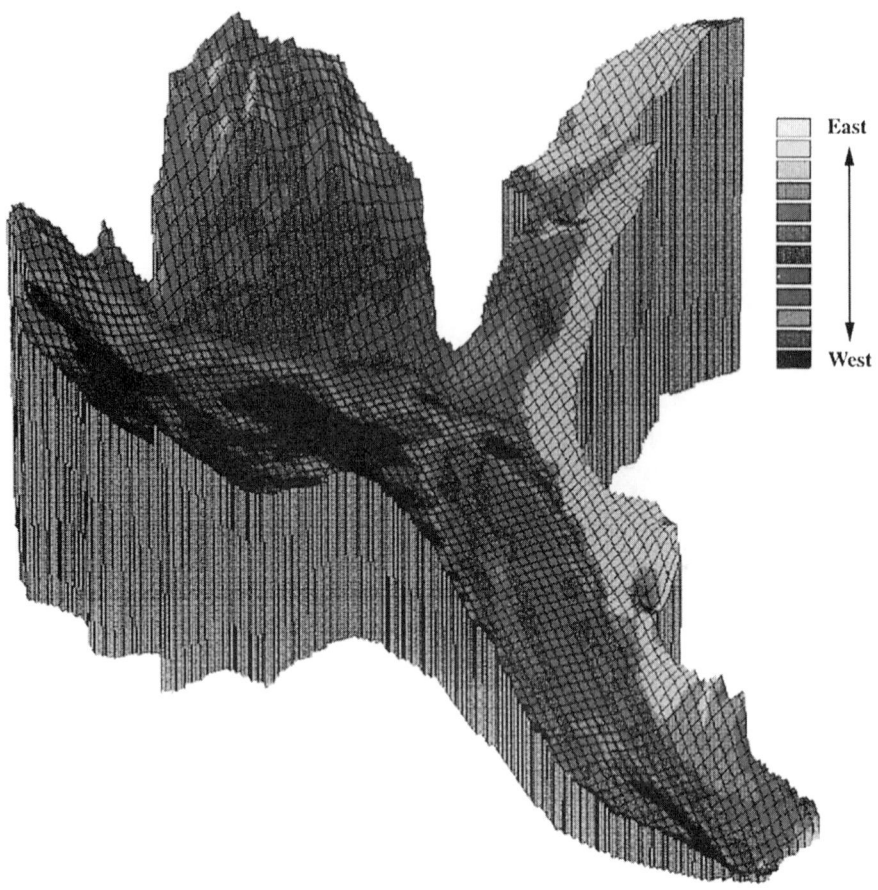

Figure 17.5 Distribution of east–west scalar aspect across Haut Glacier d'Arolla

Snow depth was recorded with a 3 m long metal avalanche probe. To minimise local irregularities, the depth at each sample point represented the mean of measurements at the central point and 2 m away to the north and south. The initial survey used the mean of five measurements at each location, but was unnecessarily time-consuming with little increase in accuracy. Sample locations were determined by electronic theodolite, and converted to map coordinates to enable direct correlation of the SWE measurements with the computed terrain parameters and zones identified by cluster analysis.

Snow density was recorded at four pits identified as A, B, C and D (Figure 17.1). It was assumed that pit A represented the glacier snout, pit B the lower glacier, pit C the glacier centre, and pit D the upper glacier. At each 20 cm increment in the vertical pit wall, 1000 cm^3 of snow was removed by a stainless steel cutter and weighed with a hand-held spring balance. Density was recorded at only a few

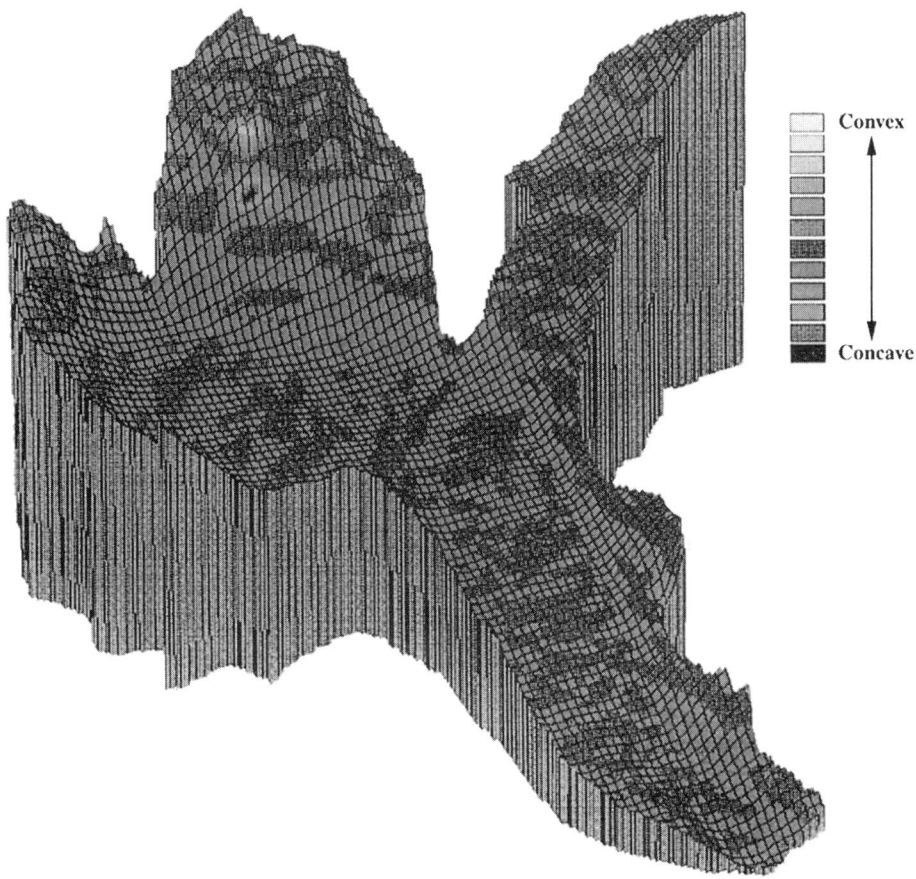

Convex

Concave

Figure 17.6 Distribution of profile (downslope) curvature across Haut Glacier d'Arolla

locations as density measurements are time-consuming and snow depth varies far more than density in alpine areas (Logan, 1973).

In addition to the glacier-wide surveys, a rapid snow depth survey was completed at 101 locations on 14 July 1993 after a major new snowfall. The rapid survey sample points were located in nine transects across the glacier between approximately 2700 m and 2850 m elevation. The rapid snow depth survey allowed analysis of the relation between terrain and snow cover on the small scale after a single accumulation event. Density measurements could not be made during this survey owing to time and safety constraints.

RESULTS

The computed terrain parameters matched well with field observations of terrain, indicating that the equations described by Zevenbergen and Thorne (1987) provide

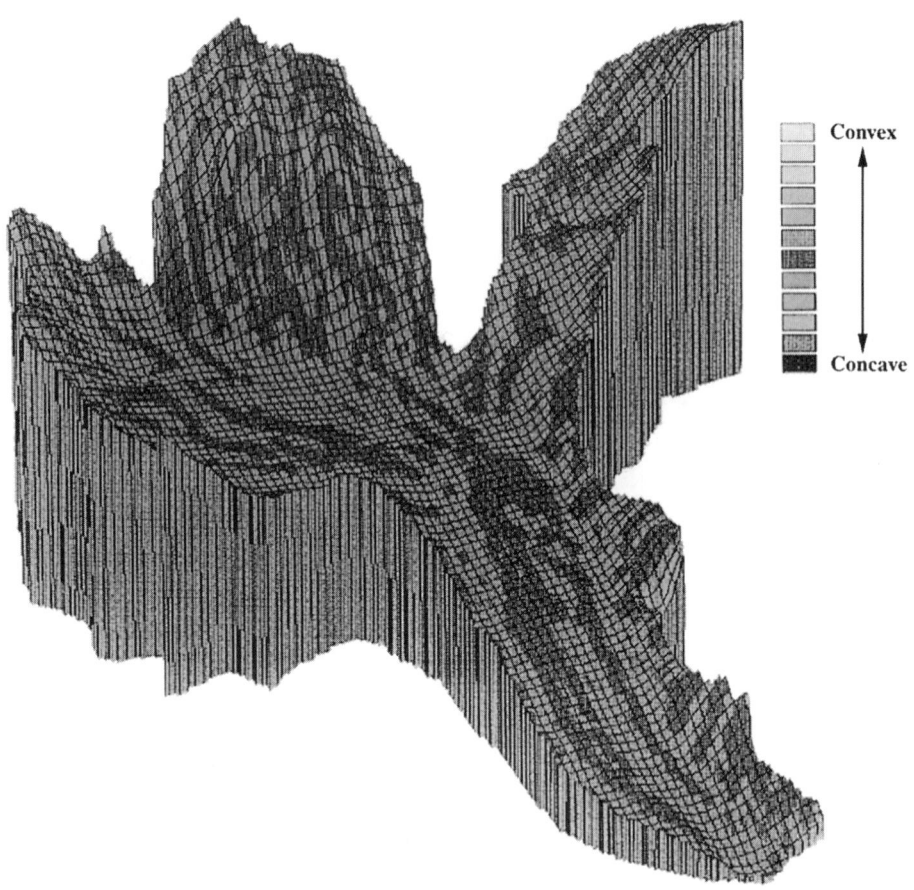

Figure 17.7 Distribution of planform (across-slope) curvature across Haut Glacier d'Arolla

an effective way of calculating slope angle, aspect and curvature from a DEM. Parametric tests were performed to relate the point SWE values to the terrain parameters. Steppuhn and Dyck (1974) noted that the distributions of snow cover data are often non-symmetric, although this was not deemed a problem as the means tend towards a normal distribution when there are a large number of samples (Goodison *et al.*, 1981). The problem of a high number of zero SWE values skewing the data in later surveys was overcome by removing from analysis any locations at which bare ice was present in the previous survey. The information provided by the SWE initially falling to zero was therefore included, but the lack of change after this was excluded. Pearson's correlation coefficient was used to determine the relation between the point SWE values and terrain parameters for each survey (Table 17.4). An optimising stepwise regression method was then used for the multiple situation to remove problems associated with collinearity (Johnston,

Table 17.1 Loadings matrix from principal components analysis on the terrain parameters. The percentage of the total variance explained by each component, and the eigenvalue of each component, are also provided

Component	1	2	3	4	5	6
Elevation	−0.741	−0.081	−0.385	0.340	0.163	−0.392
Slope angle	−0.805	−0.341	0.021	0.146	−0.037	0.462
N–S scalar aspect	−0.027	−0.214	0.878	0.393	0.094	−0.137
E–W scalar aspect	−0.448	−0.343	0.255	−0.765	0.012	−0.176
Profile curvature	−0.283	0.751	0.167	−0.138	0.545	0.105
Planform curvature	0.430	−0.654	−0.193	−0.004	0.589	0.066
Explained variance (%)	27.74	21.30	17.50	14.93	11.34	7.20
Eigenvalue	1.664	1.278	1.050	0.896	0.680	0.432

Table 17.2 Explained variance (%) and communality values (%) for significant components in principal components analysis (+/− indicates whether relation between terrain parameter and component is positive or negative)

Component	1	2	3	Communality
Elevation	−54.97	−0.66	−14.80	70.43
Slope angle	−64.75	−11.62	+0.05	76.42
N–S scalar aspect	−0.07	−4.57	+77.16	81.80
E–W scalar aspect	−20.11	−11.76	+6.49	38.36
Profile curvature	−8.03	+56.47	+2.78	67.28
Planform curvature	+18.47	−42.74	−3.71	64.92

Table 17.3 Cluster centres and data distribution within zones for 10 clusters. The centre of each zone is defined by its scores on the three components

Zone	Component 1	Component 2	Component 3	No. cases	Overall %
1	0.482	0.028	0.626	5830	49.61
2	−0.124	−6.076	−1.347	36	0.31
3	0.252	0.717	−2.381	780	6.64
4	−1.623	0.107	0.267	1204	10.25
5	−5.920	8.706	2.799	15	0.13
6	0.122	0.224	−0.753	2733	23.26
7	2.385	−5.144	−2.936	13	0.11
8	1.683	11.841	−2.972	3	0.03
9	−3.156	2.819	0.923	132	1.12
10	−0.907	−1.636	−0.125	1005	8.55

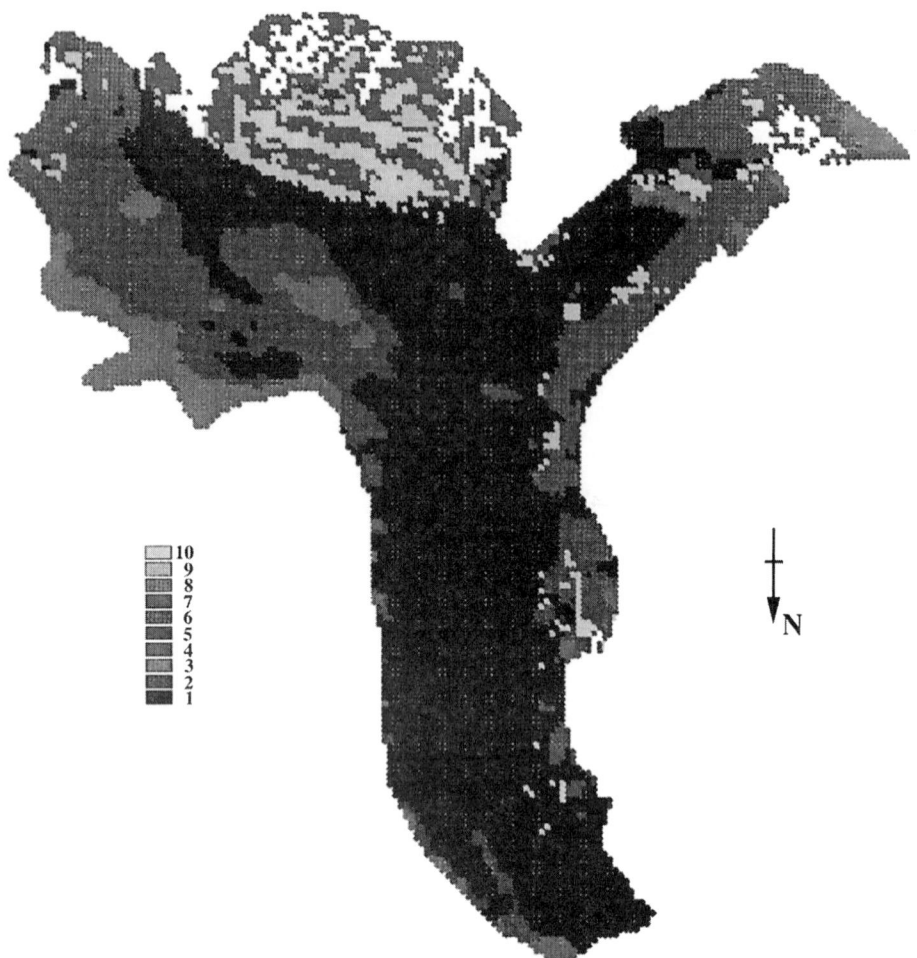

Figure 17.8   Division of Haut Glacier d'Arolla into 10 cluster zones based on terrain

Table 17.4   Pearson correlation coefficients for SWE versus terrain parameters for all surveys. Starred values are statistically significant at the 95% level

	Survey 1	Survey 2	Survey 3	Survey 4	Rapid survey
Elevation	0.727*	0.761*	0.646*	0.553*	0.197
Slope angle	−0.025	−0.034	0.162	0.303*	0.097
N–S scalar aspect	−0.233*	−0.256*	−0.355*	−0.266*	−0.322*
E–W scalar aspect	0.208	0.211	0.065	0.244	−0.232*
Profile curvature	−0.155	−0.161	−0.157	−0.085	−0.107
Planform curvature	−0.237*	−0.178	0.026	0.110	−0.124

1980). A significance level of 95% was chosen to provide a way of limiting discussion to the most important variables.

Survey 1 (20–22 June)

Snow covered the entire glacier during this survey, with a mean SWE of 0.856 m. Snow density varied little, with a mean of 0.527 g cm^{-3}, and standard deviation of ±0.031 g cm^{-3}. Correlation of SWE with terrain identified elevation, north–south scalar aspect and profile curvature as significant (Table 17.4). Stepwise regression identified elevation as the most important variable in the multiple situation, accounting for 52.9% of the SWE variance. Profile and planform curvature were also significant in stepwise regression, accounting for 3.1% and 2.9% of the variance respectively, with SWE increasing as the glacier surface became more concave.

Survey 2 (3–5 July)

There was some melt since survey 1, with a survey 2 mean SWE of 0.513 m. The snow had melted from some areas towards the glacier terminus by survey 2, meaning that density could no longer be recorded at pit A. The mean snow density at the remaining three pits remained similar at 0.533 g cm^{-3}, with a standard deviation of ±0.016 g cm^{-3}, varying much less than depth. Correlation again identified elevation and north–south scalar aspect as significant influences on SWE (Table 17.4). Stepwise regression identified elevation as the only significant variable, explaining 57.9% of the variance. This value was lower than for survey 1, although the proportion of variance explained by elevation alone was slightly higher. North–south scalar aspect was not highlighted owing to collinearity with elevation.

Survey 3 (17–18 July)

Once again there was melt since the previous survey, with a mean SWE of 0.426 m for this survey. The melt was less than between surveys 1 and 2 owing to new snowfall and freezing temperatures between 11 and 13 July. Snow density had also increased slightly since survey 3 to 0.551 g cm^{-3}, with a standard deviation of 0.059 g cm^{-3}, although it remained virtually constant over the glacier. To meet the normality assumption, the locations at which SWE was zero in survey 2 were excluded from analysis, leaving a data set with 69 values. Elevation was still the most important factor in the linear case, with north–south scalar aspect also significant (Table 17.4). Collinearity remained important as stepwise regression identified elevation as the only significant influence on survey 3 SWE, accounting for 41.7% of the variance.

Survey 4 (1–2 August)

Significant melt had occurred since survey 3, with a mean SWE loss of 0.242 m. Bare ice was now exposed over the lower part of the glacier, meaning that snow density could only be recorded at pits C and D. Mean snow density had increased slightly again to 0.554 g cm^{-3}, with a standard deviation of 0.014 g cm^{-3}. The locations at which SWE was zero in previous surveys were again removed from analysis, leaving a data set with 59 values. Correlation with SWE showed that elevation, slope angle and north–south scalar aspect were significant (Table 17.4). Although several variables were influential in the linear case, stepwise analysis identified elevation as the only significant factor in the multiple situation, accounting for 30.6% of the variance.

Rapid Survey (14 July 1993)

Correlation identified north–south and east–west scalar aspect as significant influences on snow depth (Table 17.4). Stepwise analysis identified north–south scalar aspect (10.4%) and elevation (4.9%) as significant in the multiple situation, together accounting for 15.3% of snow depth variance. This value is relatively small in comparison to the other surveys, and shows the heterogeneous nature of the snow cover on the small scale.

Terrain Clustering

The majority of sample locations (62) were in zone 1, while there was one point in zone 3, two points in zone 4, 21 points in zone 6 and one point in zone 10. Only those points located in zones 1 or 6 were analysed further as the other zones contained too little data for effective interpretation. The non-parametric Mann-Whitney U test was used to compare the SWE in different zones because the number of points in each class differed greatly, and no assumptions need to be made about the data distribution characteristics. The Mann-Whitney U test was more than 99.9% significant for all four surveys, showing that the mean SWE was significantly different between zones 1 and 6. Terrain-based zonation of the glacier therefore provided an effective division of the snow cover between zones 1 and 6, although the use of data from only two zones means that this is a preliminary test rather than a comprehensive assessment.

Comparison between the effectiveness of terrain-based zoning in predicting SWE, and SWE predictions from regression of elevation with SWE are given in Table 17.5. For terrain zoning, the predicted SWE at each sample point consists of the mean SWE for the zone in which the particular sample point was located. For regression, the predicted SWE at each sample location was determined from a line of best fit to the SWE measurements. Table 17.5 shows that regression with elevation provides the best predictions of SWE for the main surveys, although the predictions based on terrain zonation become relatively better in the later surveys.

DISCUSSION

Depth and Density

Field measurements consistently showed that snow density varied much less than snow depth, with only a small increase in density from one survey to the next. The

Table 17.5 Comparison of SWE predictions from terrain zonation and from regression of elevation with SWE. Errors are the mean deviation of predictions from observations. All figures are in metres

	Mean SWE	Zonation error	Regression error
Survey 1	0.856	0.308	0.204
Survey 2	0.513	0.303	0.188
Survey 3	0.426	0.234	0.186
Survey 4	0.184	0.166	0.135

mean snow density of 0.541 g cm^{-3}, and standard deviation of 0.030 g cm^{-3}, compares well with Elder *et al.* (1991), who found a mean of 0.520 g cm^{-3} and standard deviation of 0.044 g cm^{-3} for an alpine basin in California. This is in contrast to the mean snow depth of 1.624 m, and standard deviation of 0.760 m, for the first survey. These results support claims that more snow depth than density measurements are necessary to characterise the areal SWE variability over an alpine basin (Logan, 1973; Adams, 1976; Goodison *et al.*, 1981; Elder *et al.*, 1991). Given that depth has a strongly predictive value for SWE when the density deviation is small, accurate results can be obtained by combining many depth readings with a few density measurements. Gruzinov (1990) actually recommends abandoning density measurements altogether to increase efficiency in snow surveys.

Relation Between Snow Cover and Terrain

Elevation

The correlation coefficients consistently highlighted elevation as the single most important factor, accounting for more than 50% of SWE variation for the first two surveys, and over 30% for the last two. The rapid survey results indicated that elevation was significant, but of less importance, on the smaller scale. Elevation has been widely recognised as an important influence on variations in snow cover (McKay and Gray, 1981). The increase of SWE with elevation could be due to several factors, including the decrease in air temperature, and increase in precipitation with altitude. For the main surveys, air temperature is likely to have been of prime importance as meteorological measurements on the glacier showed a consistent decrease of at least 4°C between 2547 m and 2872 m.

The poor relationship between elevation and snow depth on the smaller scale can be explained by winds of over 8 m s^{-1} during the storm prior to the rapid survey. These wind speeds are likely to have caused redistribution, so reducing the influence of elevation on the small scale. Debris cover may also have been important as moraine was virtually free of snow due to wind scour and the higher thermal conductivity and lower albedo of rock debris compared to ice.

Aspect and Slope Angle

North–south scalar aspect was significantly negatively correlated with SWE for surveys 2, 3 and 4. This was unexpected because the lower solar radiation receipts of more northerly facing areas would suggest a positive relationship. One explanation could relate to the collinearity between north–south scalar aspect and elevation. Alternatively, the strong negative correlation between north–south scalar aspect and slope angle for the rapid survey results suggests that snow accumulation occurs against slopes that face the wind. This is because wind speeds were high and blowing in an up-glacier (southerly) direction during the rapid survey. Cline (1992) also found that slope angle had an influence on the redistribution of snow by wind.

Curvature

Profile and planform curvature were significant parameters in stepwise analysis for survey 1, accounting for 3.1 and 2.9% of the variance respectively. These values are low, but important, as determination of terrain curvature appears to help in accounting for redistribution. The higher SWE in concave areas probably relates to wind deceleration and convergence (i.e. deposition), whereas the lower SWE in convex areas probably relates to wind acceleration and divergence (i.e. erosion). The effects of avalanching could not be quantified owing to safety considerations, although observations suggest that initiation occurs in convex areas with high slope angles, and runout occurs at the base of slopes with a marked concavity. Only theoretical considerations of the relationship between terrain curvature and snow cover have been made before (Adams, 1976; Elder *et al.*, 1989, 1991; Blöschl and Kirnbauer, 1992).

Longer Term Processes

The data presented here indicate a general decline in the proportion of variance attributable to the terrain as the melt season progresses. This pattern is similar to that reported by Rychetnik (1987), where only a small proportion of the spatial variations in the time of disappearance of the snow cover could be explained by terrain. Rainfall and melt have the effect of reducing spatial contrasts in SWE, while phenomena such as redistribution are more complex. Wind is of greatest influence during accumulation and early in the melt season, when density is low and snow can be entrained from the glacier surface. This is suggested by the rapid survey results and the significance of curvature for survey 1 only.

Radiation is important in defining the spatial distribution of melt. It was hoped that the determination of aspect would account for much of the variation in radiation receipts, although it was not significant in any multiple regression calculations. This probably relates partly to shading from surrounding topography, and partly to collinearity between the terrain parameters. Collinearity becomes increasingly evident in later surveys as the sample points are restricted to the tributary glacier, which is at a higher elevation and noticeably steeper and less northerly facing than lower areas.

Overall, the results support attempts at delineating SWE according to terrain. Although the importance of the terrain declines over the melt season, multiple regression is able to account for a higher proportion of SWE variability than previous similar investigations. For an alpine basin in California, Elder *et al.* (1991) found no discernible relationship between SWE and radiation, slope and elevation for the 1986 water year, a relation of 40% in 1987, and a relation of 27% in 1988. The higher values from Arolla may be due to the sample point distribution, although the significance of several factors in linear regression, and curvature in multiple regression for survey 1, suggests that the objective evaluation of a wider range of terrain parameters may improve snow cover evaluation.

Terrain Clustering

Division of the glacier surface into 10 cluster zones according to terrain appears to have been successful, as the mean SWE was consistently higher in zone 6 than zone 1 (> 99.9% significant for all surveys). The use of only two zones in statistical analysis is limiting, although data from the other zones and visual analysis of the snow cover support clustering as a means of delineating spatial variations in SWE. Regression of elevation with SWE provided better estimates of SWE than terrain zonation, although linear interpolation does not account for the small-scale variability observed in rugged terrain (Elder and Dozier, 1990). A combination of terrain zonation and regression-based approaches may therefore provide the best estimates of SWE in alpine basins. This has important implications for snow cover monitoring as the effective division of SWE according to physically based parameters allows improved extrapolation of point values over space.

Using the predictions from terrain zonation, the volume of water held in the snowpack was estimated for each survey by adding the product of the mean SWE, grid size and number of points within each zone (Table 17.6). While this is a "rough-and-ready" calculation, it is useful as a basis against which other results can be compared. Comparison with melt model results from 1992 suggests that the general patterns are correct (Sharp *et al.*, 1993). The change in water volume between survey dates was clearly influenced by climate, with the smallest decrease between surveys 2 and 3 related to freezing temperatures and new snowfall. The importance of individual zones varies through the melt season as, for example, zone 1 dominates surveys 1 and 2, while zone 6 dominates surveys 3 and 4. This is probably because zone 6 is generally higher in the basin, meaning that terrain zonation has captured some of the influence of elevation.

CONCLUSIONS

The computer program written for this study has enabled the quantitative determination of slope angle, aspect, profile curvature and planform curvature for an entire glacier, and is applicable to any DEM with regular spacing between grid points. Using these terrain parameters, elevation has been consistently highlighted as the most important variable, accounting for 30.6% to 57.9% of the variance in

Table 17.6 Total volume of water held in the snowpack for the four main surveys (S1, S2, S3, S4) using predictions based on terrain zonation. The predicted SWE depth for each zone consists of the mean of all measurements made in that zone. The overall mean SWE was used for the zones that did not contain any sample points (zones 2, 5, 7, 8, 9)

Zone	Points	S1 SWE (m)	S1 Total (m³)	S2 SWE (m)	S2 Total (m³)	S3 SWE (m)	S3 Total (m³)	S4 SWE (m)	S4 Total (m³)
1	5830	0.757	1 765 324	0.409	953 788	0.328	764 896	0.140	326 480
2	36	0.856	12 326	0.513	7387	0.426	6134	0.184	2650
3	780	0.924	288 288	0.542	169 104	0.292	91 104	0.000	0
4	1204	0.991	477 266	0.719	346 270	0.526	253 322	0.231	111 250
5	15	0.856	5136	0.513	3078	0.426	2556	0.184	1104
6	2733	1.102	1 204 706	0.759	829 739	0.625	683 250	0.264	288 604
7	13	0.856	4451	0.513	2668	0.426	2215	0.184	957
8	3	0.856	1027	0.513	616	0.426	511	0.184	221
9	132	0.856	45 197	0.513	27 086	0.426	22 493	0.184	9715
10	1005	1.593	640 386	1.366	549 132	1.100	442 200	0.479	192 558
Total			4 444 107		2 888 868		2 268 681		933 539

SWE. There is an overall decline in the differentiation of snow cover as the melt season progresses owing to the homogenising effects of melt and rainfall.

Division of the glacier into areas of similar terrain may resolve some of the problems associated with collinearity, as cluster analysis allows the simultaneous evaluation of many terrain parameters. This more clearly represents the real physical processes as the artificial separation of variables in regression does little to account for their interaction in the field. Zonation has been attempted for one basin before (Elder *et al.*, 1989, 1991), but the problems with collinearity were ignored, and clustering was limited to radiation, elevation and slope angle. Although the conclusions from this study are preliminary owing to the statistical comparison of only two zones, it appears that cluster analysis provides an effective method of partitioning the SWE within alpine basins. The method could possibly be improved by using a multiplier to weight the importance of elevation in the clustering process.

Snow cover evaluation may be improved in the future by using pattern recognition methods such as upslope drainage area to identify groups of cells with an assemblage of characteristics favourable for avalanching and/or wind drift. The determination of slope curvature is particularly valuable because improved avalanche predictions should result. For example, cells with profile concavity which lie downslope of cells with high slope angles and downslope convexity are likely to provide a focus for deposition. This may be an important research area, as most studies (e.g. McClung and Tweedy, 1993) have focused on when, rather than where, avalanching occurs.

The results suggest that the volume of water held in the seasonal snowpack may be rapidly evaluated by taking a few depth measurements in each of the major zones identified by cluster analysis. Only one density profile for the whole study area would be necessary if the density variability is small. Stratification before sampling should increase accuracy and lower the number of samples, while also giving confidence to the areal extrapolation of point measurements for use in the calibration and testing of snowmelt models.

ACKNOWLEDGEMENTS

This work was undertaken in conjunction with the Arolla Glaciology Project, for which logistical help and permission to reproduce maps and use the DEM are gratefully acknowledged. I would like to thank Andrew McDonald, as well as all those from the Universities of Cambridge and Western Ontario, for help in the field. Francine Hughes, Martin Sharp, Ian Willis, Jon Harbor, Linda Horn, and Owen Turpin provided valuable discussion before, during and after the fieldwork. Financial assistance from the David Richards Travel Scholarship, Bedford Travel Grant, University of Cambridge Vacation Studies Fund, and Purdue University Travel Award made this work possible.

REFERENCES

Adams, W. P., 1976. Areal differentiation of snow cover in East Central Ontario. *Water Resources Research*, **12**, 1226–1234.

Alford, D., 1980. The orientation gradient: regional variations of accumulation and ablation in alpine basins. In *Geoecology of the Colorado Front Range: a Study of Alpine and Subalpine Environments*, ed. J. D. Ives, Westview Press, Boulder, Colorado, 214–223.

Arnold, N. S., Willis, I. C., Sharp, M. J., Richards, K. S. and Lawson, W. J., 1996. A distributed surface energy-balance model for a small valley glacier. I. Development and testing for Haut Glacier d'Arolla, Valais, Switzerland. *Journal of Glaciology*, **42**, 77–89.

Blöschl, G. and Kirnbauer, R., 1992. An analysis of snow cover patterns in a small alpine catchment. *Hydrological Processes*, **6**, 99–109.

Blöschl, G., Kirnbauer, R. and Gutknecht, D., 1991. Distributed snowmelt simulations in an alpine catchment. 1. Model evaluation on the basis of snow cover patterns. *Water Resources Research*, **27**, 3171–3179.

Cline, D. W., 1992. Modelling the redistribution of snow in alpine areas using geographic information processing techniques. *Proceedings, 1992 Eastern Snow Conference*, 13–24.

Elder, K. and Dozier, J., 1990. Improving methods for measurement and estimation of snow storage in alpine watersheds. In *Hydrology in Mountainous Regions. I – Hydrological Measurements; the Water Cycle*, eds H. Lang and A. Musy, International Association of Hydrological Sciences, Special Publication 193, 147–156.

Elder, K., Dozier, J. and Michaelsen, J., 1989. Spatial and temporal variation of net snow accumulation in a small alpine watershed, Emerald Lake Basin, Sierra Nevada, California, U.S.A. *Annals of Glaciology*, **13**, 56–63.

Elder, K., Dozier, J. and Michaelsen, J., 1991. Snow accumulation and distribution in an alpine watershed. *Water Resources Research*, **27**, 1541–1552.

Golding, D. L., 1974. The correlation of snowpack with topography and snowmelt runoff on Marmot Creek basin, Alberta. *Atmosphere*, **12**, 31–38.

Goodison, B. E., Ferguson, H. L. and McKay, G. A., 1981. Measurement and data analysis. In *Handbook of Snow*, eds D. M. Gray and D. H. Hale, Pergamon Press, Oxford, Chapter 6.

Gruzinov, A. V., 1990. O ratsionalizatsii rabot po opredeleniyu vodnosti snega v basseyne ledn. Abramova (On rationalising work on determining water content of snow in the basin of the Abramov glacier). *Sredneaziatskiy Regional'nyy Nauchno-Issledovatel'skiy Gidrome-teorologicheskiy Institut. Trudy*, **136**, 45–49 (Russian, with English abstract).

Johnston, R. J., 1980. *Multivariate Statistical Analysis in Geography*. Longman, London.

Logan, L. A., 1973. Basin-wide water equivalent estimation from snowpack depth measurements. In *Role of Snow and Ice in Hydrology*, International Association of Hydrological Sciences, Special Publication 107, 864–884.

McClung, D. M. and Tweedy, J., 1993. Characteristics of avalanching: Kootenay Pass, British Columbia, Canada. *Journal of Glaciology*, **39**, 316–322.

McKay, G. A. and Gray, D. M., 1981. The distribution of snowcover. In *Handbook of Snow*, eds D. M. Gray and D. H. Hale, Pergamon Press, Oxford, 153–190.

Norusis, M. J., 1990. *SPSS Base System User's Guide*. SPSS Inc., Chicago.

Rawls, W. J., Jackson, T. J. and Zuzel, J. F., 1980. Comparison of areal snow storage sampling procedures for rangeland watersheds. *Nordic Hydrology*, **11**, 71–82.

Rychetnik, J. 1987. Snow cover disappearance as influenced by site conditions, snow distribution and avalanche activity. In *Avalanche Formation, Movement and Effects*, eds B. Salm and H. Gubler, International Association of Hydrological Sciences, Special Publication 162, 355–361.

Sharp, M., Willis, I., Hubbard, B., Nielsen, M., Brown, G., Tranter, M. and Smart, C., 1993. *Water storage, drainage evolution and water quality in alpine glacial environments*. Interim Report, NERC Grant GR3/8114.

Steppuhn, H. and Dyck, G. E., 1974. Estimating true basin snowcover. In *Advanced Concepts and Techniques in the Study of Snow and Ice Resources*, eds H. S. Santeford and J. L. Smith, National Academy of Sciences, Washington, DC, 314–328.

Woo, M., Heron, R., Marsh, P. and Steer, P., 1983. Comparison of weather station snowfall with winter snow accumulation in high Arctic basins. *Atmosphere-Ocean*, **21**, 312–325.

Zevenbergen, L. W. and Thorne, C. R., 1987. Quantitative analysis of land surface topography. *Earth Surface Processes and Landforms*, **12**, 47–56.

18 Coastal Management and Sea Level Rise: a Morphological Approach

J. S. PETHICK
Centre for Coastal Management, University of Newcastle, UK

ABSTRACT

A method for the prediction of long-term morphological response of estuaries to sea level rise is described, using a hydraulic geometry approach first described by Myrick and Leopold in 1963. The estuary of the River Blackwater in Essex, England, is used as an example. Calibration and verification of the hydraulic geometry equations are achieved using a series of digital terrain models of the inter-tidal and sub-tidal estuarine morphology. Following IPCC (Intergovernmental Panel on Climate Change) predictions, a sea level rise averaging 6 mm a^{-1} was assumed over the next 50 years, which would result in an increase in tidal discharge and consequently in estuary width: changes which were predicted using the calibrated model. Results indicate that the outer estuary at this time would be between 800 m and 900 m wider than at present. Since the present estuary is entirely protected by flood embankments, these predicted changes in its morphology cannot be achieved by natural erosion and consequently flood embankments will need to be retreated if a stable estuarine form is to be attained. This process of "managed retreat" is presently under consideration as a form of coastal management in southeast England and the paper provides an initial estimate of the scale of such a retreat process.

INTRODUCTION

Managed retreat of the shoreline and its defences, both natural and artificial, is one of a number of response options which could be adopted in the face of the threat of increases in the rate of sea level rise. Alternative approaches include the somewhat academic "do-nothing" option; increasing the standard of existing defences; restoration of eroded coasts; and relocation of human infrastructures which are put at risk. Of these options, managed retreat is now regarded as one of the most powerful tools available to coastal managers, although public acceptance of the principle is likely to be a rather slow process.

The application of a managed retreat to coastal problems can take three quite distinct forms. First, the management of eroding shorelines, principally cliffs, as significant sources of sediment in order to provide inputs to adjacent or neighbouring coastal areas, is a form of managed retreat which has yet gained little

Landform Monitoring, Modelling and Analysis. Edited by S. N. Lane, K. S. Richards and J. H. Chandler.
© 1998 John Wiley & Sons Ltd.

acceptance but which may eventually be seen as an essential component of any shoreline management plan. Second, the maintenance or restoration of viable areas of supra- or inter-tidal habitat such as sand dunes, salt marshes and mudflats for their nature conservation value can be achieved by the managed retreat of existing coastal defences. Third, the provision of adequate coastal defences against flood or erosion risk can be improved by retreat of the existing line of defence, allowing natural inter-tidal landforms to develop which themselves provide an efficient and stable defence. This technique is of principal relevance to estuarine shoreline plans, although it also has applications for the open coast. It is this technique of estuarine shoreline retreat which will be examined in this paper, although all three approaches may be seen to play a part in any integrated management plan.

APPROACH

One of the difficulties of implementing a managed retreat policy in estuaries, as a response to increased flood risks due to sea level rise, is to overcome a natural tendency among coastal managers towards small-scale schemes which ignore the large-scale morphodynamics of the estuary. Since managed retreat involves the restoration of areas of the inter-tidal zone to tidal flooding, it follows that it also involves an increase in tidal prism (the total volume of water entering an estuary during the flood tide) and therefore in tidal discharge. For a relatively small managed retreat area of 100 ha this increase in tidal prism may amount to as much as 1×10^6 m^3; in a small estuary, this can give rise to a significant increase in discharge and tidal current velocity, causing erosion of the outer channel which provides tidal water to the retreat site.

In view of such estuarine-scale interactions, any managed retreat of a flood embankment within an estuary should be part of a large-scale design in which the total morphodynamics of the estuary are incorporated. The problem which faces coastal managers at the moment is that suitable long-term morphological models of estuarine development do not exist. To provide at least an interim solution to this problem, this paper reports work which uses a simple process–form model, incorporating the hydraulic geometry equations, to predict long-term changes in estuarine form.

The use of a form of hydraulic geometry model for the prediction of estuarine morphology has a long provenance. The relationship between tidal prism and the dimensions of the mouth cross-section was recognised by O'Brien (1931), who proposed that the relationship was of the form $A = aP^b$ (where A = mouth cross-sectional area; P = tidal prism; and a and b are constants). Since the discharge through the mouth section is determined by the tidal prism, this relationship is fundamentally similar to the hydraulic geometry equations which are of the form $w = aQ^m$ (where w = cross-section width and Q = discharge), and were first applied in this form to estuaries by Myrick and Leopold (1963). The relationship formulated by O'Brien (1931) has been shown to apply to a wide range of estuaries by a number of authors (e.g. Escoffier 1940; Bruun 1978; Bruun and Gerritson 1960; Gao and Collins 1994; Pethick 1994), and while van Dongeren and de Vriend

(1994) have extended the analysis to include cross sections within the estuary using partial tidal prisms. In the present paper this well known relationship is used to provide long-term prediction of changes in estuarine cross-sectional morphology and tidal discharge as a response to sea level changes.

One of the main difficulties facing any form of long-term morphological predictions is to provide verification of the models employed in their generation. Since "long term" for landforms as large as an estuary implies time scales of at least decades or centuries, it is clearly impracticable to verify models using direct observation: instead, recourse has to be made to historical data. The approach taken here is to calibrate the downstream hydraulic geometry model using a series of digital terrain models (DTM) based upon data derived from recent bathymetric surveys of the estuary. This calibration process provides the basic process–form relationship between tidal discharge and cross-section dimensions. Once such a calibration has been achieved, the model can be verified by demonstrating that it is capable of prediction over long time periods using a second series of DTMs based on historical surveys of the same estuary. A satisfactory outcome to this calibration–verification process then allows confidence to be placed in the use of the model for prediction of future estuarine morphology.

The intention of the work was to offer some preliminary guidance on the possible scale of managed retreat as a response to sea level rise within the Blackwater Estuary in Essex, England. The Blackwater has become a test area for many of the initial managed retreat experiments undertaken by Ministry for Agriculture, Fisheries and Food, the Environment Agency and English Nature. This work has involved the provision of an extensive data base including continuous measurement of waves, tides and suspended sediment and, in addition, annual topographic surveys. Using the latter information in conjunction with archival topographic information from Admiralty charts and from Ordnance Survey (OS) maps, it has been possible to provide the DTMs of the estuary which were then used in the calibration–verification process described above.

MANAGED RETREAT AND SEA LEVEL RISE

The reduction in defence standards provided by flood embankments, particularly along the east coast of England, in the face of the predicted sea level rise over the next 50 years is presently causing concern amongst coastal managers. The effect of sea level rise on these low-lying shores is two-fold. First, the joint probabilities of tide, wave and surge increase at a given surface elevation as the relative mean sea level increases. For example, the Intergovernmental Panel on Climate Change (IPCC) calculates that an increase in sea level of 30 cm over the next 30 years could result in an increase in the probability of overtopping of flood embankments on North Sea coasts from 1:250 years to 1:10 years (IPCC, 1992). A second, exacerbating effect is that increases in sea level can result in deterioration of the natural inter-tidal area fronting the flood defence. This deterioration is itself probably a result of increased probability of wave attack due to the rise in relative sea level, but the resultant loss of mudflat or salt marsh can result in a decrease in wave attenuation, thus further increasing the risk of overtopping or even breaching of the flood embankment (Owen, 1984).

Research carried out within east coast estuaries suggests that the primary response to increased sea levels is an increase in the width of the outer estuary, with a decrease in mean depth of the sub-tidal channels. This response appears to take place mainly as a result of the redistribution of sediments within the estuary. Erosion of the upper inter-tidal mudflats and the outer edge of salt marshes, probably as a result of increased wave activity as water depths increase, releases sediment which flows down the inter-tidal slope to accumulate in the sub-tidal channels. Measurements made in the Medway, Thames, Blackwater, Orwell and Humber estuaries all demonstrate that the volume of sediment eroded from the inter-tidal areas is roughly balanced by deposition in the sub-tidal channels, although in many cases this balance is offset by navigational dredging (Pethick, 1994). In addition to this internal redistribution of sediment, detailed topographic surveys of the Blackwater Estuary between 1978 and 1994 demonstrate that a small but significant net accretion has taken place, derived from marine sources, which amounts to 200 000 m^3 a^{-1}. If such a net accretion were equally deposited over the entire tidal area of the estuary, the total net accretion would be 4 mm a^{-1}, approximately equal to the annual sea level rise as recorded on adjacent tide gauges at Harwich and Sheerness (Graff, 1981). It appears therefore that the response of this estuary to sea level rise is for an increase in the flare of the plan shape of the estuary, together with an overall rise in the mean elevation of the estuary, keeping pace with sea level rise.

The morphological adjustment which results from this redistribution of sediment can be interpreted as one which provides a hydraulic cross-sectional form which has maintained its relative position in the tidal frame, as the latter rises with increases in sea level. This means that the estuary maintains its mean depth and consequently its frictional drag on the tidal flow. At the same time, however, the trapezoidal cross-section of the estuary means that an increase in mean water level results in an increase in tidal prism and consequently tidal discharge. Thus, despite the increase in bed elevation caused by deposition and the fact that the tidal frame may be expected to rise as a single unit at the same rate as sea level, the channel cross-sections, especially in the inter-tidal zone, experience a net increase in area. Since this applies to all cross-sections throughout the estuary, the overall tidal prism increases and therefore the mean discharges at each section increase, although mean velocities remain approximately constant.

The morphological response described here has been noted in most east coast estuaries only over the past 50 years, presumably due to a change in the rate of sea level rise. Since the rates of horizontal erosion of the mudflat/salt marsh boundary are rarely less than 1 m per year, this means that between 50 m and 100 m of marsh edge retreat has been observed in these estuaries over this period (Burd, 1992), in many cases resulting in the complete removal of inter-tidal salt marsh. This development is causing concern for the stability and efficiency of flood embankments, but also means that further increases in estuarine width cannot take place owing to the presence of these embankments. In those estuaries where this process of salt marsh loss is most marked, the outcome has been that inter-tidal mudflats are suffering vertical erosion at their upper margins with consequent undercutting of the flood embankments. Even more important, as the salt marshes are eroded away so the supply of sediment to the sub-tidal channels decreases so that estuary depth increases as sea level rises. Moreover, the presence of the flood embankments means that no increases in the shallow water inter-tidal area can take place, once again exacerbating the increase in depth as sea level rises and further reducing the drag on tidal flows. The result is that tidal range in these estuaries increases as sea level rises, and consequently flood risks to low-lying areas beyond the flood embankment increase accordingly.

It is clear that perhaps the only sustainable response to this mounting problem is to retreat the line of the present flood embankments, allowing the estuary space and sediments with which to respond naturally to sea level rise.

MANAGED RETREAT OBJECTIVES

Before any consideration is given to the techniques to be adopted for managed retreat in an estuary, it is imperative that clear management objectives are defined. For example, merely increasing the area of inter-tidal habitat in an estuary in the face of increased sea level rise is an important objective for nature conservation, but fails to address the problem of estuarine dynamics outlined above. It may be that the provision of areas of inter-tidal habitat can best be met by restoring tidal inputs to areas of reclaimed land adjacent to, but quite separate from, the main estuary. This can be achieved by breaching flood embankments so that the reclaimed areas are subjected to tidal flooding and deposition but are not affected by estuarine currents or wave action. However, such habitat restoration does not address the problems of estuarine width increase outlined above, so that the process of loss of inter-tidal area in the main estuary continues unabated, despite the provision of additional inter-tidal habitat. Clearly, such a limited objective may satisfy one section of estuary users but not all. Indeed, as already noted above, the increased tidal prism set up by such habitat creation schemes may actually lead to increased channel velocities and therefore cause an increase in the erosion of the inter-tidal salt marshes in the main estuary.

A more comprehensive objective for estuarine shoreline management should be to improve the hydraulic efficiency of the estuary so that inter-tidal areas are not subjected to erosion, thus allowing naturally occurring habitat to develop. In areas subjected to sea level rise, this objective can only be met by increasing the width of the estuary while at the same time decreasing the mean depth. These changes will take place naturally as long as the presence of an inhibiting flood embankment is removed. This management objective has been called bank retreat as opposed to the breach retreat of the habitat restoration schemes described above.

The suggestion that estuarine flood embankments should be setback has been extensively criticised by land owners, who see even the possibility of such a scheme as leading to a "planning blight" resulting in falling land prices. The cost–benefit issues involved in such an argument are outside the terms of reference of this paper and indeed are presently being addressed by several authorities.

THE HYDRAULIC GEOMETRY OF BANK RETREAT

The prediction of the effects of sea level rise on the dynamics of an estuary and the resultant morphological adjustments which may take place has been described above in qualitative terms, but the development of a more quantitative model has not yet been achieved. Existing hydraulic models allow the prediction of flow patterns and bed stresses as the result of input of a set of morphological boundary conditions, but are unable to achieve feedback between such hydraulics and morphology over anything but a short time span, usually measured in tidal periods rather than in months or years.

Two early papers by Langbein (1963) and Myrick and Leopold (1963), on the hydraulic geometry of small tidal estuaries, appear to provide a simple geomorphological model from which predictions of estuary response to sea level rise can be made, thus allowing some quantification of the necessary areal extent of managed retreat. Langbein's paper was based upon theoretical considerations of the hydraulic geometry of estuaries. His argument suggested that tidal discharge was related to estuarine variables such as width, depth and velocity such that, while velocity and depth showed a conservative response to changes in discharge, width was extremely responsive. His predictions for the three major hydraulic geometry equations were:

$$w = aQ^{0.71}$$
$$d = cQ^{0.24}$$
$$v = kQ^{0.05}$$

where w = width; d = depth; v = velocity; Q = discharge; a, c, k are constants.

These predictions were subsequently tested by Myrick and Leopold (1963) using data from a small tidal estuary, and were found to be substantially correct.

METHODS

The application of the hydraulic geometry approach to the problem of managed retreat requires not only a transfer of the constants to an entirely different estuarine environment, necessitating recalibration of Langbein's exponents, but also that the relationship between tidal discharge and morphology holds through time, an initial assumption that requires verification. In the present case, evidence of changes in the width of the Blackwater Estuary is available over a period of almost 150 years, using data derived from the first edition OS map (1:63 360) surveyed in 1838 and the present-day morphology which was surveyed in 1994 by the Environment Agency (Figure 18.1). Ten cross-sections were used in the analysis, located at intervals of 2 km from the mouth. The tidal prism and cross-section areas used in the analysis were calculated using data from a series of bathymetric surveys of the estuary made in 1994 by the Environment Agency (Figure 18.2). Data from these surveys, which used standard sonar techniques, were incorporated in a digital

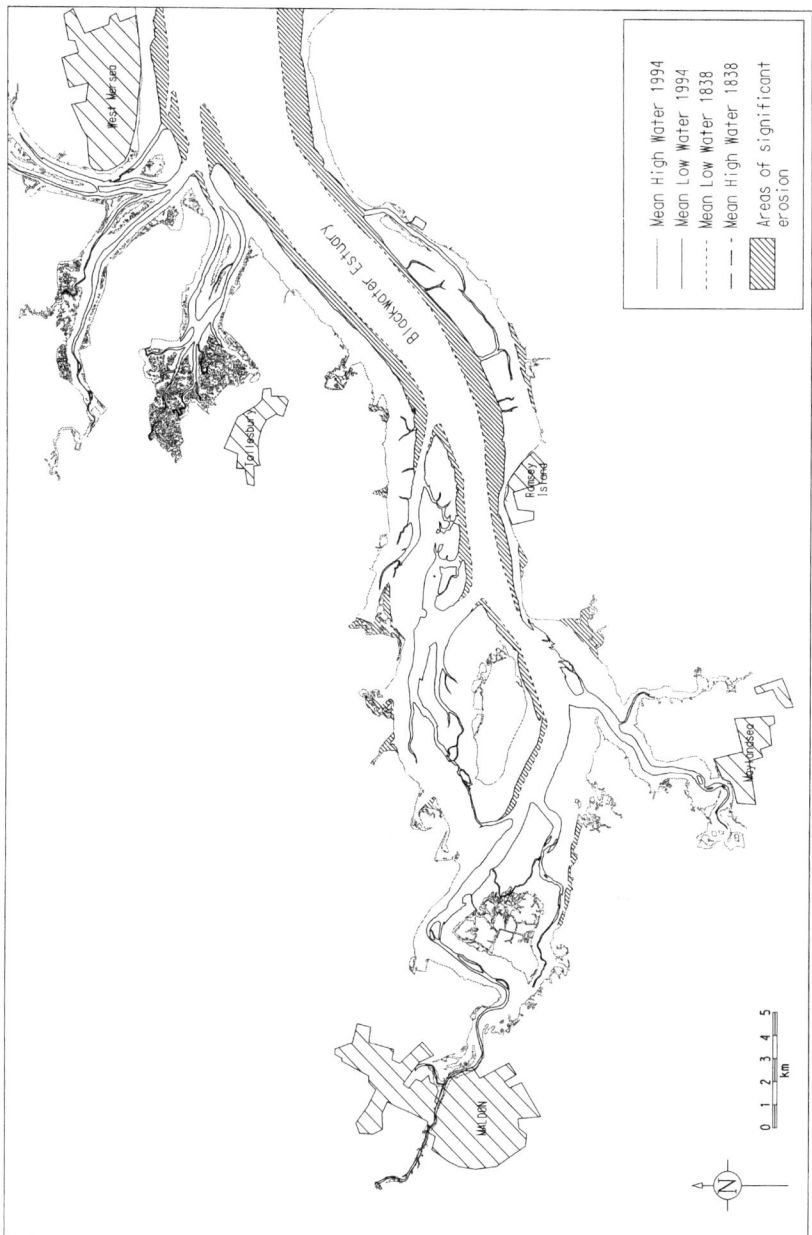

Figure 18.1 Comparison between the 1838 and 1994 surveys of the Blackwater Estuary showing changes in high and low water marks

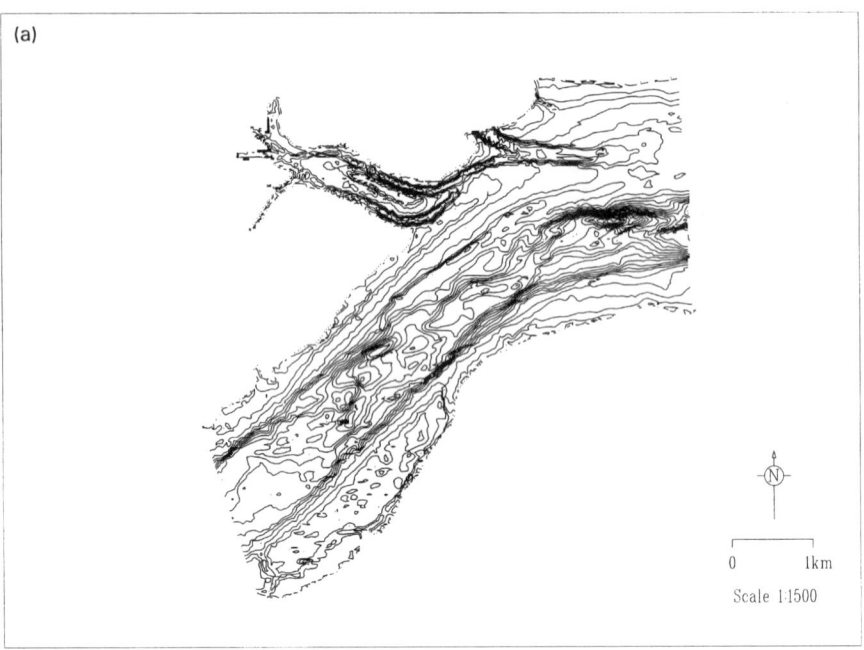

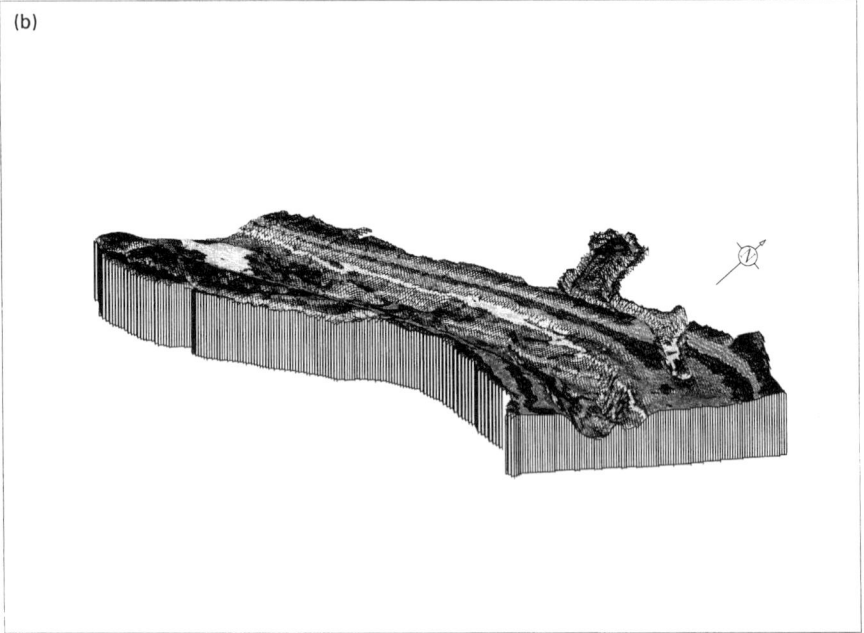

Figure 18.2 (a) Section of digital elevation model for the Blackwater Estuary compiled from bathymetric surveys undertaken in 1994; (b) isometric view of the estuary channel derived from DTM used in the analysis

elevation model. Using this DTM, calculations were then made of the total estuarine volume below high water mark spring tides (HWMST) and of the inter-tidal volume between HWMST and low water mark spring tide (LWMST). Each calculation was repeated for the 10 cross-sections to give the tidal volumes upstream (i.e. inland) of each section. Finally, cross-sectional area, width and depth were calculated for each of the 10 cross-sections.

The calculation of tidal discharge from these volumetric data presents more of a problem. Tidal discharge at each cross-section in an estuary varies throughout the tidal cycle and is determined by the rate of change in the water surface elevation (i.e. tidal stage), which in turn gives the partial tidal prism upstream of the section, that is the volume of water which is needed to fill the upstream estuary channel during each successive time interval. For the analysis, it was assumed that the process–form relationship summarised by the hydraulic geometry equations would be realised when the estuarine channel was at, or near, bankfull. Since this occurs at each high water and since at high water velocities approach zero, it was decided to use the mean discharge during the last hour of the spring tide as the input for the model. Measurement of tidal stage in the Blackwater indicates that, during the last hour of the spring tide, water elevations increase by 0.75 m and partial tidal prisms (ΔP) for each cross-section were calculated from the DTM using this figure. The mean tidal discharge over the last hour of the flood tide for each section was then calculated using $\Delta P/3600$. Changes in tidal prism and consequently tidal discharge over the past 156 years (1838–1994) were then calculated under an assumption that changes in sea level rise over this time period have resulted in a volumetric increase in the tidal prism equivalent to the product of tidal area and the change in sea level.

The analysis concentrated on the channel width of the estuary since this is the primary consideration for managed retreat. Calibration of the model was achieved by assuming that the exponent in the hydraulic geometry expression for width is constant for all estuaries, while the intercept a varies according to the local tidal and sedimentological conditions. Using the morphological data from the two surveys in 1838 and 1994, and calculating changes in discharge over this time period due to sea level changes, it is possible to provide an estimate of this intercept.

ANALYSIS AND RESULTS

For the initial calibration stage, a measure of channel width is required. In this case the width of the estuary was defined at mean sea level, that is at the mean of HWMST and LWMST. Although the low water mark has retreated landwards by a substantial amount over the past 156 years, the high water mark of the estuary is largely confined by flood embankments and has consequently remained relatively constant over this time period, thus the present-day observed width at mean sea level may be an underestimate of the natural morphology.

The tidal discharges, calculated as described above, were substituted in the hydraulic geometry equation $w = aQ^b$. Using Langbein's coefficient of $b = 0.71$, the observed width of the estuary at each time period was used with the calculated

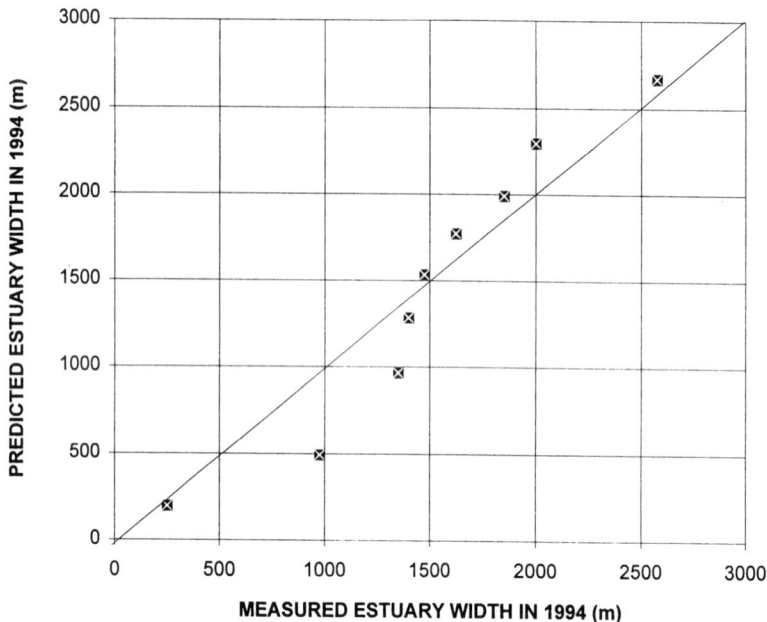

Figure 18.3 Comparison between predicted and measured estuarine channel widths in 1994, for 10 cross-sections in the Blackwater Estuary. The solid line indicates perfect correspondence and is not a best fit regression to the data points

discharge to estimate the value of the constant a. The best fit value was found to be $a = 3.8$. The predictive capability of this calibrated model is shown in Figure 18.3, which shows the comparison between observed and predicted widths for the 1994 data set. The Pearson correlation coefficient for this relationship is $R = 0.964$, $P < 0.005$. Despite the overall excellent predictive capability of the relationship, Figure 18.3 does indicate that, in the inner estuary, predicted values of width underestimate the measured values. This may be a response to the presence of two islands in the inner estuary, Northey and Osea (Figure 18.1). Division of the estuary channel around these islands appears to have resulted in two relatively shallow channels whose combined widths are greater than would have been the case for an undivided and deeper channel.

Following calibration, verification of the model requires that predictions of width are compared with a set of observations which were not included in the calibration process. For this purpose, the variations in estuarine morphology as shown in the OS survey of 1838 were compared with those predicted by the calibrated model. In order to provide such predictions, the change in tidal discharge due to changes in sea level between 1994 and 1838 was calculated. During this 156 year period, sea level in the Blackwater was assumed to have fallen by 1 mm a^{-1} (Graff, 1981). The tidal prisms for each cross-section were recalculated for the total sea level fall of 156 mm using the digital elevation models constructed from the survey data. The predicted widths at the 10 cross-sections were then compared with the observed

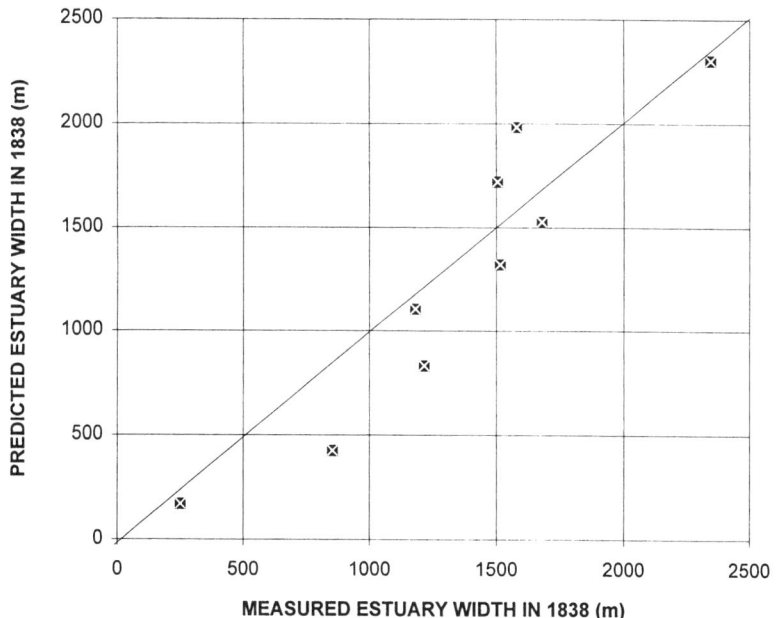

Figure 18.4 Comparison between predicted channel widths in 1838, assuming a 1 mm a^{-1} fall in sea level over the period 1994 to 1838, and measured channel widths from OS surveys, for 10 cross-sections in the Blackwater Estuary. The solid line indicates perfect correspondence and is not a best fit regression to the data points

widths (Figure 18.4). The relationship has a correlation coefficient of 0.935 ($P<0.005$) which represents a significant overall agreement, although, once again, the model underestimates the channel widths around the region of the two islands.

Despite the one localised discrepancy between predicted and observed data, it was decided that the model was sufficiently rigorous when applied over a 156 year period to allow its use as a predictor of the future changes in estuarine morphology after a 50 year period of rising sea level. This prediction assumed a sea level rise of 6 mm a^{-1} over that period (IPCC, 1992) and used as a baseline the morphology and tidal prism as surveyed in 1994. Results are shown in Figure 18.5. If the predicted widths for 2044 are compared with the actual widths in the most recent survey (1994), then an indication is given of the distance by which flood embankments should be set back in order to maintain estuarine stability (Table 18.1).

It is interesting to note that in the outer estuary the predicted widths exceed those measured in 1994 by over 800 m. This implies that, in order to maintain estuarine stability, the flood defences would need to be retreated by 400 m on each bank of the estuary. Further into the estuary this retreat distance decreases and reaches zero at approximately 11 km from the mouth as shown in Figure 18.6. In this figure, managed retreat has been indicated only on reclaimed land since this would allow restoration of an active inter-tidal zone on the estuarine side of the new flood defences. In the Blackwater, this restriction on managed retreat causes some

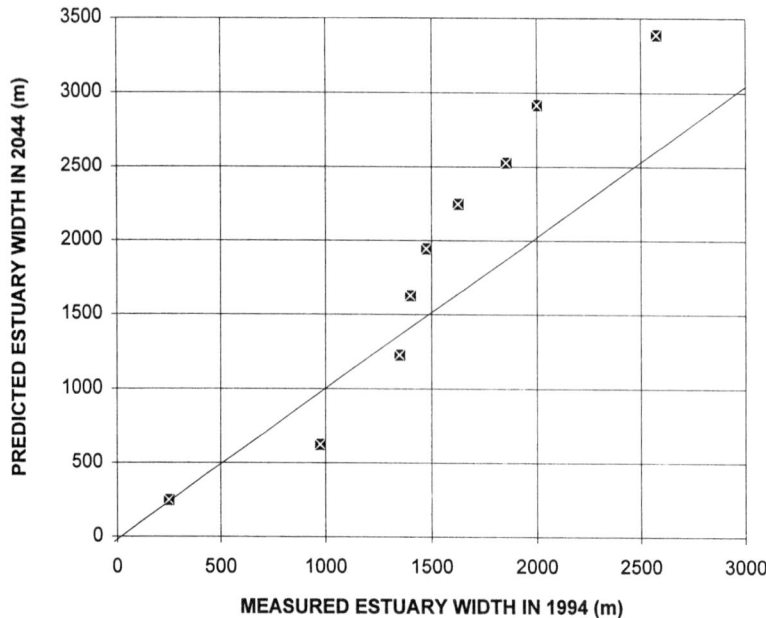

Figure 18.5 Comparison between predicted channel widths for 2044, assuming a 6 mm a^{-1} rise in sea level, and measured estuarine channel widths in 1994, for 10 cross-sections in the Blackwater Estuary. The solid line indicates perfect correspondence and is not a best fit regression to the data points

Table 18.1 Differences between measured estuarine widths in 1994 and predicted widths for 2044

Distance (m)	0	2000	4000	6000	8000	10000	12000	14000	16000
Change in width, 1994–2044	−812	−916	−675	−621	−470	−226	124	349	0

difficulties since the presence of rising ground on the estuary shores limits the amount of retreat which can be achieved in many places.

In the inner estuary, between 12 km and 14 km from the mouth, Table 18.1 and Figure 18.6 show that the predictions for estuary width are smaller than the widths measured in 1994. The area over which this decrease in width is predicted coincides with the two islands of Northey and Osea which, as noted above, cause a subdivision of the estuary channel and lead to an increase in the total measured width when compared to the predicted figure for a single channel. It is reasonable to suppose that this discrepancy applies to the predicted widths in 2044 as compared to measured widths in 1994, so that the predicted decrease in channel width shown in Table 18.1 between 12 km and 14 km may be discounted. Instead, an assumption is made that channel widths over this section of the estuary will remain constant

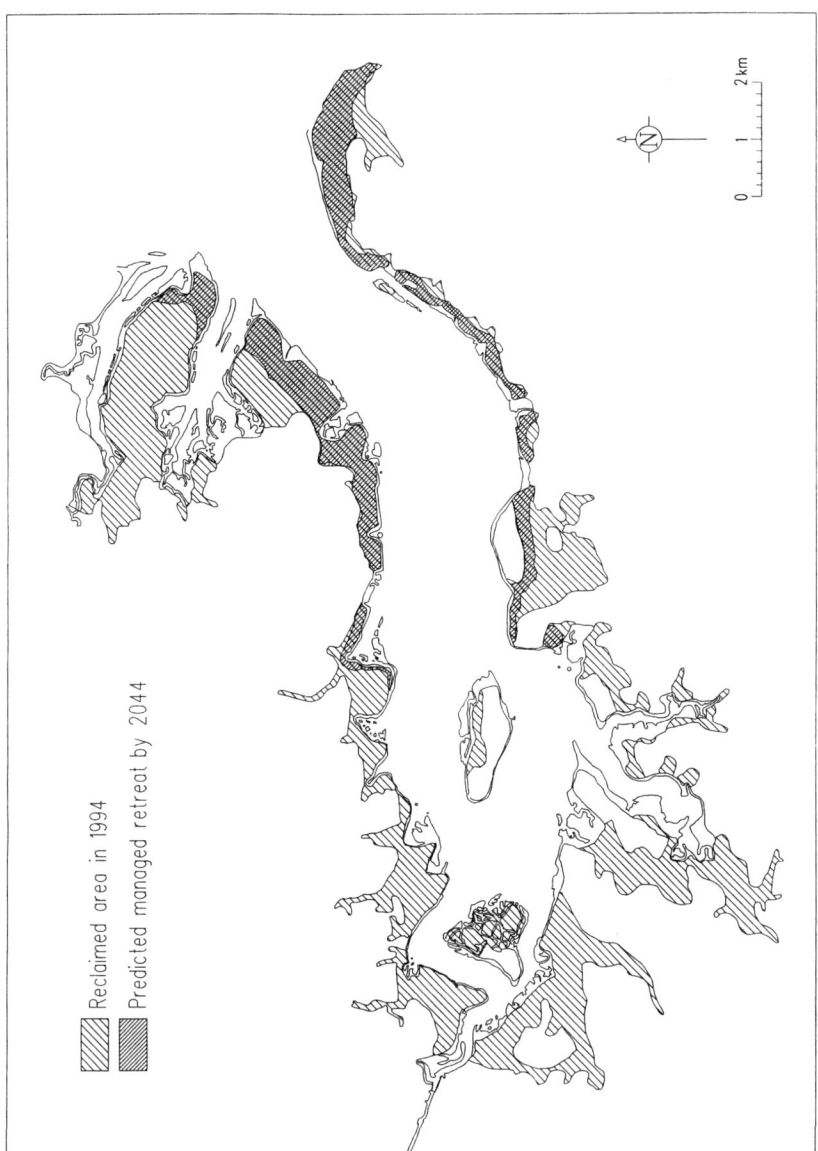

Figure 18.6 Map showing the predicted optimum managed retreat areas in the Blackwater Estuary calculated for the year 2044 and assuming a mean annual sea level rise of 6 mm over the 50 year period 1994–2044

over the next 50 year period as they are shown to do by the predictions for the cross-section immediately landward, located at 16 km from the mouth (Table 18.1).

MANAGEMENT OF MANAGED RETREAT

The calculations presented here are intended as no more than a preliminary attempt at predicting the changes in estuary outline as a result of sea level rise over the next 50 years. For example, it is clear from the inability of the model accurately to predict channel width in the sub-divided channels around the two islands that the use of a two-dimensional analysis is inadequate. Nevertheless, the results do provide an indication of the scale of changes to flood defences which may have to be made in estuaries if they are to respond to sea level changes over the next 50 years.

   The implications of the results of this work for the management of managed retreat in estuaries are of great importance. It should be stressed that the use of the Blackwater Estuary here is intended to provide no more than an example of the general pattern of retreat in estuaries and that no implications for actual retreat are intended. Nevertheless, if a shoreline management plan for an estuary such as the Blackwater were to be designed so as to accommodate a predicted 6 mm a^{-1} increase in sea level over the next 50 years, then the results given here suggest that the flood embankments on either side of the estuary should be moved landwards by a maximum of 450 m. Further landwards, however, the necessary retreat diminishes rapidly to zero at approximately 10 km from the mouth and the total retreat area involved over the whole estuary amounts to less than 200 ha or approximately 5 ha per year. The result of such a managed retreat would be the establishment of a natural inter-tidal zone throughout the estuary, reducing the probabilities of flood embankment overtopping, undercutting or breaching, and providing significant economic advantages over a policy of raising the standard of the present defences. Moreover, such a policy would allow the development of natural mudflat and salt marsh along the estuary, improving its value for wildlife and recreation and its intrinsic landscape value.

CONCLUSION

Managed retreat should be seen as a method by which estuaries are allowed to respond naturally to increases in sea level over the next 50 years. Such a policy would confer many advantages over alternatives such as do-nothing or improve standards of existing defences. Using the Blackwater as an example, the maximum retreat necessary to accommodate a 6 mm rise in sea level would be 450 m in each bank of the outer estuary, a figure which decreases to zero at 10 km from the mouth of the estuary. The loss of land to active agricultural production is estimated at 5 ha per year, distributed throughout the estuary, a loss which should be equated with the significant savings in flood defence provision and the major gains in the estuarine environment.

ACKNOWLEDGEMENT

Much of the topographic data used in this paper was provided by the Environment Agency, to whom grateful acknowledgement is made.

REFERENCES

Brunn, P., 1978. *Stability of Tidal Inlets.* Elsevier, Amsterdam.
Brunn, P. and Gerritson, F., 1960. *Stability of Tidal Inlets.* North Holland Publishing Co., Amsterdam.
Burd, F. H., 1992. *Erosion and vegetation change on the saltmarshes of Essex and north Kent between 1973 and 1988.* Nature Conservancy Council, Peterborough.
Escoffier, E. F., 1940. The stability of tidal inlets. *Shore and Beach*, **8**, 114–115.
Gao, S. and Collins, M., 1994. Tidal inlet equilibrium in relation to cross-sectional area and sediment transport patterns. *Estuarine, Coastal and Shelf Science*, **38**, 157–172.
Graff, J. 1981. An investigation into the frequency distributions of annual sea level around the coast of Britain. *Estuarine, Coastal and Shelf Science*, **12**, 389–449.
IPCC, 1992. *Global climate change and the rising challenge of the sea*, Intergovernmental Panel on Climate Change, Response Strategies Working Group.
Langbein, W. B., 1963. The hydraulic geometry of a shallow estuary. *Bulletin of the International Association of Scientific Hydrology*, **8**, 84–94.
Myrick, R. B. and Leopold, L. B., 1963. *The hydraulic geometry of a small tidal estuary.* United States Geological Society, Professional Paper 422-B.
O'Brien, M. P., 1931. Estuary tidal prism related to entrance areas. *Civil Engineering*, **1**, 738–739.
Owen, M. W., 1984. Effectiveness of saltings in coastal defence. In *Proceedings of Conference of River Engineers, Cranfield*, MAFF, London.
Pethick, J. S., 1993. Shoreline adjustments and coastal management: physical and biological processes under accelerated sea level rise. *Geographical Journal*, **159**, 162–168.
Pethick, J. S., 1994. Estuarine processes. In *Wetland Management*, eds R. Falconer and P. Goodwin, Institution of Civil Engineers, Telford, 75–87.
van Dongeren, A. R. and de Vriend, H. J., 1994. A model of morphological behaviour of tidal basins. *Coastal Engineering*, **22**, 287–310.

19 Geomorphological and Hydrodynamic Results from Digital Terrain Models of the Humber Estuary

JACK HARDISTY, RICHARD MIDDLETON, DUNCAN WHYATT and HELEN ROUSE

School of Geography and Earth Resources, University of Hull, UK

ABSTRACT

Two digital terrain models (DTMs) of the Humber Estuary have been constructed from hydrographic survey results and have been linked with tide gauge results in order to analyse longer term (decadal) geomorphological change and to model tidal discharges and tidal current velocities along the estuary. Tidal elevations along 60 km of the outer estuary, from the mouth to the Ouse–Trent confluence, have been sampled at hourly intervals from the gauge network and used to interpolate the surface elevation of the water in each of 60 segments oriented with segment boundaries and surface water "strike" along north–south axes. The results have been combined with the DTM to determine water volume, surface area and mean cross-sectional area in each segment and hence to determine volume changes and cross-sectional mean flow velocities. The results are compared with the limited field data which are presently available for spring and neap tides and the potential for real-time and longer term operation of the model is discussed.

INTRODUCTION

Digital terrain modelling techniques offer tremendous potential for the analysis and prediction of a wide range of environmental phenomena. In the present paper we explore, principally, the utility of such techniques for achieving a real-time hydrodynamic description of water movements within a large and complicated geomorphological feature such as the Humber Estuary. The paper introduces the Humber and then presents a comparison of two digital terrain models (DTMs) of the estuary. The results show that a pattern of longer term (decadal) erosion and accretion exists within the estuary, but that the changes are small and probably within the experimental errors of the field survey and analytical techniques.

Nevertheless, the general patterns revealed by this analysis require some explanation in terms of the sediment transport processes within the estuary. This question, in turn, requires a better understanding of estuary flow and of diurnal, tidal and

Landform Monitoring, Modelling and Analysis. Edited by S. N. Lane, K. S. Richards and J. H. Chandler.
© 1998 John Wiley & Sons Ltd.

seasonal variations in estuary flow conditions. Few field data are currently available with which to study flow patterns and many investigators have, therefore, turned to the use of computer-based flow simulators. In general, such models have been based upon numerical solutions to the Navier–Stokes equations and we discuss such computational fluid dynamics (CFD) programs in this paper. It is clear, however, that such programs have certain shortcomings which make them unsuitable for either real-time or longer term (decadal) operation. We therefore develop a real-time continuity-based flow simulator or time series analysis (TSA) model which runs "within" the DTM in the present paper and discuss a comparison with field results and future applications.

THE HUMBER ESTUARY

The Humber Estuary drains some 25 000 km^2 of central and eastern England and the tidal waters extend 100–120 km inland on the Trent and the Ouse (Figure 19.1) (Hardisty et al., 1995). Rising sea levels flooded the river valleys about 6000 years BP giving, in general terms, the present-day estuary shape which now flares from about 1 km wide at the Ouse–Trent confluence to over 8 km wide at Spurn Head (IECS, 1987). Freshwater inputs vary and, for example, the then regional National Rivers Authority quoted a mean annual value for 1991 of 235 m^3 s^{-1}.

North Sea waves, which penetrate the outer estuary and are responsible for sea-bed stirring and thus for increasing suspended sediment concentrations and sediment fluxes, have been recorded at the Dowsing Light. The waves are reported to have 50 year significant wave heights (H_s) of 8.23 m (using a Fisher–Tippett Type I (FT-1) distribution), 7.49 m (using a Weibull two-parameter distribution) and 6.76 m (using a Weibull three-parameter distribution (Fortrum, 1980; Bacon, 1989). More recent results from deployments of Wave Rider buoys off Holderness are reported by Bacon and Carter (1988) and suggest an FT-1 50 year return H_s of 5.99 m. Locally generated waves within the estuary are smaller, and Pethick (1987) suggests that a northwesterly wind blowing along the estuary might generate waves of 1 to 2 m at the mouth.

The Humber tides form part of the southern North Sea systems and are principally dominated by the M$_2$ amphidromic point in the eastern-central North Sea (Doodson and Warburg, 1941) with a phase lag angle of 120° with respect to lunar transit at Greenwich. In general, the tidal wave progresses in a southerly direction down the coasts of Holderness and Lincolnshire (Hardisty, 1990) and enters the mouth of the estuary some 6 h before high water Dover. The tide takes about 3 h to progress up the estuary from Spurn Head to Blacktoft and higher order harmonics are generated during the transition, causing increasing elevation and flow asymmetries. Spring tidal range in the Humber is 6.5 m at the mouth, rising to a maximum of 7.2 m at Saltend and then decreasing progressively upstream. The pattern is semi-diurnal with a pronounced neap-spring inequality. Pethick (1994) estimated the volume of the tidal prism on a mean tide as 1.2 × 10^9 m^3 and suggested that the northern Hawke Channel is ebb-dominated whereas the southern Haile Channel is flood-dominated. The level of the tide in the Humber has long

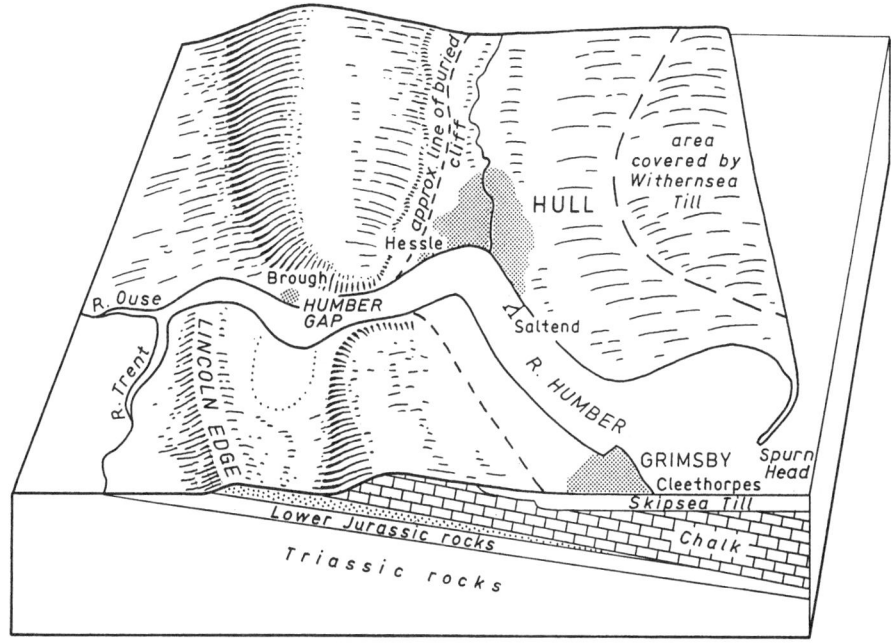

Figure 19.1   Geological map of the Humber region

been predicted for navigational and shipping purposes in the Admiralty Tide Tables, which list amplitude and phase information for the M_2, S_2, K_1 and O_1 species for a number sites in the estuary; we discuss the potential for extending these forecasts to include tidal stream predictions later in the present paper.

THE DIGITAL TERRAIN MODELS AND GEOMORPHOLOGICAL RESULTS

The first stage in the development of a real-time estuary flow simulator was the construction of new DTMs of the Humber bathymetry using the techniques described in, for example, Petrie and Kennie (1987), Moore *et al.* (1991) and Lane *et al.* (1994). These DTMs form the bathymetric basis for the implementation of both the CFD and the TSA flow models described below. Two DTMs have been constructed, based upon point depths digitised at irregular spacings from the Associated British Ports (ABP) charts of the estuary for the annual surveys of 1980 and 1989. The two DTMs also include additional data from the Admiralty chart east of Spurn Head. All depths have been converted from the various charts to Ordnance Datum Newlyn (ODN) using linear and planar interpolations based upon six segments of the estuary and following consultations with the hydrographic surveyors at ABP Hull. The datum corrections varied from –3.9 m at Spurn Head

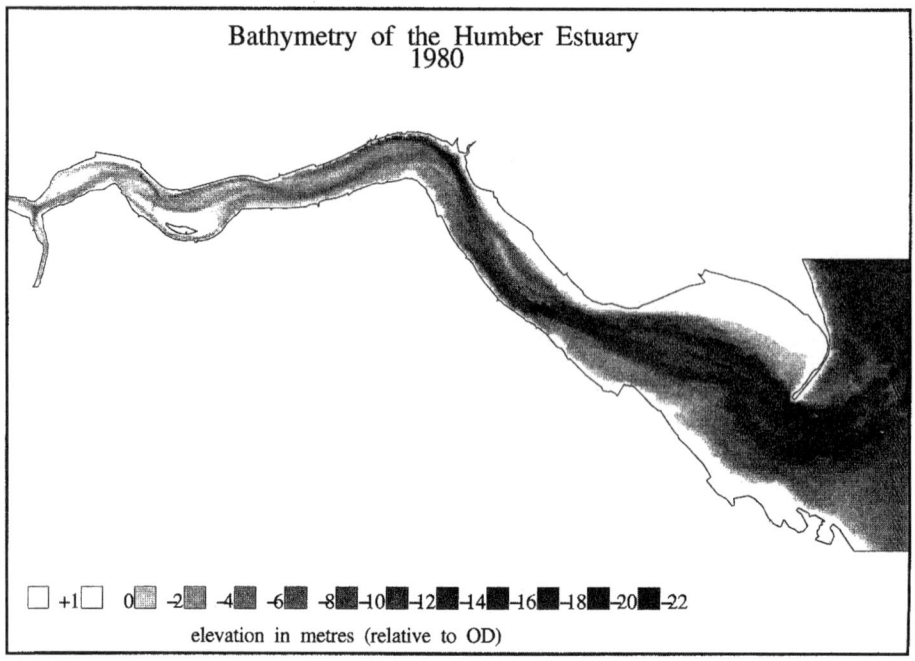

Figure 19.2 DTM of the Humber Estuary based upon the 1980 ABP (Humber) data

to –1.1 m at Burton Stather. The first model was constructed from the 1980 surveys and charts (Figure 19.2) and the second from the 1989 surveys and charts.

The estuary margins were digitised from local 1:500 000 Ordnance Survey maps and used to generate pseudo-depths set to 1 m above high water spring tides, particularly in regions shoreward of the ABP surveys. In addition, the shoreline was used to constrain the digital terrain model construction area which was applied to the digital terrain model. The data were processed in the ARC-INFO Geographical Information System to generate a DTM using the Delaunay triangulation method (Petrie and Kennie, 1987) which is increasingly accepted for modelling natural topography (Lane *et al.*, 1994). This model was sampled on a regular 100 m grid which, after masking with the shoreline data, produced the cell model which was used in the CFD and the TSA calculations below.

The errors involved in the generation and utilisation of the DTMs were assessed in two stages: there are errors in the original depth determinations and there may then be further errors introduced in the construction of the DTMs. The hydrographic surveys are carried out on a more or less continuous basis by ABP Humber and for the 1980 charts involved Hi-Fix position fixing with a quoted error of +/–2.5 m. The system was upgraded during the 1980s and the 1989 chart was based upon a Microfix network with a location accuracy of +/–1 m. Depths were measured with Atlas Deso hydrographic sounders which are calibrated on a daily basis for changes in the celerity of sound using the standard immersed reflector technique and corrected for tidal elevation from the gauge network described below. Depths are determined to a combined error of +/–0.05 m (Holmes, ABP Hydrographer, *pers. comm.*). It is noted that there are no cumulative errors in either the depth or the position data. The DTM was constructed using linear triangulation. The stability of the solution was examined by comparing the estuary volume estimated from a 500 m grid based upon the 100 m grid data, and using three different interpolation strategies. Firstly, the spring tide volume was estimated using the DTM depth at cell centres as 5.347 km^3, whilst using the mean depth of all non-land cells produced an estimated 5.366 km^3, and finally using the median depth of all non-land cells produced an estimate of 5.378 km^3. The similarity of these three estimates was taken as an indication that the interpolation strategies were not introducing cumulative errors and the DTM was therefore used for further analysis.

The models were first used to investigate geomorphological change in the estuary. The cell model was used to calculate the volume required to fill the estuary up to integer increments of elevations with respect to Ordnance Datum (OD) from an eastern boundary defined as a line running north–south from the point of Spurn Head to the Ouse–Trent confluence. In addition, it was clear that much of the estuary change was concentrated in the upper estuary and the analysis was therefore repeated separately for the lower and upper estuary sections either side of a north–south line at Hull. This type of analysis allows the location of the changes to be identified even though there may be no net change in the total estuary volume. The results for the 1989 DTM data can be summarised by considering the estuary volume at the high and low water marks for different tidal ranges, such as mean high water spring (MHWS), mean high water neap (MHWN), mean low water spring (MLWS) and mean low water neap (MLWN) as below. For example, taking tidal elevations with respect to OD at the mid-estuary site at Hessle, the results are:

	MHWS	MHWN	MLWN	MLWS
ODN	3.9	1.8	−1.2	−3.0
Volume (km^3)	2.7	2.1	1.4	1.0

The estuary volume for a spring tide is thus 1.7 km^3 and for a neap tide is 0.7 km^3. There are, of course, some errors in these estimates due principally to the fact that the water surface is not, at any state of the tide, represented by a geoid and is not, therefore, represented by a surface which is parallel to the OD at all locations. Neither are high water or low water simultaneously attained in all parts of the estuary. Also, future attention may need to be given to the digitisation and conversion to lattice procedures. However, with these considerations and the errors detailed above, the two figures represent a reasonable estimate of tidal volumes.

Analysis of the change in volume of the estuary by comparison of the two digital terrain models reveals a complicated geomorphological response. The results are represented as change in volume below OD levels in Figure 19.3 and as the volume change for each 1 m OD slice in Figure 19.4. It is clear that four quite different responses can be identified: (i) between –20 m OD and about –10 m OD the estuary volume is decreasing (that is, accretion is occurring in the deeper water areas) with a total of some 0.02 km^3 volume change between the two DTMs; (ii) between –10 m OD and about –5 m OD the change in estuary volume remains

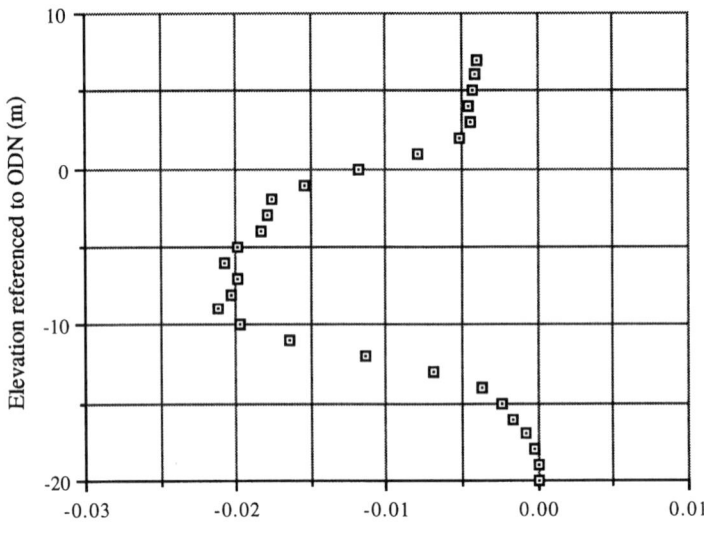

Figure 19.3   Integrated estuary volume changes, 1980 to 1989

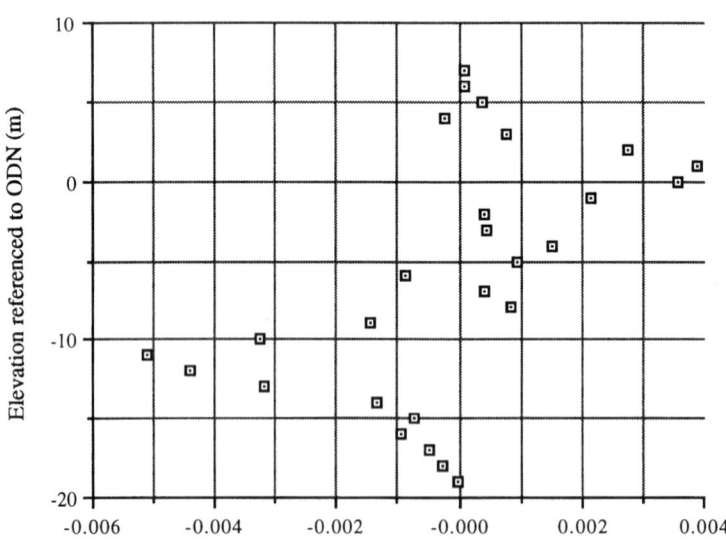

Volume change by slice 1980 to 1989 (km³)

Figure 19.4   Estuary volume changes by slice, 1980 to 1989

constant so that neither net accretion nor net erosion is occurring; (iii) between −5 m OD and about 2 m OD (approximately High Water on neap tides at Hessle) the estuary volume is increasing (that is erosion is occurring in the shallower water areas) with a total of some 0.015 km^3 volume change between the DTMs; and (iv) above 2 m OD there is no volume change.

Overall, there was some 0.005 km^3 of accretion between 1980 and 1989 and this appears to have principally occurred in the deeper water areas. Since there were approximately 6300 tides over the nine year period then, on average, the accretion figure represents annual accretion of 0.56 × 10^6 m^3 or some 800 m^3 of accretion on an average tide and, therefore, a tidal influx of some 1.28 × 10^6 kg or 1280 tonnes of sediment. Although these results are preliminary and do not identify diurnal, tidal and seasonal variations, they do provide first estimates of net geomorphological change within the estuary. Any explanation of the rate and location of the accretion must depend upon a better appreciation of instantaneous and residual flow patterns than is presently available and therefore the DTMs have also been utilised to develop tidal flow models, as detailed below.

ESTUARY FLOW MODELLING USING TIME SERIES ANALYSIS

There are two stages in the implementation of any tidal flow simulator based upon flow continuity methods: first, the tidal elevations are determined and second mass continuity is used to convert tidal elevations into flow vectors. This approach was introduced as the "modified tidal prism method" by Ketchum (1951) and is described, with particular reference to the determination of estuary flushing characteristics, by Dyer (1973, p.110). Time series analyses are often used to determine and to predict tidal elevations (see Doodson and Warburg (1941) and the discussion below) but, for the present purposes, we obviate the first stage by using the observed tidal elevations directly from the ABP tide gauge network and then progress to use the DTM to determine flow vectors. At the present time some 15 gauges are operational and data are constantly revised and transmitted via a VHF radio telecommunications link to ABP control at Hull King George's Dock. The Humber Observatory receives and decodes these signals and data from seven gauges are utilised in the present analysis. These gauges are Spurn, Grimsby, Saltend, King George Dock, Albert Dock, Humber Bridge and Blacktoft.

The tidal flow analysis is based upon the DTM and tidal elevation data described above, and can be explained with reference to the schematic shown in Figure 19.5.

The estuary is made up of sixty segments numbered from west to east. Consider a time t (in hours) referenced to low water Spurn Head and a segment N wherein the depth of water is η. The volume change, δV, during time T to $T+\delta T$ can be calculated by two independent methods. Firstly, the volume change δV, is given by the change in water depth $\delta \eta$ multiplied by the cross-sectional area, which is approximated by the width of the segment (in metres), $(Y_N + Y_{N+1})/2$, multiplied by 1000 because each segment is 1 km in length. Thus the volume change, δV, is given by:

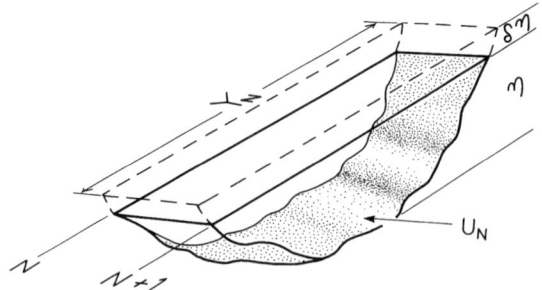

Figure 19.5   Schematic of tidal flow model

$$\delta V = 1000 \; \delta\eta \; (Y_N + Y_{N+1})/2 \tag{19.1}$$

Secondly, from mass continuity, the volume change can be related to the difference between the flow through the upstream segment boundary and the downstream segment boundary given by the product of the appropriate flows and cross-sectional areas. If the cross-sectional mean velocities at the downstream and upstream boundaries of segment N at time T are U_N and U_{N+1} respectively, then the volume change is:

$$\delta V = U_N \; (\eta_{N,t} - \eta_{N,t+\delta t}) \; Y_N - U_{N+1} \; (\eta_{N+1,t} - \eta_{N+1,t+\delta t}) \; Y_{N+1} \tag{19.2}$$

Simple algebra allows these two expressions to be equated and the cross-sectionally averaged mean flow velocities through each segment boundary to be solved, provided that the depth of water at each step can be superimposed upon the digital terrain model to determine the segment geometries. A short program was written in PASCAL to interrogate the DTM and hence solve the equations for particular tidal elevation data from the gauges for the various examples shown below.

RESULTS AND COMPARISON WITH FIELD DATA

The model was run to compare typical spring and neap tidal conditions as shown in Figures 19.6 and 19.7. The tide gauge data for the spring tide of 19 January 1995 and for the neap tide on 1 August 1994 were sampled at hourly intervals and linked with the DTM to solve equations (19.1) and (19.2) for cross-sectionally averaged flow in each of the 60 segments as described above. The spring tidal flows are plotted in Figure 19.6 for 12 h from low water Spurn Head. The flood tide (hours 1–6) runs for approximately 6 h with peak velocities of 1.9 m s^{-1} being attained in the middle estuary at Hull Roads between Kingston upon Hull and Hessle, whilst peak flows at the mouth of 0.8 m s^{-1} are attained between 3 and 4 h after low water. Ebb velocities (hours 7–12) are of similar magnitude with a pronounced asymmetry and again maximum velocities are attained in the middle estuary. The neap tide results are plotted in Figure 19.7 and show a similar overall pattern but

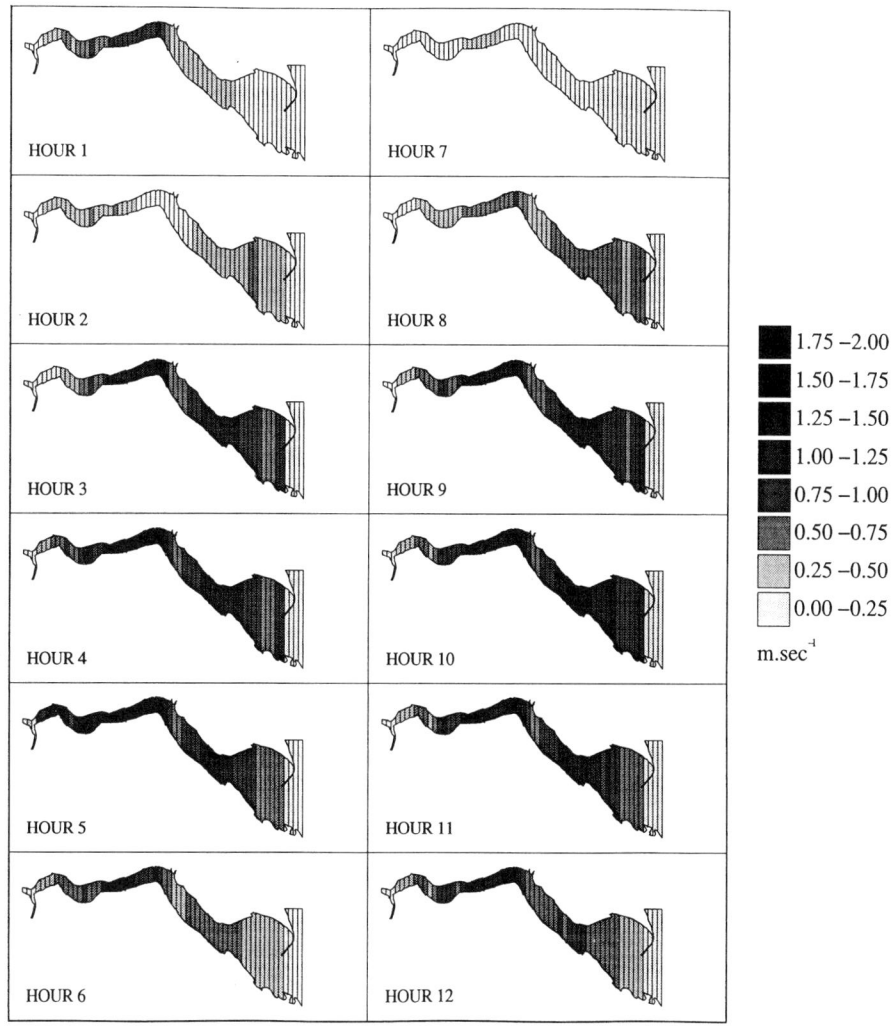

Figure 19.6 Hourly maps of flow from the TSA model for the spring tide of 19 January 1995

with rather lower velocities, the peak ebb in Hull Roads being 1.2 ms⁻¹. Comparison of these results with the limited amount of field data is encouraging. For example, peak ebb flows at tidal diamond R in Hull Roads on Admiralty Chart 109 (River Humber and the Rivers Ouse and Trent) are listed as 3.4 knots and 2.3 knots (1.73 and 1.17 ms⁻¹) for the spring and neap tides respectively, which compare reasonably well with the model results highlighted above.

A more detailed comparison was carried out with tidal flow velocities recorded during a *Sea Vigil* research cruise at an anchor site adjacent to the Bull Float

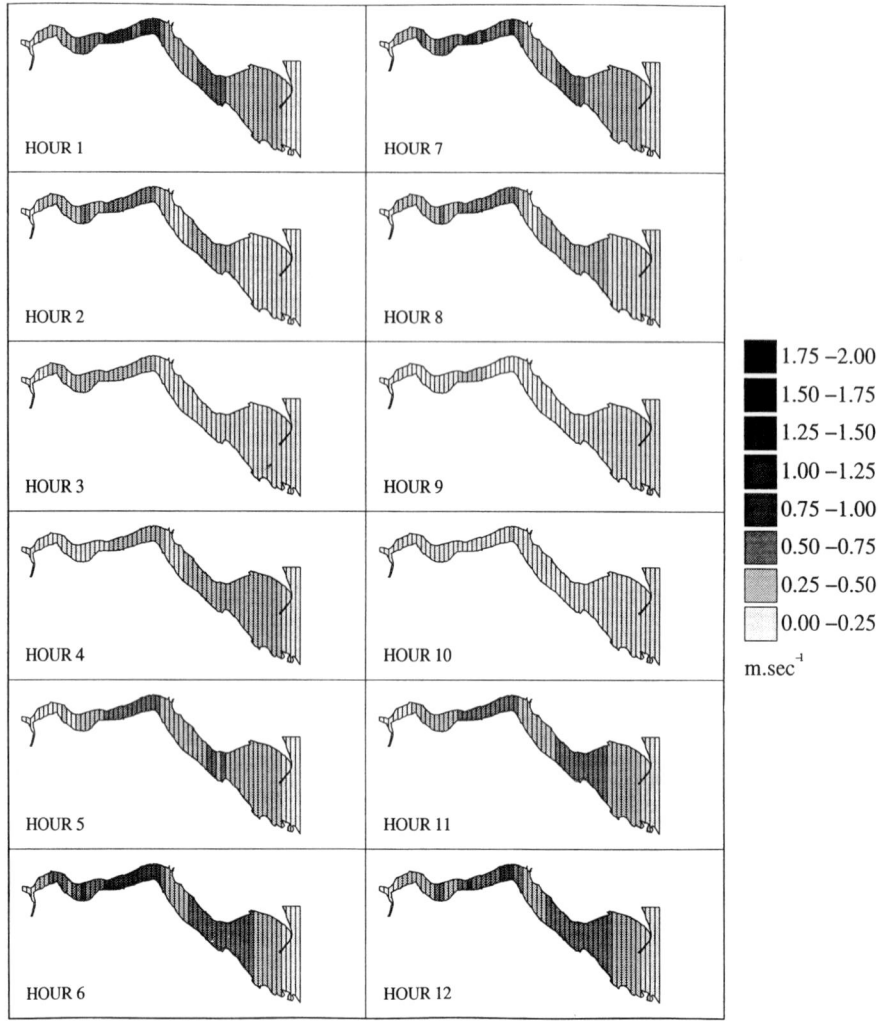

Figure 19.7 Hourly maps of flow from the TSA model for the neap tide of 1 August 1994

midway across the estuary mouth (in segment 55) for the 24 November 1994. These field data are depth but not cross-sectionally averaged. The TSA model was run with tide gauge data from the seven gauges described above and the results are shown as hourly flow velocities in Figure 19.8. These results are also encouraging, and differences between the tidal flow model and the prototype data are attributed to the difference between the cross-sectionally averaged predictions of the model and the actual location of the current meter at one point in the wide estuary mouth.

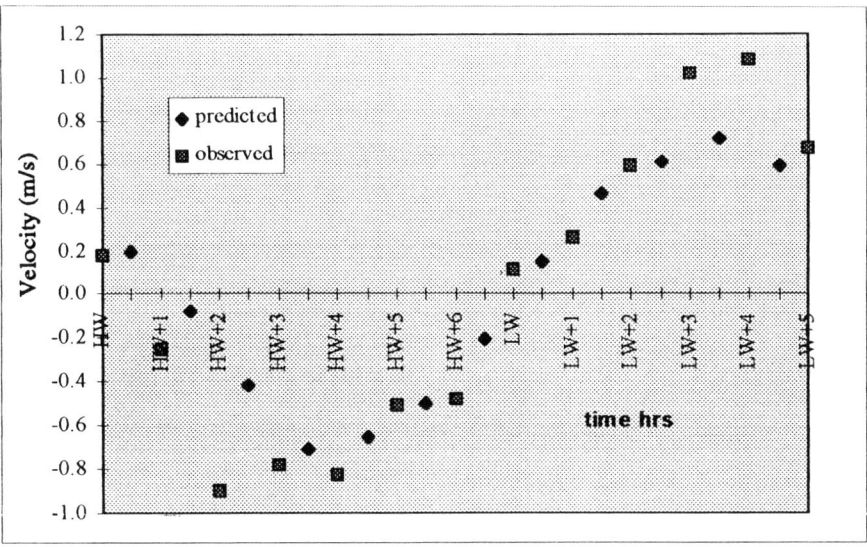

Figure 19.8   Observed versus TSA predicted flows for the Bull Float, 24 November 1994

DISCUSSION AND CONCLUSIONS

This paper has presented some geomorphological and hydrodynamic results for the Humber Estuary which have been obtained from digital terrain models constructed from hydrographic survey data. There are, perhaps, three significant results:

(i)   The comparison of the DTMs has provided first estimates of the longer term (decadal) evolution of the estuary. The analysis described here suggests that there is a mean annual accretion of some 0.56×10^6 m^3 in the estuary and that this occurs principally within the deeper channels, whilst erosion dominates within the shallower areas. However, these results must be treated with caution since small errors in the field surveys and in the analytical techniques could easily account for the relatively small changes in the cross-sectional and volumetric results.

(ii) It is also apparent from the results presented here that the TSA approach can usefully predict flow velocities in near real time throughout the estuary and is a viable alternative to full numerical solutions based upon the dynamical equations. The approach is being developed into an operational model since model run-time can be significantly shortened to produce near real-time predictions, particularly with the incorporation of the tide gauge data as described above. Again, however, there are particular problems which require further work since, for example, the model predicts cross-sectionally averaged velocities which cannot be compared in a fully quantitative sense with the existing field data. We have recently completed transects across the mouth of the estuary using acoustic Doppler current profilers and these data will allow more accurate comparison with the model results than has been possible here.

(iii) Finally, although, as described above, the model works with water levels determined from the tide gauge measurements, it is clear that a fully predictive capability can be developed. This is because, for many years, harmonic analyses of tide gauge data have been undertaken to determine amplitude, angular velocity and phase of the major tidal species as functions of astronomical data (see Doodson and Warburg, 1941). The results are routinely used in tide tables and such tidal predictions can be used to replace the gauge data in the TSA analyses. It should then be possible to examine the relationships between longer term estuary flow dynamics (for example, many years of tidal conditions could quickly be run in the simulator) and the geomorphological changes described above.

ACKNOWLEDGEMENTS

The authors acknowledge the financial support of the Natural Environment Research Council (Grants AAPS/GR3/8228 *Modelling and in situ measurement of seabed gravel dynamics* and LOIS/GST/02/747 *Modelling and in situ measurement of sediment flux in the lower Humber with particular reference to freshwater controls, tidal and storm forcing*), the National Rivers Authority, and the Research Support Fund of the University of Hull. Nick Hughes and Richard Hirst have assisted greatly in the field. The assistance of Peter Sergeant and Tim Rhodes on the NRA's *Sea Vigil* is gratefully acknowledged. Useful discussions with many other LOIS collaborators are acknowledged including the other PIs on the Flux Curtain: David Huntley, Sarah Metcalfe and Terry Marsden.

REFERENCES

Bacon, S., 1989. *Waves recorded at Dowsing Light Vessel 1970–1985*. Institute of Oceanographic Sciences Deacon Laboratory, Unpublished Report No. 262. 60pp.

Bacon, S. and Carter, D. J. T., 1988. *Waves and wave spectra recorded at two sites off the Holderness coast*. Institute of Oceanographic Sciences Deacon Laboratory, Unpublished Report No. 260, 88pp.

British Transport Docks Board (BTDB), 1971. *Collection of base data from the Humber Tidal Model*. Report H4, Humber Estuary Research Committee.

Doodson, A. T. and Warburg, H. D., 1941. *Admiralty Manual of Tides*. HMSO, London, 270pp.

Dyer, K. R., 1973. *Estuaries: a Physical Introduction*. Wiley, London, 140pp.

Falconer, R. A., 1993. An introduction to nearly horizontal flows. In *The Coastal, Estuarial and Harbour Engineers Reference Book*, ed. M. B. Abbott, E&FN Spon Ltd, 27–36.

Falconer, R. A. and Owens, P. H., 1990. Numerical modelling of suspended sediment fluxes in estuarine waters. *Estuarine, Coastal and Shelf Science*, **31**, 745–762.

Fortrum, B. C. H., 1980. *Analysis of waves recorded at the Dowsing Light Vessel between 1970 and 1979*. Institute of Oceanographic Sciences, Unpublished Report 126.

Hardisty, J., 1990. *The British Seas: an Introduction to the Oceanography and Resources of the North-West European Continental Shelf*. Routledge, London.

Hardisty, J., Rouse, H. L., Middleton, R., Hirst, R. S. and Scott, T., 1995. The Humber Observatory: Concepts, Instrumentation and Software Systems. *Earth Surface Processes and Landforms*, **20**, 859–880.

IECS, 1987. *The Humber Estuary Environmental Background*. The Institute of Estuarine and Coastal Studies, University of Hull, for Shell UK, 59pp.

Ketchum, B. H., 1951. The exchange of fresh and salt water in tidal estuaries. *Journal of Marine Research*, **10**, 18–38.

Lane, S. N., Chandler, J. H. and Richards, K. S., 1994. Developments in monitoring and modelling small-scale river bed topography. *Earth Surface Processes and Landforms*, **19**, 349–368.

Moore, I. D., Grayson, R. B. and Ladson, A. R., 1991. Digital terrain modelling: a review of hydrological, geomorphological and biological applications. *Hydrological Processes*, **5**, 3–30.

Pethick, J. S., 1987. Physical characteristics of the Humber Estuary. In *The Humber Estuary: Environmental Background*, Institute of Estuarine and Coastal Studies, University of Hull, 6-20.

Pethick, J. S., 1994. *Humber Estuary: Coastal Processes and Conservation*. Unpublished Report to English Nature by the Institute of Estuarine and Coastal Studies, University of Hull, 56pp.

Petrie, G. and Kennie, T. J. M., 1987. Terrain modelling in surveying and civil engineering. *Computer Aided Design*, **19**, 171–187.

Author Index

Subject Index

Website Index

Data sources for DTM construction, global, NOAA at NCAR Boulder
   http://www.ngdc.noaa.gov
Data sources for DTM construction, summary list
   http://www.geo.ed.ac.uk
Data sources for DTM construction, UK, Ordinance Survey
   http://www.open.gov.uk/ordsurv
Data sources for DTM construction, USA, Data Clearinghouse
   http://www.nsdi.usgs.gov
Data sources for DTM construction, USA, Eros Data Centre
   http://edcwww.cr.usgs.gov

GPS information
   gopher://unbmvsl.csd.unb.ca:70/hPUB.CANSPACE.GPS.INTERNET.SERVICES.HTML
GPS information
   http://www.inmet.com:80/~pwt/gps_gen.htm
GPS information
   http://wwwhost.cc.utexas.edu/ftp/pub/grg/gcraft/notes/gps/gps.html
GPS information
   http://galaxy.einet.net/editors/john_beadles/introgps.htm
GPS Information, Relevant Journal Contents Pages
   http://www.geod.emr.ca/~craymer/tcg/
GPS World Home Page
   http://www.advanstar.com/GEO/GPS/

Satellite Imagery, AirSar
   http://www-airsar.jpl.nasa.gov
Satellite Imagery, Pharus
   http://TUDEDV.ET.TUDELFT.NL/www/ttt/rs/pharus/pharus_home.html
Satellite Imagery, Planetary Exploration
   http://www.jpl.nasa.gov/mip/planet.html
Satellite Imagery, Remote Sensing Society
   http://axp10.iend.wau.nl/sar/sig/rad_sig.htm
Satellite Imagery, Canadian Space Agency, Sun-synchronous, Radarsat
   http://radarsat.espace.gc.ca/welcome.html
Satellite Imagery, European Space Agency (ESA), near circular, polar and sun-synchronous
   orbits, ERS-1
   http://gds.esrin.esa.it/ERS1.1
Satellite Imagery, European Space Agency (ESA), near circular, polar and sun-synchronous
   orbits, ERS-2
   http://sloth.esrin.esa.it:80/specers2.htm
Satellite Imagery, NASA(USA), various orbits, SEASAT
   http://www.jpl.nasa.gov/mip/seasat.html
Satellite Imagery, NASA(USA), various orbits, SIR-A
   http://www.jpl.nasa.gov/mip/sira.html
Satellite Imagery, NASA(USA), various orbits, SIR-B
   http://www.jpl.nasa.gov/mip/sirb.html

201906 ✓

is to be returned ... re